Noise and Vibration Control on Ships

Raymond Fischer • Leonid Boroditsky

Noise and Vibration Control on Ships

Understanding and Cutting Through the Noise

ASA PRESS

Springer

Raymond Fischer
(Retired)
Billerica, MA, USA

Leonid Boroditsky
(Retired)
Canton, MA, USA

ISBN 978-3-031-55169-7 ISBN 978-3-031-55170-3 (eBook)
https://doi.org/10.1007/978-3-031-55170-3

This Springer imprint is published by the registered company Springer Nature Switzerland AG
The registered company address is: Gewerbestrasse 11, 6330 Cham, Switzerland

Paper in this product is recyclable.

The ASA Press

ASA Press, which represents a collaboration between the Acoustical Society of America and Springer Nature, is dedicated to encouraging the publication of important new books as well as the distribution of classic titles in acoustics. These titles, published under a dual ASA Press/Springer imprint, are intended to reflect the full range of research in acoustics. ASA Press titles can include all types of books that Springer publishes, and may appear in any appropriate Springer book series.

ASA PRESS

The Acoustical Society of America

On 27 December 1928, a group of scientists and engineers met at Bell Telephone Laboratories in New York City to discuss organizing a Society dedicated to the field of acoustics. Plans developed rapidly, and the Acoustical Society of America (ASA) held its first meeting on 10–11 May 1929 with a charter membership of about 450. Today, ASA has a worldwide membership of about 7000.

The scope of this new Society incorporated a broad range of technical areas that continues to be reflected in ASA's present-day endeavors. Today, ASA serves the interests of its members and the acoustics community in all branches of acoustics, both theoretical and applied. To achieve this goal, ASA has established Technical Committees charged with keeping abreast of the developments and needs of membership in specialized fields, as well as identifying new ones as they develop.

The Technical Committees include acoustical oceanography, animal bioacoustics, architectural acoustics, biomedical acoustics, computational acoustics, engineering acoustics, musical acoustics, noise, physical acoustics, psychological and physiological acoustics, signal processing in acoustics, speech communication, structural acoustics and vibration, and underwater acoustics. This diversity is one of the Society's unique and strongest assets since it so strongly fosters and encourages cross-disciplinary learning, collaboration, and interactions.

ASA publications and meetings incorporate the diversity of these Technical Committees. In particular, publications play a major role in the society. The Journal of the Acoustical Society of America (JASA) includes contributed papers and patent reviews. JASA Express Letters (JASA-EL) and Proceedings of Meetings on Acoustics (POMA) are online, open-access publications, offering rapid publication. Acoustics Today, published quarterly, is a popular open-access magazine. Other key features of ASA's publishing program include books, reprints of classic acoustics texts, and videos. ASA's biannual meetings offer opportunities for attendees to share information, with strong support throughout the career continuum, from students to retirees. Meetings incorporate many opportunities for professional and social interactions, and attendees find the personal contacts a rewarding experience. These experiences result in building a robust network of fellow scientists and engineers, many of whom become lifelong friends and colleagues.

From the Society's inception, members recognized the importance of developing acoustical standards with a focus on terminology, measurement procedures, and criteria for determining the effects of noise and vibration. The ASA Standards Program serves as the Secretariat for four American National Standards Institute Committees and provides administrative support for several international standards committees.

Throughout its history to present day, ASA's strength resides in attracting the interest and commitment of scholars devoted to promoting the knowledge and practical applications of acoustics. The unselfish activity of these individuals in the development of the Society is largely responsible for ASA's growth and present stature.

Without a doubt, we would like to thank our wives for supporting and aiding in decisions and choices made over our careers. This often meant putting up with extensive overtime and time away from home. They found ways to keep the home front running and bringing up our families. Additionally, our wives were very encouraging and supportive while we were writing this book. We could not have done all this without their support.

Preface

> Even in the most beautiful music, there are some silences, which are there so we can witness the importance of silence. Silence is more important than ever, as life today is full of noise. We speak a lot about environmental pollution but not enough about noise pollution.
>
> ~ Andrea Bocelli

Quiet is proving to be a desirable goal with respect to both humans and animals. Creating a quiet ship is a complex process. Every ship type from small craft to large tankers can be engineered to be quieter than their previous types. How to achieve this given noisy, powerful machinery placed in the restricted space of a thin shell that accepts and radiates both airborne and underwater noise is the objective of this book. How to deal successfully with these contradictive requirements is laid out in terms of criteria, program planning, acoustic modeling, engineering of optimal treatments, and testing for compliance.

This book should prove useful to those involved in ship acoustics: designers, ship builders, ship owners/operators, marine engineers, and naval architects. They need a practical guide that should clarify the process of developing reasonable acoustic goals and the method of meeting them. Currently, there are no books in the technical literature that combine existing theoretical and practical experience in this field. This book is the first attempt to systematize this information. To the extent possible, the use of sophisticated mathematics was avoided; physical models are used to explain acoustical constructs.

The book contains multiple references that can be used to extend one's knowledge on a particular topic. With this approach, the reader should have the opportunity to solve acoustic problems from beginning to end. Specifically, how to meet acoustic criteria while designing optimal noise and vibration controls for a complex ship. Cutting through the noise can be simplified with a better understanding of the various facets.

Billerica, MA, USA — Raymond Fischer
Canton, MA, USA — Leonid Boroditsky

Acknowledgment

We would like to take the opportunity to acknowledge all the outstanding engineers and staff members who aided us in creating a quiet marine environment.

Contents

About the Authors

Raymond Fischer Fischer received his MS in Ocean Engineering from the University of Massachusetts, Amherst, in 1976. He was employed by Bolt, Beranek, and Newman (BBN) for 12 years and by Atlantic Applied Research for 5 years. In 1990, he started his own company, Noise Control Engineering, Inc. (NCE). He retired in 2016.

Fischer was extensively involved with the prediction of radiated noise, sonar self-noise, habitability noise, aural detectability, and structural vibrations on over 400 different vessels. He has also directed many shipboard noise and vibration measurements for demonstration of compliance with habitability criteria, for diagnostic testing and to verify noise modeling techniques. His areas of specialty include design, development, and noise control engineering to solve noise and vibration problems in small high-speed crafts, yachts, work boats, ice breakers, naval surface ships, and other nonconventional vessels. He was the Technical Leader for a US Navy Small Business Innovative Research (SBIR) program to provide more accurate shipboard airborne noise predictions. He coauthored the original and a Supplement to SNAME's Design Guide for Shipboard Airborne Noise Control, T&R 3-37. He has also contributed to the development of acoustic goals and several design guides for the US Navy and the Canadian Navy.

Fischer was the Principle Investigator for the Office of Naval Research's (ONR) Noise Induced Hearing Loss (NIHL) program. This multiyear effort started in 2010. This resulted in a new and verified acoustical engineering tool – DesignerNOISE®. New management and programmatic approaches for Navy acquisitions were also developed. Acoustic modeling tools were established for composite material constructions, underwater radiated noise, and return on investment.

Fischer was the project manager for the habitability and radiated noise control efforts on NOAA's Fisheries Research Vessel (under contract with VT Halter Marine), Fast Missile Craft (FMC), the Alaska Regional Research Vessel, and the development/optimization of spray-on damping treatments. Under his guidance, NCE conducted radiated noise trials on SEAFIGHTER (ONR's X-Craft), research vessels, seismic vessels, and commercial vessels.

He has 84 published papers on ship acoustics. He has over 400 technical reports or technical memos to his credit over his career.

Leonid Boroditsky Boroditsky graduated from Leningrad Shipbuilding Institute in 1959 with a master's in science degree. From 1959 to 1994, he was employed by the Acoustical Lab at the Central Research Institute of Shipbuilding Technology.

Boroditsky defended his PhD thesis about ship structure sound vibration insulation in 1966. His areas of special interest were icebreaker noise, flow noise, and propeller noise on fast craft. Boroditsky coauthored, with V. Spiridonov, a book entitled *Structureborne Noise Reduction in Ship Compartments* (1974).

Boroditsky joined NCE in 1996 and retired in 2021. He was a major contributor to the development of DesignerNOISE®, 3-D noise prediction software. He saw the implementation of this software into an ongoing design practice. His latest areas of interest were the acoustic characteristics of composite materials, aircraft carrier noise reduction, damping coating optimization, and on-deck noise control.

Boroditsky participated in numerous government and commercial projects to abate noise levels on ships. He performed multiple diagnostic tests for different kinds of vessels and developed treatments to mitigate mechanical and hydrodynamic noise.

Boroditsky has published papers in *Sudostroenie*, *Sound and Vibration*, *Noise Control Engineering Journal*, *American Acoustical Society*, *International Congress on Sound and Vibration*, *Inter Noise*, and *Naval Engineers Journal*.

Chapter 1
Introduction

If your goal is to achieve a quiet vessel, be it a ship, a small boat or anything in between, this generates the questions of 'how quiet' and 'for whom'? Are crew and/or marine life to be considered? Are noise and vibration both to be studied? What is the operating profile of the vessel, and how much time, effort and money will be invested to accomplish a quiet vessel? In the long run, designing, building and operating a quiet vessel will have significant impacts. These include weight, scheduling, space and cost. Other considerations are the impacts on the level and complexity of the design effort. The only way to minimize these sometimes 'adverse' impacts is to ensure that valid and implementable controls are utilized. These controls will need to be optimized with respect to the acoustic goals, vessel operation, and environment being protected. This means that this control process needs to be integrated into the overall ship design.

Ships have many intensive noise and vibration sources in close proximity to compartments that are noise sensitive – berthing, work and watch stations, and other manned positions. Source-path-receiver modeling is often used to predict this noise environment. Depending on the accuracy of the modeling used, one can determine and identify the optimal controls. Controls consist primarily of additional materials, noise and/or vibration treatments, and/or operational changes. In addition to the material/treatment costs, scheduling, testing and labor become factors affecting decisions on optimal procedures to meet goals or criteria. One assumption here is that the acoustic goals are appropriate and achievable.

The authors believe that necessary and sufficient noise and vibration control on ships should be based on detailed acoustical studies conducted during the design stage. The most successful projects show that acoustical design should be incorporated starting from the Concept or Preliminary Design stage. Fixing noise problems after Builder Sea Trials usually cost an order of magnitude more than if those treatments and material were considered during a proper design process. Unfortunately, consideration of acoustic goals is omitted or ignored in many designs due to their "usual" impact on space, weight and cost of treatments or just plain lack of interest.

R. Fischer, L. Boroditsky, *Noise and Vibration Control on Ships*,
https://doi.org/10.1007/978-3-031-55170-3_1

The core part of any acoustical design is based on realistic noise and vibration predictions: airborne and structureborne induced noise level calculations. These calculations must reflect the actual ship construction and systems. This book is an attempt to cover these issues in a comprehensible and concise manner. The noise prediction process includes knowledge of three parts: acoustic[1] source description, acoustic path analysis, and receiver compartment or location acoustic characteristics.

There is an enormous variety of acoustic sources on ships, more so than for most land-based factories. The most well-known are mechanical sources: main and auxiliary engines, gearboxes, pumps, compressors, fans, and other equipment basically located in machinery rooms. A noise source may be located not only inside but also externally to the ship. Aircraft landing and taking off may be a source of intensive on-deck noise, but they also influence the noise inside a ship. Main engine intake and exhaust systems very often generate high noise levels at on-deck stations and, sometimes, at inner compartments such as the Pilothouse.

Extensive ventilation systems may create excessive noise levels very distant from the machinery room. Ventilation noise is a combination of aerodynamic and mechanical noise. Practice shows that at some distance from engine rooms, ventilation (air conditioning) may be the main contributor to overall noise levels in compartments.

Propellers, thrusters, water jets and other propulsors are another category of noisy sources. Underwater radiated noise is basically connected with excessive noise levels from propulsors when they cavitate. Icebreaking and waves and inflow interactions with the hull are very important for noise in the bow compartments of some ships. The preponderance of these sources are considered in this book.

The acoustic path analysis requires information concerning hull structure, joiner panels, insulation, coating structures, machinery location, and locations of interest relative to critical sources. Noise path analysis includes analytical consideration of sound waves spreading through air and structure and analysis of a structure's ability to insulate, absorb, and radiate sound. Considering and analyzing this information can be a labor-consuming segment of the noise prediction process. The process is complicated by the strong frequency dependance of the acoustic sources, path effects, and receiver characteristics. For airborne noise, for instance, there are usually nine octave bands to be considered from 31.5 Hz to 8000 Hz, the typical range of human hearing. In addition, there are frequency weightings such as A-weighting for human hearing or vibration frequency weightings to account for vibrations affecting the human body along different axes. For underwater radiated noise, the frequency span can be much lower and higher – 1 Hz to 50,000 or 100,000 Hz.

Notwithstanding the sophistication of the noise prediction process for a ship's environment, one major principle must be formulated. That is, the separation of airborne noise from structureborne noise. This terminology basically reflects the different media related to the spreading of sound. Airborne noise is transmitted through the air. Structureborne noise is vibration spreading through ship structural

[1] The term "acoustic" as used herein generally includes both noise and vibration.

components with eventual noise radiation from this structure into air in the compartment of interest. Each of these components needs to be analyzed with different methods and are processed differently. These methods for airborne and structureborne noise calculation are discussed in detail in this book.

When excessive noise levels are expected and the relative contributions of structureborne noise and airborne noise have been determined, noise and vibration control approaches can be developed. This would include consideration of operating conditions such as speed and machinery line-ups, administrative controls such as allowed exposure time, and application of physical treatments. The treatments may insulate noise or vibration. They may absorb noise and vibration or may block it. In each case, it is important to qualify and quantify the effectiveness of each of the recommended treatments versus frequency. The treatments may have "global" (whole ship) or "local" (one structure or compartment) effectiveness; they may reduce only airborne or structureborne noise or both. For example, resilient mounts under machinery equally reduce the structureborne noise of this source everywhere in a ship; therefore, this treatment has a "global" effect. A joiner panel may reduce radiation from a bulkhead in one particular room where this joiner panel is installed. This is an example of a "local" effect. Damping tiles, depending on location and area, may have a combination of local and global effects. The effectiveness of salient noise and vibration treatments will be considered in this book individually and in combination. Other factors, such as the frequency range over which the treatment is effective, its weight and space impact, and that it meets marine regulatory requirements, need to be factored into this process.

The authors worked as acoustical consultants for over 40 years each. This book reflects their combined experience in marine acoustical consulting and the corporate experience of the consulting company Noise Control Engineering (NCE) established in the USA in 1991. Other papers and reference material by the authors are noted as needed. Prominent works by others in the field are also included.

The authors believe that it is not reasonable to provide the basics of acoustics, vibration theory, and wave spreading in elastic media, as these are widely published (Beranek 1971; Beranek and Ver 1992; Junger and Feit 1986). It is also assumed that the reader has basic knowledge regarding acoustics: frequency, decibel, sound pressure, acceleration, resonance, etc. A simplified engineering approach sometimes prevails over extensive and complicated scientific considerations. If an empirical approach leads to the goal faster than an analytical approach, the authors will use the first with a reference to any limiting assumptions.

The objective of this book is to educate ship owners/operators, naval architects/ marine engineers, ship builders, and engineering students on how to understand and address relevant factors involved in designing, building, delivering and operating a 'quiet' vessel. A vessel herein is defined as any mechanical system operating in an enclosed volume in a marine environment – fixed, floating, or submerged. 'Quiet' in this case addresses both the noise and vibration of a vessel and how it influences the operating environment both inside and outside the vessel.

A comprehensive discussion of the terms used in the field is given in several basic texts on acoustics or noise control (Fischer et al. 1983; Fischer 2020; Fischer and Boroditsky, 2001; Beranek 1971; Beranek and Ver 1992; Crocker 2007; Pierce 2019; Kinsler and Frey 1962; Fahy and Thompson 2015; Loeser 1999; Junger, 1986; Bies and Hansen 1988; Boroditsky and Spiridonov 1974; Fischer and Bahtiarian, 2017a, 2017b; Humes, 2005; Nikiforov, 2005; Plunt, 1980; Ross, 1983; Southall 2019; Urick, 1983; Discovery of Sound in the Sea - www.dosits.org; and Junger and Feit 1986). These references provide a discussion of relevant acoustic terminology and general principles that are needed as a basis for understanding the acoustic environment in and around a ship. This includes a discussion of the primary units for both noise and vibration – decibels, dB, and decibel math. The concepts of airborne noise and structureborne noise as important components are introduced.

References

Beranek, L. (ed.): Noise and Vibration Control. McGraw-Hill Book Co., New York (1971)

Beranek, L., Ver, I. (eds.): Noise and Vibration Control Engineering, Principles and Applications. John Wiley & Sons, Inc, New York (1992)

Bies, D., Hansen, C.: Engineering Noise Control. UNWIN HYMAN, Boston (1988)

Boroditsky, L., Spiridonov, V.: Structureborne Noise Control in Ship Compartments. Sudostroenie, Leningrad (1974) (in Russian)

Crocker, M. (ed.): Handbook of Noise and Vibration Control. Wiley, NY (2007)

Discovery of Sound in the Sea – http://www.dosits.org

Fahy, F., Thompson, D.: Fundamentals of Sound and Vibration, 2nd edn. CRC Press, Boca Raton (2015)

Fischer, R.: Sound and noise control. In: *The Marine Engineering Series*, 4th edn. The Society of Naval Architects and Marine Engineers (SNAME), Alexandria VA (2020)

Fischer, R., Bahtiarian, M.: Chapter 38: Ship noise control. In: Encyclopedia of Maritime and Offshore Engineering. Wiley (2017a)

Fischer, R., Bahtiarian, M.: Ship noise control. In: Finch, D. (ed.) Encyclopedia of Maritime and Offshore Engineering. Wiley (2017b). Online ISBN: 9781118476406

Fischer, R., Boroditsky, L.: Technical and Research (T&R) Bulletin 3–37 (Supplement), Supplement to the Design Guide for Shipboard Airborne Noise Control. The Society of Naval Architects and Marine Engineers (SNAME), Alexandria VA (2001)

Fischer, R., Collier, R.: Noise prediction and prevention on ships. In: Crocker, M. (ed.) Handbook of Noise and Vibration Control. Wiley, New York (2007)

Fischer, R., Nelson, D., Burroughs, C.: Technical and Research (T&R) Bulletin 3–37, Design Guide for Shipboard Airborne Noise Control. The Society of Naval Architects and Marine Engineers (SNAME), Alexandria VA (1983)

Humes, L., Joellenbeck, L., Durch, J. (eds.): Noise and Military Service: Implications for Hearing Loss and Tinnitus. National Academies Press (2005) http://www.nap.edu/catalog/11443.html

Junger, M., Feit, D.: Sound, Structures, and their Interactions, 2nd edn. The MIT Press, Cambridge, MA (1986)

Kinsler, L., Frey, A.: Fundamentals of Acoustics, 2nd edn. Wiley, NY (1962)

Loeser, H.: Fundamentals of Ship Acoustics, Acoustical Phenomena in and Around Ship Hulls. The Society of Naval Architects and Marine Engineers (SNAME), New Jersey (1999)

Nikiforov, S.: Acoustic Design of Naval Structures, D. Feit (English translation), Office of Naval Research, ADA 463069 (2005)
Pierce, A.: Acoustics. In: An Introduction to its Physical Principles and Applications, originally published in 1981, Reprinted 1989 in Physics Today, **34**(12) (2019).
Plunt, J.: Methods for Predicting Noise Levels in Ships. Chalmers University of Technology, Goteborg, Sweden (1980)
Ross, D.: Mechanics of Underwater Noise, 3rd edn. Peninsula Publishing, Los Altos. CA (1983)
Southall, B.L., Finnean, J., et al.: Marine mammal noise exposure criteria: updated scientific recommendations for residual hearing effects. Aquat. Mamm. **45**(2), 125–232 (2019)
Urick, R.: Principles of Underwater Sound, 3rd edn. Peninsula Publishing, Los Altos. CA (1983)

Chapter 2
Basic Acoustic Principles – Noise and Vibration

This chapter discusses the basic acoustic parameters used throughout this book. This includes information on the speed of sound and vibration in various media, relationships between acoustic power and sound pressure, sound propagation, direct v. reverberant fields, and transmissibility of sound through ship structures. Other parameters are the radiation efficiency of air-backed and water-backed plates and structure/fluid interaction. Finally, noise and vibration units are defined. Unless otherwise noted, the mks system is utilized throughout this text.

Basic parameters used in acoustics are covered quite well by Beranek (1971), Thumann and Miller (1990) and Engineering Acoustics from Wikibooks.[1] Some basic concepts, especially applicable to ships, are as follows:

- Some of the sound impinging on a bulkhead, deck or deckhead will be reflected, transmitted, or absorbed. The controlling factors are the Young's modulus, density, and thickness of the solid material and the amount of absorption on the source/receiver sides.
- The noise in each ship compartment is affected by the amount of internal noise due to any sources in that compartment, generally HVAC, and the amount of noise transmitted from the adjacent compartment(s).
- The best way to minimize noise transmitted throughout a ship is to isolate the vibration source from the structure before it can excite the compartment boundaries/foundations (as structureborne noise).
- The best way to reduce airborne transmitted noise is to vibration isolate joiner work and/or add mass and decoupling to the solid surfaces. Limp mass on a resilient underlayment such as fiberglass will provide this "decoupling."
- Any finite-sized structure on the ship has a resonant frequency – whether it is equipment, structure, piping or HVAC systems – which, if excited, may become a sound short or way for acoustic energy to be easily transmitted and likely cause a noise and/or vibration problem.

[1] https://en.wikibooks.org/wiki/Engineering_Acoustics

R. Fischer, L. Boroditsky, *Noise and Vibration Control on Ships*,
https://doi.org/10.1007/978-3-031-55170-3_2

- Ships, no matter how well designed, can become noisy due to sound shorts, poor maintenance, incorrect installation of treatments, and the fact that acoustic energy, such as heat, will take the shortest and path of least resistance to any compartment of concern.

Unfortunately, and this is the bane of acoustical consultants specializing in shipboard noise, the typical treatments discussed above:

- Add weight
- Take away space
- May be costly in terms of implementation and maintenance.

2.1 Decibel Addition/Subtraction and Units for Noise and Vibration

The decibel units used in acoustics are not linear. Going from 60 dB to 120 dB, such as going from a temperature of 10 °C to 20 °C, is not twice as loud or twice as hot. One rule of thumb to remember is that an increase of 5 dB in sound is usually *perceived* as being twice as loud. Thus, going from 60 to 65 dB is considered by most as twice as loud. However, some people are more sensitive to noise (and vibration), and a change of even 2 dB can be noticeable and problematic.

A majority of the calculations carried out in this book are simple arithmetic operations, except in regard to adding and subtracting decibel values. In this case, it is necessary to 'combine' the values logarithmically. When two or more sound pressure levels, vibration levels, or power levels are combined, the following steps need to be taken:

- Take the antilogs of each dB value
- Arithmetically sum these antilogs
- Determine the log value of this sum

For example, the combination of three acoustic sources of 90, 95, and 96 dB re 20 μPa (or 2×10^{-5}N/m^2) would be:

$$Combination = 10\ \mathrm{Log}\left(\sum\left(10^{90/10} + 10^{95/10} + 10^{96/10}\right)\right) = 99\ \mathrm{dB\ re\ 20\ \mu Pa} \quad (2.1)$$

It is slightly more complicated when trying to determine the contribution of only one source when all that is measured is the combined level of two sources and the level with just one of the sources secured. Then, the math is as follows. For the case where the combined level from two sources is, L_{comb} is 90 dB, and the measured level with just one source L_{p1} is 86 dB. Then, the unknown source L_{p2} is

$$L_{p2} = 10\ \mathrm{Log}\left(\sum\left(10^{L_{comb}/10} - 10^{L_{p1}/10}\right)\right) = 88\ \mathrm{dB\ re\ 20\ \mu Pa} \quad (2.2)$$

When pressure levels, vibration levels, or power levels are added to any type of transfer function,[2] the values are treated arithmetically. For example, if a source level of 90 dB re 20 μPa is attenuated by spreading losses by 6 dB, the received level is 90–6 = 84 dB re 20 μPa.

All values in decibels should be rounded up to the nearest whole number. Fractions of a dB do not amount to anything of value. Along the same vein, values within +/−2 dB can be considered almost equivalent and sometimes are with certain criteria.

2.2 Units for Noise and Vibration

One should note that the primary quantities of interest are sound pressure (p), acoustic power (W), and vibration – as either velocity (v) or acceleration (a).[3] One should always bear in mind that sensors and instrumentation that measure these parameters always use squared quantities for pressure and velocity; hence, '20 Log' is used in the equations. Power is not reported as a squared quantity – hence 10 Log is used in the power level equations.

Therefore, when computing a dB value for a physical quantity, the following equations apply:

Pressure, p, measured in N/m^2 or Pa, and using a standard reference of $p_o = 20$ μPa (or 20×10^{-6} Pa) is converted to a sound level value in decibels as:

$$L_p \text{ dB re } 20\ \mu\text{Pa} = 20\ \text{Log}\left(p/_{20 \times 10^{-6}} \right) \tag{2.3}$$

This book uses acceleration whenever possible. Thus, a given value of acceleration, a (m/sec^2), can be converted into an equivalent level dB re 1 μG (1 G = 10 m/sec^2 and 1 microG[4] is 10×10^{-6}m/sec^2):

$$L_a \text{ dB re } 1\ \mu\text{G} = 20\ \text{Log}\left(a/_{10 \times 10^{-6}} \right) \tag{2.4}$$

Using the relationships between rms acceleration, a(m/sec^2), velocity, v(m/sec); and displacement, d(m); at frequency f (Hz):

$$a = 2\pi f \nu \text{ and } \nu = 2\pi f d \tag{2.5}$$

One can develop the following equations to convert to velocity in dB re 1×10^{-8} m/sec or displacement in dB re 1×10^{-11} m:

[2]Transfer functions relate an input quantity to an output quantity.

[3]Unless otherwise noted the metric system is used throughout.

[4]A microG is technically 9.8×10^{-6}m/sec^2; the G value for convenience is rounded to 10.

$$L_\nu \text{ dB re } 1 \times 10^{-8} \frac{\text{m}}{\text{sec}} = L_a \text{ dB re } 1\ \mu\text{G} + 44 - 20\ \text{Log}(f) \tag{2.6}$$

and

$$L_d \text{ dB re } 1 \times 10^{-11}\text{m} = L_a \text{ dB re } 1\ \mu\text{G} + 88 - 40\ \text{Log}(f) \tag{2.7}$$

In an historical note, the space/time averaged square-velocity on a surface of a structure is proportional to the total energy stored in a resonant system. This is proportional to the radiated sound power. In this case, the reference velocity can be chosen so that the velocity level numerically equals the sound pressure level on the surface of the structure radiating into air. Thus:

$$L_p = 20\ \text{Log}\ {}^{p}\!/_{p_0} = L_\nu = 20\ \text{Log}\ {}^{\nu}\!/_{\nu_0} \tag{2.8}$$

and

$$\nu = {}^{p}\!/_{(\rho_0\ c_0)} \tag{2.9}$$

where $p_o = 20\ \mu\text{Pa}$, $\rho_0 c_0 \sim 400$ mks rayls, and rayls – the sound specific impedance – is defined as the ratio between the sound pressure and the particle velocity it produces.

In this case, $\nu_0 = {}^{p_0}\!/_{\rho_0\ c_0} = 5 \times 10^{-8}$ m/sec rather than the 1×10^{-8} m/sec reference currently utilized. This implies that using the standard v_o value of 1×10^{-8} m/sec would over predict the equivalent sound pressure by a factor of 14 dB.

Along this same line, the sound power, W (watts), radiated by a plate of surface area, S (m^2), is related to the velocity, v (m/sec), of the plate as:

$$W = \langle \nu^2 \rangle\ \rho_0\ c_0 S\ \sigma_{rad} \tag{2.10}$$

and

$$\nu = {}^{p}\!/_{\rho_0\ c_0} \tag{2.11}$$

where the brackets $\langle\ \rangle$ denote the root mean square value.

Then, the radiated power, L_w dB re 1×10^{-12} W, is

$$L_w \text{ dB re } 1 \times 10^{-12}\ \text{W} = L_\nu \text{ dB re } 1 \times 10^{-8}\ \text{m/s} + 10\ \text{Log}(S) + 10\ \text{Log}(\sigma_{rad}) - 14 \tag{2.12}$$

where σ_{rad} is the radiation efficiency of the ship plating (see also Sect. 2.7). At higher frequencies, typically above 2000 Hz, the radiation efficiency term goes to zero, and the radiated power is directly related to the sound pressure, as modified by the plate area.

The sound pressure squared, p^2, at a distance, r, m, from a source is equal to:

$$p^2 = (\rho_0\, c_0)^2\, Q\, \langle v^2 \rangle\, S\, \left(\sigma_{rad}/4\pi r^2\right) \tag{2.13}$$

and substituting acceleration for velocity yields

$$p^2 = (\rho_0\, c_0)^2\, Q\, \left(\langle a^2 \rangle/(2\pi f)^2\right)\, S\, \left(\sigma_{rad}/\left(4\pi r^2\right)\right) \tag{2.14}$$

Thus,

$$\begin{aligned} L_p \text{ dB re } 20\ \mu\text{Pa} = {} & 19 + 10\,\text{Log}(Q) + L_a \text{ dB re } 1\ \mu\text{G} + 10\,\text{Log}(S) + 10\,\text{Log}(\sigma_{rad}) \\ & - 20\,\text{Log}(f) - 20\,\text{Log}\, r \end{aligned} \tag{2.15}$$

where Q is the directivity factor and r is the distance source to the receiver.

In the hemi-spherical acoustic nearfield, for a 1 m^2 area and above the coincidence frequency, this equation reduces to

$$L_p \text{ dB re } 20\ \mu\text{Pa per } 1\ \text{m}^2 = L_a \text{ dB re } 1\ \mu\text{G} - 20\,\text{Log}(f) + 22 \tag{2.16}$$

2.3 Speed of Sound in Air, Fluids, and Structures

The speed of sound in any fluid is temperature-, pressure-, and density dependent. Thus, the speed of sound in a gas turbine exhaust is different than that in cabins. That in air is different than that in water. The speed of sound in air is $c_a = \sqrt{\gamma P/\rho}$, where γP is the bulk modulus, P is the absolute air pressure (Pa), γ is the ratio of specific heats (1.4), and ρ is the mass density (kg/m^3). This is equal to 343 m/s for air. In water, the speed of sound is $c_w = \sqrt{B/\rho}$, where B is the fluid's bulk modulus. For water, the speed of sound is approximately 1500 m/s.

Always, bear in mind that the pressure, intensity, or power level in air is different than the pressure, intensity or power level in a fluid by the ratio of their acoustic impedance, Z, which is equal to ρc.

The speed of a flexural wave in a thin plate in vacuo is $c_p = \sqrt{Eh^2\omega^2/12\rho}$, where E is the Young's modulus of the material, h is the plate thickness, ω is $2\pi f$, and ρ is the mass density of the material. Note that this speed is frequency dependent, whereas the speed of sound in a fluid or air is independent of frequency. This speed is slower if the plate is water loaded on one side or both sides (Loeser 1999). This dissimilarity leads to the 'critical frequency' component of radiation efficiency discussed below.

The wavelength, λ(m), of sound in air at 21 °C at a frequency, f, is:

$$\lambda = c/f \tag{2.17}$$

where c = 343 m/sec, the speed of sound.

2.4 Sound from a Source of Known Acoustic Power

Sound, L_p, at a distance r from a small point source of known acoustic power, L_w dB re 1 pW, is simply computed as:

$$L_p(r) \text{ dB re } 20\ \mu\text{Pa} = L_w - 20\ \text{Log}(r) + 10\ \text{Log}(Q) - 11 \tag{2.18}$$

where r is the distance from the source and Q is the source directivity (the ratio of the acoustic power radiated to the receiver to the average radiated in all directions). This equation is correct for free space, showing that the received level decreases by 6 dB for a doubling of distance. Often, the sound will come from a 'finite' sized source of dimensions a and b, with $a > b$. In this case, the received level will depend on the distance r relative to the a and b dimensions of the plate as:

$$\begin{aligned} r < b/\pi &: L_p \text{ dB re } 20\ \mu\text{Pa} = L_w + 10\ \text{Log}(\pi/4ab) \\ b/\pi < r < a/\pi &: L_p \text{ dB re } 20\ \mu\text{Pa} = L_w - 10\ \text{Log}(r) - 10\ \text{Log}(4a) \\ r > a/\pi &: L_p \text{ dB re } 20\ \mu\text{Pa} = L_w - 20\ \text{Log}(r) - 11 \end{aligned} \tag{2.19}$$

If the source is in a fluid such as water, the equation equivalent to [2.18] is[5]:

$$L_p(r) \text{ dB re } 1\ \mu\text{Pa} = L_w - 20\ \text{Log}(r) + 10\ \text{Log}(Q) + 51 \tag{2.20}$$

using the typical reference of 1 μPa for underwater noise.

It is worth noting that the supposed conversion rate between the radiated acoustic power in water and the associated mechanical power is 1×10^{-8}. In air, this ratio is between 10^{-2} and 10^{-4}. Thus, a mechanical power of 1 kW would equate to an acoustic power of 70 dB re 1 pW in water and 110 dB re 1 pW in air. The equivalent radiated pressures would be 99 dB re 20 μPa−1 m in air and 121 dB re 1 μPa−1 m in water. In essence, the same mechanical input causes a higher pressure reading in water than in air for the referenced values. This is something to be very much aware of when comparing pressure and power levels in decibels in air and in water.

[5] For underwater noise, the reference p_o would be 1 μPa. The reference pressure in water is equated to −120 dB while that in air is −94, a difference alone that adds 26 dB to the same pressure in air.

2.5 Direct Versus Reverberant Field

Engine or equipment rooms typically have a variety of acoustic sources. If the acoustic power level, L_w dB re 1 pW, is given for each item, the resulting noise in the compartment can be calculated. A "small" source will radiate uniformly in all directions, and the pressure in a free field decays at a rate of 6 dB per doubling of distance from the source. Placing this source inside an enclosure will result in a reverberant field, consisting of reflections from the enclosure walls. The pressure will no longer decay at a rate of 6 dB per doubling of distance; however, it will decay to a set level depending on the type and quantity of absorptive materials inside the enclosure or room. This results in the following equation used to determine the direct and reverberant sound level, L_p, in a compartment due to a source in the compartment that emits a known sound power level, L_w:

$$L_p \text{ dB re } 20\ \mu\text{Pa} = L_w + 10\,\text{Log}\left(Q_\theta/\left(4\,\pi\,r^2\right) + 4/R\right) \tag{2.21}$$

where Q_θ is the directivity factor depending on the position of the source in the room (+3 dB if placed in the floor, +6 dB is placed at the intersection of two surfaces, and +9 dB if placed in a corner); r is the distance from the source, m, and R is the Room Constant, m^2, of the room.

The room constant, R, a measure of the absorption in a room, is computed as:

$$R = S\,\alpha_{sab}, m^2 \tag{2.22}$$

where S is the surface area of the compartment, m^2, and α_{sab} is the average Sabine absorption coefficient.

The closer the Sabine absorption coefficient is to one, the lower the reverberant pressure level in the room due to lower reflections from the boundaries. The $1/r^2$ term in Eq. 2.21 denotes the 6 dB reduction per doubling of distance in the free field portion of the room.

For a 'typical' berthing compartment with a modicum of absorption on the overhead and bulkheads, the reverberant field exists within approximately 1 meter of a point source, as shown in Table 2.1. In this case, the sound pressure would only decrease by 2 dB going from 0.5 to 1 m from the source rather than by 6 dB if the source were in a free field. If this 90 dB re 1 pW source were in the free field, the pressure level at 1 m would be 79 dB re 20 μPa versus the reverberant level of 89 dB. This point, where the reverberant field equals the direct field, is called the Hall Radius, r_H, and is equal to $r_H = R/(16\,\pi)$.

As shown in Table 2.2 for an engine room, the reverberant field starts approximately 2 meters from a point source. In this case, the sound pressure would only decrease by 3 dB going from 0.5 to 1 m from the source rather than by 6 dB if the source were in a free field. This is due to the larger room dimensions. Note that at low frequencies, the absorption value significantly decreases, resulting in a higher received pressure both in the free field and reverberant fields, as shown in Table 2.2.

Table 2.1 Representative free versus reverberant sound calculations – Berth and Engine Room

	Berth Dimensions, m^2	α_{sab} [a]	Distance r, m	Received L_p dB re 20μPa @ r
Deck	16.7	0.02	0.5	87.8
Overhead	16.7	0.7	1	85.7
Bulkheads	12.5	0.15	2	
			4	84.7
	Engine Room Dimensions, m^2	α_{sab}	Distance r, m	Received Lp dB re 20 μPa @ r
Deck	41.8	0.02	0.5	86.4
Overhead	41.8	0.7	1	83
Bulkheads	19.7	0.15	2	81.5
			4	81

[a]For mid- to high-frequency

Table 2.2 Free versus Reverberant Sound for Engine Room with Little Absorptive Treatment

	Engine Room Dimensions, m^2	α_{sab}	Distance *r*, m	Received Lp dB re 20 μPa @ r
Deck	41.8	0.01	0.5	90
Overhead	41.8	0.1	1	89
Bulkheads	19.7	0.05	2	88.7
			4	88.6

Note that in the free field, for $r < 1$ m, the received pressure for the same source is decreased by 5 dB, and the reverberant field is increased by 5 dB due to the decreased absorption over the case shown in Table 2.1.

The absorption coefficient is affected not only by the absorption on the boundaries but also by the type and extent of furnishings in the compartment. Additionally, the bulkheads, decks, and overheads themselves affect how much energy is transmitted into adjacent compartments through the boundaries. This 'apparent absorption' is significant at low frequencies (Fischer and Boroditsky 2001). Finally, in very large rooms, atmospheric absorption[6] may be important above 2000 Hz; atmospheric absorption is critical with respect to propagation in the free field – such as from exhaust stacks to nearby port facilities.

For an on-deck or free-field acoustic source, Eq. 2.22 can be used without the *4/R* term. In this case, the Q_θ term accounts for the directivity of the source. Furthermore, ducted openings are often highly directive[7].

Representative absorption coefficients are provided in Table 2.3 (Fischer et al. 1983). Included in this table is the NRC – Noise Reduction Coefficient, which is an arithmetic average of the absorption coefficient, α_{sab}, between the 125 and 4000 Hz

[6]ANSI/ASA S1.26-2014 Method for Calculation of the Absorption of Sound by the Atmosphere

[7]See SNAME T&R 3-37, Table 7.1 and SNAME Supplement, Appendix E: Design Guide Errata

Table 2.3 Representative absorption coefficients

Description	Octave Band Freq., Hz									
	31.5	63	125	250	500	1000	2000	4000	8000	NRC
Steel or Aluminum Plate	0.01	0.01	0.02	0.03	0.03	0.03	0.02	0.02	0.02	0.03
Glass	0.30	0.20	0.16	0.04	0.03	0.02	0.02	0.02	0.02	0.05
Carpeted deck	0.02	0.04	0.08	0.10	0.15	0.20	0.25	0.20	0.15	0.16
Acoustic Ceiling Tile	0.03	0.10	0.20	0.25	0.50	0.70	0.75	0.75	0.70	0.52
50 mm thick fiber-glass with vapor barrier	0.10	0.20	0.28	0.92	0.99	0.79	0.53	0.37	0.25	0.63
50 mm Rockwool			0.21	0.69	1.13[a]	1.15[a]	1.16[a]	1.18[a]	1.14[a]	0.95
50 mm thick light-weight polyimide foam			0.35	0.59	0.95	1.03[a]	0.96	0.87	0.94	0.73

[a]Values greater than 1 are an artifact of the standard testing procedure (ASTM C423 – www.astm.org) and should be set to a maximum value of 1

octave bands. The NRC is the value measured by the majority of vendors. Standardized test methods are provided in ASTM C384–04 (2022) and ISO 354:2003.

It is worth noting that the sound pressure level of a source in an enclosed space, $p_{enclosed}$, will increase at low frequencies over that in a free field, p_{free}, by the following ratio:

$$p_{enclosed}/p_{free} = \left(c^2/\omega^2\right)\left(4\pi r/V_0\right) \tag{2.23}$$

where c is the speed of sound, $\omega = 2\pi f$, r is the distance in the free field, and V_O is the enclosed volume.

Putting $V_O = (2r)^3$ for a cube and $r = 1$ m, one finds that the $p_{enclosed}$ is greater than p_{free} for frequencies below 68 Hz. Thus, enclosing a source in a small enclosure typically increases its apparent source level at low frequencies.

2.6 Transmissibility

The above section considers the resulting airborne noise in a 'source' space on a vessel. The primary phenomenon of interest with respect to airborne noise transmission within the ship is that of the transmission loss of decks, bulkheads, or overheads – considered 'homogeneous' structures. In general, ribbed panels have a transmission loss value below that of the 'mass law'[8] and a negative slope for frequencies below the stiffened panel resonance (controlled by the rib spacing and plate thickness), a positive slope above the resonant frequency up to a plateau

[8]Mass law transmission loss, $TL = 20\ log\ (\rho_s f) - 48$; where ρ_s is the mass per unit area, kg/m^2, f is the frequency

(controlled by surface mass), a dip at the coincidence frequency, and a positive slope above that (Beranek and Ver 1992).

When the transmission loss, *TL* dB, is known for a particular construction, the received noise level $L_{p\ rec}$, in an adjacent compartment for a given known pressure level in the source room, $L_{p\ source}$, is:

$$L_{p\ rec}\ \text{dB re } 20\ \mu\text{Pa} = L_{p\ source} - TL + 10\ \text{Log}(S_p/R) \tag{2.24}$$

where S_p is the area of the interface and R is the room constant in the receiver space. Note that the difference in sound in the source and receiver spaces not only depends on the TL of the interface but also depends on the area of the interface and absorption of the receiver space.

At low frequencies, the transmission loss is generally controlled by the surface mass, ρ_s (kg/m^2), of the interface as $TL \cong 20\ \text{Log}\ \rho_s$. Doubling the surface mass increases the *TL* by 6 dB. Doubling the interface area increases the received level by 3 dB, while doubling the room constant reduces the noise by 3 dB. If the interface of interest has absorption and/or joiner paneling, then the *TL* is that of the composite construction. Due to resonances of ribbed panels, there may be a dip in the *TL* at low frequencies (Beranek and Ver 1992). Additionally, due to the mass-air-mass spring resonance between a joiner panel/air gap/bulkhead, there is likely a dip in the *TL* at this frequency. Insulation in this gap usually minimizes this dip. Bear in mind that theoretically the *TL*, as predicted by the mass law as being proportional to $f\rho$, means the *TL* increases with increasing frequency. In reality, this value is capped at approximately 55 dB by realistic constructions and flanking paths – especially around penetrations or break-in/breakout ducting. Vendor data measured in a lab setting should also be capped at roughly 55 dB.

If the total interface is made up of n different constructions, each area S_n, the composite *TL*, dB, can be computed as:

$$TL, \text{dB} = 10\ \text{Log}\left(\sum\nolimits_n S_n / \sum\nolimits_n S_n \tau_n\right) \tag{2.24}$$

where $\tau = 10^{(-TL/10)}$

Finally, any hole in the construction from the source to the receiver side greatly diminishes the *TL*. As an example, a 150 m^2 bulkhead rated with a *TL* of 40 dB would have a composite *TL* value of 20 dB if it had just a 1% opening. This cuts the *TL* in half!

The transmission loss of some typical constructions is given in Table 2.4 (adapted from Fischer et al. 1983). Bear in mind that Noise Reduction, *NR*, is not the same as transmission loss, *TL*. *NR* is a measure of the difference in noise levels in the source and receiver spaces. However, the TL takes into account the intervening area:

$$TL, \text{dB} = NR + 10\ \text{Log}(S_p/R) \tag{2.25}$$

Table 2.4 Representative transmission loss values, dB

	Octave Band Frequency, (Hz)								
Material	31.5	63	125	250	500	1000	2000	4000	8000
4.8-mm, Steel	14	20	26	29	34	37	39	35	47
4.8-mm, Steel ribbed	12	18	24	28	33	36	37	35	47
4.8-mm, Steel ribbed, 12 kg/m^2 sheathing	13	19	30	34	39	44	55	32	57
4.8-mm, Al	5	11	17	23	26	31	33	28	37
4.8-mm, Al ribbed	3	9	16	22	25	30	32	28	37
6.4-mm, Steel (by Mass Law)	16	22	28	34	40	46	52	58	64
6.4-mm, Steel	16	22	26	31	36	40	37	42	51
6.4-mm, Steel ribbed	10	18	25	30	35	38	34	42	51
6.4-mm, Steel ribbed, 12 kg/m^2 sheathing	10	18	29	40	49	54	50	52	61
6.4-mm, Al ribbed	3	9	17	23	27	31	28	32	42
6.4-mm, Al ribbed, 12 kg/m^2 sheathing	3	9	21	33	41	47	42	42	52
9.5-mm, Steel ribbed	18	24	26	32	36	39	36	47	56
9.5-mm, Al ribbed	10	15	18	24	28	30	28	38	47
12.7-mm, Steel ribbed	20	27	28	34	39	35	41	51	60
12.7-mm, Steel ribbed, 12 kg/m^2 sheathing	20	27	32	44	53	51	57	61	70
16-mm, 12 kg/m^2 sheathing	6	12	18	24	29	31	25	25	30
Double Panel, St 6.4-mm thick, 100-mm air gap	22	29	36	48	60	60	60	60	60
Double Panel, St 6.4-mm thick, 50-mm air gap	22	29	36	45	57	60	60	60	60
Double Panel, St 3-mm thick, 50-mm air gap, with absorptive material in gap.	15	21	27	47	62	75	75	75	75
Double Panel, Al 6.4-mm thick, 100-mm air gap	13	19	25	38	38	38	38	46	60
Double Panel, Al 3-mm thick, 50-mm air gap, with absorptive material in gap.	8	14	20	31	46	54	54	57	60
50-mm Softcore Joiner Lining (on 6.4-mm Bulkhead)	16	24	32	41	51	57	52	59	66
50-mm Softcore Joiner Lining (on 6.4-mm Bulkhead)	13	21	29	38	48	54	49	56	63
50-mm Floating Floor on 9.5-mm Steel Deck	18	24	33	46	52	58	63	70	78
75-mm Floating Floor on 9.5-mm Steel Deck	18	24	36	47	52	59	65	73	82

2.7 Structure/Fluid Interactions and Radiation Efficiency

Of primary interest aboard ships is the interaction between structureborne vibrational energy in the hull structure and its coupling, via radiation, as airborne (or fluidborne) noise. This relationship is defined as:

$$L_w \text{ dB re 1pW} = L_a \text{ dB re 1 }\mu\text{G} + 10\,\text{Log}(\sigma) + 10\,\text{Log}(\rho c) + 10\,\text{Log}(S) - 20\,\text{Log}(f) - 4 \tag{2.26}$$

where

L_w is the radiated acoustic power for a given acceleration level, L_a, of the structure,
S is the radiating area, m^2, σ is the radiation efficiency,
f the frequency of concern, Hz, and
$\rho\ c$ the surrounding fluid density and sound speed in rayls (kg/(m^2–s)).

This coupling is controlled by the frequency-dependent 'radiation efficiency' of the structure. The controlling parameters for the radiation efficiency are the plate thickness[9], frame spacing, hull material Young's modulus, E, and density, ρ.

For the same vibration level on a hull plate and above its critical frequency, the plate will radiate acoustic power by 36 dB more into water than it will into air based on the difference in ρc: ρc for water is 1,500,000 mks rayls, while ρc for air is 406 mks rayls. Thus, radiation into the ocean from a hull plate is often critical. The impact of the difference in radiation efficiency for a 6.2 mm metal plate is addressed below in Table 2.5.

Going one step further, the resulting pressure level L_p dB re 20μPa at a distance r from a source in air with a given vibration level L_a dB re 1μG is given in Eq. (2.15).

Table 2.5 Radiation efficiency of air- and water-backed plates and impact on radiated power

Octave Band Freq.		31	63	125	250	500	1000	2000	4000	8000	16,000
Air loaded, Plate 1	10 Log σ, dB	−15	−16	−16	−16	−15	−13	7	−2	−1	0
Water loaded, Plate 1	10 Log σ, dB	−59	−50	−42	−33	−26	−19	−12	−7	−6	−5
Delta radiation	dB	−44	−34	−26	−18	−11	−6	−19	−6	−5	−4
Delta acoustic power, L_w	dB	−10	0	8	16	23	28	15	28	29	30

[9] At low frequencies, an equivalent plate thickness can be used to account for the 'smearing' of the ribs, which will effectively increase the apparent plate thickness.

In water, the equivalent equation for the resulting sound pressure level, L_p dB re 1μPa, is:

$$L_p \text{ dB re } 1\mu\text{Pa} = -20\,\text{Log}(r) + L_a \text{ dB re } 1\ \mu\text{G} + 10\,\text{Log}(\sigma) + 10\,\text{Log}(S) - 20\,\text{Log}(f) + 117 \quad (2.27)$$

The radiation efficiency describes the frequency-dependent coupling of bending[10] waves in a plate to the adjacent fluid – air or liquid; basically, the ratio of the different wavelengths in each component. The wave speed in the plate changes with frequency, while that in the fluid is independent of frequency. When the speed in the plate is lower than that in the fluid, the coupling is low. The coupling is best when the speed in the plate is equal to that in the fluid; this is called the critical frequency. The coupling is slightly lower above the critical frequency.

As noted above, the critical frequency, f_c, is the point at which the plate radiates most efficiently. The critical frequency for a plate is $f_c = c^2/(1.8\ c_l\ h)$, where c is the speed of sound in the fluid, c_l is the longitudinal wave speed in the plate and h is the plate thickness. The longitudinal wave speed in steel is approximately 5960 m/s and that in aluminum is 6420 m/s. Note that the thicker the plate is, the lower the critical speed; thus, a thick plate radiates more efficiently at lower frequencies, contrary to one's intuition. The critical frequency of a steel or aluminum plate in water is:

$$f_{c\ metal/water} = 24{,}750/h(\text{cm});\ \text{and a steel or aluminum plate in air:}$$
$$f_{c\ metal/air} = 1{,}290/h(\text{cm})$$

There are three regions of concern for radiation efficiency. Above the critical frequency, the radiation efficiency of a ribbed plate has a value of approximately 1, or *10* Log(σ) = 0 dB. Around the critical frequency, the radiation efficiency is hay-stacked with an approximate value between 3 and 7 dB. Below the critical frequency, the radiation efficiency decreases with lower frequencies. In the lower frequency bands, 'modal radiation' dominates at the interframe resonant frequencies (Beranek, 1992). These radiation values are listed below:

$$\text{Above critical frequency, } f_c\text{: } \sigma = 1\Big/\sqrt{\left(1 - {}^{f_c}\!/_{f}\right)}$$

$$\text{At the critical frequency, } f_c\text{: } \sigma = \left({}^{l_x f_c}\!/_{c}\right)^{0.5} + \left({}^{l_y f_c}\!/_{c}\right)^{0.5} \quad (2.28)$$

[10]Throughout this book, longitudinal and shear waves are ignored. The authors believe there is little loss in accuracy by doing so. More exact models, such as Statistical Energy Analysis (SEA), could address these wave types. The basic coupling from longitudinal waves to bending wave occurs by scattering at structural intersections.

Below the critical frequency, f_c: $\sigma = \left({c^2}/{Sf_c^2}\right) g_1(\alpha) + \left({Pc}/{Sf_c}\right) g_2(\alpha)$ (2.29)

where l_x and l_y are the rib spacings, S is the area between the ribs, and the perimeter, $P = 2\ (l_x + l_y)$ and

$$g_1(\propto) = \left[{8}/{\pi^4} \frac{1 - 2\alpha^2}{\alpha(1-\alpha^2)^{.5}}\right] \text{ for } f < fc/2 \text{ and equals 0 for } f > fc/2 \quad (2.30)$$

$$g_2(\propto) = \left({1}/{4\pi^2}\right)\left(\frac{(1-\alpha^2)\left(\ln\left({(1+\alpha)}/{(1-\alpha)}\right)\right) + (2\alpha)}{(1-\alpha^2)^{1.5}}\right) \quad (2.31)$$

where $\alpha = \sqrt{{f}/{f_c}}$

Around the first panel resonant frequency, the radiation efficiency is approximately equal to:

$$\sigma = 4Sf^2/c^2 \quad (2.32)$$

Typical radiation efficiencies for a 1.5 × 0.6 m by 0.63 cm (Plate 1) and a 1.27 cm (Plate 2) thick plate are shown in Table 2.6. The in air critical frequency is 2000 Hz for Plate 1 and 1000 Hz for Plate 2.

A numerical procedure combining finite element methods with Rayleigh integrals to determine radiation efficiency is provided in Knuth et al. 2023.

Note that the thicker plate radiates more efficiently, by approximately 4 dB, below its critical frequency. In a simply supported plate, the 1,1 mode is 46 Hz for a 0.63 cm plate thickness and 93 Hz for a 1.27 cm thick plate. The radiation efficiency for this mode is −12 dB for the thinner plate and −6 dB for the thicker plate. Note that edge conditions – simply supported, clamped, or free – will all affect the natural frequency and radiation efficiency (Beranek, 1992).

As mentioned above, with respect to the radiated power into air or water, the acoustic power radiated is higher by 36 dB due to adjusting for the different $\rho c's$. The other factor is the difference in air-loaded versus water-loaded radiation

Table 2.6 Representative radiation efficiencies, ribbed plate in-air

	Octave Band Freq, Hz	31	63	125	250	500	1000	2000	4000	8000
Plate 1	10 Log σ, dB	−15	−16	−16	−16	−15	−13	7	−2	−1
Plate 2	10 Log σ, dB	−11	−12	−12	−11	−10	5	−2	−1	0
Plate 1-Plate 2	dB	−4	−4	−4	−4	−5	−18	8	−1	0

efficiencies, as shown in Table 2.5. The critical frequency is approximately 36,000 Hz for a water-loaded 0.635 cm thick plate; below this frequency, it is a poorer radiator than an air-backed plate. However, the combination of changes in radiation efficiency and ρc results in a significantly higher radiated power, L_w (Eq. (2.28)), as illustrated in the bottom row of Table 2.5.

2.8 Acceptance

An important part of noise prediction algorithms is the calculation of hull structure vibration levels induced by incident airborne noise from equipment sources (Boroditsky and Fischer 2012). For example, if a noisy piece of equipment is placed on effective mounts, airborne noise excitation is not a significant factor with respect to habitability and underwater noise. Examples are an aircraft noise when an aircraft lands or takes off aircraft carrier decks.

2.8.1 Multi-Modal Approach

Smith 1962 and Maidanik 1962 developed equations to model the modal vibration response to an incident diffusive airborne sound field.

$$\langle \nu_m^2 \rangle = {}^{2c\pi^2}\!/_{M\rho\omega^2}\ S_p(\omega_m)\ \mu_m \tag{2.33}$$

where v_m –modal velocity, c-speed of sound in air, M- mass of subsystem, ρ-air density, $S_p(\omega_m)$ – sound pressure spectral density at the natural frequency ω_m, and $\mu_m = \frac{\rho\sigma c}{\rho\sigma c + \eta\omega m}$, where σ –radiation efficiency, m-mass per area unit, c is the speed of sound, and η – loss factor.

If every vibration mode in the frequency band of interest Δf equally contributes to the overall vibration level, one can assume:

$$\langle \nu^2 \rangle_\Sigma = \langle \nu_m^2 \rangle\ \Delta N \tag{2.34}$$

where ΔN – number of modes in the frequency band of Δf width.

Using the expression for plate modal density (Cremer and Heckl 1973), it is possible to derive:

$$\Delta N = \left(\frac{\pi A f_c}{c^2} \right)\ \Delta f \tag{2.35}$$

where A is the panel area.

If the sound spectral density is uniform in a frequency band, the overall sound pressure in the band is equal to:

$$\langle p^2 \rangle = 2\pi \Delta f S_p(\omega) \tag{2.36}$$

Combining the Eqs. (2.33, 2.34, 2.35, and 2.36) and considering acceleration instead of velocity, one can come to the transfer function:

$$L_p - L_a = 10 \text{ Log}\left(\frac{m\rho c}{4\mu f_c \pi^2}\right) \tag{2.37}$$

where L_p is the noise level near a plate, dB re 20 μPa; L_a is the acceleration level dB re μG; m is the mass of a plate unit area, kg/m^2; and the resistance ratio μ is calculated for each frequency band.

Equation (2.37) can be easily confirmed under laboratory and field conditions. Figures 2.1, 2.2, and 2.3 demonstrate the measured and calculated responses of a hull structure. The common tendency is a significant discrepancy between measured and calculated values at the lower frequencies of the audible frequency range. The heavier the structure is, the lower the natural frequency of the plating, and the greater the disagreement in the low-frequency range. When there are multiple resonances in a frequency band, Eq. (2.37) gives satisfactory results for ship hull structures.

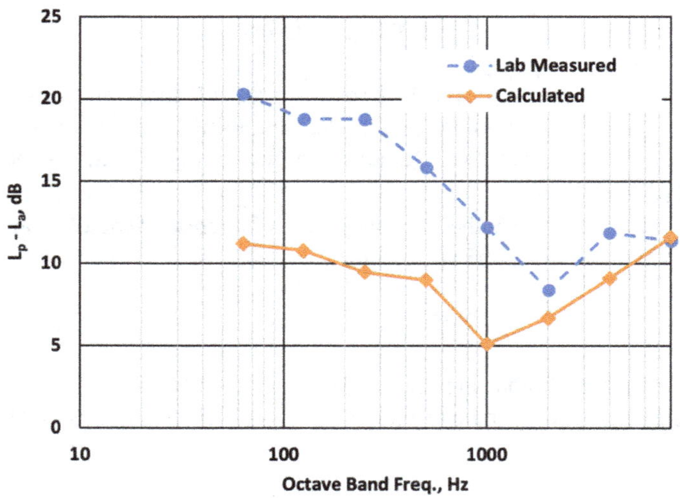

Fig. 2.1 Transfer function for a 10 mm thick steel plate

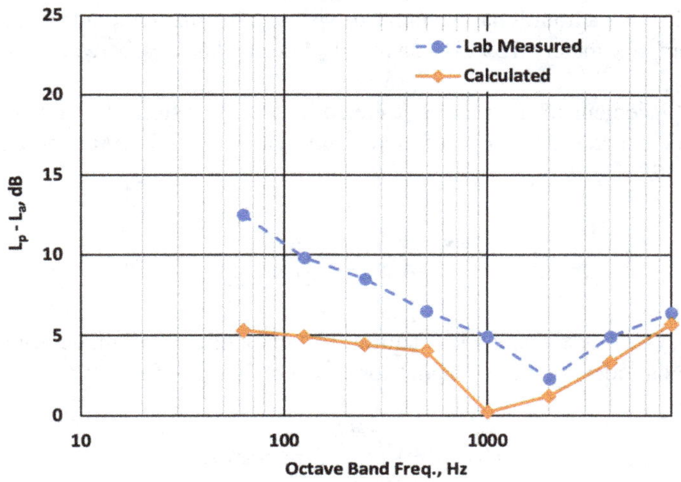

Fig. 2.2 Transfer function for a 10 mm thick aluminum plate

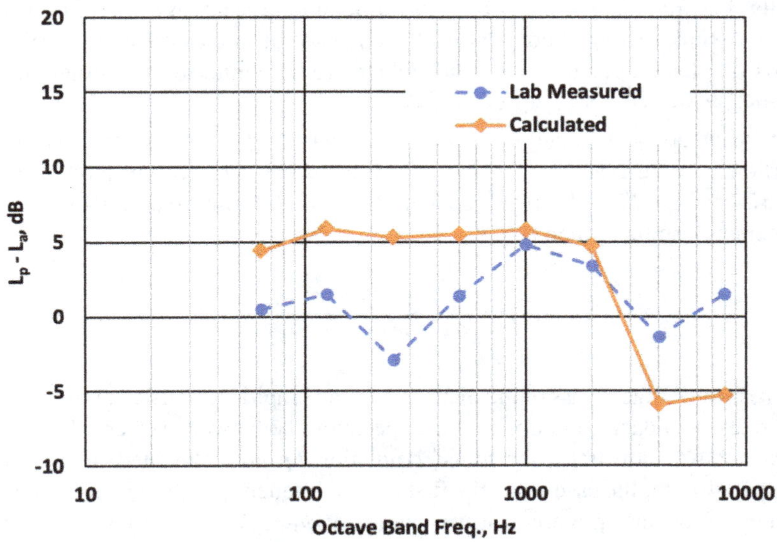

Fig. 2.3 Transfer function for a 3 mm thick aluminum plate

2.8.2 Low-Frequency Approach

Equation (2.37) and measurements demonstrate a weak dependence on frequency in the frequency range where the vibration is multi-modal in the frequency band. However, this is not the case for some hull structures at low audible frequencies.

For example, with a traditional 0.6 m frame spacing, the first natural frequency of the plating is approximately 125 Hz for metal plate thicknesses between 15 mm and 25 mm.

The displacement of the plate at frequency ω =2 π *f* from distributed pressure *p(x,y)* can be expressed in terms of modal summation (Fahey, 2007 and Boroditsky and Fischer 2012):

$$w(x,y)=\sum\nolimits_{r=1}^{\infty}\frac{\phi_r(x,y)\widetilde{F}_r}{M_r\left[\omega_r^2(1+j\eta)-\omega^2\right]} \tag{2.38}$$

where $\phi_r(x,y)$ – mode shape, ω_r – natural frequency of the *r*-th mode, η-loss factor, $r(r_1,r_2)$-modal index, and $\widetilde{F}_r$ – modal force, which is equal to:

$$\widetilde{F}_r=p(\omega)\iint_0^{l_xl_y}\phi_r(x,y)dxdy \tag{2.39}$$

If the frequencies of interest are below the first natural frequency, the sum sign may be omitted, and only the first term in Eq. (2.38) may be counted. The sound pressure typically does not change much over the structural element area (deck, bulkhead). One can assume *p(x,y)* = const at each frequency of interest. This integral value depends on plate boundary conditions, which are indefinite for real structures. For supported edges: $\iint_0^{l_xl_y}\phi_r(x,y)dxdy=\frac{4l_xl_y}{\pi^2}$

For stiffer boundary conditions, this value will be less; however, it is directly proportional to the plate area. For a two-dimensional system, the modal mass is $M_r = M/4$, where *M* is the plate total mass. With all these assumptions, the plate displacement can be written as:

$$w(x,y)=\frac{16\phi_1(x,y)p(\omega)}{\rho_s\pi^2\left[\omega_1^2(1+i\eta)-\omega^2\right]}\gamma \tag{2.40}$$

where ρ_s is the surface mass of the plate and γ is an empirical coefficient that depends on the plate boundary condition and is expected to be between 0.5 and 1.0.

Figure 2.4 demonstrates the transfer function *Lp-La,* calculated in accordance with Eq. (2.40), for the case when the first natural frequency of plating is less than the frequency of the airborne noise. It was assumed: $\phi_1(x,y)=\alpha=1$ One can see that the qualitative frequency dependence of this transfer function is close to the measured values for relatively thick plates (Figs. 2.3 and 2.4). The order of value also corresponds to measured data in the low-frequency range. The 12 dB per octave decay is more or less close to the experimental data for thick plating.

However, finite element and field structural analysis show that at least one to two resonances in an audible octave band are expected. Let us assume that at least one resonance exists in the octave band and that a resonance frequency is in the middle of

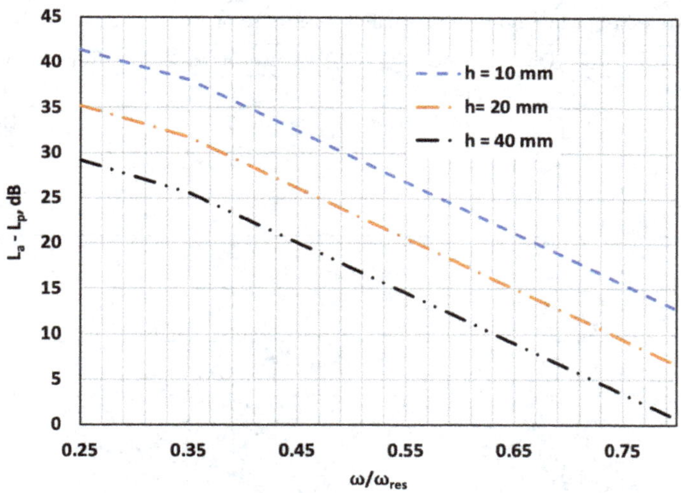

Fig. 2.4 Transfer function for steel plating (frequency below natural frequency)

an octave band. In this case, the average in octave band acceleration squared may be derived from Eq. (2.40) as:

$$\langle a^2 \rangle_{oct.} = \frac{16^2}{\rho_s^2 \pi^4 \Delta\omega_{oct.}} \int_{\omega_{res}/\sqrt{2}}^{\omega_{res}\sqrt{2}} \frac{p^2(\omega)\omega^4 d\omega}{\left(\omega_{res}^2 - \omega^2\right)^2 + \omega_{res}^2 \eta^2} \tag{2.41}$$

Approximate value of this integral for diffusive field:

$$\langle a^2 \rangle_{oct.} = \frac{64 p^2(\omega)}{\rho_s^2 \pi^5 \eta \sqrt{2}} \tag{2.42}$$

The transfer function in this case is:

$$Lp - La = 10 \text{ Log} \frac{\rho_s^2 \pi^5 \eta \sqrt{2}}{16^2} \tag{2.43}$$

Table 2.7 shows the values of this transfer function for three plate thicknesses for $\eta = 0.005$ (bare steel hull in the low-frequency range) for three cases: one resonance in the octave band, two resonances and three resonances in the octave bands.

The values in Table 2.7 are well correlated with the measured data in the frequency range between 250 Hz and 500 Hz shown in Figs. 2.1, 2.2, 2.3, and 2.4.

Table 2.7 Transfer function Lp-La for a small number of resonances in octave bands, dB

Thickness, mm	One resonance	Two resonances	Three resonances
10	14	11	9
20	20	17	15
40	26	23	21

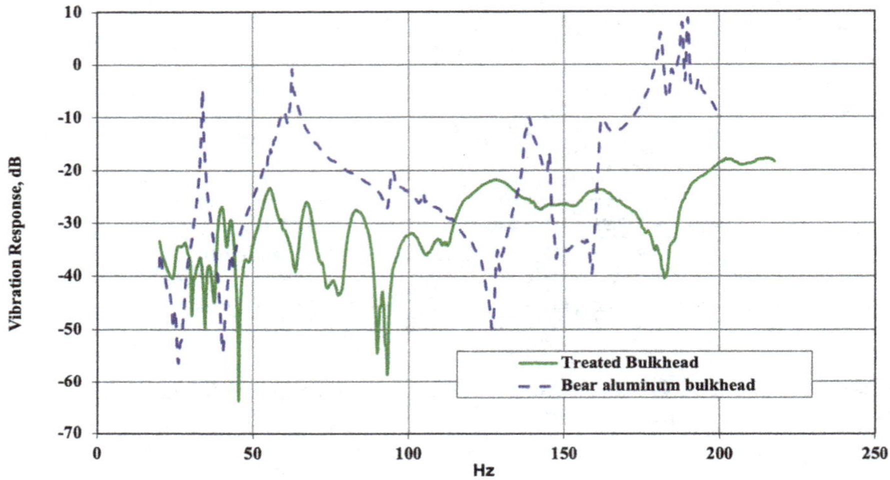

Fig. 2.5 Swept sine response of panels treated and untreated

2.9 Damping Effectiveness

Damping materials have been found to be very effective in reducing both the airborne and structureborne response of framed structures, especially on high-speed, aluminum naval craft (Boroditsky and Fischer 2008). Damping treatments generally require thicker layers of damping on steel and, in some cases, a constraining plate. This treatment not only reduces machinery-induced noise but is also effective in abating flow-induced noise, whether it is excited by waves or foils. Any type of damping generally reduces the resonant response of framed structures, especially when the thickness of the damping material is equal to that of the substrate. As shown in Fig. 2.5, the resonant responses at 70 and 180 Hz are significantly reduced. Thus, at low frequencies, the damping may significantly reduce the structureborne noise, acceptance, and airborne transmission loss.

Damping treatments have significant frequency- and temperature-dependent properties (Nashif et al. 1985). Vendors often supply ‘reduced-frequency curves’ showing the relationships between damping or Young’s Modulus and these two parameters – see Fig. 8.27. The loss factor and modulus are on the right-hand side, and the reduced frequency is on the lower x-axis. Other treatment types are not

Table 2.8 Loss factor values – 3 mm thick aluminum ribbed bulkhead

Bulkhead Material	Coating Type	Plate Thickness, mm	Coating Thickness, mm	Octave Band Damping Loss Factor				
				63	125	250	500	1000
Aluminum	No coating	3	0	0.007	0.006	0.007	0.0047	0.0032
Aluminum	Damping Material	3	2	0.063	0.055	0.057		
Damping Material Loss Factor, dB				9.5	9.6	9.1		

generally temperature dependent. Another critical parameter is the material-to-base plate thickness ratio; the higher it is, the better. Optimal areas to receive treatment should be analyzed by acoustic models capable of considering all the potential sources and the acoustic transmission paths to the receivers of interest. The measured loss factors of the composite bulkhead/treatment are provided in Table 2.8. The authors found that the real effectiveness of a damping treatment is always less than 10 Log (η/η_0), where η is the loss factor of a coated structure and η_0 is the same for an uncoated structure.

The application of damping treatments can impact airborne transmission loss and acceptance. The following examples are for a 2 mm (80 mils) thick damping layer on a 3 mm thick aluminum bulkhead. As shown in Fig. 2.6, damping increases the airborne *TL* by approximately 5 dB across the audible frequencies. More importantly, the damping treatment reduces the acceptance of the panel to incident airborne noise by 8 to 12 dB. This would have a significant impact on secondary structureborne noise (Boroditsky and Fischer 2008). Since acceptance is the inverse of radiation, damping can also have a significant impact on this parameter. Figure 2.7 shows that the acceptance of an aluminum bulkhead is greatly decreased by the application of 2 mm (80 mils) of damping material.

Based on before and after testing regarding bow thruster-induced noise, it was found that the airborne noise in compartments around the thruster room was reduced by 3 to 5 dB, while damping of the room's surfaces reduced the vibration by approximately 10 dB (Spence 2011).

2.10 Fluid Loading

Water loading on a stiffened plate will impact multiple acoustic characteristics. Water or fluid loading will shift plate natural frequencies lower. It will also decrease the vibration response of a stiffened plate and increase the damping loss factor in the

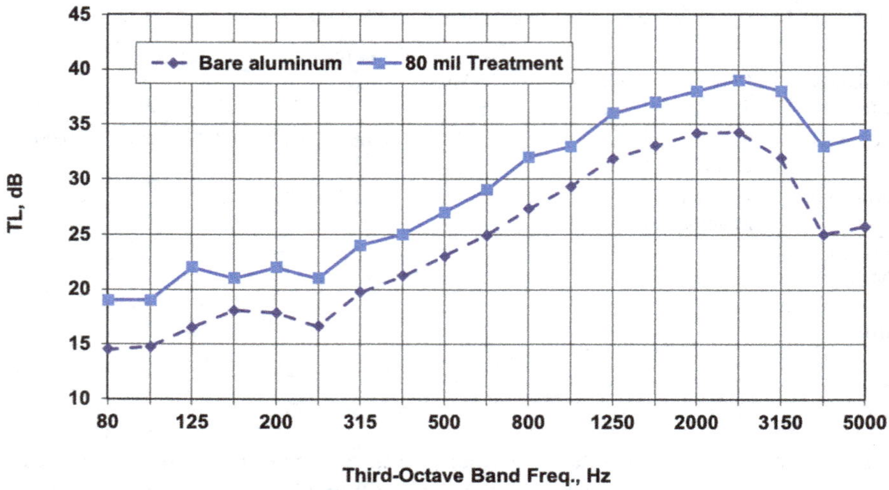

Fig. 2.6 Airborne TL, bare and with damping

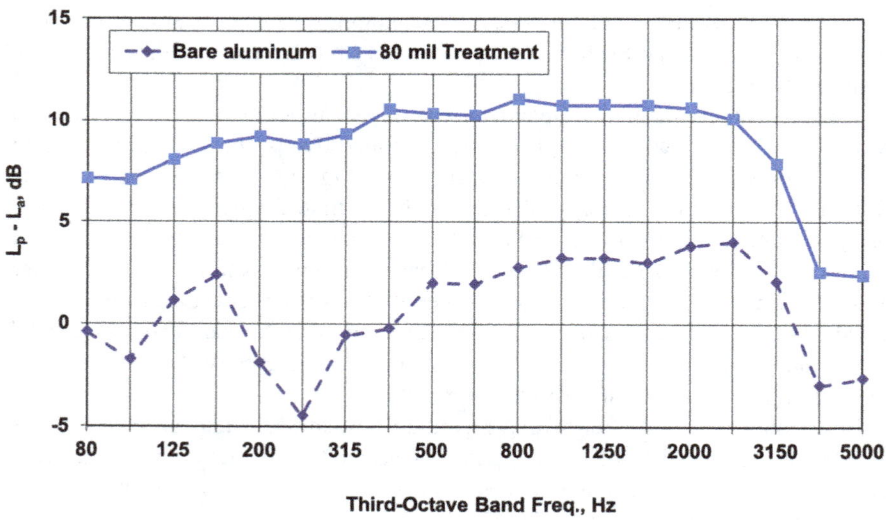

Fig. 2.7 Plate acceptance, bare and with damping

form of 'radiation damping.' In one study, a 50% difference was found in the fundamental natural frequency of a plate coupled to infinite or finite fluid loading (Kim et al. 2007). Finally, water loading will impact the radiated power.

The reduction in modal frequencies of a stiffened plate due to water loading is a direct impact of the 'added-mass', the water going along for a ride with the vibrating system. The rule-of-thumb is that the added mass is equal to a depth of approximately a sixth of the bending wavelength. This means that the effect will be significant at low frequencies and much less significant at high frequencies. This fluid loading parameter, β, is computed as (Rumerman 2007):

$$\beta = \rho c / \omega \rho_m t = \rho c_c k_b^2 / \sqrt{12}\, \rho_m c k^2 = \rho c_c / \sqrt{12}\, \rho_m c \Omega \tag{2.44}$$

where ρ_{m} is the mass density of the plate, t is the plate thickness, c_c is the speed of the compressional wave in the plate, k_b is the wavenumber of the bending waves in the plate in-vacuum for the frequency of interest, ρ is the mass density of the fluid medium, c is the speed of sound in the fluid (frequency independent), k is the wavenumber in the fluid, and $\Omega = k_b/k$.

For instance, the bending wavenumber, k_b, of a 0.008 m thick plate at 31.5 Hz is 4 m^{-1}. The fluid wavenumber at the same frequency is 0.13 m^{-1}, and k_b, including the fluid loading, is 6 m^{-1}.

The parameter Ω is a ratio of the bending wavenumber, k_b, in the plate to the fluid wavenumber or a fraction of the 'coincidence frequency' ratio. When this value is below 0.5, the power, Π, radiated by a force on the plate becomes:

$$\Pi = F^2 \left(\rho / 2\omega (\rho_m t)^2 \right) \left(1 - \left(\beta / \sqrt{1+\beta^2} \right) \right) \tag{2.45}$$

When β is small, the power, Π, radiated by a force on the plate becomes:

$$\Pi_{\beta\ small} \sim F^2 \left(\beta \rho / 2\omega (\rho_m t)^2 \right) \tag{2.46}$$

and when β is large, the power, Π, radiated by a force in the plate becomes

$$\Pi_{\beta\ large} \sim F^2 \left(\rho / 4\beta^2 \omega (\rho_m t)^2 \right) \tag{2.47}$$

Therefore, the difference in radiated power due to water loading is proportional to $1/2\beta^2$. For example, for a β of 2, the radiated power decreases by 9 dB for the same F input.

The impact of water loading on radiation efficiency, σ_{rad}, is covered by Sect. 9.3.2. The significant increase in the critical frequency, f_{cw}, due to water loading over that of a dry plate, f_{ca}, effectively lowers the water-loaded radiation efficiency for all frequencies below f_{cw}. Thus, for the same vibration level on a water-loaded versus dry plate, the radiated power will be less due to this lower radiation efficiency. For instance, a 0.001 m thick plate will have a critical frequency of 12 kHz in air and 223 kHz in water.

Radiation damping can be considered as the capacity of the plate to radiate acoustic energy. It is defined as (Choi et al. 2004):

Table 2.9 Radiation loss factor – wet versus dry aluminum plate

Freq., Hz	Wet loss factor	Dry loss factor	Difference factor
100	0.002526	0.000257	9.8
160	0.00249	0.000211	11.8
250	0.002424	0.000176	13.8
500	0.002373	0.000145	16.4
800	0.002317	0.000114	20.3
1250	0.002226	8.83E-05	25.2
2000	0.00215	7.27E-05	29.6
3000	0.002074	6.21E-05	33.4
5000	0.001952	6.11E-05	31.9
6300	0.001861	8.61E-05	21.6
10,000	0.00179	0.000237	7.6
Min	0.00179	0.0001	7.6
Max	0.002526	0.000257	33.4

$$\eta_{rad} = {}^{\rho}\!/_{\rho_m}\ {}^{1}\!/_{kta}\ \sigma_{rad} \tag{2.48}$$

and

$$\alpha = \left(1 \quad if\ f > f_c \middle| \left(1 + {}^{\rho}\!/\!\left(\rho_m t \sqrt{k_b{}^2 - k^2}\right)\right)\ if\ f < f_c\right) \tag{2.49}$$

where k is the fluid wave number: $k = 2\pi f/c$, k_b is the plate bending wave number $= \omega/(D\omega^2/t\rho_m)^{0.25}$, and D is the plate bending stiffness $= Et^3/12(1 - \nu^2)$ where E is the Young's modulus and ν is Poisson's ratio.

As shown in Table 2.9, the radiation loss factor for a plate in water is typically orders of magnitude higher than that in air over most of the frequency range of concern. The loss factor of a water-backed plate is up to 33 times that of a dry plate.

Note that the water loading effect also applies in the determination of the blade natural frequencies (see Sect. 5.3.5.2). Water loading is in the range of 25 to 30% of the blade mass. A more precise estimate of the added mass to a propeller blade, M_a, is given by Lewis 1977 and Slater 1984, as:

$$M_a = (0.21\ \rho n D^3)(MWR^2)/((1 + (P/D)^2)\ (0.3 + MWR)) \tag{2.51}$$

where D is the propeller diameter, P is the nominal propeller pitch, MWR is the mean width ratio (ratio of blade width to propeller diameter), n is the number of blades, and ρ is the fluid density.

The above equation applies to the axial motion of the propeller 'with rotational restraint.' For an unrestrained shaft, the added mass M_{au} is:

$$M_{au} = M_a / \left(1 + (P/2\pi)^2\, M_a / I_p\right) \qquad (2.52)$$

The added polar mass moment of inertia, I_a, is

$$I_a = 0.8\, M_a \ \ (P/2\pi)^2 = 0.02\, M_a\, P^2 \qquad (2.53)$$

where I_p is the polar mass moment of inertia of the propeller in air.

References

Beranek, L. (ed.): Noise and Vibration Control. McGraw-Hill Book Company, NY (1971)

Beranek, L., Ver, I. (eds.): Noise and Vibration Control Engineering, Principles and Applications. Wiley, NY (1992)

Boroditsky L., Fischer, R.: Acoustical Effectiveness of Damping Coatings. Acoustics '08, Paris, June 2008

Boroditsky, L., Fischer, R.: Ship Hull structure response on airborne noise excitation. In: 19th International Congress on Sound and Vibration, Lithuania, July 2012

Choi, K.K., Dong, J., Vlahopoulos, N., Wang, A., Zhang, W.: Design sensitivity analysis and optimization of high frequency radiation problems using energy finite element method and energy boundary element method. In:10th AIAA/ISSMO Multidisciplinary Analysis and Optimization Conference, NY, August 30-September 1, 2004

Cremer, L., Heckl, M.: Structure-Borne Sound, pp. 452–453. Springer-Verlag, NY (1973)

Fahey, F., Gardenia, P., Sound and Structural Vibration, Elsevier (2007)

Fischer, R., Boroditsky, L.: Supplement to SNAME T&R 3-37- Shipboard Airborne Noise Control. Society of Naval Architects and Marine Engineers (SNAME), Jersey City, NJ (2001)

Fischer, R., Nelson, D., Burroughs, C.: SNAME T&R 3-37 – Shipboard Airborne Noise Control. Society of Naval Architects and Marine Engineers (SNAME), Jersey City, NJ (1983)

Kim, K., Choi, S., Cho, J.: Theoretical and numerical study on hydroelastic vibration of plate for various bounded fluid field. ICSV14. (2007)

Knuth, C., Squicciarni, G., Thompson, D.: An engineering approach for estimating the radiation efficiency of orthogonally stiffened plates. J. Vib. Acoust. **145**(3), 031006 (2023)

Lewis, F.M.: Hull vibration of ships. In: Crocker, M. (ed.) Principles of Naval Architecture. SNAME NY (1977)

Loeser, H.: Fundamentals of Ship Acoustics. The Society of Naval Architects and Marine Engineers (SNAME), Jersey City, NJ (1999)

Maidanik, G.: Response of ribbed panel to reverberant acoustics fields. J. Acoust. Soc. Am. (JASA). **34**(6), 809–826 (1962)

Nashif, A.D., Jones, D.I.G., Henderson, J.P.: Vibration Damping. Wiley, NY (1985)

Rumerman, M.: Radiation efficiencies for steel plates in water. Shock. Vib. **14**, 81–88 (2007)

Slater, J.: A Review of Hydrodynamic added mass inertia of vibration submerged structures. Defense Research Atlantic, TM 84/F, April, 1984

Smith, P.: Response and radiation of structural modes excited by sound. J. Acoust. Soc. Am. (JASA). **34**(5), 640–646 (1962)

Spence, J.: Measurement of spray-on damping effectiveness and application to bow thruster noise on ships. Noise Control Eng. J. **59**(2), 126 (2011)

Thumann, A., Miller, R.: Fundamentals of Noise Control Engineering, 2nd edn. The Fairmont Press, Inc, Liburn GA (1990)

Chapter 3
Acoustic Criteria

This chapter starts to answer the question of "How Quiet"? A noisy craft affects humans – both onboard the vessel and ashore or nearby (e.g., in the case of man-overboard or in-port). Ship-induced noise also affects sea life, sometimes over large distances. Acoustic criteria, or in some cases only 'guidance', have been established to limit these impacts. For crew and passengers, noise impacts hearing loss, health, comfort, and performance as a minimum. For warfighters, noise can affect stealthiness and situational awareness. For sea creatures, noise can affect feeding, communication, health, and reproduction.

Any crew subjected to high noise may incur hearing loss. Furthermore, some crew members are more sensitive to noise and suffer losses not registered by the 'average' person. High noise and vibration levels can cause sleeplessness, adverse operational performance, and other health issues. It is also important that any acoustic goals be fully and comprehensively defined to limit these consequences. This includes vessel and machinery operating characteristics, noise/vibration level in well-defined units and frequency bandwidth and range, measurement locations, recognized test standards, reporting requirements and how any 'overages' or excesses are to be resolved. Salient government and regulatory acoustic requirements need to be understood, so these requirements are not misapplied or used in circumstances where the requirement is inappropriate or, even worse, unachievable.

Most hearing criteria have been developed with respect to a frequency weighting that heavily discounts low frequencies and emphasizes high frequencies – called A-weighting. Human responses to vibration levels also have various frequency weightings that are used to determine the magnitude of their impact. These frequency weighting and the length of time of exposure apply to both humans and sea

R. Fischer, L. Boroditsky, *Noise and Vibration Control on Ships*,
https://doi.org/10.1007/978-3-031-55170-3_3

creatures.[1] Differences between times of exposure at sea and those of ashore in factories are presented to highlight the impact of operation at sea.

Similar to the requirement to reduce noise impacts on the crew, there is current interest in reducing underwater noise radiated by vessels to lessen the impact on ocean mammals and other wildlife. Most of the effort carried out to assess and control the noise impact on the crew would be directly applicable to noise coupling to the ocean. Sonar systems are also specialized marine systems with underwater acoustic requirements. There are usually signal-to-noise (or sonar self-noise) limits for these systems to work effectively.

Each of these noise and vibration criteria can have different limits, frequency ranges and bandwidths, and distances associated with the impact. The operating profile of the vessel should play a large part in establishing the system requirements. A high-speed vessel that only operates at high speed for a very limited time may only have a noise limit based on normal operational speeds. Achieving any type of acoustic goal or requirement compounds the need to thoroughly understand, predict, and control noise and vibration from all types of vessels.

Note that some noise limits are provided as 'guidance' and some as 'criteria.' Guidance documents generally have no compliance implications. Criteria documents have a set compliance level set with potential consequences if not met.

The following books and papers address additional details: Beranek and Ver 1992; Boroditsky and Spiridonov 1974; Fischer 2001, 2007, 2017 and 2020; Humes et al. 2005; and Loeser 1999; Fischer 1983.

With respect to the marine field, there are a multitude of noise and vibration criteria and 'guidance' developed by international, national, and various marine organizations or non-government organizations (NGOs). Noise criteria exist for

- hearing conservation
- crew/passenger comfort and fatigue
- communications – including hearing warning signals
- annoyance
- privacy
- health (impact on sleeping, and the human body/organs[2])
- impact on the environment, both above and below the waterline.

Vibration criteria address:

- human comfort and fatigue
- acceptable machinery and equipment vibration
- structural fatigue

[1] Hearing sensitivity and exposure limits for mammals are beyond the scope of this book. The reader is referred to Southall et al. 2009 and IMO – https://www.imo.org/en/MediaCentre/HotTopics/Pages/Noise.aspx

[2] http://textbook.maritimemedicine.com; Tim Carter, Textbook of Maritime Medicine, 2nd Ed., 2013.

Table 3.1 Hazardous noise limits

Noise limit	Exposure time	Comment
85 dB(A)	8 h	OSHA (1971, rev. 1983)
87 dB(A)	8 h	EU
87 dB(A), 140 dB(C)	Daily or weekly	EC directive 2003/10/EC; Merchant Shipping and Fishing Regulation 2006; lower exposure action values – 80 dB(A) daily or weekly
85 dB(A)	8 h	Upper limit for 'required action level' from OSHA
82 dB(A) $L_{eff}(24)$	24 h	US Coast Guard Navigation and Vessel Inspection Circular No. 12–82; Time weighted average over 24 h
80 dB(A) $L_{eff}(24)$; 135 dB(C)	24 h	Code on noise levels on board ships, MSC 91/22/Add.1, Annex 1; applies to existing ships above 1600 gross tonnage, new ships below 1600 'as far as reasonable and practical;
85 dB(A)	8 h	Department of Defense Instruction 6055.12, Hearing Conservation Program (HCP)
85 dB(A)	Hazardous noise compartment	Appendix F, Shipboard Compartment Noise Requirements, MIL-STD 1474E
70 dB(A)	24 h	Protection for most sensitive person, EPA

EC (2003) https://osha.europa.eu/en/legislation/directives/82
OSHA Occupational Safety and Health Admin.; https://www.osha.gov/noise
EU European Union; https://osha.europa.eu/en/legislation/directives/82
EPA (1974) Environmental Protection Agency; https://www.epa.gov

For the military, airborne criteria are developed for aural detectability and 'situational awareness' (Casali et al. 2011). Underwater noise with respect to detectability and operation of sonar systems for the military are other concerns. All of these criteria are frequency- and, in some cases, time dependent.

The earliest criteria typically addressed hearing conservation. High noise levels are also known to increase lost work time and decrease productivity (Avery and Gray 2009). Early hearing conservation limits were compiled by Guignard (1973). The same year, the US Environmental Protection Agency (EPA) issued criteria for noise limits, EPA 550/9-73-002. A 'brief historical overview and perspective' of occupational hearing loss is provided by Berger et al. 2003.

In Technology for a Quieter America, National Academy of Engineering, 2010, the following hazardous limits are defined (see Table 3.1):

There are two common hearing conservation exchange rates involving the time of exposure:

- 3 dB per halving (or doubling) of exposure recommended by NIOSH, ANSI 200a, ISO 1999:2013, e.g.,
  - 85 dBA for 8 h same as 88 dBA for 4 h
- OSHA and MSHA use a 5 dB exchange rate
  - 90 dBA for 8 h same as 95 dBA for 4 h

On board ships, noise may be quantified as steady-state, fluctuating,[3] intermittent,[4] or impulsive.[5] Each may have their own limits for level and exposure parameters. Steady-state noise is considered 'continuous' as the noise in an engine room or from HVAC systems in accommodations. 'Fluctuating' noise is continuous, but its level rises and falls, typically by more than +/−5 dB. Intermittent noise is not steady, having starts and stops between periods, such as an air compressor that cycles on and off. Impulsive noise is considered intense with a short duration.

Research on non-Gaussian noise exposure aboard ships is exploring Kurtosis as a metric of understanding hearing loss. The difference on ships is that there may be a unique impulse exposure in a reverberant space. It has been proposed that the kurtosis value is a metric for determining the impulsiveness of a signal, thereby eliminating the difficulties of establishing relative peak levels and durations (Erdreich 1986). Additional research has been conducted to understand the relationship between kurtosis and hearing loss (Hammernik 2003). These studies have shown that high-level non-Gaussian noise (impulsive) routinely causes greater levels of hearing damage than energy equivalent (i.e., equivalent dose) Gaussian exposures. Damage increases with increasing kurtosis, and the spectrum of hearing loss follows spectral kurtosis (Lei et al. 1994).

While qualitative levels exist based on various parameters or studies, not all people respond the same way to sound. Their subjective response to noise and hearing impact depends on hearing acuity, individual characteristics and attitude, and level and characteristics of the sound itself, such as frequency and length of exposure (Webster 1975). Thus, most criteria are not truly 'universal' but generally define a norm. Some people are ultra-sensitive to noise, while others respond adversely to low-frequency noise. The spectrum and levels of sound in the auditory field are illustrated in Fig. 3.1. The lower blue line in the plot represents the threshold of hearing.

Carter (2013) considers the following possible impacts of high noise on humans: sleep/alertness, cardiovascular, and vision. These all depend on one's physiological and psychological make-up and sensitivity to noise. High vibration adds the following potential effects: back pain, respiratory, and balance/sea sickness. Vibration criteria exist for impact on human comfort and fatigue, as well as damage to structures and machinery balance and condition. Other shipboard stressors, such as heat and light, may add to the above concerns.

Community noise criteria – land side – exist for operations dockside or underway, say along a river.

Quantitative noise criteria do not yet exist with respect to marine life; however, they are being considered by various regulatory agencies, particularly the International Maritime Organization (IMO).[6] Anthropogenic noise has measurable impacts

[3]Noise levels that rise and fall (rapidly or gradually) more than +/−5 db(A)

[4]Noise that is discontinuous

[5]Noise that is transient of short duration and rapid growth of more than 40 dB.

[6]https://www.imo.org/en/MediaCentre/Pages/WhatsNew-1818.aspx

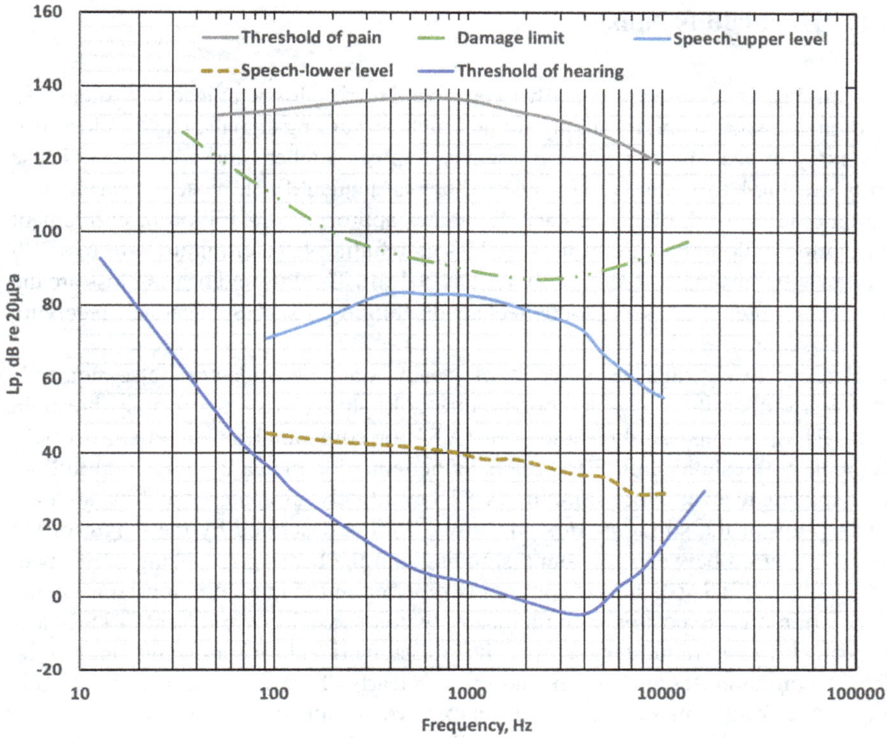

Fig. 3.1 Human auditory sound field

on fish and mammals, affecting their communications and health. There is some evidence that fish are also affected by vibration in the fluid – especially with respect to pile driving or explosive demolition. Research agencies have developed underwater radiated noise (UWRN) guidance to reduce the impact on sea creatures; the military has created limits based on the detectability of the vessel.

Many of these marine life criteria overlap with the types of criteria developed for humans, e.g., hearing conservation, communications, and health. Significant differences between land-based and marine-based criteria for hearing loss revolve around the time of exposure (8 h vs. 24 h), type of exposure (mid- to high-frequency vs. low frequency), steady-state noise vs. impulsive or fluctuating, and exposure to a vibrations existing on the marine platform.

This chapter is as crucial as the remaining chapters on predicting and controlling noise since the criteria and their implications need to be understood by the designer/operator/builder. These need to be translated into the ship's specification by the owner/operator, or there is little hope of delivering the vessel in an expeditious manner. Criteria need to be reasonable to be enforced. Bear in mind that any specification that says the vessel "is to be vibration free" will never be met. A quantitative value is always needed. Similarly, for a "noise free" ship. This type of breakdown will lead to increased cost and significant delays.

3.1 Hazardous Noise

Hearing loss is socio-economically expensive. Hearing loss can lead to difficulty in recognizing speech and localizing sounds such as warning signals. Hearing loss also adversely impacts other people, especially family members and coworkers. These costs are hard to quantify. However, there is a quantifiable cost – hearing loss compensation for the U.S. military. Estimates approach $1.2 billion of entitlement for compensation and care of hearing loss and auditory system injuries in over 1.8 million Veterans in 2012 alone (Yankaskas 2013b). Tinnitus and hearing loss are the two most prevalent service-connected disabilities for US veterans receiving compensation.

Hearing loss, either as a result of steady state or impact/impulse noise, is commonly quantified by a temporary threshold shift (TTS), permanent threshold shift (PTS), asymptotic threshold shift (ATS) or compound threshold shift (CTS). Temporary threshold shift (TTS) refers to the reduction in hearing threshold following exposure to an auditory stimulus. A TTS eventually dissipates sometime after the interruption of the stimulus. In some cases, TTS is described by the length of the recovery period between the sound stopping and the hearing test being performed. For example, TTS_2 is a metric commonly reported in the literature, representing the TTS 2 min after exposure. From a disability compensation standpoint, TTS is less critical. However, in military settings, it is mission critical for service members to be able to communicate and understand orders clearly. Therefore, minimizing TTS is imperative from a mission readiness perspective or 'situational awareness.'

Permanent threshold shift (PTS) describes a hearing threshold shift that will never recover. Permanent threshold shifts resulting from noise exposures are sometimes referred to as noise-induced permanent threshold shifts (NIPTS). Researchers have at times tried to correlate TTS to PTS resulting from impact and impulse noise. However, others have shown that TTS recovery patterns resulting from impact/impulse exposures are more complex than the linear time recovery pattern observed with continuous noise exposures. The complex recovery process has led some to question the assumptions made regarding TTS and PTS for impact and impulse noise (Henderson and Hamernik 1986).

Compound Threshold Shift (CTS) is a metric used within the Army's Auditory Hazard Assessment Algorithm for the Human (AHAAH) (Price 2010). It is the threshold shift measured immediately following noise exposure. The measured threshold shift will contain both temporary and permanent components; thus, the term compound.

In the United States, one of the earliest noise criteria was aimed at hearing loss prevention. The US Air Force developed hearing conservation considerations in *Air Force Regulation (AFR) 160-3 "Precautionary Measures Against Noise Hazards"* in 1956. This was followed by the *Walsh-Healy Act* [41-CFR 50-204.10], which defined occupational noise limits for government supply contracts. It took till 1970 to create the Occupational Safety and Health Administration (OSHA) and the National Institute for Occupational Safety and Health (NIOSH), both of which provide noise limits and institute noise control programs for hearing exposure. This became the well-known limit of 90 dB(A) for 8 h of exposure with a 5 dB exchange rate[7] based on data from industrial surveys (29-CFR-1910.95). The basic assumption was that the other 16 hours of the day were spent in a 'quiet' environment to allow the ear to recuperate. At sea, this quiet time is not generally achieved. One of the more complete hearing 'policies' is contained in *Dept. of the Army, Headquarters, 101ST Airborne Div., "Army Hearing Program", CAM Regulation 40-5, Jan. 2008 (revised Jan. 2015*).

In terms of marine community limits, OSHA claimed to regulate noise limits per 29-CFR 1910.95 on 'uninspected vessels', such as fishing vessels, towboats, and tugs (under 300 Gross Tons – GT). For inspected vessels, in 1982, the USCG developed *Navigation and Vessel Inspection Circular (NVIC) 12-82* with the following stated purpose:

> PURPOSE. This Circular contains the Coast Guard's recommended guidelines to the U.-S. maritime industry for addressing conditions of high noise. The guidelines were developed in consideration of the need for protecting crewmembers from noise exposures, which may produce permanent noise-induced hearing loss; for providing crewmembers with suitable conditions for recuperation from the effects of exposure to high noise levels; and for providing a safe working environment by giving consideration to the need for effective speech communication and for hearing audible alarms and warnings. Amplifying information in attached enclosures is provided as guidance in addressing key aspects of noise control.

Section 3.d provides the limits as follows:

> d. The exposure limit recommended herein (82 dB(A)) is, like almost all exposure limits, based upon an evaluation of a certain degree of risk to some personnel and does not insure that all personnel will not incur hearing damage. For this reason, a hearing conservation program containing a system for periodic audiometric testing of personnel is necessary to detect those susceptible persons at the initial outset of a hearing impairment before any appreciable damage is accrued. Since such a program would generally not be necessary if exposure levels (computed without regard to attenuation contributed by hearing protective devices) were further reduced to 77 dB(A), the Circular suggests that new vessels be designed so that the 77 dB(A) level may be achieved.

It is important to note that 82 dB(A) reflects the longer exposure time expected on a vessel. In this case, OSHA limits based on 24 h of exposure and a 5 dB exchange rate were used. [For an 85 dB(A) limit and 3 dB exchange rate this would be 80 dB(A).] It is also important to note that the NVIC provided 'guidelines' and 'suggestions' – not requirements. In 1981, the International Maritime Origination (IMO) established a guideline for shipboard airborne noise (*IMO Resolution A468XII-1981*). This guideline covered sound measurement, acoustic insulation, hearing protection and maximum recommended sound pressure levels by ship compartment type. The *IMO Code of Practice for Noise Levels in Ships* and other regulatory organizations have moved to an Leq-24 (Equivalent 24-hour noise exposure) of 80 dB(A) to protect the ship's crew.

---

[7] The exchange rate provided limits for other half orders of time; for example, 100 dB(A) for 4 h of exposure or 85 dB(A) for 16 h of exposure.

As noted above for the USCG, the purpose was to reduce mariner exposure from noise to provide a safe working environment (i.e., the ability to hear audible alarms), reduce noise-induced hearing loss (NIHL), and provide an acceptable degree of comfort for the passengers and crew on the ship. The USCG NVIC 12–82 hearing loss limit is suggested to be 77 dB(A) for vessels built after 1985 and over 1600 gross tons (GT).

The marine industry is required to place warning signs on spaces having hazardous noise levels, typically above 85 dB(A). This includes diesel machinery spaces and most auxiliary machinery spaces. In these spaces, hearing protection is to be worn. The US Navy in *OPNAVINST 5100.19F, 5May 2019, Navy Safety and Occupational Health Program Manual for Forces Afloat*, requires single hearing protection in spaces with noise levels between 85 to 96 dB(A) or 140 to 165 dBP (peak pressure impact or impulsive noise) and double hearing protection in spaces with noise levels above 96 dB(A) or 165 dBP. The Navy has a hearing conservation program for personnel exposed to continuous and intermittent noise that "has an 8-hour time-weighted average (TWA[8]) noise level of 85 dB(A) or greater or impact or impulse sound levels of 140 dBP or greater. Note that weapon firing on naval vessels can produce impulsive levels as high as 161 dB(C) on the bridge wings and 123 dB(C) on the bridge (Paddan 2016).

IMO, in Annex 1, Resolution MSC.337(91), Adoption of the "Code on Noise Levels on board Ships," dated 30 Nov. 2012, sets a hearing loss criterion at an Leq-24 of 80 dB(A). It also sets a compartment noise limit of 110 dB(A) for machinery spaces. In compartments with levels over 85 dB(A), 'suitable hearing protection should be used.' The IMO peak level is 135 dB(C).[9] Furthermore, it is stated that no seafarer, even with hearing protection, should be exposed to levels exceeding 120 dB(A) or an Leq-24 of 105 dB(A).

3.1.1 Time Weighted Average (TWA) Noise

Time weighted averaging (TWA) is one method of determining the noise exposure of crewmembers. TWA is based on the concept that when crewmembers are exposed to noise, they receive a 'noise dose'. This dose accumulates over time, and the TWA can be used to assess the risk of hearing damage. Spending short periods of time in high noise environments can result in a dose that is equivalent to spending longer periods of time in lower noise environments. The TWA calculation is typically based on some reference timeframe, such as 8 or 24 h, with an associated limit for that timeframe.

[8]See Industrial Hygiene Field Operations Manual, Technical Manual NMCPHC-TM6290.91–2 Feb 2021 for DoD computation; https://www.med.navy.mil/Portals/62/Documents/NMFA/NMCPHC/root/Industrial%20Hygiene/7IHFOM_CH5.pdf?ver=b0w4KMBeoYE1ojy0mRHsMA%3d%3d

[9]C-weighting is essentially flat between 31.5 Ha and 8000 Hz, and looks more at the impact of low frequency sound on hearing.

Table 3.2 Parameters for time weighted average calculations

Parameter	OPNAVINST 5100.19D	BUMED Notice 6260
Exchange Rate	4 dB	3 dB
Base Hours	8	8
Limit	84 dB(A)	85 dB(A)

The Time Weighted Average noise is calculated using the equation:

$$TWA = q \, \mathrm{Log}\left[\frac{1}{T}\sum t_i \, 10^{L_i/10}\right] \tag{3.1}$$

where t_i is the time in compartment i, L_i is the noise level in compartment i, T is the total number of 'Base Hours' (i.e., the reference timeframe), and q is determined from the equation:

$$q = \frac{E}{\mathrm{Log}(2)} \tag{3.2}$$

where E is the exchange rate in decibels. For a 'standard' summation and logarithmic averaging of noise levels, the factor q would be 10 for a 3 dB exchange rate and 16.6 for a 5 dB exchange rate.

Different references require the use of different values for the exchange rate, base hours, and exposure limit for those base hours. Table 3.2 provides a summary of how these parameters are prescribed in *OPNAVINST 5100.19D*[10] (2000) and *BUMED Notice 6260,* (2012).

The impact of hearing protection devices would be accounted for by reducing the L_i by the Noise Reduction Rating (NRR) of the device. Note that with OSHA's derating of the NRR by 25% for ear muff and 50% for formable ear plugs, even the highest rated combinations of muffs and plugs are not expected to provide real-world reductions much greater than 20 dB (NIOSH 1998). NIOSH identifies the following factors affecting the likelihood of wearing hearing protection:

- Convenience and availability
- Belief that the device can be worn correctly
- Belief that the device will prevent hearing loss
- Belief that the device will not impair a worker's ability to hear important sounds
- Comfort
- Adequate noise reduction
- Ease of fit
- Compatibility with other personal protective equipment

[10]Not included in current OPNAV M-5100.23, dated 5 June 2020.

Table 3.3 Hearing protection insertion loss values

Percent wearing earplugs		47% not wearing		7%	25%	15%	7%	
Insertion %		n/a		25	50	75	100	
Insertion depth, mm		n/a		5–9	10–14	15–19	>20	
Insertion Loss, dB								
OB Freq, Hz	250	500	1000	2000	3000	4000	6000	8000
25% insertion	2	2	1	1	6	11	12	2
50% insertion	17	14	15	17	26	31	28	17
75% insertion	24	25	27	27	31	36		39
100% insertion	29	32	35	34	33	40		45
Plug A	25	18	26	30	31	35	47	31
Plug B	29	32	37	39	41	39	47	28
Plug C	58	54	56	40	40	37	46	22

Some typical hearing protection insertion loss values are provided in Table 3.3 (adapted from Yankaskas, no date). This table shows the effectiveness of foam plugs at various insertion values from 25 to 100%. Remark that for this study 47% never wore earplugs. Models A through C were custom molded, deep insertion plugs evaluated in the study. Note that at some point, bone conduction becomes the limit for the effectiveness of in-ear plugs. This is true for launch crews on aircraft carriers; hence, the wearing of full helmets for additional hearing protection.

Another concern after hearing loss is tinnitus. This is generally described as ringing or hissing heard without any real noise stimulus present. Research is continuing on the pervasiveness of tinnitus and hearing loss among veterans (*National Academies Press* 2006).

This research also includes the development of kurtosis of a signal as a means of capturing the hazards of non-Gaussian noise exposure. The US Army has developed the *Auditory Hazard Assessment Algorithm for the Human Ear (AHAAH),* which is a theoretical model of the human ear developed to evaluate the hazard of high-level impulsive sounds (Price 2010). The tool is accepted by the Society of Automotive Engineers (SAE) for the evaluation of airbag noise and is under consideration to be the tool employed by military communities to assess military impulse exposures via the Department of Defense – Design Criteria Standard Noise Limits, MIL-STD-1474E (2015). Alamgir et al. 2016, presented a paper on how the 'economic burden of hearing loss' should be considered by the US military.

Kurtosis is likely the most promising for applications of interest to the US Navy because kurtosis-based metrics may be incorporated into existing noise prediction software such as *DesignerNOISE*®. This noise prediction software is acceptable per *Mil-Std-1474E, Department of Defense Design Criteria Standard Noise Limits, 12 Feb. 1997*. Furthermore, kurtosis-corrected equivalent levels are similar to the existing energy-based criteria used to assess hearing damage risk.

The noise limit values and level of performance needed from Hearing Protection Devices are provided in *BUMED Notice 6260* (2012), reflecting the latest information for the Department of Defense Hearing Conservation Program.

3.2 Noise Limits – Compartments

Many organizations, both government and non-government, have developed some simple and some complex noise requirements for both crew and passengers. As stated in the Introduction, it is necessary to understand the criterion imposed on any particular project to meet the assigned levels in an optimum way with respect to all the other vessel requirements – speed, tonnage limits, layout, thermal, use of marine qualified materials, etc. These are covered in the following sections.

3.2.1 IMO/SOLAS Noise Limits

As of 2012, ships covered by the International Convention for the Safety of Life at Sea (SOLAS) (MSC 1974) are **required** to meet noise and vibration requirements set in the Code Noise On Board Ships (2012). The Code provides mandatory maximum noise level limits for machinery spaces, work spaces, control rooms, accommodations, mess, and other compartments. The SOLAS Regulation 3-12, Protection against noise applies to the following:

1. Ships of 1600 gross tonnage and above:
    - For which the building contract is placed after 1 July 2014; or
    - In the absence of a building contract, the keels that are laid or that are at a similar stage of construction on or after 1 January 2015; or
    - The delivery of which is on or after 1 July 2018,
    - unless the Administration deems that compliance with a particular provision is unreasonable or impractical.
2. On ships delivered before 1 July 2018 and:
    - Contracted for construction before 1 July 2014 and the keels of which are laid or which are at a similar stage of construction on or after 1 January 2009 but before 2015; or
    - In the absence of a building contract, the keels are laid or are at a similar stage of construction on or after 1 January 2009 but before 1 January 2015.

    Measures shall be taken to reduce machinery noise in machinery spaces to acceptable levels as determined by the Administration. If this noise cannot be sufficiently reduced, the source of excessive noise shall be suitably insulated or isolated, or a refuge from noise shall be provided if the space is required to be manned. Ear protection shall be provided for personnel required to enter such spaces, if necessary.

3. Ships shall be constructed to reduce onboard noise and to protect personnel from the noise in accordance with the *Code on noise levels on board ships*, adopted by the Maritime Safety Committee by resolution MSC.337(91), as may be amended by the Organization, provided that such amendments are adopted, brough into force and take effect in accordance with the provisions of Article VIII of the present Convention concerning the amendment procedures applicable to the annex other than chapter I. For the purpose of this regulation, although the Code on noise levels on board ships is treated as a mandatory instrument, recommendatory parts as specified in chapter I of the Code shall be treated as non-mandatory, provided that amendments to such recommendatory parts are adopted by the Maritime Safety Committee in accordance with its Rules of Procedure.
4. Notwithstanding the requirements of paragraph 1, this regulation does not apply to the types of ships listed in paragraph 1.3.4 of the Code on noise levels on board ships[11].

The 'Code on noise on board ships' provides allowable crew exposures to hazardous noise and to noise in living and working spaces (Badino 2012b; Bahtiarian 2014). Additionally, it addresses 'verification of acoustic insulation between accommodations.' This is a required 'sound insulation index' between cabins, messrooms, and corridors. The sound reduction index[12] varies between R_w values of 30 and 45 depending on which compartments are adjacent. Guidance on noise control approaches is provided along with procedures for determining noise exposure. The Code defines testing for compliance and testing methods, e.g., during thruster operations. The noise level limits for various spaces are shown in Table 3.4 (IMO Resolution MSC.337, 2012).

3.2.2 U.S. Navy Equipment and Compartment Noise Limits

U.S. Navy (USN) based its compartment noise limits largely on a Naval Ocean Systems Center (NOSC) report by Lambert 1981. The main purpose was to ensure "that noise does not interfere with ship's missions or with the health of personnel." This report addressed the fact that 'Navy ships tend to be noisy because of factors such as propulsion and ventilation requirements, as well as size, weight, cost and schedule constraints.' Those last factors still impede the delivery of 'quiet' vessels. One way the government has dealt with meeting noise criteria is approval of "changes or waivers to ship specifications which would permit higher levels than

[11]Code does not cover dynamically supported or high-speed craft, fishing vessels, pipe-laying or crane barges, mobile offshore drilling units, pleasure yachts, ships of war and troopships, ships not propelled by mechanical means, pile driving vessels and dredgers.

[12]ISO Standard 717-1:1996 as amended (1:2006), part 1.

Table 3.4 Code on noise levels on board ships, limits, dB(A)

Designation of rooms and spaces	Ship size 1600 up to 10,000 GT	> 10, 000 GT
Work spaces		
Machinery spaces[a]	110	110
Machinery control rooms	75	75
Workshops other than those forming part of machinery spaces	85	85
Non-specified work spaces[b] (other work areas)	65	65
Navigation spaces		
Navigation bridge and chartrooms	65	65
Look-out posts, incl. navigating bridge wings[c] and windows	70	70
Radio rooms (with radio equipment operating but not producing audio signals)	60	60
Radar rooms	65	65
Accommodation spaces		
Cabin and hospitals[d]	60	55
Messrooms	65	60
Recreation rooms	65	60
Open recreation areas (external recreation areas)	75	75
Offices	65	60
Services spaces		
Galleys, without food processing equipment operating	75	75
Serveries and pantries	75	75
Normally unoccupied spaces		
Spaces referred to in section 3.14 of the code (locations with unusually high noise levels where seafarers may be exposed, even for relatively short periods, and at intermittently used machinery locations.	90	90

[a]If the maximum levels are exceeded when machinery is operating (only permitted if dispensation is granted in accordance with paragraph 1.3.6 of the Code), stay should be limited to very short periods or not allowed at all. The area should be marked according to section 7.4 of the Code
[b]Examples are open deck workspaces that are not machinery spaces and open deck workspaces where communication is relevant
[c]Reference is made to the *Recommendation on methods of measuring noise levels at listening posts* (resolution A.343(IX)), which also applies
[d]Hospitals: treatment rooms with beds

might otherwise be optimal." A new standard, *MIL-STD-1474E, Department of Defense Design Criteria Standard, Noise Limits, 15 April 2015* "provides specific noise limits and other requirements to equipment designers and manufacturers. It is intended to cover typical operational conditions. Required noise limits shall not be exceeded if the material is to be acceptable to the procuring activity." The document lists the preferred method of noise control as "source reduction", followed by engineering controls. Furthermore, it states that **administratively limited exposure** (authors emphasis) is the "least preferred method."

Table 3.5 US Navy compartment definitions

Category	Definitions
A-3 & A-12	Spaces where direct speech communication must be understood with minimal error and without need for repetition. Category A-3 applies to spaces where maximum talker-listener distance is less than 6 feet (2 m). Category A-12 applies to spaces where maximum talker-listener distance is 6 feet (2 m) or greater.
B	Spaces where comfort of personnel is the primary consideration; where speech communication considerations are secondary.
C	Spaces where it is essential to maintain especially quiet conditions.
D	Spaces where high noise levels exist, where speech communication is not mandatory; where ear protection is not provided; and where the preventions of hearing loss is the primary consideration.
E	Spaces where high noise levels exist, where speech communications are required over short distances but can be accomplished with high vocal effort and where speech amplification and amplified telephones are normally available.

Table 3.6 Maximum Steady State Airborne Noise Levels for Compartments and On-Deck Stations

Airborne noise category	Steady State A- weighted levels	Octave Band Center Frequencies (Hz) (will apply only if A-weighted levels are not met)								
		32	63	125	250	500	1000	2000	4000	8000
A-3	70	81	78	75	72	69	66	63	60	57
A-12	60	72	69	66	63	60	57	54	51	48
B	65	78	75	72	69	66	63	60	57	54
C	60	72	69	66	63	60	57	54	51	48
D	85	98	95	92	89	86	83	80	77	74
E	80	88	85	82	79	76	73	70	67	64

Appendix E of the MIL-STD states: The objective here is to use equipment with lower source noise levels than the criteria for compartment in which they are located. The equipment requirements include "instrumentation and procedures for measurement and analysis of, and maximum acceptable sound pressure level criteria," and acceptability for ALL shipboard equipment." It also covers equipment mounting – both hard and resilient.

The equipment airborne noise limit corresponds to the assigned noise Category of the compartment in which it will be located (see Table 3.5). It also requires an octave band noise level that is 6 dB below the values given in Table 3.6. Multiple measurement locations are identified for each equipment type from small to medium to large. Structureborne noise criteria, if needed, are developed as part of the Noise Control Program Plan.

Appendix F of the Standard 'establishes in-air noise limits and imposes testing requirements, measurement procedures, administrative requirements and reporting requirements for demonstrating conformance to steady-state and impulsive airborne noise limits for compartments accessible by shipboard personnel and for on-deck work stations.' Airborne limits apply to steady-state noise, noise from intermittently

Table 3.7 Maximum allowable A-weighted limit for intermittent noise

Noise category	Level (dBA)
A-3	80
A-12	70
B	75
C	70
D	85
E	85

operated machinery and equipment, peak sound pressure and peak sonar transmission noise. The Standard applies to all surface ships, platforms, new constructions and modifications, and Service Life Extension Programs (SLEP).

This document sets up compartment categories A through E for vessels, as defined in Table 3.5.

Compartments and on-deck stations are assigned the A-weighted noise limits listed in Table 3.6. If the A-weighed noise exceeds the limit, then the un-weighted octave band limits in the table will apply.

For noise contributions from intermittently operated machinery and equipment that is experienced within Category A-3, A-12, B and C spaces, airborne noise levels shall not exceed the A-weighted limits shown in Table 3.7. Intermittently operated implies a short duration interval, consisting of a noise burst lasting from one second to 15 minutes that interferes with speech communication. Such operations include but are not limited to bow thrusters, ballasting and de-ballasting pumps, compressors, distributed piping systems, flight operations, or workshop tools.

The maximum peak sonar transmission noise limit in any compartment is 84 dB at the sonar transmission frequency.

The MIL-STD also requires the following components:

- Noise Control Program Plan (NCPP)
  - Assignment of noise limits to compartments and on-deck work stations
  - Compartment airborne noise analysis
  - Identification of material types for
    - insulations, damping materials, resilient mounts, resilient pipe hangers, flexible pipe connections, floating floors, and acoustic enclosures
  - Identification of conceptual machinery and equipment items and air moving devices supplied by contractor that are to be preinstallation noise tested
  - Identification of Government-Furnished Equipment (GFE) that generates noise, assignment of noise performance limits, provision for pre-installation airborne and structureborne noise testing, and
  - Identification of special requirements that are to be used to achieve compliance with assigned noise limits, such as:
    - Enclosures
    - Equipment and machinery to be resiliently mounted
    - Internal combustion engine and turbine muffles and silencer acoustic performance

- Pre-installation testing of the acoustic performance properties of acoustic insulation, floating floors, damping treatments, and joiner systems, and
- Procedures for controlling propeller-induced noise

The MIL-STD references the SNAME Design Guide T&R Bulletin 3-37 (Fischer, 1983) or a 3-D acoustical design tool called "*DesignerNOISE*®"[13] for performing the noise analysis. Other components include:

- Compartment airborne noise analysis
- Noise control program plan
- Design history booklet
- Compartment noise survey testing and reporting
- Identification of noise excesses
- Identification of hazardous compartments
- Equipment pre-installation test plans and reports
- Sound test failure notifications and recommendations
- Noisy impact studies
- Consultations with subject matter experts (within U.S. Navy)
- Waivers

3.2.3 Miscellaneous and Specialized Noise Criteria

In addition to crew/passenger hazardous and compartment noise limits, there are other types of noise and vibration criteria. These are discussed in the following sections.

3.2.3.1 Speech Interference Level (SIL)

Speech Interference Levels is a single numerical value that is the arithmetic average of the ambient or background sound in the 500, 1000, and 2000 Hz (and sometimes 4000 Hz) octave bands. It is used as a metric regarding how well-spoken communications can be understood with minimal mis-understanding. The SIL is the average of the background noise over which a speaker needs to be effectively heard by a listener at a set distance. These are the bands that typically control vocal communications. Some of the regulatory suggestions are given below.

[13]Developed by Noise Control Engineering, LLC as part of a US Navy sponsored Small Business Innovative Research (SBIR) program.

Table 3.8 Talker-listener distance vs. speech effort (SIL)

Talker-listener distance (m/ft)	Speech interference level, dB			
	Normal	Raised	Very Loud	Shouting
0.15/0.5	71	77	83	89
0.3/1.0	65	71	77	83
0.6/2.0	59	65	71	77
0.9/3.0	55	61	67	73
1.2/4.0	53	59	65	71
1.5/5.0	51	57	63	69
1.8/6.0	49	55	62	67
3.7/12.0	43	49	55	61

Mil-Std-1747D[14] provided a standard table for allowable SIL (see Table 3.8). The computed SIL levels for Category A-12 and A-3 are 53 and 60 dB, respectively. This means communications should be effective with a normal voice over 1.2 m and over greater distances with a raised voice in a Category A-12 space. For Category A-3 space, a raised voice is needed to communicate for distances greater than 0.6 m.

The Naval Ocean Systems Center (NOSC), in its Airborne Noise Limits for Merchant Ships (Lambert 1979), recommends speech interference as follows:

- Normal voice, distance 2–5 m, 60 dB(A)
- Raised voice, 1–2 m, 70 dB(A)
- Very loud voice, distance <1 m, 75 dB(A)

The World Health Organization (WHO), (WHO 1980) also recommends SIL bands of 500 to 2000 Hz. For communications beyond a few meters, interference starts at approximately 50 dB. Speech with more vocal effort can be understood at background noise levels of 65 dB(A).

3.2.3.2 Effect of Noise on Sleep

A research project by the US Coast Guard (Calhoun and Lamb 2006) and an article by Tamura et al. 1997 address the effects of shipboard noise on sleep. Studies directed toward marine craft recommend noise levels between 65 and 70 dB(A) to limit the impact of noise on sleep. Sleep disturbance can start as low as 40 to 50 dB (A). 70 dB(A) is loud enough to significantly change the sleep patterns of most people. It is possible that some crew become de-sensitized to this over time at sea.

For comparison, the A-weighted equivalent of the recommended octave band requirement for Category B lies within the recommended range of 65 to 70 dB(A). Crew cabins with Category B have an equivalent A-weighted level of 69 dB(A). Medical cabins, covered by Category C, have a more stringent equivalent level of 63 dB(A).

[14] Since replaced by Mil-Std 1747E

WHO Europe (2009) has found that measurable effects on sleep start at a background level as low as 40 dB(A). The adverse effect of noise was found to partly depend on the nature of the source. High noise levels at low frequency, such as those found onboard ships, were found to require a lower guideline to avoid adversely impacting sleep. On any ship, it would not be possible to meet the WHO criterion for community noise sleep disturbance effects.

The compartment limits were set at 40 dB(A) for one project on which the authors were consulted. This goal was achieved through the application of extensive and costly treatments. The crew actually felt the low level adversely impacted their sleep since they could hear every change in sound due to variations in machinery line-up, voices, footstep, etc. These were previously masked by higher ambient noise levels.

3.2.3.3 Community and Shoreside Noise Guidance

The operation of vessels dockside, in port, or along waterways with adjacent residential properties are often subject to community noise limits. This is often true for ferries that are continually moving in and out of ports. Community criteria vary among countries and even cities. Community criteria are often set as A-weighted levels. In addition to A-weighted levels, there are limits based on time of day and 'equivalent energy' noise levels. Two of the more stringent criteria often invoked are the day-night average noise level. This is defined as a 24-hour average L_{eq} with a 10 dBA "penalty" added to noise during the hours of 10:00 P.M. to 7:00 A.M. to account for increased nighttime noise sensitivity. L_{eq} is an equivalent energy noise level where the average acoustic energy content of noise is distributed over any chosen exposure time, usually 12 to 24 h. L_{eq} is the constant noise level that delivers the same acoustic energy to the ear as the actual time-varying noise over the same exposure time. In addition, there may be a Community Noise Equivalent Level (CNEL). The CNEL usually adds a supplemental noise penalty for evening hours, typically between 7 and 10 PM. For calibration, communities with a $L_{eq\ 24\ hr}$ below 45 dB(A) are considered quiet. Noisy urban areas have a CNEL of approximately 70 dB(A).

The World Health Organization (WHO) has developed Guidelines for Community Noise (WHO 1999). This guidance suggests that sound pressure levels outside living spaces do not exceed 45 dB L_{night} and 60 dB L_{Amax} (NoMEPorts 2005).

U.S. Environmental Protection Agency (U.S. EPA) under the Noise Control Act of 1972 has developed guidelines for acceptable noise level limits for residential land uses. The limits set forth for residential uses are 55 dBA L_{dn} for outdoors and 45 dBA L_{dn} for indoors. The Federal Transit Administration has also developed guidelines for the allowed increase in noise over the already existing noise shoreside. For instance, an area with a baseline level between 60 and 65 dB(A) could only stand an increase of 5 dB before the impact was deemed severe.

Dockside noise from ferries is due mainly to diesel exhaust and engine room ventilation fans. Ferries usually do not have on-deck limits for these sources unless an upper deck has an exposed passenger seating area. Therefore, these same sources impact dockside receivers. Murphy and King 2014 concluded that "that environmental noise at shipping ports has the potential to be a significant public health concern unless it is tightly regulated and appropriate noise management and mitigation measures are implemented to prevent long-term exposure." This study also found significant low-frequency noise during port handling activity, which may also lead to a greater health concern.

Mammal are also impacted by vessels operating shoresides or underway along rivers. Hearing impacts are better understood for terrestrial mammals than for ocean animals. Mammals have a significant bandwidth – from 1 Hz to over 150,000 Hz. Noise impacts vary from increased alertness to avoidance. In many mammals, hearing loss, once it occurs, is permanent. The impacts of anthropogenic noise on species and ecosystems are summarized in Sordello et al. 2020.

3.2.3.4 Other Criteria

Some criteria have been developed for specific types of vessels or for particular operations, such as naval versus commercial. Some of these criteria[15] from regulatory societies are listed below, e.g.,

- American Bureau of Shipping (ABS) – *Guide for Crew Habitability on Ships, Feb. 2016*; with HAB and HAB+ A-weighted limits for noise and maximum weighted root mean square (r.m.s.) acceleration level and sound insulation indexes, R_w; others include Guide for:
    - Compliance with the *ILO Maritime Labour Convention, 2006 Title 3 Requirements*
    - Passenger Comfort on Ships
    - Crew Habitability on Workboats
    - Crew Habitability on Mobile Offshore Drilling Units
    - Crew Habitability on Offshore Installations
    - Habitability of Industrial Personnel on Accommodation Vessels
    - Comfort on Yachts
    - Naval Ship Rules
- Det Norske Veritas (DNV) – *Rules for Classification of Ships, Part 6, Chapter 33, Newbuildings Special Equipment and Systems* – Additional Class; Comfort Class, Jan. 2014; A-weighted noise limits by compartment, sound insulation indexes, R_w, and frequency weighted overall r.m.s. vibration velocity limits.
- Germanischer Lloyd SE – 24 *Guidelines for Compliance with MLC 2006 Noise and Vibration Requirements*
- Lloyd's Register (LR) –*Rules and Regulations for the Classification of Ships, Part 7, Other Ship Types and Systems, July 2011*

[15] Since Regulatory criteria tend to change with time, it is recommended that appropriate web sites be reviewed for the latest criteria.

- Lloyd's Register (LR) – Naval Ship Rules and Ship Vibration and Noise Guidance Notes, July 2006
- Bureau Veritas (BV) – *Part E, Additional Class Notations, Ch. 6, Comfort on board (COMF), July 2014, amended Jan. 2016*
- UK – *The Merchant Shipping and Fishing Vessels (Control of Noise at Work) Regulations 2007*; hazardous noise exposure of 85 dB(A).

Many regulatory bodies invoke the IMO "*Code on noise levels on board ships*" adopted by resolution MSC.337(91) – such as the Indian Registry of Shipping, ClassNK, and Norwegian Maritime Authority (MNA).

There are also limits on the airborne emissions from ships while in-port or underway (Lloyd's Register (LR), for instance) to reduce their impact shore-side.

3.3 Habitability, Structural, and Machinery Vibration Criteria

Similar to habitability noise limits, there are a multitude of habitability vibration standards. Many of the noise standards covered above also incorporate habitability vibration limits. In general, habitability limits lie below 200 Hz. There are three major axes of concern – vertical, longitudinal, and lateral.

Machinery and structural vibration limits tend to cover higher frequencies than those used for habitability. These criteria are aimed at reducing the failure of machinery and equipment or structural failure caused by excessive vibration. The excitations in this case can be:

- 'forced' by direct machinery excitation
- flow induced by bow waves or HVAC
- hydro-dynamic – propeller excitation
- ice impacts
- or even the free or forced resonant response of some structure or system.

Requirements also exist for the 'quality' of machinery vibration and their expected life with those vibrations.

Vibration limits are varied in that some use displacement, some use velocity, and others use acceleration. Many vibration requirements are given as physical quantities, such as velocity (m/sec), rather than a decibel quantity. The other important variable to remember is that all values given in dB are 'squared' quantities. Thus, determining the physical quantity from a dB value requires the following mathematics for 'squared' variables for acceleration, velocity, and displacement (similar to that for pressure):

$$physical\ quantity = reference\ quantity \times 10^{dB\ value/20}$$

Table 3.9 Overall frequency-weighted r.m.s. acceleration values, habitability guidelines for different areas on a ship

Area classification	A		B		C	
	mm/sec^2	mm/sec	mm/sec^2	mm/sec	mm/sec^2	mm/sec
Values above which adverse comments are probable	143	4	214	6	286	8
Values below which adverse comments are not probable	71.5	2	107	3	143	4

ISO 6954 -2000

3.3.1 Human Vibration Limits

As with noise criteria, there are a multitude of possible human vibration limits. A few are directly applicable to the marine field. One such criterion is *ISO Standard 6954, 2nd edition, 2000 – Mechanical vibration - Guidelines for the measurement, reporting and evaluation of vibration with regard to habitability on passenger and merchant ships*. This standard sets 'frequency-weighted' vibration limits in the three principal axes – vertical, longitudinal, and transverse or athwartships. One-third octave band frequencies between 1 and 80 Hz are measured at two locations on each deck.

As noted in the introduction, high vibration can affect one's performance and health. Some allowable values are provided in Table 3.9.

For guidance, Classification A can be passenger cabins, Classification B can be crew accommodations, and Classification C can be working areas.

The *ABS Guide for Crew Habitability on Ships (2016)* has three ratings – Hab, Hab+, and Hab ++. The vibration guideline covers a weighted acceleration from 1 to 80 Hz. The limits for Hab+ and Hab++ are more stringent than those listed in Table 3.9 for 'adverse comments are probable'.

Norsok 2004 also has a noise and vibration standard for Working Environment developed by the Norwegian petroleum industry.

3.3.2 Machinery/Equipment Vibration Limits

There are two standards of primary interest. The first considers the forced response of equipment and whether it will withstand shipboard excitations. This is covered by International Standard *ISO 10055, Mechanical Vibration – Vibration Testing Requirements for Shipboard Equipment and Machinery Components, 1996*. Acceptability is based on whether the equipment/machinery is functional after being subjected to the specified vibration and time of exposure[16].

[16]Vibration acceptance criteria are also given in *ANSI S2.26, Vibration testing requirements and acceptance criteria for shipboard equipment.*

The other guideline considers measurements made on parts to 'characterize their running conditions with respect to trouble free operation.' There are several parts to *International Standard ISO 10816, Mechanical Vibration – Evaluation of Machine Vibration by Measurements on Non-rotating Parts, 1995:*

- Part 1: General guidelines,
- Part 2: Large land-based steam turbine generator sets,
- Part 3: Industrial machines with nominal power above 15 kW and nominal speeds between 120 r/min and 15,000 r/min when measured in situ,
- Part 4: Gas turbine driven sets excluding aircraft derivatives,
- Part 5: Machine sets in hydraulic power generating and pumping plants, Part 6: Reciprocating machines with power ratings above 100 kW

The ABS (American Bureau of Shipping) provides both 'Guidance Notes' and vibration criteria/limits in their Guidance Notes on Ship Vibration, 2006. Specific vibration (and noise) criteria are contained in ABS *Rules for Building and Classing Naval Vessels* (NVR), jointly developed by the US Naval Sea Systems Command (NAVSEA) and ABS.

ANSI S2.28 The guide for the measurement and evaluation of broadband vibration of surface ship auxiliary rotating machinery also has measurement procedures and acceptable machinery source levels.

Lloyd's has Guidance Notes on Ship Vibration and Noise (2006), see Fig. 3.2. Section 3.7 of this document covers machinery vibration, including assessment

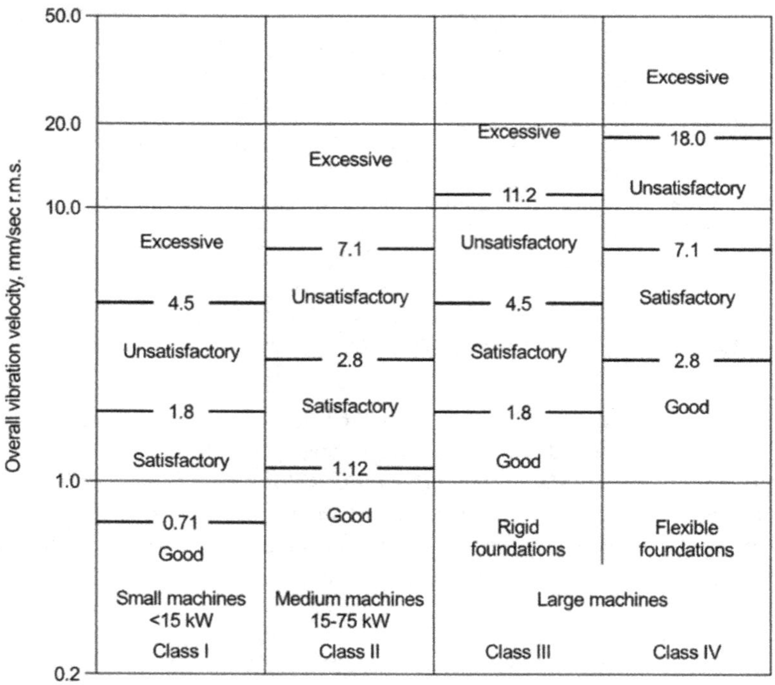

Fig. 3.2 Acceptable vibration levels for machinery

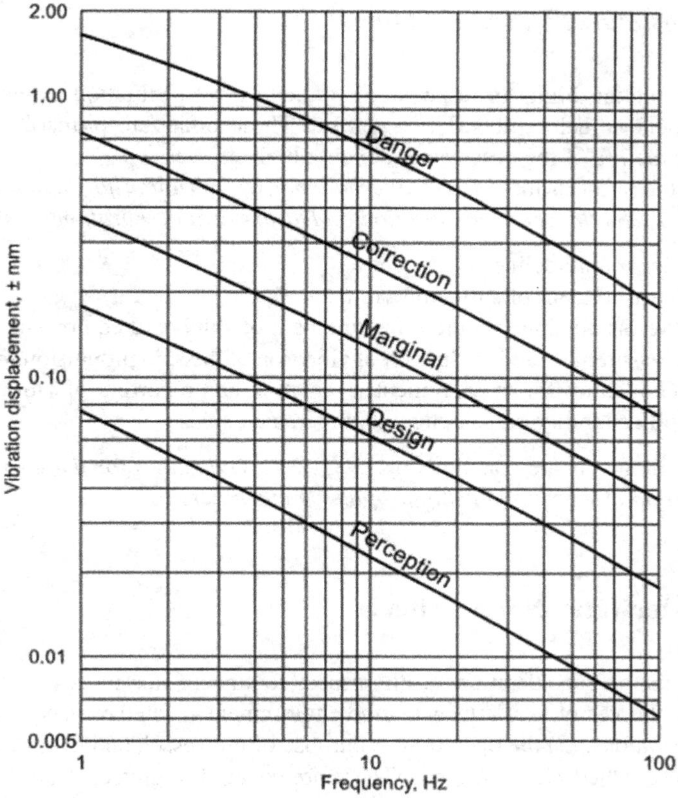

Fig. 3.3 Acceptable vibration limits for piping systems

criteria for determining whether vibration is considered excessive, unsatisfactory, satisfactory, or good for small machines, medium machines, and large machines on rigid or flexible foundations. 'Good' is defined as typical of new machines, 'Satisfactory' as okay for long-term operation, 'Unsatisfactory' not okay for long-term use, and 'Excessive' probable damage to the machine.

While not replicated herein, Lloyd's Guidance Notes also address vibration limits for diesel alternator sets, turbochargers, reciprocating compressors, gearboxes and rolling element bearings. Lloyd's Guidance also provides a way to access the vibration of pipelines (Lloyd's 2006). This chart is provided in Fig. 3.3. Note that these limits are in terms of displacement in mm.

Torsional shafting analyses are usually needed in most specifications. In addition to ensuring that the shafting system is acceptable to various torsional standards, any torsional, axial, or lateral resonance should be reflected in the Campbell resonance avoidance curve discussed in Sect. 10.1.4.9.

3.3.3 Structural Vibration Limits

Structure vibration limits in response to free and forced vibration excitation also cover frequencies below 200 Hz, in general. These limits are primarily aimed at prolonging the life of the vessel and avoiding high stresses.

One of the latest of these is the *International Organization for Standardization (ISO), ISO 20283, Mechanical vibration – Measurement of vibration on ships.*

- Part 1: General guidelines
- Part 2: Measurement of structural vibration
- Part 3: Pre-installation vibration measurement of shipboard equipment
- Part 4: Measurement and evaluation of vibration of the ship propulsion machinery
- Part 5: Guidelines for measurement, evaluation and reporting of vibration with regard to habitability on passenger and merchant ships

Similar requirements are contained in *ANSI S2.27 Guidelines for the measurement and evaluation of vibration of ship propulsion machinery.*

3.4 Underwater Noise Criteria

The three major considerations with respect to underwater noise are the noise radiated by the vessel as it affects (1) the environment – whether fish/mammals or military operations, (2) the operation or purpose of the vessel, and (3) its own sonar system – also called platform noise. These are covered in greater detail below. As noted by Bahtiarian 2013, "Underwater noise is not just the military's problem anymore."

3.4.1 Vessel Signature

The earliest underwater noise requirements were developed for naval vessels with respect to detectability and/or stealth and possible interference with active/passive sonar systems. Military systems are generally classified and not covered by this text.[17] Commercial multi-beam systems are addressed in Sect. 3.4.2.

In addition to the reference pressure of 1 μPa in fluids rather than 20 μPa in air, these standards reference a distance of 1 m from the 'acoustic' center of the object.' In general, this is obtained by scaling from the distance of the measurement back to the 1 m reference (which would be inside the hull for almost any vessel). The scaling law usually lies between spherical – 20 Log r/r_0 to cylindrical – 10 Log r/r_0 depending to a large degree on the depth of the water. ISO is developing (draft)

[17] For a recent discussion of warships and underwater noise regulations see Papanicolopulu 2011.

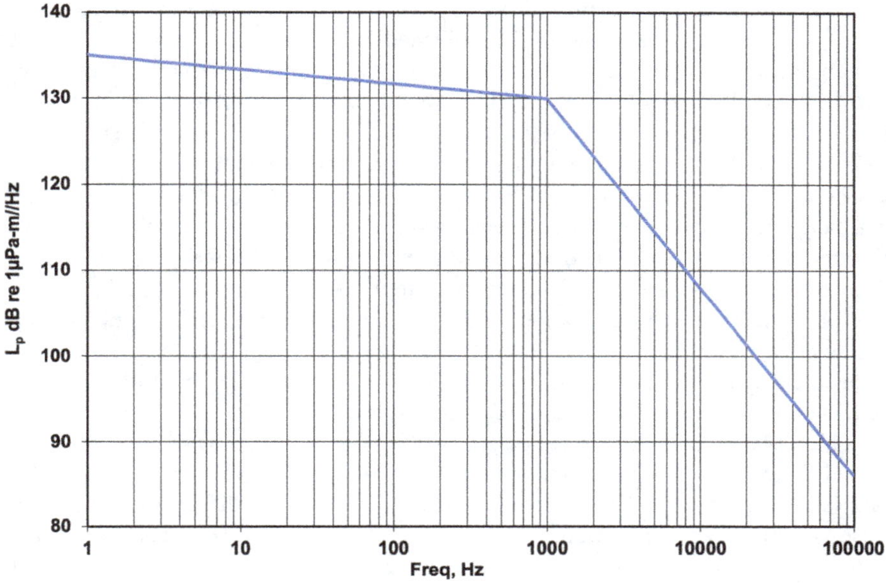

Fig. 3.4 ICES Underwater Noise Limit for Fisheries Vessel, 11 knots

test standards for both depths – *ISO 17208-1, Underwater Acoustics – Quantities and procedures for description and measurement of underwater sound from ships – Part 1: Requirements for precision measurements in deep water used for comparison purposes* and *ISO 17208-2, Underwater Acoustics – Quantities and procedures for description and measurement of underwater noise from ships – Part 2: Determination of source levels from deep water measurements.* A similar standard is being developed for shallow water. It is necessary to realize that the ship's signature is also directional – being different for abeam, astern, bow, and keel aspects.

One of the earliest commercial and most stringent underwater noise criteria was developed by the International Council for the Exploration of the Sea (ICES) for fisheries research vessels (Mitson 1995). This criterion, set for operation at 11 kn, was nominally to lower the vessel signature to a point where it would not adversely impact fish counts conducted by specialized sonar systems (see Fig. 3.4). The frequency range of concern ranges from 1 Hz to 100,000 Hz; the level below 1000 Hz based on possible 'fish reaction' to the ship signature and above 1000 Hz on meeting fish sonar self-noise requirements. The fish 'reaction' criterion is an open issue being investigated by the marine bio-acoustic community. While the criterion given in the reference is in narrow bands, the measurement procedures require the use of one-third octave bands.[18] The arithmetic conversion from a broadband

[18]Measurement below 10 Hz is difficult due to motions of the hydrophone. Several different hydrophones may be needed to cover the entire range from 1 Hz to 100,000 Hz.

measurement to an equivalent narrowband criterion is inherently problematic. In reality, the measurements need to be performed in narrow bands to detect relevant tonal noise since the standard is given as a narrow band requirement. In essence, since it is usually easier to measure, third-octave band measurements are collected. These data, assumed to be flat across each third band, are scaled to an equivalent 1-Hz bandwidth. This procedure misses any strong tonal signature.

The only ships that are quiet enough to pass the ICES standard have had significant amounts of engineering and acoustic controls. These ships typically have diesel generators mounted on very soft, two-stage isolation systems that supply electricity to low-vibration DC motors that turn custom-designed propellers. The controlling factor is having a propeller that is cavitation free up to 11 knots. Diesel engine noise makes the stringent low-frequency section of the ICES criteria difficult to meet. Generally, hydro-acoustic sources, namely, the propeller, control the higher frequency section of the ship's radiated noise spectrum.

A broad-based consensus on allowable anthropogenic noise does not exist. In the United States (U.S.), nominal mammal harassment levels have been developed by the National Marine Fisheries Service (NMFS). Behavioral disruption for impact noise is 160 dB and for continuous noise is 120 dB (*NOAA Technical Guidance, 2016)*. This is also the overall noise level. For example, a vessel producing an overall signature of 190 dB re 1 μPa–m would meet this harassment threshold for a mammal 3162 m away {spherical spreading = 20 Log (3162/1) = 70 dB; needed propagation loss is 190–120 = 70 dB}.

Not only is the overall noise of concern with respect to mammals and other organisms in the ocean. It is believed that the frequency range, duration, hearing acuity of the species, particle velocity (Popper and Hawkins 2018) and/or energy density could play roles in the response of ocean life. The acoustic impacts on underwater life include communication, navigation, and threat/hunting responses. These interactions are not fully understood at this point (NOAA 2016; Erbe 2013).

The European Union [E.U. 2008], United Nations General Assembly, Convention on Migratory Species[19] [UN Convention, 2008] and daughter agreements [ACCOBAMS 2004 and ASCOBANS], International Whaling Commission (IWC)[20] Scientific Committee, Oslo and Paris Conventions for the protection of the marine environment of the North–East Atlantic (OSPAR) and International Maritime Organization (IMO 2008a, b, 2009),[21] ISO TC 8 SC2 – Marine Environmental Protection[22] are all reviewing and considering the impact of anthropogenic noise and what limits, if any, should be considered. In the United States, the Marine Mammal Commission on 'A Sound Approach to Research and Management –

[19] https://www.cms.int/#:~:text=The%20Convention%20on%20Migratory%20Species%20(CMS)%2C%20also%20known%20as,migratory%20animals%20and%20their%20habitats.

[20] https://iwc.int/commission/commission-sub-groups/scientific-committee

[21] IMO Correspondence Group, "Noise from Commercial Shipping and Its Adverse Impacts on Marine Life."

[22] https://www.iso.org/committee/45816.html

Marine Mammals and Noise' has made a report to the US Congress. The objective would be to reduce the noise from ships "by 10 dB per decade." Approaches being considered by the marine community include the design of quieter ships, routing of ships away from environmentally sensitive areas, and operational measures, including speed and path changes. It is believed that the noisiest 10% of the vessels contribute the greatest majority of the overall noise pollution (Leaper et al. 2009).

Detailed discussions of potential regulations for underwater noise are provided in McCarthy 2004. This reference covers the following topics: Scientific Aspects of Underwater Noise, Focusing Events, Policy Development, Politics, Potential Solutions and Obstacles, and Use of Marine Protected Areas.

Starting in 2004, IMO began to address the issue of underwater noise and its impact on marine mammals given that continuous anthropogenic noise is primarily caused by shipping. One of the first steps of the IMO Marine Environment Protection Committee (MEPC) was to develop non-mandatory technical guidelines (*MEPC.1/Circ.833, Annex, Guidelines for the reduction of underwater noise from commercial shipping to address adverse impacts on marine life, April 2014*) and measurement standards. The IMO document considers that propeller noise is a prime contributor to anthropogenic noise from ships and provides some general guidance on design methods to mitigate this source (see also Sect. 9.1.1). The document also assesses current gaps and future focus areas.

The Government of Canada, Transport Canada, is actively interested in underwater radiated noise. This includes workshops, developing tools to predict and implement effective quiet designs, developing new, real-time tracking tools, and developing a tool to detect marine mammals. The European Union (EU) also has a collaborative program, called SILENV,[23] to review the "current state of scientific knowledge.., review current national and international guidelines and regulations with regard to marine mammals and fish noise exposure, levels and criteria; define parameters of underwater noise measurements that need to be collected from ships. . .".

The guidelines recognize that underwater noise from shipping may have both short-term and long-term negative effects on marine mammals and other aquatic life. The design guidance is for ship designers, shipbuilders and ship owners/operators. It also mentions that hull form and operational aspects such as speed can be factors as well. Similar types of information are contained in the *Marine Vessel Environmental Performance Assessment Guide: General Measures: Ocean Health & Aquatic Life – Underwater Noise, SNAME Technical & Research Bulletin 6-01 GM-1, 2014.*

ABS, DNV, and other non-governmental organizations (NGOs) have developed underwater noise criteria by vessel type and purpose, e.g., acoustic research versus dredging. Det Norske Veritas (DNV) has implemented underwater radiated noise criteria for ships based on the actual classification or utilization of the ship. The "Silent Class Notation" consists of five classifications of vessels: Acoustic (A),

[23] Ships Innovative soLutions to rEduce Noise, https://cordis.europa.eu/project/id/234182

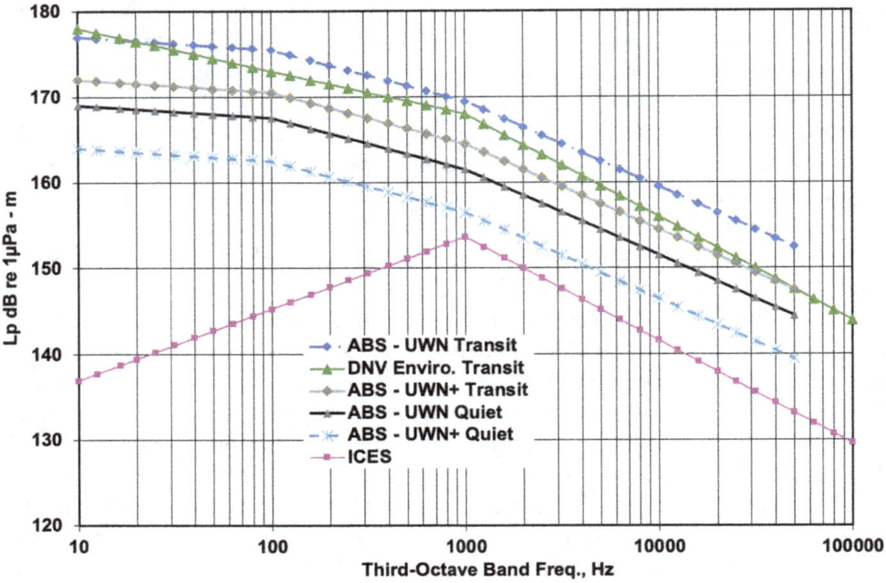

Fig. 3.5 Examples of ICES, DNV and ABS class underwater radiated noise Criteria. (Note: UWN – underwater noise; Enviro. – Environmental)

Seismic (S), Fishery (F), Research (R), and Environmental (E). Representative examples of radiated noise criteria are presented in Third-Octave Bands (TOB) in Fig. 3.5, normalized to 1 meter.

For comparative purposes, typical radiated noise criteria for seismic, oceanographic research, and sonar research vessels and ICES are provided in Fig. 3.6. These criteria only apply to slow speed operation, typically below 12 knots, i.e., non-cavitating propellers. The North Atlantic Treaty Organization (NATO) operates a very quiet research vessel, the *R/V Alliance* (Bonetti and Calcagno 1986). The *Alliance* is supposedly one of the quietest surface vessels in the commercial world. It has extensive treatments, as listed in Table 3.10. This vessel has criteria from a high of 12 knots down to drifting (on battery power). These criteria are compared to the ICES requirement in Fig. 3.7. The ICES criterion is higher than that for the NATO vessel except below 40 Hz.

3.4.2 *Sonar Self-Noise*

Ambient noise in the ocean and the noise generated by the ship itself combined account for the 'sonar self-noise.' Ambient noise due to a high sea state can often mask sonar signals. This self-noise affects the potential classification of a target signal via a sonar system – whether onboard or towed as shown in Fig. 3.8. It is

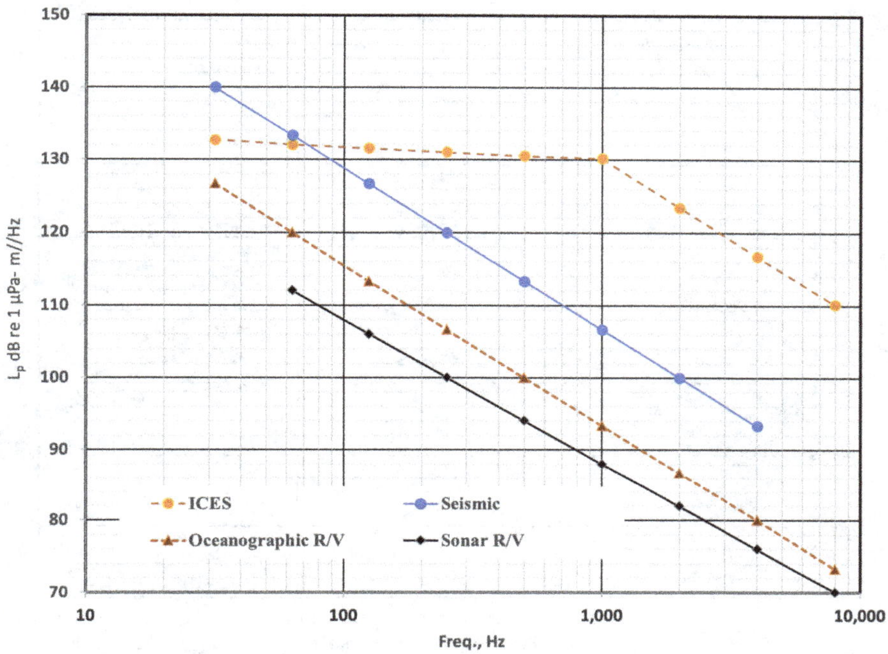

Fig. 3.6 Radiated noise criteria – for other' research vessels'

Table 3.10 NATO R/V design parameters and treatments

General Characteristics		Treatment
LOA	93.0 m	
Breath, moulded	15.2 m	
Displacement	3000 t	
Max. Speed	17 kn	
Equipment	*Rating*	
(2) Electric motors	6 pulse SCR	Smoothing choke
(2) Main Prop. Diesel	2040 kW @ 1200 rpm	Double stage isolation mount, Enclosure, 20-ton raft
(3) Ship Service (SS) Gensets		Double stage isolation mount, Enclosure
(2) Controllable Pitch (CP) Propellers	130 rpm, 3.35 m dia., 5 Bld.	Design, high tip clearance, skew

important to distinguish between 'platform noise' and 'sonar self-noise.' Platform noise considers only the ship's contribution to an omni-directional hydrophone located in the position of the sonar array. The sonar self-noise is the ship's contribution as measured after signal processing through the array and processing of the signal. Thus, array directivity and array gain need consideration.

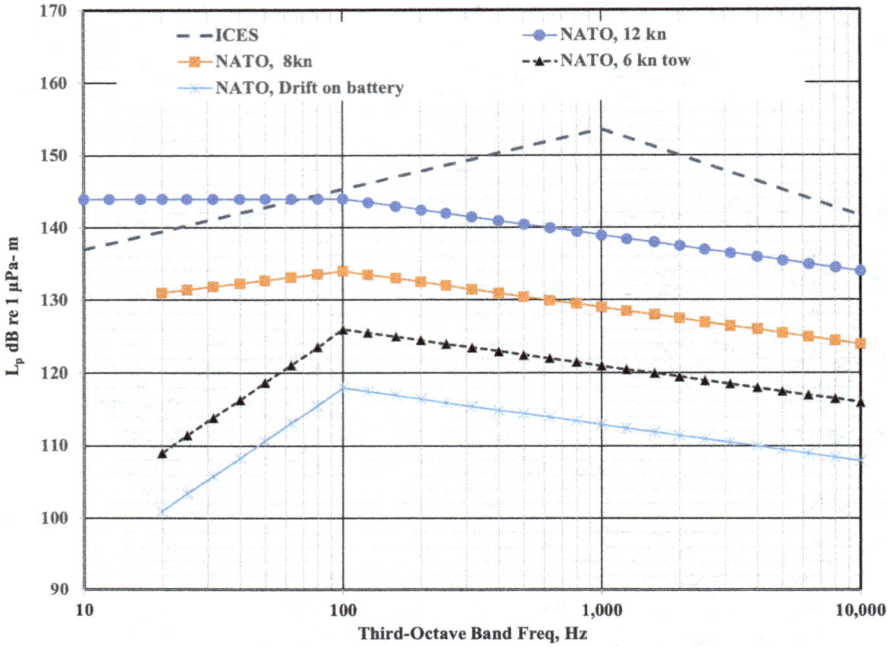

Fig. 3.7 Radiated noise criteria – ICES v. NATO R/V

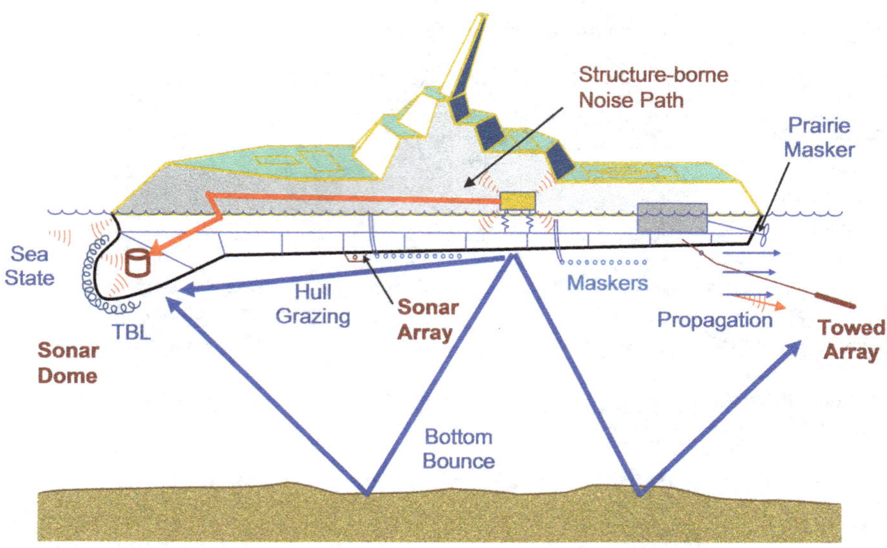

Fig. 3.8 Components for sonar self-noise (Fischer et al. 2008)

Table 3.11 Typical platform self-noise level requirements

Equipment	Frequency of band, 1 Hz band width	Spectrum level, dB re 1 μPa
Multibeam Survey System (Note 1 – Depending on bottom type)	70–100 kHz	35
	30 kHz	40
Single Beam Survey System	12 kHz	49
	38 kHz	40
	120 kHz	32
	200 kHz	32
Sub-Bottom Profiler	4 kHz	60
	15–20 kHz	47
Acoustic Doppler Current Profiler (ADCP)	38 kHz	40
	75 kHz	32
	150 KHz	32
Acoustic Navigation & Tracking System	20–30 kHz	47

The ship's platform noise is controlled by propulsion and auxiliary machinery-induced airborne and structureborne noise, propulsor radiated noise, sea-connected piping systems, and flow or turbulent noise along the hull plate and sonar face or from control surfaces. The platform noise is quite speed- and operational line-up dependent. It will also change significantly during turns. Certain sources, such as propellers and sea chest opennings, will also have a directivity component that needs consideration. Finally, the sonar generally cannot operate if the array is in the path of any bubble sweep-down from the bow.

Generally active sonar systems work best with a Signal-to-Noise (SNR) ratio of 10 dB or better within their operational bandwidth. Most active sonars, especially multi-beam systems, operate above 2000 Hz. In many cases, propeller cavitation dominates in this high-frequency range. The allowed sonar self-noise level and frequency range is nominally established by the sonar vendor. This value is controlled by the source level and directivity of the sonar transducers, pre-amplifiers, array parameters, baffles, processing algorithms, and location of the sonar relative to the hull and major equipment. Some vessel specifications quote an upper speed and specific sea state for the sonar system to operate. In this case, a judgment needs to be made on the relative contribution of the platform noise relative to that from sea-state ambient noise. A possible conservative assumption would be for the platform noise to be 10 dB below the sea state noise. Examples of sonar platform self-noise limits are shown in Table 3.11.

One implicit assumption is that sonar platform-noise limits will not be met with the operation of bow thrusters. This is due to their expected cavitation noise. Another critical factor is ship speed. Flow-excited Turbulent Boundary Layer (TBL) vibration of the sonar dome has been found to be a significant component of self-noise at moderate speeds. The noise is proportional to the seventh power of the free stream velocity over the dome, meaning that a doubling of the velocity can increase the platform noise by 21 dB.

3.5 Summary

As shown in Fig. 3.9, any ship design loop is centered on the 'Ship Specification' (Fischer and Yankaskas 2011). Any specification should consider what criteria are achievable and reasonable from an overall systems viewpoint. The selection of the acoustic criteria should depend on the type and function of the vessel and how the vessel will be operated, i.e., full speed, station keeping, etc. As noted above, hearing conservation, comfort, and communications are just a few of the noise types and criteria to be considered. Habitability vibration criteria need the same functional considerations listed above.

A good acoustic specification generally defines the acoustic limits in unambiguous terms, how to verify those limits are achieved, and under what conditions. Terms such as "there shall be no vibration" are unacceptable since every ship will vibrate – the question is at what level. Terms such as 'noise will be limited' are too indefinite. Crew surveys have shown noise almost always annoys.

Two general types of specifications exist – the hardware specification and the performance specification. If the owner/designer has performed a noise and vibration analysis and knows the treatment types to be installed, a hardware specification should be released. If the owner wants the designer/builder to design and select treatments, then a performance specification is more appropriate. Problems almost always occur if the ship's specification is a combination of performance and specified treatment approaches. In this case, it is often difficult to optimize the treatments, and the design is stymied from the beginning. Settling conflicts can be costly and delay construction.

Meeting noise and vibration criteria does not necessarily imply that there are no complaints or even hearing loss. Some people have higher sensitivity to acoustic inputs than those used to develop various criteria (Biot and De Lorenzo 2007). Annoyance, in particular, is subjective and even harder to quantify (Badino et al. 2012a). The impact of noise and vibration on sleep depends not only on the person's sensitivity but also on the background noise and the existence of intermittent noise. In some specifications, a defined 'overall' noise excess (usually 2 to 5 dB) is allowed in a percentage of the spaces. Similarly, there may be an allowable excess defined over a band of frequencies.

In some cases, the owner/operator may choose to substitute another 'simpler' acoustic requirement for a more complex requirement, such as using a hull vibration criterion in lieu of an underwater radiated noise criterion or basing underwater noise criteria on ocean ambient noise (Stocker 2005). In this case, the owner/operator has to have accurately modeled and analyzed the ship's systems.

Another approach that has been used successfully is for the owner/operator to provide incentives or penalties with respect to meeting the acoustic requirements (Bahtiarian 2001). The incentives would need to spell out the monetary benefits of meeting noise and vibration levels a specific amount below the original criteria. The penalties would need to define the cost of not meeting the acoustic requirements in a timely manner and/or being a specific amount over the original requirements.

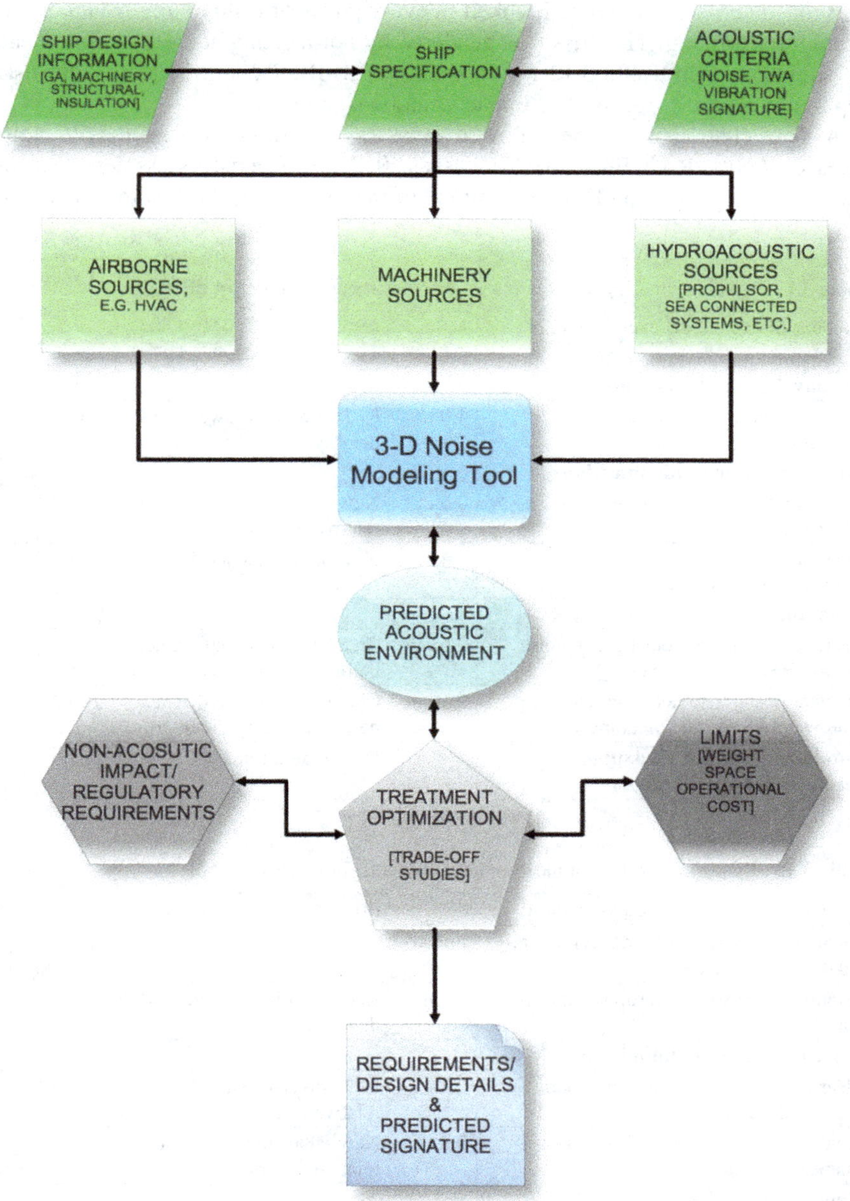

Fig. 3.9 Ship design loop

Acoustic criteria should be invoked in every phase of a ship's life from concept through retro-fitting. The "risks" engendered by not meeting acoustic requirements are summarized in Table 3.12 for an excess of 4 dB. Table 3.13 shows the risks implied by an 8 dB excess. The impacts can be significant.

A flow chart that can be followed to assess the risks associated with noise excesses is provided in Fig. 3.10 (Yankaskas 2013a, b). The risk assessment decision tree is usually based on (1) the magnitude of the excess, (2) the treatments already

Table 3.12 Risks implied by not meeting in-air noise requirements by 4 dB

Amount of excess, dB excess amount	4	dB	
Hazardous noise wrt 85 dB(A) per day 3 dB exchange rate	Impact		% change
Potential impact on hazardous exposure time	4.8 h	Reduction in allowed work per day	60.3%
Impact on crew communications			
Clear communications at 1 m talker-listener distance	0.6 m	New talker-listener distance relative to 1 m	36.0%
Clear communications at 2 m talker-listener distance	1.4 m	New talker-listener distance relative to 2 m	32.2%
Situational awareness			
Relative distance to detect given noise (one-way spherical spreading)	0.63 m	Change in relative distance before detecting given noise	36.9%
Relative distance to detect given noise (one-way cylindrical spreading)	0.40 m	Change in relative distance before detecting given noise	60.2%
Non-hazardous noise (assume accommodations 70 dB(A)		Discernable impact	

Table 3.13 Risks implied by not meeting in-air noise requirements by 8 dB

Amount of excess, dB excess amount	8	dB	
Hazardous noise wrt 85 dB(A) per day 3 dB exchange rate	Impact		% change
Potential impact on hazardous exposure time	6.7 h	Reduction in allowed work per day	84.3%
Impact on crew communications			
Clear communications at 1 m talker-listener distance	0.3 m	New talker-listener distance relative to 1 m	68.0%
Clear communications at 2 m talker-listener distance	0.8 m	New talker-listener distance relative to 2 m	58.6%
Situational awareness			
Relative distance to detect given noise (one-way spherical spreading)	0.4 m	Change in relative distance before detecting given noise	60.2%
Relative distance to detect given noise (one-way cylindrical spreading)	0.16 m	Change in relative distance before detecting given noise	84.2%
Non-hazardous noise (assume accommodations 70 dB(A)		Discernable impact	

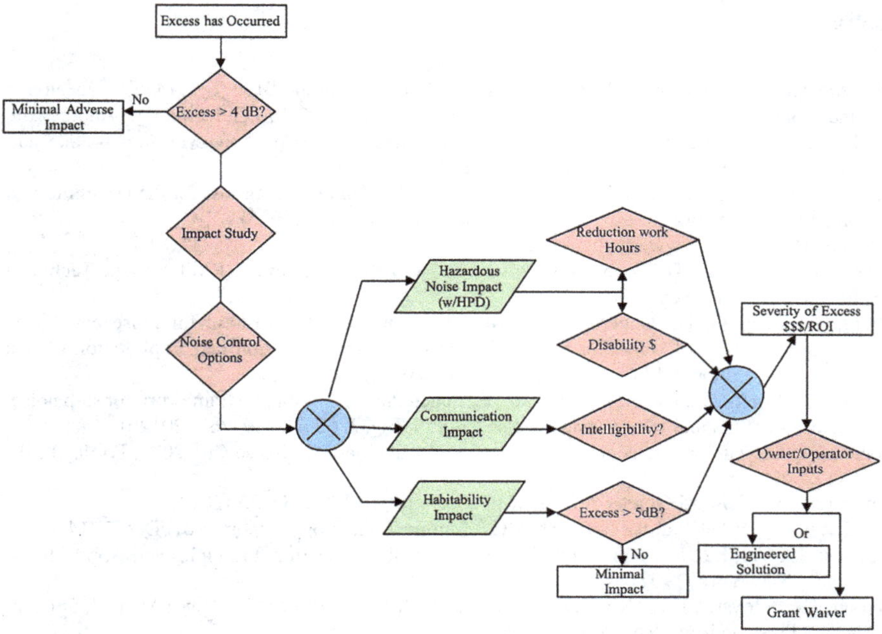

Fig. 3.10 Risk assessment flow chart. (NIHL Noise Induced Hearing Loss; US Navy Office of Naval Research program)

utilized, (3) whether the impact is on hazardous noise, communications, or habitability, and (4) the severity of the excess and its impact on cost to rectify and return on investment. Based on the information built from following the tree, the owner/operator would make a decision to either accept the excess or come up with another engineered solution.

If an owner or regulatory society decides to impose acoustic criteria, selection of the criteria should address the following components:

- Purpose/objective of the criteria (hearing, comfort, environmental, sonar system operation, research, etc.)
- Operation of the vessel – speed and equipment line-up (all equipment or only selected equipment operating)
- Definition of allowable acoustic criteria in terms of band-width, lower and upper frequency range, and amplitude of the allowed sound pressure level or vibration level
- Type of criteria – hardware specification or performance specification
- Types of equipment (including propulsor); control treatments or operating conditions likely needed to attain the criteria – are these feasible or achievable
- Verification methods – analytical and/or measurements
- Possible incentives for meeting the criteria or penalties for not meeting them, including risk assessment

References

ACCOBAMS, Agreement on the Conservation of Cetaceans of the Black Sea, Mediterranean Sea and Contiguous Atlantic Area (ACCOBAMS), Resolution 3.10 "Guidelines to Address the Impact of Anthropogenic Noise on Marine Mammals in the ACCOBAMS Area", 2007 and Resolution 2.16 "Impact Assessment of Man-Made Noise", 2004

Alamgir, H., Tucker, D., Kim, S.-Y., et al.: Economic burden of hearing loss for the US military: a proposed framework for estimation. Mil. Med. **181**, 301–306 (2016)

ASCOBANS, https://www.ascobans.org

Avery, L., and Gray, D.: Shipboard Noise Control, Lessons Learned, BMT Group, Technical Instruction # 043, Sept. 2009

Badino, A., Borelli, D., Gagero, T., Rizzuto, E., Schenone, C.: Criteria for noise annoyance evaluation on board ships. In: Sustainable Maritime Transportation and Exploitation of Sea Resources. Taylor and Francis Group, London (2012a)

Badino, A., Borelli, D., Gagero, T., Rizzuto, E., Schenone, C.: Normative framework for ship noise: present situation and future trends. Noise Control Eng. J. **60**(6), 740–762 (2012b)

Bahtiarian, M.: The precise acoustic design of a pilot station boat. Noise-Con 2001, Portland ME, Oct. 2001

Bahtiarian, M.: Toward Quieter Seas. ASME Mech. Eng. **135**(6), 32–35 (2013)

Bahtiarian, M.: What's all the Noise at IMO?. Marine Technology Report, Nov./Dec. 2014

Beranek, L., Ver, I. (eds.): Noise and Vibration Control Engineering, Principles and Applications. Wiley, New York (1992)

Berger, E.F., Royster, L.H., Royster, J.D., Driscoll, D.P., Layne, M.: The Noise Manual, 5th edn. AIHA Press, Fairfax, VA (2003)

Biot, M., De Lorenzo, F.: Noise and vibrations on board cruise ships; are new standards effective? In: 2nd International Conference on Marine Research and Transportation (ICMRT '07), Naples, Italy, 2007

Bonetti, E., P. Calcagno: Low noise as the first task of a vessel – a description and some remarks about Acoustic Quieting Design Criteria and Features of SACLANT ASW Centre Oceanographic Vessel. In: Shipboard Acoustics, Nartinus Nijhoff Publishers, Dordrecht, Netherlands, 1986

Boroditsky, L., Spiridonov, V.: Structureborne Noise Control in Ship Compartments. Sudostroenie, Leningrad (1974)

BUMED – Department of the Navy, "Interim Navy Medicine Hearing Conservation Program Guidance, BUMED Notice 6260, December 13, 2012

Calhoun, S., and T. Lamb, *Human Factors in Ship Design: Preventing and Reducing Shipboard Operator Fatigue,* Univ of Michigan, Dept. of Naval Architecture and Marine Eng., US Coast Guard Research Project, 2006

Carter, T.: Textbook of Maritime Medicine, 2nd edn. Norwegian Centre for Maritime Medicine (2013) http://textbook.maritimemedicine.com

Casali, J., et al.: Warfighter auditory situational awareness: locating the shooter with and without hearing protection. In: Human Systems Integration Symposium, Vienna, VA Oct. 2011

Code on noise levels on board ships, IMO Resolution MSC.337, adopted on 30 Nov. 2012

Department of Defense (DoD), Design Criteria Standard Noise Limits, MIL-STD-1474E, 15 April 2015

E.U. 2008. The European Union Marine Strategy Framework Directive (Adopted in June 2008). Available on http://ec.europa.eu/environment/water/marine/index_en.htm

EC: Directive 2003/10/EC of the European Parliament and of the Council of 6 February 2003 on the Minimum Health and Safety Requirements Regarding the Exposure of Workers to the Risks Arising from Physical Agents (Noise) (2003)

EPA (U.S. Environmental Protection Agency). 1974. Information on Levels of Environmental Noise Requisite to Protect Public Health and Welfare with an Adequate Margin of Safety. Document 550/9-74-004

Erbe, C., "International regulation of underwater noise," Acoust. Aust., 51(1), April 2013
Erdreich, J.: A distribution based definition of impulse noise. J. Acoust. Soc. Am. **79**(4), 990–998 (1986)
Fischer, R.: Sound and noise control. In: *The Marine Engineering Series*, 4th edn. The Society of Naval Architects and Marine Engineers (SNAME), Alexandria VA (2020)
Fischer, R., and M. Bahtiarian, "Ship Noise Control," *Encyclopedia of Maritime and Offshore Engineering*, Online ISBN: 9781118476406, (D. Finch *ed), John Wiley & Sons, 2017
Fischer, R., Boroditsky, L.: Technical and Research (T&R) Bulletin 3–37 (Supplement). The Society of Naval Architects and Marine Engineers (SNAME), Alexandria VA, Supplement to the Design Guide for Shipboard Airborne Noise Control (2001)
Fischer, R., Collier, R.: Noise prediction and prevention on ships. In: Crocker, M. (ed.) Handbook of Noise and Vibration Control. Wiley, New York (2007)
Fischer, R., Yankaskas, K.: Noise Control on Ships – Enabling Technologies. ASNE Intelligent Ship IX, May (2011)
Fischer, R., Boroditsky, L., Spence, J.: Acoustic design tools – present and future integrated software. In: 9th International Naval Engineering Conference & Exhibition. IMarEST, London (2008)
Fischer, R., Nelson, D., Burroughs, C.: SNAME T&R 3-37 – Shipboard Airborne Noise Control. Society of Naval Architects and Marine Engineers (SNAME), Jersey City, NJ (1983)
Guignard, J.: A Basis for Limiting Noise Exposure for Hearing Conservation. Aerospace Medical Research Lab AMRL-TR-73-90, Wright-Patterson Air Force Base, Ohio, July 1973
Hammernik, R.: The effects of amplitude distribution of equal energy exposures on noise-induced hearing loss: the kurtosis metric. J.Acoust. Soc. Am. **114**(1), 386–395 (2003)
Henderson, D., Hamernik, R.P.: Impulse noise: Critical review. J. Acoust. Soc. Am. (JASA). **80**(2), 569–584 (1986)
Humes, L., Joellenbeck, L.M., Durch, J. (eds.): Noise and Military Service: Implications for Hearing Loss and Tinnitus. National Academies Press (2005)
IMO 2008a. Minimizing the introduction of incidental noise from commercial shipping operations into the marine environment to reduce potential adverse impacts on marine life. Paper MEPC 58/19 submitted by the United States to 58th session of IMO Marine Environment Protection Committee. October 2008
IMO 2008b. Report of the 58th Session of IMO Marine Environment Protection Committee. October 2008
IMO: Noise from Commercial Shipping and its Adverse Impacts on Marine Life, Report from the Correspondence Group (MEPC59/19), April 09 2009
IMO Resolution MSC.337, Code on noise levels on board ships, adopted on 30 Nov. 2012
Lambert, D., NOSC Technical Document 243, Airborne Noise Levels on Merchant Ships, Naval Ocean Systems Center, San Diego, CA, April 1979
Lambert, D., NOSC Technical Report 712, *Setting Airborne Noise Limits for Navy Shipboard Compartments*, Washington DC, 29 June 1981
Leaper, R., Renilson, M. R., Frank, V. and Papastavrou, V. "Possible steps towards reducing impacts of shipping noise", Presented to the IWC Scientific Committee, June 2009, Madeira, Portugal, SC/61/E19
Lei, S., Ahroon, W., Hammernik, R.: The application of frequency and time domain kurtosis to the assessment of hazardous noise exposures. J. Acoust. Soc. Am. **96**(3), 1435–1444 (1994)
Lloyd's *Guidance Notes, Ship Vibration and Noise,* Lloyd's Register, July 2006
Loeser, H.: *Fundamentals of Ship Acoustics, Acoustical Phenomena in and Around Ship Hulls*. The Society of Naval Architects and Marine Engineers (SNAME), New Jersey (1999)
Marine Vessel Environmental Performance Assessment Guide*: General Measures: Ocean Health & Aquatic Life – Underwater Noise,* SNAME Technical & Research Bulletin 6-01, GM-1, 2014
McCarthy, E.: International Regulations of Underwater Sound. Kluwer Academic Publishers, Boston (2004)

Mitson, R. B. "Underwater noise of research vessels: Review and recommendations," ICES Cooperative Research Report 209, 1–61, Copenhagen, 1995

MSC 91/22/Add.1, Annex 2, Page 1, Resolution MSC.338(91), *Adoption of Amendments to the International Convention for the Safety of Life at Sea,* 1974 As Amended and Code on noise levels on board ships

Murphy, E., King, E.: An assessment of residential exposure to environmental noise at a shipping port. Environ. Int. **63**, 207–215 (2014)

National Academies Press, Institute of Medicine of the National Academies, "Noise and Military Service", Washington DC (2006)

NIOSH – National Institute of Occupational Safety and Health. 1998b. Criteria for a Recommended Standard: Occupational Noise Exposure. Publication 98–126, 1998

NOAA Technical Guidance for Assessing the Effects of Anthropogenic Noise on Marine Mammal Hearing, July 2016. https://www.fisheries.noaa.gov/west-coast/endangered-species-conservation/esa-section-7-consultation-tools-marine-mammals-west

NoMEPorts, *Good Practice on Port Areas Noise Mapping and Management*, 2005

NORSOK Standard, Working environment, S-002, Rev. 4, Lysaker, Norway, August 2004

OPNAVINST 5100.19D, "Occupational Safety and Health, Program Manual for Forces Afloat," Volume I, Appendix B4-A, October 5, 2000

OSHA: Occupational noise exposure (general industry). In: 29 CFR 1910.95. Federal Register (1971)

Paddan, G.S.: Occupational noise exposure on a Royal Navy Warship during Weapon Fire. Noise Health. **18**(84), 266–273 (2016)

Papanicolopulu, I.: Warships and noise regulation: the international legal framework. Mar. Pollut. Bull. **63**, 35–39 (2011)

Popper, A.N., Hawkins, D.: The importance of particle motion to fishes and invertebrates. J. Acoust. Soc. Am. **143**, 470–488 (2018)

Price, G.R.: Executive Summary of the Development and Validation of AHAAH. https://arl.devcom.army.mil/resources/ahaah/ahaah-executive-summary/, September 10, 2010

Sordello, R., Ratel, O., DeLachapelle, F., et al.: Evidence of the impact of noise pollution on biodiversity: a systematic map. Environ. Evid. **9**, 20 (2020)

Southall, B., J. Berkson, D. Bowen, et al.: Addressing the effects of human-generated sound on marine life: an integrated research plan for U.S. *Federal Agencies.* A Report of the Joint Subcommittee On Ocean Science & Technology (JSOST) (2009)

Stocker, M.: Comments of on the National Marine Fisheries Proposed Action for Ocean Noise Criteria, Jan. 2005

Tamura, Y., Tawada, T., Sasazawa, Y.: Effect of ship noise on sleep. J. Sound Vib. **205**(4), 417–425 (1997)

Technology for a Quieter America, Committee on Technology for a Quieter America, National Academy of Engineering, ISBN: 0-309-15633-5, 2010

Webster, J.C.: Psychoacoustic Problems with Noise Control. Presented at 51st Convention, Human Factors Technology Div., Naval Electronics Laboratory Center, San Diego, CA May 1975

World Health Organization: Guidelines for Community Noise, (B. Berglund, et al. Ed.), 1999

World Health Organization, Environmental Health Criteria 12, Noise.: https://apps.who.int/iris/bitstream/handle/10665/39458/9241540729-eng.pdf (1980)

World Health Organization Europe: Night Noise Guidelines for Europe (2009)

Yankaskas, K.: Private Communication Re NIHL Research Program – Overview Presentation, Office of Naval Research (2013a)

Yankaskas, K.: Prelude: noise induced tinnitus and hearing loss in the military. Hear. Res. **295**, 3–8 (2013b)

Chapter 4
Noise and Vibration Control Program Plan

There is a price to be paid to achieve compliance with the acoustic requirements imposed by marine regulatory agencies or owner/operators (Fischer and Pettit 2014). These requirements, which may appear as airborne noise, underwater radiated noise, sonar self-noise, and/or vibration limits, are spelled out in ship specifications. They are becoming more prevalent as detrimental impacts from noise are receiving greater recognition. The impacts range from hearing loss claims by humans, lower retention of crew, and the need to control sound levels experienced by marine mammals. Making the correct decisions during the ship design and build process will minimize cost and will improve Return On Investment (ROI). Costs include the monetary price paid for materials, labor, and maintenance. Cost should also incorporate the 'cost' of added weight, less space, and potential adverse impact on the construction schedule and operation of the vessel. An example of the hidden cost of noise control is the increase in fuel cost over the life of the vessel due to added weight (Fischer and Pettit 2014). The Program Plan described herein will help develop the steps needed for a successful design while minimizing the cost paid to achieve compliance.

Commercial off-the-shelf noise and vibration control treatments and labor to install them are significant cost-drivers. The authors' rule of thumb, developed over the last 40 years, is that the engineering and the cost of acoustic control treatments for the lead ship of a conventional marine vessel should be 1–3% of the vessel cost. A specialized vessel, such as a fisheries research vessel, would likely require 15–20% of the ship cost. Engineering design could be 10–25% of this cost, depending on the complexity and stringency of the requirements (Fischer and Pettit 2014; Verboom 2004).

This chapter addresses the steps to be considered in making decisions on how best to meet noise and/or vibration limits for marine vessels. At the same time, costs will be minimized through optimization of noise control treatments (see Chap. 10). It is important to note that this is an iterative process – often requiring multiple passes through the design process.

R. Fischer, L. Boroditsky, *Noise and Vibration Control on Ships*,
https://doi.org/10.1007/978-3-031-55170-3_4

The objective of any Noise and Vibration Control Program Plan (NVCPP) is to deliver an acoustically complaint vessel to the owner in the most efficient and effective manner possible. This NVCPP should cover design, construction, and testing. It also describes the procedures through which the designer and builder will ensure the following:

- Satisfactory and optimal acoustic measures are incorporated into the project to meet the stated acoustic requirements within the construction schedule.
- Acoustic measures are properly implemented during construction
- Proper testing is conducted to verify performance
- A plan is in place to diagnose and achieve compliance for any noted excesses.

The key elements in a NVCPP are identified in Table 4.1 (Fischer 2006).

This Plan should also provide design guidance and ways to anticipate difficulties and limitations early on. This will result in quicker resolution of problems. Thus, the plan describes engineering, administrative, management, and ship construction functions that will be performed to ***integrate*** the acoustic design into the overall ship design. This plan should provide an organized approach, describing how acoustic performance requirements are reviewed, assessed, and translated into design, construction and testing of the ship. The extent of the plan depends to some extent on the stringency of acoustic requirements and the experience and capabilities of the design agent and shipyard.

Consideration needs to be given to underwater noise, habitability noise, structural and habitability vibration, and/or sonar self-noise. The primary intention is to identify, quantify, and control the dominant sources and acoustic transmission paths. The process to select, optimize, and apply control treatments should be laid

Table 4.1 Key Elements of a Noise & Vibration Control Program Plan (NVCPP)

Identify managerial, technical, procurement and administrative authorities.
Provide clear identification of noise and vibration limits and ship acoustic test requirements (i.e., machinery operating modes) that apply when data are acquired for the purpose of demonstrating compliance with acoustic limits.
Recognize potential risk factors and possible reduction approaches
Identify acoustic inputs for critical cost drivers, long-lead control materials, and design and construction margins
Specify clearly hardware or goal driven requirements cited in the ship specifications.
As necessary, include acoustic requirements (software, hardware, test, etc.) in procurements of:
Machinery and equipment. Noise and vibration control hardware and materials. Subcontractor services such as propulsion systems, HVAC, joiner systems, etc.
Require and identify extent of noise and vibration modeling, analyses, and reporting.
Detail range of noise and vibration control design, integration, review, and verification
Prepare and review of affected detail design drawings
Compel quality assurance and inspection of acoustic constructions (provide training as necessary)
Specify test and trials of the as-built ship
Provide full documentation of acoustic program, including any excesses and remedial efforts

out. Finally, the NVCPP should define deliverables to be furnished in support of the acoustic control effort. An outline of the steps in the design process is contained in Figs. 4.1 for noise and 4.2 for vibration. The vibration control plan needs to address broadband machinery vibrations, structural response, and possible limits to impose on source vibration. Predicting and controlling vibration levels goes hand-in-hand with controlling noise levels. Items such as avoidance of structural resonances and acceptable foundation impedances need to be considered as part of the vibration plan. Finite Element Analysis (FEA) is often the appropriate tool to address these later items.

The acoustical plan and resulting design are part of an ***iterative*** process; e.g., sometimes proposed noise treatments may contradict cost/weight/safety issues. Safety issues include meeting regulatory requirements pertaining to fire, smoke, and toxicity. These factors need to be resolved, and inconsistencies jointly worked out during the design phase by the various parties – including any acoustic consultant, the project manager, material and equipment vendors, and other relevant subcontractors.

Current technology, in terms of prediction tools and control methods, can handle most marine noise and vibration issues. However, without a comprehensive program plan and attention to details, these tools and treatments often prove ineffective or are not applied appropriately. On commercial vessels, the owners may have an intense

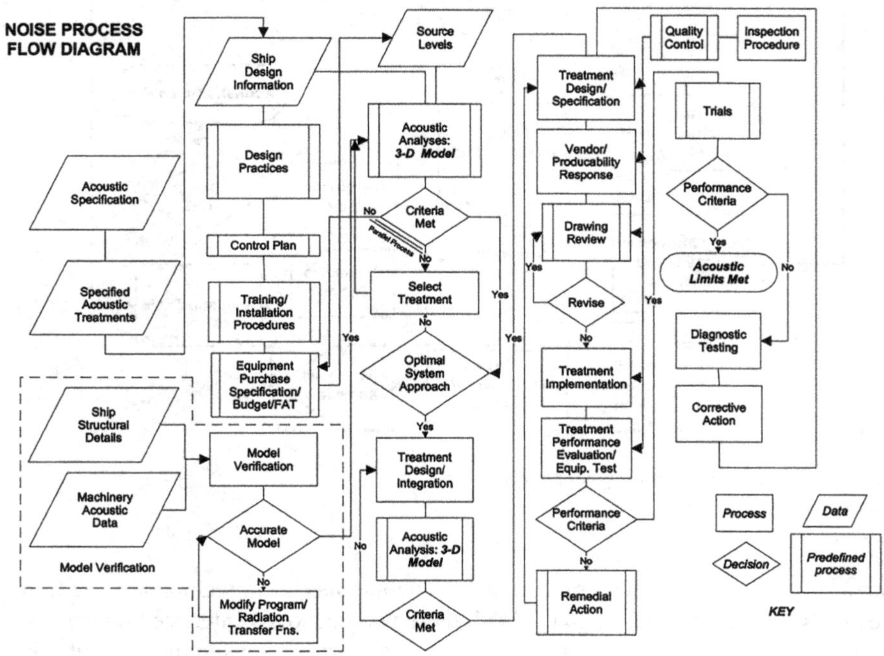

Fig. 4.1 Noise control program plan

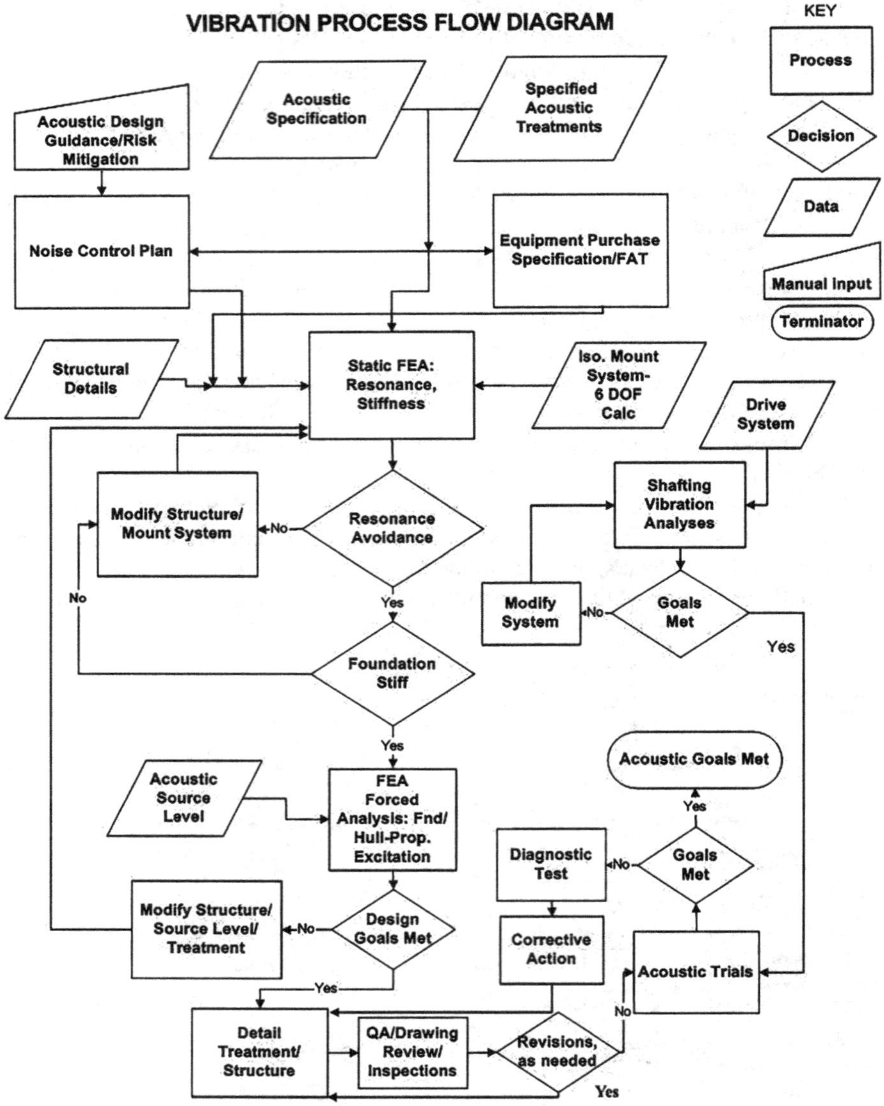

Fig. 4.2 Vibration control program plan

interest in obtaining a quiet product, and the builder has his reputation at stake if the vessel fails to meet acoustic goals. With these two anchors – interested owners and builders – and for a plan that is executed properly, there are significant paybacks. The victors are the ocean environment, on-board acoustic environment, and the total cost of ownership (Fischer and Yankaskas 2011).

From its inception, any vessel is designed to achieve a particular set of objectives – speed, displacement, endurance, etc. Deriving the hull form, propulsion equipment, and layout is a relatively well-known and straightforward engineering process. However, the same type of 'systems' design needed to achieve a quiet vessel is usually relegated to a very low priority, if considered at all. In reality, achieving a quiet vessel cuts across all standard ship design engineering functions – Hull, Mechanical & Electrical (HM&E), insulation, propulsor, etc. – since almost every component on the vessel can contribute to the acoustic environment. Thus, a Noise and Vibration Control Program Plan provides a methodology necessary to ensure that a comprehensive 'systems' approach is considered early and throughout the design, construction, and delivery of the vessel. The Program Plan needs to be understood by the designers and builders and then implemented and followed through to a successful conclusion.

While controlling habitability noise is relatively straightforward, controlling underwater radiated noise requires a higher level of design and handling. Radiated noise levels can be reduced in an optimal manner on any vessel but only with the proper attention to detail. Furthermore, attention to these system details can also result in fuel efficiencies, longer periods between maintenance, and lower adverse impacts on weight, space and cost.

In cases where there is a stringent acoustic requirement and the design agent and/or shipbuilder lack relevant experience, it may be advantageous for the owner or builder to hire a 'consultant.' A person with the correct experience can provide their experience and feedback directly to the various parties.

The status of these elements shown in Figs. 4.1 and 4.2 should be monitored and tracked to assess progress of the noise control program. The point would be to identify and incorporate suitable cost-effective noise control treatments into the ship design and to ensure that the ship design does not include any non-acoustic features that inadvertently compromise acoustic performance.

The specification typically includes a deliverable document, which often identifies contract data requirements for the vessel. The format, schedule, interfacing with any electronic management information system, and items to be included in the submittals should be detailed as part of the plan. Another critical part of the plan is a milestone schedule, identifying critical decision points, deliverables, construction inspections and compliance testing.

4.1 Management, Organization Structure and Scheduling

This first part of the NVCPP includes a description of the technical, management, and organizational structure. It needs to identify actions that will be taken by team personnel to implement the plan. This is important to ensure incorporation of the specified noise requirements into the design and construction of the vessel. An organizational chart should be provided that gives the authority to impose the full set of orderly actions needed to meet the noise requirements. This should include,

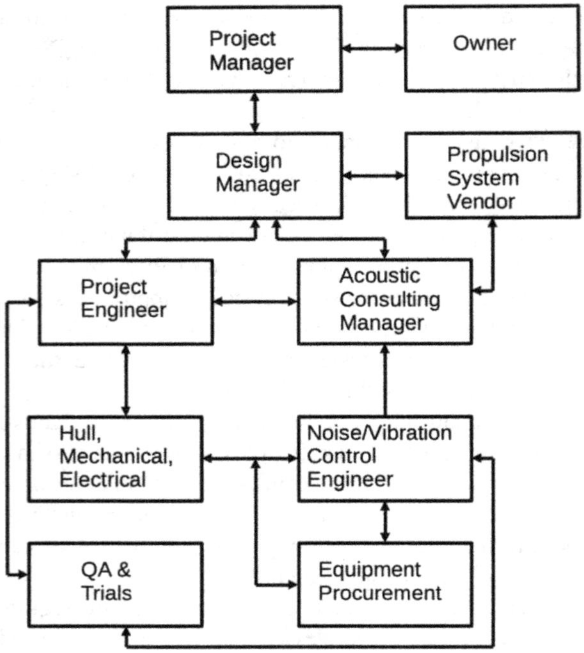

Fig. 4.3 Organization structure

but not be limited to, requirements on managerial responsibility, design, procurement, fabrication, assembly, construction, test, training, and quality control functions (see Fig. 4.3). The program manager, engineering manager, and key noise control engineer should be identified by name. Experience and resumes should be provided for key personnel. Where there are stringent or unfamiliar acoustic requirements, the team should include a marine acoustic control expert or a subcontract to a marine acoustic expert. These personnel will serve as members of the noise control team. Interfaces with the owner and regulatory agencies should be identified.

A team that does not understand and execute the Plan and have the needed relationships in place can spend a significant amount of time chasing and rectifying incorrect steps.

4.1.1 Procurement Specification

Achievement of acoustic goals is often critically dependent on machinery noise and vibration levels. Fluidborne source levels may be important (Fischer 2006), especially for quiet vessels. Source-level 'budgets' are often established early in the design process. Either a preliminary prediction process or 'reverse engineering' can be used to establish the frequency-dependent noise and vibration limits for critical equipment. If a similar vessel/criteria exists, equipment levels from that vessel could be used. In some cases, the ship's specification lists acceptable acoustic source levels for specific equipment, e.g., propulsion and major auxiliary machinery.

Occasionally, a preliminary analysis or ship specification requires machinery and/or equipment to be purchased to a particular acoustic level. In this case, a procurement specification needs to be developed and submitted to equipment vendors. Potential acoustic 'equivalence', if allowed by the specification, should be considered as part of this effort. Factory Acceptance Testing (FAT) may be invoked as part of any equipment purchase specification. In addition, to avoid potential problems, all rotating machinery should meet regulatory agency balancing criteria, for example, ISO 21940-11:2016. It is also useful to select 'quiet' types of bearings for the equipment and shafting.

For any 'equivalence', the vendor should provide the following:

1. The machinery vendor should provide sound and/or vibration data measured on similar units. This should consist of the following:
    a. Vibration on the mounting feet of the equipment with units of acceleration (g's, inches/second^2, mm/second^2, dB relative to 1 μg, etc.) as a function of frequency from 25 Hz to 10,000 Hertz.[1] This may be supplied as narrowband (from 25 to 10,000 Hertz) or 1/3 octave band (from 25 to 10,000 Hertz bands). A description of the test methods and conditions used to acquire the data should be provided. The overall vibration level is not a useful metric.
    b. Sound pressure level (in dB re 20 μPa) at distances of 3 to 5 feet from the equipment or sound power levels (in dB re 1 pico-Watt) of the equipment. The sound data are measured at four sides and over the top of the equipment. The sound data are to be provided as a function of frequency in octave band or 1/3 octave band formats (from 31.5 to 8000 Hertz octave bands). A description of the test methods and conditions used to acquire the data should be provided. A-weighted noise is not a useful metric in this case.

The shipyard's Request for Quotations for noise and/or vibration critical equipment should contain a 'fact sheet' addressing potential acoustic requirements. Vendors should be requested to supply 'historical' source level data with their quotation. From this, a representative Purchase Description, PD, should be developed for use on all critical equipment. As part of the PD, a generalized Factory Acceptance Test (FAT) procedure should be defined, if utilized. The main part of this PD should address:

- Scope
    - Identify equipment and machinery to be tested
    - Operating conditions
    - When and where tests are to be conducted
    - Acceptability of prior tests on similar units

[1]This range would cover the octave bands from 31.5 Hz to 8000 Hz; starting in the lowest third-octave band of 25 Hz, to the highest third-octave band of 10,000 Hz.

- Requirements
  - Source airborne noise limits
  - Source vibration limits
- Testing
  - Acoustic levels and bandwidth
  - Invoke test standards or define test standards for noise and vibration
- Measurement locations
  - For airborne sound, usually at 1 m on 5 sides; often in third-octave or octave bands and A-weighted level
  - For vibration levels, usually on top of resilient mounts in the vertical direction for units placed on a suitably stiff foundation; often in third-octave and narrow bands (1 Hz)
  - Whether measured at vendor's plant, test laboratory, or on vessel
- Instrumentation
  - For airborne noise, Sound Level Meter (SLM) is acceptable
  - For vibration, accelerometers and a multi-channel spectrum analyzer
- Measurement procedures
  - Usually defined in regulatory standards (ISO, ANSI, DNV, etc.) or military specifications
- Documentation
  - Graphs showing test conditions, sensor location, units, gains, frequency range, and bandwidth. These graphs should also identify criteria to quickly show compliance or excess. Calibration sheets are to be witnessed and included in the report.
- Shipping instructions
  - Equipment should be packaged and shipped to protect and prevent any degradation of acoustic characteristics.

This procedure will identify instrumentation, calibration, mount type, factory foundation construction, factory ambient noise/vibration levels, test methods, test standards, data to be collected, and reporting requirements. Scheduling is to be provided with sufficient notice to the owner/shipyard and other members for potential witnessing. These data are to be supplied for approval to the contractor prior to vendor shipping the unit. The shipyard's purchase technical specifications should require vendors to exercise particular care when handling, packing, storing, and shipping noise tested machinery.

4.1.2 Risk Reduction

Team members should work closely to identify key design and construction challenges early in the program. The objective would be to identify and minimize risks. One of the first steps would be to review the ship's specification to identify those areas and criteria requiring significant noise or vibration controls. The second would be to review 'shipyard standard design and construction techniques' to determine if they reflect good acoustical practices that match the acoustic requirements. Risk reduction also means working with machinery vendors to ensure that their equipment is compliant with required or anticipated source-level requirements. Finally, this review should identify potential 'errors' in the specification as provided. These errors (or professional differences in understanding the criteria) should be adjudicated before beginning the control process.

As needed, acoustic impact studies should be carried out as early as possible to identify approaches to meet the various criteria. The same acoustic goals can often be met by differing control approaches. In our experience, no two shipyards have taken the same approach to identical acoustic issues. These early studies would assess the status of the engineering work needed to meet the specified criteria, provide initial and updated noise/vibration predictions, determine if additional testing is needed for new or unusual conditions, and examine the available noise and vibration control approach alternatives. These studies can examine potential conflicts between acoustic requirements and non-acoustic considerations imposed by other sections of the ship's specification or other regulatory requirements.

4.2 Acoustic Analysis and Control

This part of the Program Plan would address procedures for performance, documentation, and review of airborne and structureborne noise analyses (predictions). These would cover the impact of propulsors, propulsion machinery, auxiliary machinery, pumps and piping systems, and wave-induced noise. The analyses methods and computational computer codes used to guide the noise control design should be specified in the plan. Requirements for documenting the analyses for submittal to the responsible owner or government authorities should be included.

Using information extracted from contract drawings and information available from the ship's specifications, a *baseline* noise prediction should be developed. The analysis should address all salient shipboard noise producers. Transmission over structureborne, airborne, and fluidborne paths need to be analyzed. Propeller contributions at the proper operating speed(s) should also be included in the analysis. Technology exists, with proper Computer Aided Design (CAD) processes, to accurately predict noise during any stage of ship design (van den Berg 2015). When this technology is applied early in the design process and updated as the design progresses, the results are such that the treatments are optimized in every respect. This will also minimize cost and produce a quiet vessel. The goal is to provide a highly accurate prediction so that the optimal controls can be selected during trade-off studies. For additional information, see Chap. 12 on cost and Return On Investment (ROI).

The results of these models, compared to the criteria, will determine any acoustical requirements for equipment procurements, if utilized. These can be airborne and/or structureborne noise source level performance limits. This will also identify potential noise control treatments for inclusion in ship design and construction. The output would be treatment details, including areas of coverage for acoustic insulation materials and damping materials, specific material types, candidate vendors, typical treatment cross-sections, and special installation requirements, if needed.

The shipyard would typically incorporate these recommendations into detail design drawings. The noise analyses and noise control treatment recommendations need to be documented in technical reports. These reports would be updated as design information, Factory Acceptance Test, if needed, and ship noise testing results become available. Reports should include detailed assumptions, machinery source-level data, developed acoustic transfer functions, predicted noise levels, recommendations for additional noise control, as needed, and other pertinent information. A revision history should be required for each document.

The shipyard's Project Engineer should take each recommendation for noise control treatments identified in the noise reports under advisement and assess the performance, schedule, technical risk, and cost ramifications of the recommendations. Recommendations should also be screened for compliance with other engineering, functional, regulatory, and contractual requirements. If the change significantly affects cost or schedules, the Project Manager may request alternatives to be considered. The final decision with respect to the inclusion of additional noise control measures will then be documented with appropriate justifications. The earlier the analyses are started, the sooner any significant potential contractual or design problems can be identified.

4.3 Design Reviews

The Program Plan needs to identify a schedule and procedures for design reviews to ensure that specific noise performance requirements are incorporated in the detail design of the vessel. Internal and programmatic design reviews should be considered. Key design parameters that are incorporated into the ship design to achieve design compliance with specified acoustic performance requirements should be summarized. Critical path decisions should be presented along with resolutions and ways forward. Decision makers should be involved in Critical Design Reviews.

The Program Plan needs to address methods to be followed for submission of contract deliverables to ensure they are complete and adequate. The Plan needs to address owner/regulatory comments and resolve any feedback.

4.4 Engineering Change Proposals

An important part of the Plan is the establishment of procedures and actions to be taken to assess the impact of engineering change proposals or requests for waivers. These are bound to happen. Changes to the shipboard machinery and/or design may be requested by the owner or may be initiated by the shipyard. The shipyard Program Manager would notify the owner and noise team of any impending Engineering Change Proposals (ECP's) that involve the following:

(i) Relocation or addition/deletion of machinery or sonar systems,
(ii) Any changes to the noise critical machinery items, those having a direct impact on the noise generated or planned treatments,
(iii) Any changes to foundation designs or mounting system designs for noise critical machinery items,
(iv) Any additional noise control design measures suggested by the owner or by the shipyard that are not specified in the ship's specification,
(v) Any changes to the general arrangement of the ship, or
(vi) Any changes to the ship's hull, appendages, insulation, sheathing or propulsion system designs.

The noise control team, to determine their acoustic impacts, will evaluate ECPs. A noise expert, if utilized, would inform the shipyard of the overall impact of any ECP on noise and vibration performance. Any impact on the suite of noise control measures and treatments to eliminate excessive noise and vibration would also need to be assessed. The impacts should be documented.

4.5 Noise Control Treatment Material

Similar to equipment procurement, the Program Plan should address methods of obtaining and installing noise control materials. Procedures should be included for the development and review of actions to ensure that any specified requirements, delineating the acoustic performance of critical noise control materials and hardware, are invoked in procurement documentation. These need to be validated and verified by the material supplier. Again, this can include requests for historical performance data, hopefully based on a recognized military or IMO, ANSI, etc., acoustical test standards.

Conventional noise treatments include low noise producing machinery and equipment, isolation mounts, acoustic insulation, damping, high transmission loss materials, silencers, etc., when combined can have a significant impact on:

- Schedule, especially for long-lead items such as isolation mounts
- Installation cost, for example – installing acoustic insulation or gluing damping tiles in place

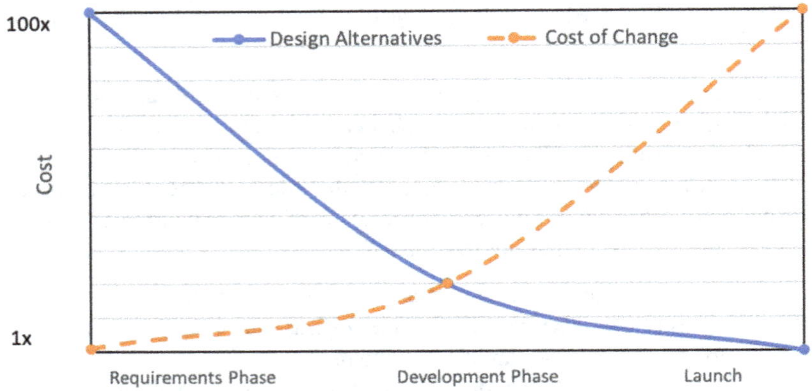

Fig. 4.4 1:10:100 rule applied to ship requirements, development, and launch

- Weight, installing high-density mineral wool in lieu of low-density fiberglass (or foams if allowed by regulatory agencies)
- Space, thicker insulation and floating floors can have large gaps above the structural deck
- Material costs, some in the neighborhood of $100 to $200/sq. meter (Fischer and Pettit 2014).

It is beneficial to optimize the noise and vibration mitigation treatments during the ship design stage rather than after the construction phase is completed. After the initial construction is complete, the retrofit of noise control treatments may cost as much as ten to one hundred times more than the costs incurred had treatments been selected by implementing an evaluation and optimization process during the design phase – see Fig. 4.4. It is also possible that obtaining expert advice can more than pay for itself by avoiding poor design choices, especially for vessels with stringent criteria.

4.6 Drawing Review

Details should be included in the Program Plan as to how drawings will be created and reviewed. Detail design drawings produced by the shipyard should be reviewed by a noise control expert to ensure that the intent of the treatment recommendations is included in the design. They should ensure that non-acoustic features do not compromise acoustic performance. Engineering drawings addressing acoustic features should be listed on the ship's Drawing Index, which should be the central control document available to all team members. The Program Manager has the responsibility to direct and review the work of each section within engineering to ensure that day-to-day design efforts incorporate all salient noise control features. It is the responsibility of the Project Engineer to ensure that the design staff is informed of the noise control features that are incorporated into the overall ship design.

Drawing reviews are an integral part of the noise control effort, providing needed "insurance." The results of these reviews should be documented in the form of letter reports and marked drawings, as appropriate, and submitted as per the Noise and Vibration Control Program Plan.

Each design stage has a certain level of development with respect to noise control issues. A systematic review of drawings during each stage by a noise control expert should be considered a good practice.

During the detail design stage, the incorporation of controls should be verified for each appropriate drawing. The areas of interest at this stage are as follows:

- resilient mount arrangement/details
- acoustic and thermal insulation
- sheathing
- foundations for resiliently mounted machinery
- HVAC and/or piping system arrangements (in noise sensitive areas only)
- details of joiner bulkheads
- intake and exhaust arrangements for internal combustion engines or gas turbines
- flexible hose, bellow, and duct connection details
- shafting supports and bearing, if appropriate
- noise and vibration treatment construction details

4.7 Inspection of Acoustic Constructions

A critical component of the Program Plan is the inspection of acoustic constructions. All noise control hardware should be inspected during and subsequent to installation. The shipyard's Quality Assurance (QA) lead should have the responsibility to oversee and inspect installations. QA should certify that installations are in accordance with design. A noise expert, who can certify that installations are in accordance with design, should assist him. The status of all acoustic constructions should be tracked, maintained, and reported to the noise control team. To facilitate this, a construction inspection reporting spreadsheet should be developed (see Chap. 13). An inspection spreadsheet should contain field entry positions for recording items inspected. It should be noted when the noise control installation is complete and ready to undergo operation.

The shipyard, owner (or owner's representative), and QA personnel should perform general and specific inspections of installations with acoustic content. These inspections, for which many non-government and Mil-Standards exist, should include the following:

- Resiliently mounted machinery, including mounts – type, deflection height, and alignment, piping and electrical connections and the foundation

- Piping and exhaust systems supported with resilient pipe hangers
- Thermal, fire, and acoustic insulation systems
- High Transmission Loss (HTL) constructions
- HVAC and piping silencers and duct linings, internal and external
- Damping materials
- Hull (missing location of cathodic anodes within flow paths to propeller, orientation of grating to sea chests and bow thruster)
- Other special treatments such as floating floors, enclosures, etc.

4.8 Training

The Program Plan needs to define the training to be provided to the design and construction teams. If the shipyard has not had experience with a ship built to the specified acoustic criteria, a noise professional with appropriate experience should provide this training. The team members receiving this training should be identified. This typically includes HM&E[2] engineers, propulsion designers, lead foremen, key trade personnel, and quality assurance inspectors. Training for design personnel should occur early on in the project and prior to training for production and QA personnel. The major design factors influencing the acoustic quality of the ship should be presented. As needed, owner personnel should also be invited.

4.9 Testing and Verification

The Program Plan should identify all required acoustic testing, including purpose, test details, location, schedules, and personnel needed to attend or conduct the test (also see Chap. 13 for details). The object of testing is to provide acoustic source levels, demonstrate compliance with some parts of the project or specification, or perform diagnostic testing. Any test plan should identify the needed acoustic instrumentation; equipment, ship or machinery operating conditions; calibration certificates; field calibrations to be provided; witness requirements; and level of detail for test reports. Dates for delivery of the draft test plans and final plan responding to owner comments should be identified.

Test procedures will need to comply with the requirements of the ship's specification concerning ship operating conditions, water depth and positions for acoustic sensors (microphone or hydrophone) and vibration measurements. Compliance testing is usually performed during Builder's or Acceptance Trials. Mechanisms in place to address any measured noise or vibration excesses should be included in the plan.

---

[2] Hull, Mechanical and Electrical

Test procedures and acceptance requirements are critical for contracts having 'acoustic' incentives or penalties. In some cases, intermediate testing may be needed to reduce risks. This may include using an artificial source, cycling equipment as it comes on-line, or doing overside testing with a hydrophone.

4.10 Deliverables

The following lists the deliverables that should be required for a comprehensive Noise Control Program Plan, as appropriate:

- **Design Reports**
  - Noise Control Program Plan (Initial Draft and Final Accepted)
  - Airborne Noise Analysis Report
  - Habitability Vibration Analysis Report
  - Hull Vibration Analysis Report
  - Shaft Vibration Analysis Report
  - Resilient Mount/Foundation Vibration/Resonant System Design and Analysis Report
  - Sonar Self-Noise Analysis Report
  - Underwater Radiated Noise Analysis
  - List of Acoustic Isolation Systems
  - Inspection and Corrective Action Report
- **Test/Trials Reports**
  - Factory Acceptance Noise Test Report
  - Airborne Noise Survey Report
  - Vibration Survey and Equipment Structureborne Noise Survey Reports
  - Sonar Self-Noise Survey Report
  - Underwater Radiated Noise Report

To augment the planning for noise critical designs, the following 'internal reports' should be maintained at set intervals:

- Progress reports
- Design guidance
- Review of Purchase Descriptions
- Review of vendor acoustic data and vendor supplied noise/vibration controls
- Updated predictions and recommended noise/vibration controls
- Review of 'noise critical drawings'
- Resolution of conflicts between the specification/regulatory requirements and control approaches

References

Fischer, R.: Elements of a successful ship noise control project. In: Marine Environmental Engineering Technology Symposium (MEETS).Arlington, VA, Jan. 2006

Fischer, R., Pettit, L.: A noise control decision making process for marine vessels. In: SNAME Maritime Convention, Houston TX, Oct. 2014.

Fischer, R., Yankaskas, K.: A program approach to reducing shipboard noise on naval vessels. In: Noise-Con 2011, Portland, OR, July 25, 2011

ISO 21940-11:2016 Mechanical Vibration. Rotor balancing procedure and tolerances for rotors with rigid behaviour

van den Berg, N.G.: A Prediction Method for the Underwater Acoustic Signature of Steel Surface Vessels, Allowing for a Quick Approximation During the Design Stage. Marine Technology, Delft University of Technology, Oct 2015

Verboom, W. C.: Radiated Ship Noise Reduction Presentation. TNO TPD, March 2004

Chapter 5
Shipboard Noise and Vibration Sources

Noise, whether habitability or underwater radiated, is most often dominated by the acoustic levels associated with propulsion sources. This includes the prime movers and propulsors. Next items of concern are the auxiliary equipment. High vibrations exist at their attachment point to the ship's structure. Furthermore, this equipment is often well distributed throughout the length and width of the vessel. Vibrations induced by this equipment and traveling through and along the vessel structure are called "structureborne noise." The other significant path starts as high noise levels in machinery spaces. This noise impinges on the surrounding ship structure, which has a measurable vibration response. This vibration, as it is distributed along over the vessel, is considered "secondary structureborne noise". The remaining path of concern is airborne transmission from one compartment to the next or from an airborne noise source to an on-deck station. It is critical to understand both these noise and vibration components. This allows one to rank sources of importance by their frequency-weighted acoustic levels.

The propulsor is the dominant source with respect to underwater noise. Once cavitation occurs, it commonly dominates the ship's underwater signature. This cavitation-induced noise can also be heard inside the ship, so it becomes another source that eventually contributes to airborne noise inside compartments.

This chapter considers the acoustic characteristics of sources – both in terms of level and frequency. Some acoustic sources are well known and understood. Others, including novel sources, may require further investigation to reduce risk and ensure accurate acoustic predictions.

There are generally six methods of obtaining acoustic source levels to be used in ship acoustical predictions; in order of preference, these are as follows:

R. Fischer, L. Boroditsky, *Noise and Vibration Control on Ships*,
https://doi.org/10.1007/978-3-031-55170-3_5

1. The fundamental goal is to have vendor[1] measured noise and vibration data on the particular machinery item being considered.
2. The second choice is to specify the machinery be subject to Factory Acceptance Tests (FATs), where the machinery being used on the vessel is tested for its noise and vibration characteristics according to an acceptable regulatory standard.
3. Measurement taken previously on as close to an identical piece of machinery (vendor, horsepower, rpm, type of machinery, etc.) as possible. These levels should be 'scaled' by knowing the relationships between power, rpm, or other operating parameters (see SNAME T&R 3-37)
4. A database of measured noise and vibration for as close to an identical piece of machinery (horsepower, rpm, etc.) as possible; again, scaled by operating parameters.
5. The next to last, from an accuracy or risk point of view, is using empirical scaling laws or algorithms based on machinery type and scaled by certain controlling parameters, such as horsepower, rpm, and operating pressure (as supplied by the SNAME Design Guide and Supplement, 1983, 2001).
6. The least appropriate is maximum or overall (OA) machinery source level limits by generic type/power as provided by regulatory agencies and government sources (ref Mil-Std, ISO std. by machinery grade), as the max data do not usually have any frequency content.

The first is usually available from vendors who have or use acoustic test facilities. The second is a method usually utilized on specialized vessels, research and military. For the third, one depends on vendors or acoustic consultants building up a database of previous measurements. The fourth is generally maintained by large shipyards or acoustic consultants. The fifth approach is available from various design guides, many of them maintained by professional societies (Fischer et al. 1983 and Fischer and Boroditsky 2001, Heckl 1988, and a list of Databases found at the end of this chapter). Finally, the last approach only provides the maximum allowed or overall level by machinery type. Thus, it is the least conservative approach. This may lead to maximizing the noise/vibration controls to be applied.[2] The least desirable approach is to not use any acoustic data, which may lead to an objectionably noisy vessel. This approach is often taken by shipyards that 'just know how to build quiet ships.'

For many ship projects, the acoustical characteristics of major sources are determined by the Factory Acceptance Test. The usual acoustical test parameters are the sound pressure level or sound power and vibration level spectra. For many noise prediction methods and acoustical modeling software, the sources are characterized

[1] These source level data should be measured and reported in accordance with accepted ISO or other accepted regulatory body standards for acoustic testing. It is not uncommon for vendor acoustic data to be optimistic (providing low levels) since it is often based on a small sampling and the variability in tolerances and manufacturing quality are not captured appropriately.

[2] Vendor provided 'balance' requirements are not suitable since they are only valid at a single frequency. Source levels used in the prediction process should be provided in a minimum of octave bands. Third-octave bands data provide additional information on the frequency content.

Table 5.1 Diesel engine source vibration level corrections for hard-mounted engines

Octave Band Freq.	31.5	63	125	250	500	1000	2000	4000	8000
Free – Hard Mounted Vibration, dB	5	3	2	1	0	0	0	0	0

by sound power[3] and vibration energy, which impinge on the foundation (ship structure). Most machinery vendors in the United States do not, in practice, measure vibration energy and/or sound energy. In the European Union (EU), the opposite is becoming true due to regulatory requirements such as EU Machinery Directive 2006/42/EC.

Noise and vibration levels are usually measured at the manufacturer's facility or shipbuilder's facility. There are multiple standards[4] that determine the conditions needed for accurate measurements. All standards have many common features, such as accuracy, testing location, allowed background levels, sensor locations, instrumentation, reporting, and calibration.

Airborne noise level measurements are usually performed at several positions one (1) meter from each of the four sides and over the top surface of the machinery. Usually, direct field sound pressure prevails at this distance. Therefore, low ambient noise is a precondition for informative noise measurements.

The vibration levels of the source feet strictly depend on the stiffness of the support or foundation (Aatola 2002). To have an opportunity to compare vibration levels from different sources, many standards require measurements with machinery installed on resilient (isolation) mounts. This would provide the so-called 'free vibration' level of the equipment.[5] Free vibration is usually used in the noise prediction process. Adjustments from 'free vibration' to that measured for a hard-mounted diesel are provided in Table 5.1 (Fischer et al. 1983). Here, it is seen that the attachment to a high stiffness or impedance foundation lowers the expected free vibration in the lower frequency bands.

The parameters of interest are the acoustic noise or vibration levels over the frequency range of interest, bandwidth, and reference units. For hearing and most habitability noise considerations, this range is typically the octave bands from 31.5 to 8000 Hz referenced to 20 μPa for noise and 1 μG for structureborne noise. For human responses to vibration and structural fatigue, this range is usually narrow bands or one-third octave bands from 1 to 200 Hz. For machinery vibration, the range can go from 1 Hz to 20,000 Hz. Underwater radiated noise can be either narrow, third-octave, or octave bands between 1 Hz and 100,000 Hz referenced to 1 μPa.

[3]While sound power is the preferred metric, sound pressure is the most common type of data provided. See Eq. (5.1).

[4]ASNI S2.26, S2.27, and S2.28; ISO 4867, ISO 6954, ISO 7919, ISO 10055, ISO 10816, ISO 13332, ISO/CD 20155 N330, ISO 20283, ISO 8528; ABS Guidance Notes on Ship Vibration.

[5]At low frequencies, the equipment is never truly free of the attachment method and environment.

In this book, the sound pressure and vibration levels are the main 'source' parameters of interest. The relationship between power and pressure for a small source is given in Sect. 7.1:[6]

$$L_p \text{ dB re } 20\ \mu\text{Pa} = L_w \text{ dB re } 10^{-12}\ \text{W} - 20\ \text{Log}(r) - 11 \quad (5.1)$$

where r is the distance from the source, m.

"Empirical" source noise and vibration formulations are provided in SNAME T&R 3-37 and its Supplement (Fischer et al. 1983 and Fischer and Boroditsky 2001). Other source references are Ashcroft 2018; Thumann and Miller 1990; Plunt and Odegaard 1983; Bies and Hansen 2009; Nilsson and Tyvand 1981, 1983; and Heckl 1988. For the SNAME Design Guide, an empirical approach was based on curve fitting a large naval database of information for each machinery type. These formulas typically involve mechanical power and rpm. Other variables may include mass, flow rate, and pressure. Since extensive documentation exists in the above references, these formulas will not be repeated herein except for a few important sources, such as propulsion equipment.

The following is a list of the types of equipment typically considered in addressing noise and vibration environments onboard vessels:

1. Propulsion drivers
   (a) Diesel
      (i) Direct drive
      (ii) Geared drive
      (iii) Diesel electric
   (b) Gas Turbine
      (i) Geared drive
      (ii) Gas turbine electric
   (c) Hybrid – diesel/battery
   (d) Gearbox
   (e) Propeller
      (i) Fixed Pitch
      (ii) Controllable Pitch
      (iii) Bow/stern thrusters
      (iv) Water Jets
      (v) Cyclodial

[6]The same equation for the sound pressure in water in L_p dB *re* 1 μPa would substitute +51 dB in lieu of the −11 dB.

2. Auxiliary equipment

   (a) Ship Service Generator

      (i) Diesel generator
      (ii) Gas turbine generator

   (b) Compressor – Air and Refer

      (i) Rotary
      (ii) Reciprocating
      (iii) Refrigeration/Air Conditioning
      (iv) De-ballast

   (c) Pump -

      (i) Fluid
      (ii) Air

   (d) Hydraulic Systems
   (e) Motors
   (f) Fans/Unit Coolers/Fan Coil Units

      (i) Forced Draft Blowers
      (ii) Ventilation Fans

         1. Centrifugal
         2. Airfoil
         3. Forward/Backward Curve
         4. Vane and Tube Axial
         5. Propeller
         6. Collective Protection System (CPS)

      (iii) Electronic cooling

   (g) Transformers/SCRs
   (h) Distributed systems

      (i) Diesel intake/exhaust (including turbo)
      (ii) HVAC

         1. Flow regenerated – vanes, turns, etc.

      (iii) Hydraulic

         1. Flow – through piping system

   (i) Miscellaneous

      (i) Boilers/Steam Turbines
      (ii) Valves
      (iii) Impulsive sources
      (iv) Tools

(v) Bow wave impact
(vi) Breaking ice
(vii) Cavitation from flow over appendages or propulsor
(viii) Vibration resonances – natural or forced, including vortex shedding excitation

Based on the authors' experience, the repeatability of source level measurements is usually +/−2 dB at best. At low frequencies, the variance could be as high as +/−7 dB. The variability or repeatability in a laboratory setting is usually in the 1 to 3 dB range.

5.1 Airborne and Structureborne Machinery Source Levels

Table 5.2 lists the expected range of A-weighted noise levels for some typical shipboard propulsion equipment. For reference, the table also provides the allowable 'exposure' times, without hearing protection and with double hearing protection.

Note that without hearing protection, the exposure time before incurring hearing damage is measured in seconds to minutes.

5.1.1 Propulsion and Ship Service Engines

In many cases, the propulsion engine is the main source. It controls noise levels in the vicinity of the source and is sometimes remote from the source. Various diesel propulsion options exist – direct drive, diesel/geared, diesel electric, or hybrid combination with other types of drives such as gas turbine. There are three major components used to assess diesel-induced airborne noise: casing, intake, and exhaust. The intake/exhaust are not usually of concern if completely ducted and

Table 5.2 Sound level ranges for noise sources and estimated exposure times with and without double-hearing protection

Source	Low Level dB(A)	High Level dB(A)	Allowed Worst Case Unprotected Exposure Time[a]	Estimated Exposure Duration With Double Hearing Protection[b]
Diesel	98	120	9 s	2.5 h
Gas Turbines	85	101	12 min	Unlimited
Shipboard Machinery/ Equipment	84	114	36 s	6 h

[a]Based on 85 dB(A) limit for 8-h and 3 dB exchange rate
[b]Estimate using 30 decibel (dB) reduction for double hearing protection, realizing this may be a conservative best-case scenario

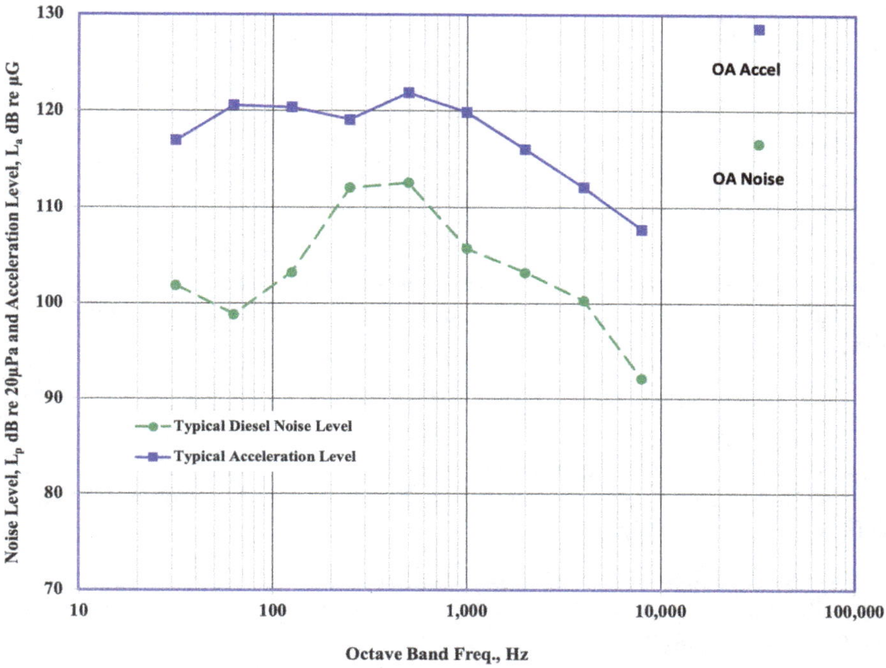

Fig. 5.1 Typical diesel source acoustic levels. OA Overall

silenced. A turbocharge should be considered a significant noise source. The casing noise source levels are typically dependent on whether it is a low-speed (below 700 rpm) or medium- to high-speed diesel (above 700 rpm). For the same power, a low-speed diesel is noisier below 500 Hz and quieter above 500 Hz; however, the A-weighted noise level is still higher for the low-speed diesel.

The noise level may also depend on the operating power/rpm versus the rated power/rpm. The type of intake and diesel mass are also parameters that usually affect the airborne source level. Typical diesel source levels are shown in Fig. 5.1. 'OA' is the overall acceleration level[7] or the overall noise level. Note that the diesel vibration level contains high levels at low frequencies, below 500 Hz. Low-frequency noise is usually dominant on diesel-propelled vessels. Hence, the need for low-frequency isolation mounts on diesel, whether propulsion or generator drives.

The diesel source levels are computed as follows (Fischer and Boroditsky 2001).

[7] Vibration levels are provided in acceleration as L_a, dB re 1μG; to convert to velocity levels, L_v, dB re 10^{-8} m/s, use the following equation: L_v, dB re 10^{-8} m/s $= L_a - 20 \log (f) + 44$.

***Diesel Engine Exhaust*:**
One approach for predicting the octave band exhaust source levels in dB re 1 pW is:

$$L_w \text{ dB re 1pW} = 77 + 10 \text{ Log}\left(P \times R_{rpm}\right) + 30 \text{ Log}\left(\frac{O_{rpm}}{R_{rpm}}\right) + 10 \text{ Log}\left[\frac{1}{\left[\left(\frac{f_r}{f}\right)^3 + \left(\frac{f}{f_r}\right)\right]}\right] \tag{5.2}$$

where R_{rpm} is the rated rpm, P is the rated power in kW (hp = kW/0.7457), f is the octave band center frequency, O_{rpm} is the operating rpm, and fr is the firing rate. The firing rate is determined as:

$$fr = \text{Number of cylinders} \times O_{rpm} \times 2/(60 \times \text{No.Strokes}) \tag{5.3}$$

Diesels are 2 or 4 stroke, depending on the number of compression cycles. In general, Detroit Diesel and EMD are 2 stroke engines, while Caterpillar, Wartsila and Cummins are 4 stroke engines.

***Diesel Casing Radiated* (Fischer and Boroditsky 2001):**
The following equation is for the casing radiated source level of diesel operating below and above 700 rpm:

For diesels operating over 700 rpm, the power level in dB re 1 pW is:

$$L_w \text{ dB re 1pW} = 57 + 10 \text{ Log}\left[\frac{R_{rpm} \times P \times (1 + (P/m))}{\left(\frac{f}{1000} + \frac{1000}{f}\right)}\right] + 20 \text{ Log}\left(O_{rpm}/R_{rpm}\right) \tag{5.4}$$

where m is the mass in kg of the engine (kg = weight, lbs.*0.4536). If the mass is not known, the term P/m can be estimated as $P/m = 0.1\ \sqrt{(P/1000)}$.

For diesels below 700 rpm:

$$L_w \text{ dB 1 pW} = 3 + 4.5 \times \text{Log}(P) + A\left[5.5 \text{ Log}(P) + 10 \text{ Log}\left(R_{rpm}\right) - 43\right] + B \tag{5.5}$$

where the A and B coefficients are listed in Table 5.3 (Fischer and Boroditsky 2001):

Table 5.3 Octave Band coefficients, dB – Diesel airborne noise

OB Freq, Hz.	31.5	63	125	250	500	1000	2000	4000[a]	8000[a]
A	1.6	1.5	1.4	1.3	1.2	0.8	0.8	0.8	0.8
B	93	96	98	100	102	102	96	91	86

[a] Estimated level due to lack of a reliable database

Table 5.4 Octave Band adjustments, dB – Diesel structureborne noise

	Octave Band Center Frequency (Hz)								
Description	31.5	63	125	250	500	1000	2000	4000	8000
Adjustments for Baseline Structureborne Noise, dB	0	5	11	16	21	27	29	27	22

Table 5.5 Octave Band adjustments, dB – L_{sB1}

OB Freq, Hz.	31.5	63	125	250	500	1000	2000	4000[a]	8000[a]
Δ	−17	−19	−20	−21	−24	−28	−34	−43	−50

[a] Estimated level due to lack of a reliable database

Diesel Source Structureborne Noise Levels **(Fischer and Boroditsky 2001):**
High-speed diesel, isolation mounted

The baseline structureborne noise source level is calculated as:

$$L_{sb} \text{ dB re } 1\,\mu\text{G} = -20\,\text{Log}(m) + 20\,\text{Log}(P) + 30\,\text{Log}\left(R_{rpm}/O_{rpm}\right) \tag{5.6}$$

The octave band levels are determined by adding the coefficients in Table 5.4 to the baseline level.

Low to medium speed, hard mounted diesel (below 1000 rpm) (Fischer and Boroditsky 2001):

Step 1 – Determine L_{sB1}:

$$L_{sB1} \text{ dB re } 1\mu\text{G} = 55 + 5.5\,\text{Log}(P) + 10\,\text{Log}\left(R_{rpm}\right) + 30\,\text{Log}\left(O_{rpm}/R_{rpm}\right) + \Delta \tag{5.7}$$

Add adjustments, Δ, provided in Table 5.5 to determine octave band levels of L_{sB1}.
Step 2– Determine L_{sB2}:

$$L_{sB2} \text{ dB re } 1\mu\text{G} = A + B\,\text{Log}(P/1000) \tag{5.8}$$

Add adjustments provided in Table 5.6 to determine octave band levels of L_{sB2}:
Step 3: Compute

$$L_{sB} = 0.5\,(L_{sB1} + L_{sB2}) \tag{5.9}$$

and add the adjustments of Table 5.7

Table 5.6 Octave Band adjustments, dB – L_{sB2}

OB Freq, Hz.	31.5	63	125	250	500	1000	2000	4000[a]	8000[a]
100–250 rpm									
A	74	73	73	63	63	60	58	56	54
B	4	3	2	10	7	6	1	1	1
250–600 rpm									
A	86	82	80	78	80	75	69	64	59
B	1	3	3	8	6	5	1	1	1
600–1000 rpm									
A	91	86	85	83	85	83	79	78	77
B	6	6	8.5	5	7	4.5	4	3	2

[a] Estimated level due to lack of a reliable database

Table 5.7 Octave Band adjustments, dB – L_{sB}

OB Freq, Hz.	31.5	63	125	250	500	1000	2000	4000	8000
Adj, dB	0	6	12	18	24	30	36	42	48

Table 5.8 Octave Band adjustments, dB – L_{sB}

OB Freq, Hz.	31.5	63	125	250	500	1000	2000	4000[a]	8000[a]
A	81	81	77	76	77	69	70	68	63
B	13	5	4	9	2.5	18	13	10	10
Δ	0	6	12	18	24	30	36	42	48

[a] Estimated level due to lack of a reliable database

Table 5.9 Adjustments for hard-mounted diesel-generators located between decks

OB Freq, Hz.	31.5	63	125	250	500	1000	2000	4000	8000
Adj, dB	7	8	10	10	11	15	18	18	18

Hard mounted Diesel Generator (Fischer and Boroditsky 2001):

Structureborne source levels for a hard mounted genset are computed as:

$$L_{sB}\ \text{dB re } 1\ \mu\text{G} = A + B\,\text{Log}\left(\frac{P}{250}\right) + \Delta, \tag{5.10}$$

where the values for A, B, and Δ are provided in Table 5.8.

These values apply for diesels located on an integrated tank top structure. The vibration levels would increase by the adjustments given in Table 5.9 for hard mounted diesel generators mounted on higher levels in the vessel.

For isolation mounted diesel-generators, use the structureborne noise levels for the appropriate diesel. If the diesel and generator are mounted on a common skid, reduce the diesel vibration levels by the ratio of the diesel to that of the entire genset package, i.e., 10 Log (diesel mass/genset mass).

It is important to note that 4 cycle diesels also generate tones at multiples of half the rotation rate. Diesels have a forest of tones that are easily detectable in the underwater radiated signature of a commercial vessel.

5.1.2 Gas Turbines

Gas turbine-driven vessels are generally quieter than their diesel counterparts. This is primarily because this machinery is rotary vice reciprocating. The vibration source levels – both tonal and broadband – are lower for comparable weight-to-power levels. Furthermore, the tones produced by the gas turbine are much higher in frequency due to their higher rotation rate, as high as 3600 rpm or 60 Hz. The rotor tones are usually in the thousands of Hertz due to the high number of blades on the turbine compared to the firing rate associated with the number of cylinders in diesel. A comparison of representative diesel and gas turbine vibration levels is provided in Fig. 5.2. The actual vibration on any vessel will depend on the vendor's

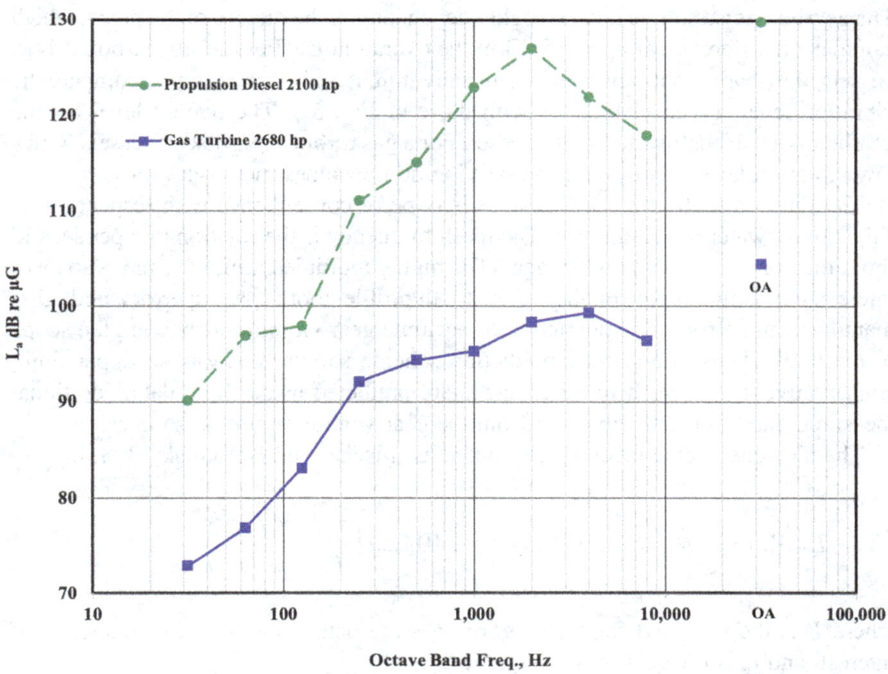

Fig. 5.2 Representative source vibration levels – gas turbine versus diesel

equipment source level, the type of ship foundation (mechanical stiffness or dynamic impedance), and the attachment method.

Gas turbine inlet and exhaust sound power levels can be estimated by the following equations (Tupov and Semin 2008):

Intake source power level, $L_{w\text{-}intake}$:

$$L_{w-intake} \text{ dB re 1pW} = 10\, m_1 \, \text{Log}(P) + 137 \quad (5.11)$$

where P is the gas turbine power, MW, and m_1 is a factor determined to be 0.74 (by measurements of several gas turbines).

Exhaust source power level, $L_{w\text{-}exhaust}$:

$$L_{w-exhaust} \text{ dB re 1pW} = 10\, m_1 \, \text{Log}(P) + 137 \quad (5.12)$$

where m_1 is a factor determined to be 0.71. Note: these two equations show the intake and exhaust levels to be very close.

5.1.3 Gearboxes

The gearbox noise source level is highly dependent on the quality of the gears, which can make a difference of up to 15 dB in the overall noise. For diesel/gearbox driven vessels, the gearbox vibration level is usually immaterial – for example, compare the vibration levels given in Fig. 5.1 with those of Fig. 5.3. The overall level for the gearbox is 18 dB below that of the diesel. For a gas turbine/gearbox or vessels with a low vibration level driver, the gearbox vibration becomes more relevant.

The tones at tooth mesh frequencies may be of consequence with respect to the ship's underwater radiated noise. The mesh frequency is the rotation rate per second times the number of teeth on that gear. The higher multiple harmonics may also be of interest, depending on the quality of the tooth profile, tooth spacing errors, and other manufacturing errors. The prediction of accurate gear vibration is beyond the scope of the book. Thus, the generic prediction scheme shown below is an upper limit, conservative approach. If noise excesses are predicted using these data, additional measured data should be obtained from the gear vendor or similar sources.

The airborne source level is determined as (Fischer and Boroditsky 2001):

$$L_w \text{ dB re 1pW} = 68 + 10\, \text{Log}(P) - 10\, \text{Log}\left(\frac{f_g}{f} + \left(\frac{f}{f_g}\right)^2\right) + \Delta \quad (5.13)$$

where P is the kW load for the gearbox, f is the octave band center frequency of interest, and f_g is the gear mesh frequency.

Δ represents the influence of the manufacturing tolerance of the gearbox. These values are provided in Table 5.10. Gears produced to a high tolerance (after VDI

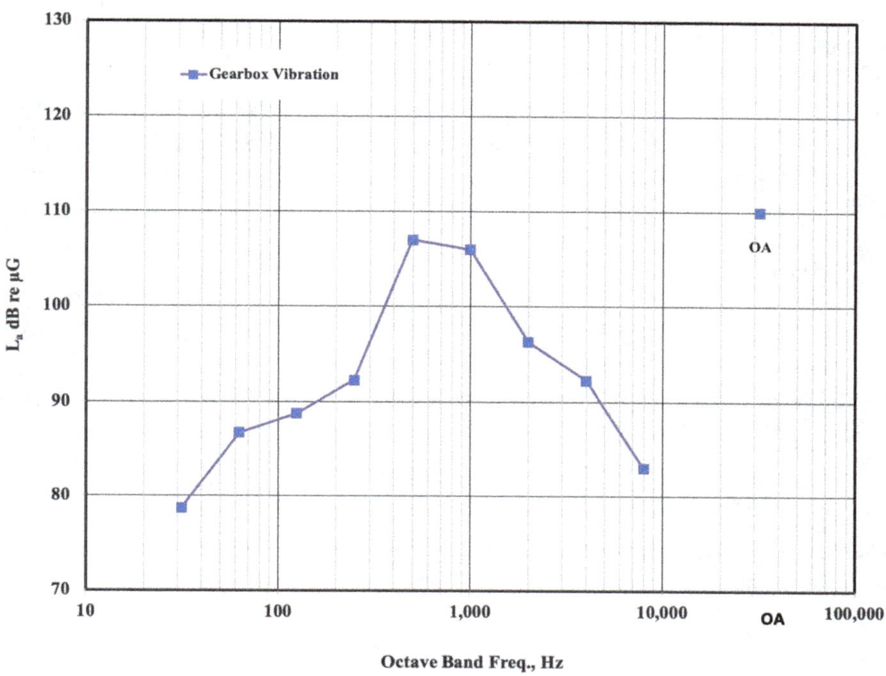

Fig. 5.3 Typical gearbox source structureborne level

Table 5.10 Gear manufacturing tolerance adjustment, dB

Class	B3	C1	C2	C3	D1	D2	D3
Δ, dB	0	2.5	5	7.5	10	12.5	15

2159 (1985)) are rated Class B. Those produced to normal tolerances are in Class C. Class D gears are manufactured to low tolerances.

If the manufacturing tolerance in accordance with VDI 2159 (1985) is known, Eq. (5.13) above provides the most accurate approach.

For reduction gears operating at full speed and load, the baseline structureborne noise source level is given by: (Fischer et al. 1983)

$$L_{sB} \text{ dB re } 1\ \mu\text{G} = 48 + 10\ \text{Log}\ (P) \tag{5.14}$$

where P is the rated load in kW. To determine the source levels in each octave band, the adjustments given in Table 5.11 are added to the baseline level (adjustments are always added to the baseline level).

The airborne power level of smaller spur, spiral bevel, and planet system gears can be computed as follows (Bowes 1979):

Table 5.11 Octave band adjustments for the baseline gearbox structureborne noise source level, dB

	Octave Band Center Frequency (Hz)								
Description	31.5	63	125	250	500	1000	2000	4000	8000
Adjustments for Baseline Structureborne Noise Source Level	0	9	3	8	23	33	33	28	18

Spur Gears, $L_{w\text{-}spur}$:

$$L_{w-spur}\ \text{dB re 1 pW} = 20\ \text{Log}(\tau) + 37.8\ \text{Log}(f) - 72 - D_{sg} \tag{5.15}$$

where τ is the torque, N-m, f is the frequency, and D_{sg} is -5 for 'Harmonic No.[8] = 2'.

Spiral Bevel Gears, $L_{w\text{-}sbg}$:

$$L_{w-sbg}\ \text{dB re 1 pW} = 20\ \text{Log}(\tau) + 37.8\ \text{Log}(f) - 81 - D_{sbg} \tag{5.16}$$

where τ is the torque, N-m, f is the frequency, and D_{sbg} is +7 for 'Harmonic No. = 2'.

Planet System Gears, L_{w-ps}:

$$L_{w-ps}\ \text{dB re 1 pW} = 12.8\ \text{Log}(\tau) + 37.8\ \text{Log}(f) - 39 - D_{ps} \tag{5.17}$$

where τ is the torque, N-m, f is the frequency, and D_{ps} is -10.5 for 'Harmonic No. = 2'.

Information on the quality of gearbox vibration is as follows (Nayak 1973):

- Helical gears have flatter spectra than spur gears; lines at tooth contact frequency and harmonics tend to dominate the spectra of helical gears to a smaller extent than they do for spur
- Axial vibration may be more important for helical gears than for spur gears
- Helical gears are quieter than spur gears by 0 to 10 dB overall
- Well lubricated gears are much quieter than poorly lubricated gears (by as much as 20 dB)
- Bearing type makes a difference of as much as 6 dB overall radiated noise – (Optiz 1967) ranks bearing types[9] as follows:
  - Radial ball bearing
  - Angular contact ball bearing with preload – 2 dB quieter than above
  - Tapered roller bearing – 3dB quieter than radial ball bearing
  - Tapered roller bearing with axial preload – 4dB quieter than radial
  - Analysis suggests that decreased radiation is due to increased damping; thus, 'fluid-film' bearings are the quietest

[8]The harmonic number is not defined, however, these 3 equations are useful in determining the relative source level of each gear type.

[9]ISO has multiple standards for various bearing types

- Cylindrical housing is quieter than rectangular housing
- Increasing the face-width or diameter or helix angle reduces noise
- A total contact ratio (nominal number of teeth in contact) near an integer results in a quieter gear (reduced stiffness variation)
- Lines at tooth contact frequency and harmonics are caused mainly by mesh stiffness variation and by errors other than tooth spacing errors.
- Using a higher quality gear decreases overall noise, but only if a) lines at tooth contact frequency and harmonics are not strong or b) these lines are caused by errors other than tooth-spacing errors.
  - Extremely high manufacturing accuracy: $1 \times 10^{-8} < \eta_{ac} < 5.7 \times 10^{-8}$
  - High manufacturing accuracy: $5.7 \times 10^{-8} < \eta_{ac} < 3.2 \times 10^{-7}$
  - Normal manufacturing accuracy: $3.2 \times 10^{-7} < \eta_{ac} < 2.5 \times 10^{-6}$
  - Poor manufacturing accuracy: $2.5 \times 10^{-6} < \eta_{ac}$
  - where η_{ac} = overall radiated power/transmitted power
- Tip relief, profile modifications, and crowning may all be used to reduce noise and vibration

5.2 Propulsor Noise Source

The metric of interest with respect to habitability noise is the vibration induced on the hull by propellers, thrusters, or waterjets. Some measured hull vibrations are given in Tables 5.12, 5.13, and 5.14.

Table 5.12 Vibration levels on hull plates above or around propulsors (measured by authors)

Frequency, OB Hz	31	63	125	250	500	1000	2000	4000	8000
Free run, 12 knots	80	82	105	100	98	99	105	113	104
Towing, 4 knots	102	107	123	112	114	113	125	126	113

Support Boat – vibration level above propeller plane, dB re 1 μG

Table 5.13 Vibration on plate near Waterjets

Frequency, OB Hz	31	63	125	250	500	1000	2000	4000	8000
Aluminum catamaran, 35 knots	86	92	101	112	120	121	128	124	122

Fast Catamaran – Vibration on plates near waterjets, dB re 1 μG

Table 5.14 Vibration on Waterjet components, dB re 1 μG

Frequency, OB Hz	31	63	125	250	500	1000	2000	4000	8000
Transom	108	107	108	115	122	130	128	128	131
Duct	133	130	127	130	132	133	132	132	134

This chapter addresses both aft hull vibrations induced by a cavitating propeller and hull vibrations induced by a bow tunnel thruster.

5.2.1 *Propeller-Induced Hull Vibrations*

The vibrations induced on the hull above the propeller can vary greatly. To a large extent, cavitation is the main parameter controlling hull-induced vibrations. Other factors are propeller tip clearance, hull plate thickness, and type of cavitation. Some measured data are shown in Tables 5.12, 5.13, and 5.14. Note that towing can significantly increases the hull vibration – see Table 5.12.

Propulsor-induced noise can be the greatest source of habitability noise in aft compartments on ships and is usually the prime contributor to the ship's underwater radiated noise. This noise is generated via several phenomena, as follows in order of significance:

- Cavitation – most severe, usually above 10 knots, broadband noise with tones at blade rates and multiples
- Vortex shedding – off trailing edge and leading edge, due to pressure differences across face, narrower hump of energy – can also excite 'singing'
- Turbulence ingestion – inflow not laminar due to hull form, shafting, and appendages; the more turbulent the flow, the greater the broadband noise
- Shaft and bearing forces – propeller excited longitudinal, bending, and torsional forces imparted to hull structure; particularly at harmonics of blade rate.

These phenomena were investigated as early as 1957 by Saunders.

Propeller-induced noise is often one reason for unacceptable noise levels in the aft compartments. Propeller noise control is a serious problem for combat ships, research vessels, pleasure craft, and cruise ships. The propeller generates considerable noise when it starts to cavitate – also known as the cavitation inception speed. For onboard compartments, the propeller is considered a source of structureborne noise. From a habitability noise prediction and noise control point of view, this noise source would be fully determined if the vibration levels on the hull plating could be accurately calculated.

The response of the hull plate, as compiled by the authors, is based on algorithms from the SNAME Design Guide Supplement (Fischer and Boroditsky 2001) and Brown 1986. The vibration induced by the propeller depends on two phenomena: the underwater radiated sound pressure from the propeller and the response of the hull to this pressure field. The current procedure considers only the case where the propeller is cavitating. In the non-cavitating state, propeller-induced vibration is not normally of concern.

The propeller's radiated pressure level is a sum of two components: noise from tip vortex cavitation and noise from blade surface cavitation. Other cavitation phenomena are shown in Fig. 5.4; however, these are not usually as significant if the propeller is properly designed.

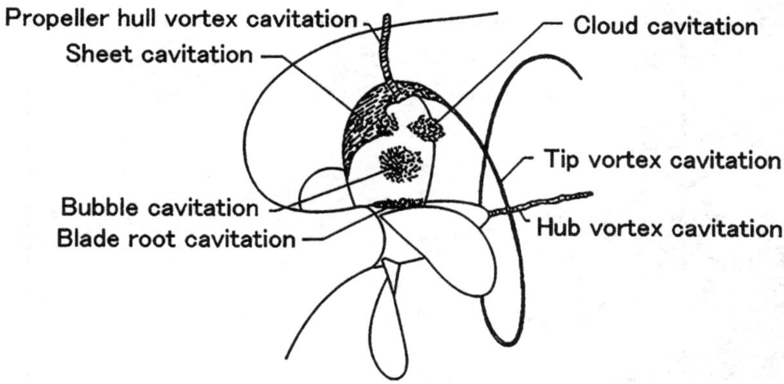

Fig. 5.4 Propeller cavitation phenomena (Ishii et al. 2020)

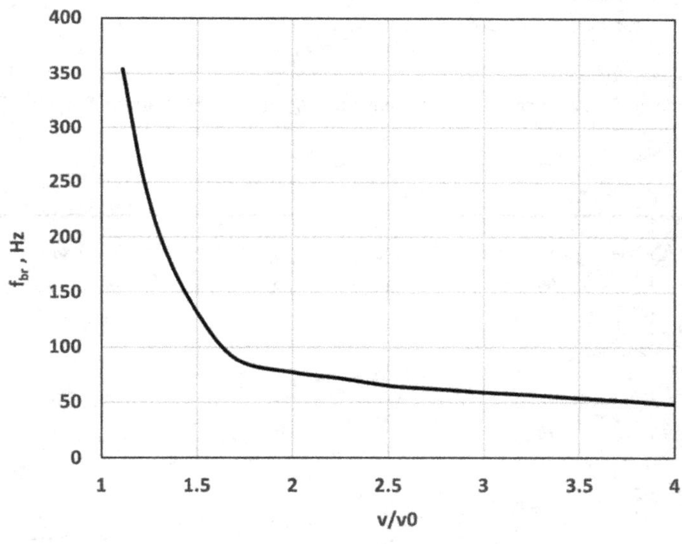

Fig. 5.5 Tip vortex break frequency, f_{br} Hz

The tip vortex and blade surface cavitation components are determined graphically for each octave band by the following steps (taken from Fischer and Boroditsky 2001):

1. Determine the break-frequency – the frequency where the sound pressure spectrum peaks. This is done using the graphs in Fig. 5.5 for the tip vortex and Fig. 5.6 and 5.7 for surface cavitation, depending on the v/v_o ratio.
2. Determine the sound pressure levels at this break-frequency for both tip vortex, Eq. (5.18), and blade cavitation, Eq. (5.19), and mark their locations on an octave band graph prepared by the user.

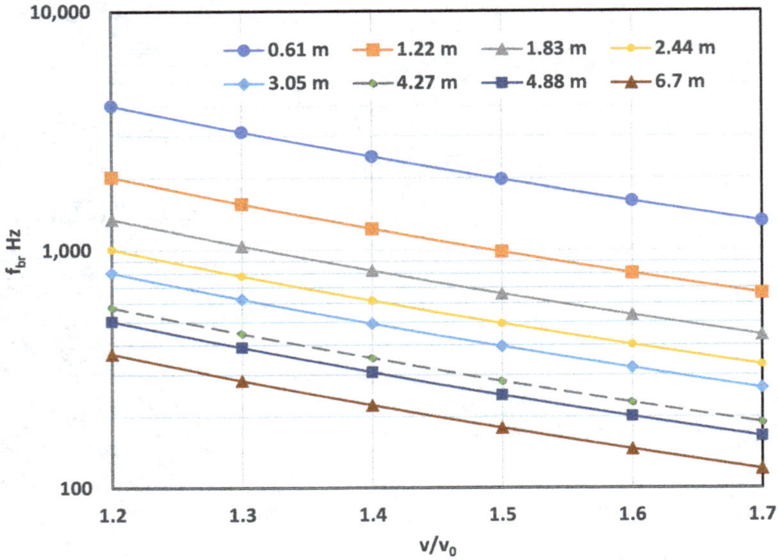

Fig. 5.6 Blade surface cavitation – break frequency, f_{br} vs. v/v_o for Various Propeller Diameters, D ($v/v_0 < 1.7$)

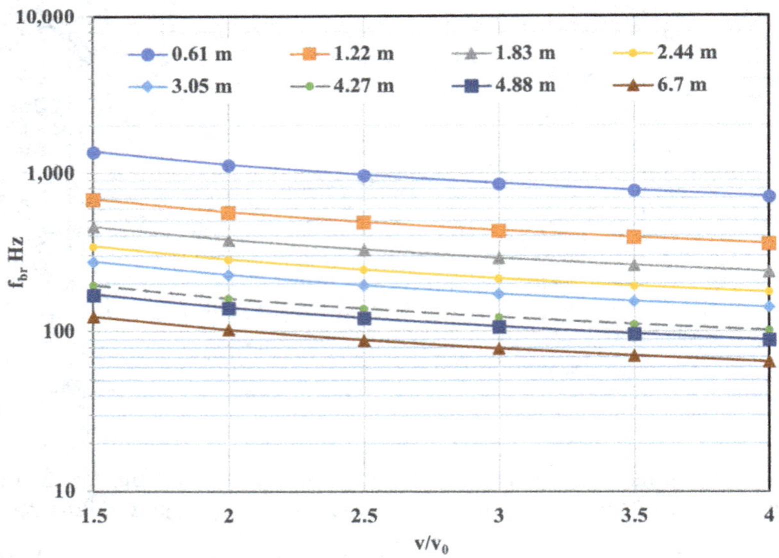

Fig. 5.7 Blade surface cavitation – break frequency, f_{br} vs. v/v_o for Various Propeller Diameters, D ($v/v0 > 1.7$)

3. On this graph, draw a line to the right with a slope of −3 dB/octave until the 8000 Hz octave band is reached for each type of cavitation – tip and blade surface.
4. On this graph, draw a line to the left with a slope of −3 dB/octave until the 31.5 Hz octave band is reached for each type of cavitation – tip and blade surface.
5. The total propeller sound pressure level is determined by combining the tip and blade surface sound pressure levels, as shown in Eq. (5.20).

The sound pressure at 1 m from the propeller, L_{pt}, in the octave band containing the 'tip-cavitation break frequency' is:

$$L_{pt} \text{ dB re 1 μPa–m} = 10 \text{ Log}\left[0.23 n_b \left(\frac{N_p}{60}\right)^3 (D)^4 \left(\frac{1}{f_{br}}\right)\right] + 155 + C_t \quad (5.18)$$

where f_{br}, tip cavitation break-frequency, Hz (from Fig. 5.5); n_b, number of blades; N_p, propeller RPM; D, propeller diameter, m; C_t, correction, which depends on ratio v/v_O (v, speed of ship, v_O, inception speed[10]); and for $v/v_O < 2.35$, $C_t = 10$ Log [0.67 $((v/v_O) - 1)$]; for $v/v_O > 2.35$, $C_t = 0.0$

The blade cavitation sound pressure level at 1 m from the propeller, L_{pb}, in the octave band containing the 'blade surface-cavitation break-frequency' is equal to

$$L_{pb} \text{ dB re 1 μPa–m} = 10 \text{ Log}\left[0.23 n_b \left(\frac{N_p}{60}\right)^3 (D)^4 \left(\frac{1}{f_{br}}\right)\right] + 168 + C_b \quad (5.19)$$

where the notations are the same as for Eq. (5.18) except $C_b = 20$ Log $(1-(v_O/v))$. Figures 5.6 and 5.7 show the break-frequency versus propeller diameter and inception speed.

These equations form Steps 2 through 4. The last step is to combine the results of Eqs. (5.18 and 5.19) to determine the total propeller signature, L_{1m}, as:

$$L_{1m} \text{ dB re 1 μPa–m} = 10 \text{ Log}\left(10^{L_{pt}/10} + 10^{L_{pb}/10}\right) \quad (5.20)$$

Once the propeller's pressure level is known, the induced acceleration on the hull plating above, L_a, is calculated using Eq. (5.21), combining the propeller source level with a 'transfer function':

$$L_a \text{ dB re 1 μG} = L_{p@hull} + \Delta L_{pa} \quad (5.21)$$

[10]Inception speed-speed of ship where the propeller begins to cavitate, v_o (kn) = 7+ 0.005 L_s, where L_s is the ship length, ft. This value is very conservative, cavitation inception speed would be much higher with a competently designed propeller and minimized wake. Another formulation is that $v_O = min\ \{max[(1.42 - 1.2\ c_b)V_d; 9] : 14\}$, $c_B$is the block coefficient (Jalkanen et al. 2018). Per this equation all ships cavitate by 14 kns.

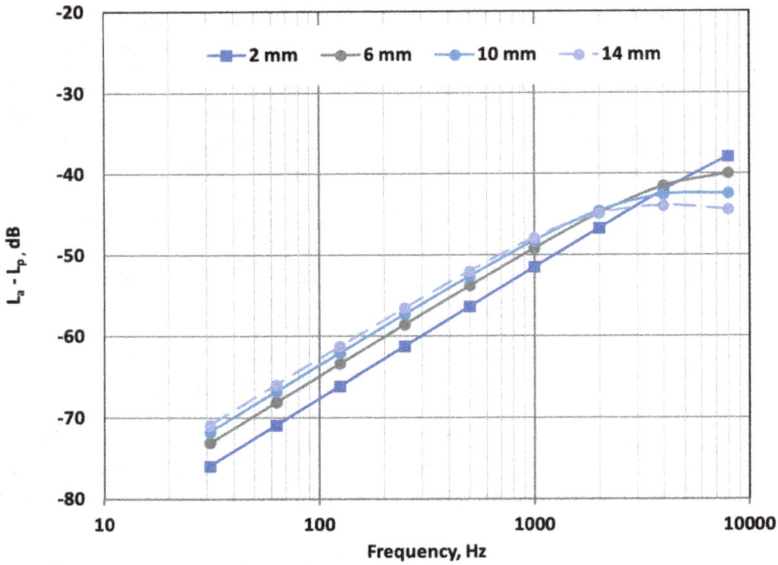

Fig. 5.8 Transfer function ($L_a - L_p$) vs. frequency for steel hulls of various thickness

where ΔL_{pa} is the transfer function relating the incident pressure, $L_{p@hull}$, to the hull response. This transfer function is provided in Fig. 5.8 for steel hulls and Fig. 5.9 for aluminum. The sound pressure incident on the hull is computed as:

$$L_{p@hull} \text{ dB re 1 } \mu\text{Pa} = L_{1m} - 10 \text{ Log} \left[\frac{0.33D + 3.3d_t}{0.33D + 3} \right] \tag{5.22}$$

where D is the propeller diameter, m; d_t is the distance from the propeller tip to the hull plate, m; and L_{1m} is the propeller source level per Eq. (5.20).

The acceleration levels computed in Eq. (5.21) are used as the vibration source levels for propeller structureborne induced noise. In the octave band containing the blade passage rate, $n_b \times N/60$, 5 dB is added to the acceleration level. This vibration would be attenuated by path effects – damping and intersection losses – as is done with other vibration sources, such as propulsion diesels.

5.2.2 *Bow Tunnel Thruster-Induced Noise and Vibration*

The A-weighted noise in a berth near a tunnel thruster with a controllable pitch propeller can be predicted by the following empirical equation (Carlton, 2007 and Buiten and de Regt n.d.):

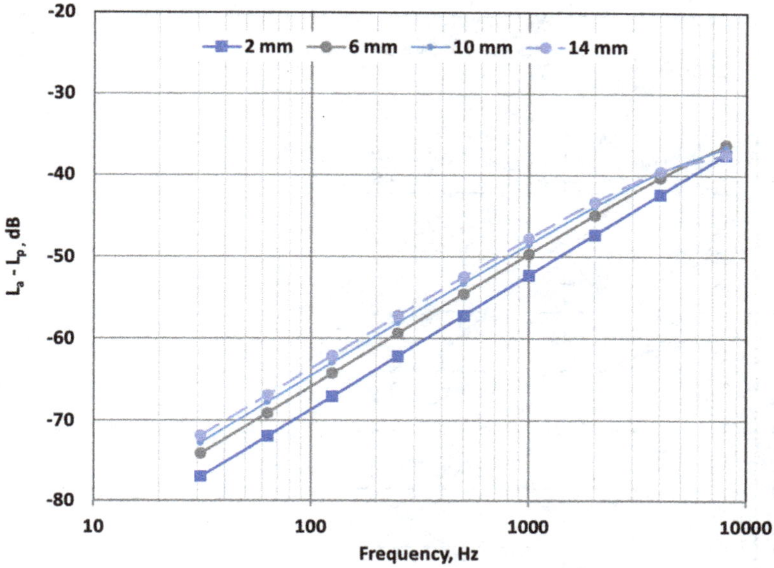

Fig. 5.9 Transfer function ($L_a - L_p$) vs. frequency for aluminum hulls of various thickness

$$L_p \text{ dB(A)re } 20\mu\text{Pa} = L_{p0} - L_p(\Theta) - L_p(L/D) - L_p(path) \tag{5.23}$$

where

L_{pO} is the A-weighted base level noise in a cabin located near the thruster room with the thruster operating at full power, P, kW,
$L_p(\Theta)$ is the change associated with operating at partial power.
$L_p(L/D)$ is a factor depending on the tunnel length, L, to the tunnel diameter, D, and
$L_p(path)$ accounts for the number of frames, Fr, and decks, D_K, to the receiver space of interest.

The additional equations are as follows:

- $L_{pO} = 108 - 7.07\ K + 10.029\ K^2 + (24.058 - 4.689\ K + 0.615\ K^2) \times \text{Log}(P_B)$, where K is the tunnel centerline immersion-to-duct diameter ratio, $1 < K < 3$ and P_B is the power rating of the thruster.
- $L_p(\Theta) = 26.775 - 13.387\ \text{Log}(\Delta P_B)$, where ΔP is the percent of full power
- $L_p(L/D) = -4.393 + 14.593\ \text{Log}(L/D)$, where $1 < L/D < 10$
- $L_p(path) = 10^{\left(0.904827 + 0.968977\ \text{Log}(D_K) - 0.348142\ (\text{Log}\ (D_K))^2\right)} + F_r 10^{\left(-0.22233 - 0.126009\ (D_K) + 0.007657 (D_K)^2\right)}$ where $1 < D_K < 5$ ($D_K = 0 =$ tanktop) and $1 < Fr < 50$

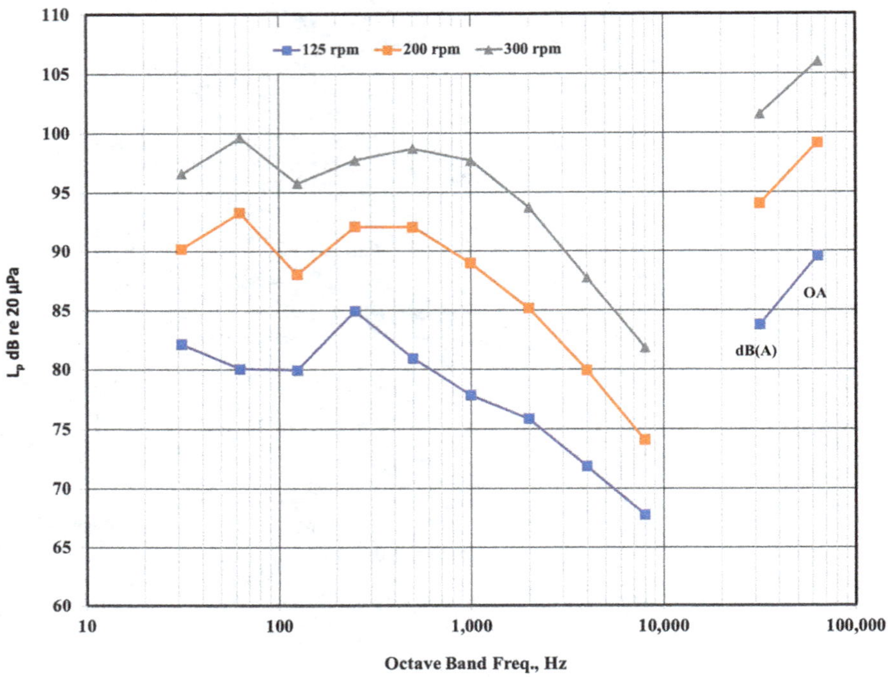

Fig. 5.10 Tunnel-induced noise vs. operating speed

Path losses based on the number of frames, *Fr*, and decks, *Dk*, from the thruster to the receiver space are determined by this equation.

These equations are partly derived from Carlson and Buiten. Figs. 5.10 and 5.11 provide typical noise levels measured in a bow thruster room and in a cabin over the thruster room for various operating speeds. Table 5.15 provides a survey of noise levels in various compartments for several operating speeds and when thrusting to either port or starboard. In this case, the directional thrusting differences are not too significant.

The vibration on the tunnel bow thruster, L_a, is computed as (Heckl 1988):

$$L_a \text{ dB re 1 } \mu\text{G} = 20 \text{ Log}(P) + \Delta_1 \tag{5.24}$$

where P is the rated power of the thruster in kW and the adjustments, Δ_1, are provided in Table 5.16.

Table 5.17 provides the predicted propeller source levels in dB re 1 µPa in octave bands, the measured tunnel wall vibration levels in dB re 1 µG, the measured wall vibration levels and the difference between the two. This method can be accurate if all the bow thruster design parameters are known.

Measured vibration levels on various ship structures are provided in Fig. 5.12. Note the low-frequency amplification on the hull plate and bulkhead over the vibration on the top of the duct wall.

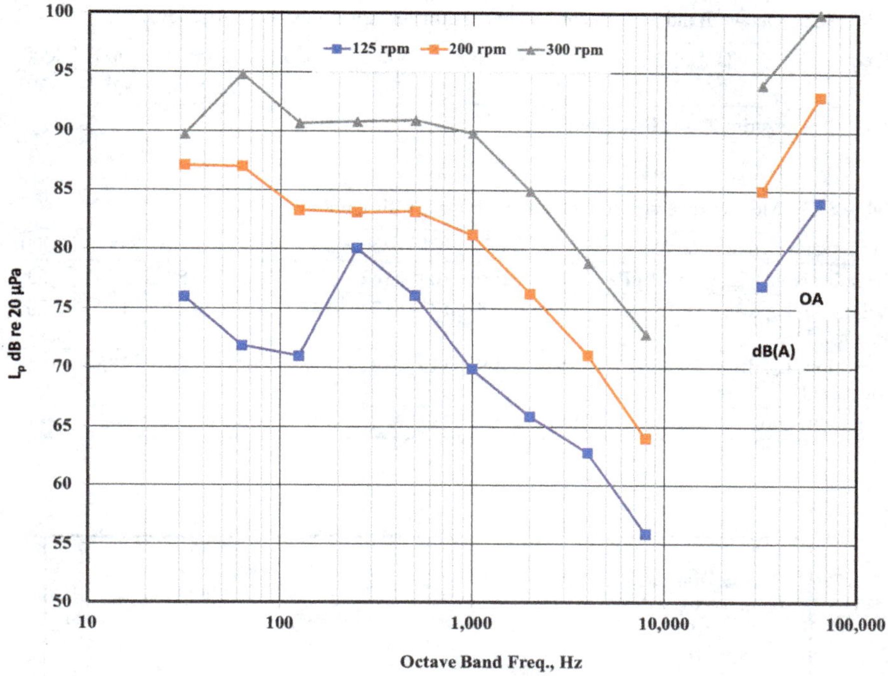

Fig. 5.11 Tunnel-induced noise – room over

Table 5.15 Noise Survey, dB(A); CPP Bow Thruster

	Pitch =	0%	30%	30%	70%	70%
	Direction =	None	Stbd.	Port	Stbd.	Port
Deck	*Condition* = compartment	Non-Cav.	Non-Cav.	Non-Cav.	Cav.	Cav.
Hold	Bow Thruster Room	98	97	97	104	102
Main	Mess	83	78	81	90	89
	Lounge	79	79	78	87	86
	Galley	81				
01 Level	Forepeak	85	85	87	95	95
	Laundry Room		72	74	84	82
	Fwd. Stbd. Inboard Berth	67	62	65	n/a	80
02 Level	Fwd. Stbd. Berth	66				
	Fwd. Port Berth	63				

Two 600 kW bow thrusters operating. Note that at 0% pitch, the noise in the mess exceeds that when the thruster operates at 30% load. This is caused by hydraulic control valve noise for no-load operation.

Table 5.16 Octave Band adjustments – Tunnel thruster structureborne source level

Freq.	31.5	63	125	250	500	1000	2000	4000	8000
Δ_1	35	44	82	84	52	51	50	53^E	53^E

Note: E superscript is for 'Estimated'

Table 5.17 Measured versus predicted bow thruster wall vibration

Freq, Octave Band, Hz	31	63	125	250	500	1000	2000	4000	8000
Impeller Noise, L_p dB re 1 μPa	152	160	163	162	161	160	158	156	153
Vibration, L_a, dB re 1μG, Predicted	108	118	123	124	125	126	124	122	120
Vibration, L_a, dB re 1 μG, Measured	101	109	115	121	122	124	125	125	123
Delta Vibration, dB	7	9	8	3	3	2	−1	−3	−3

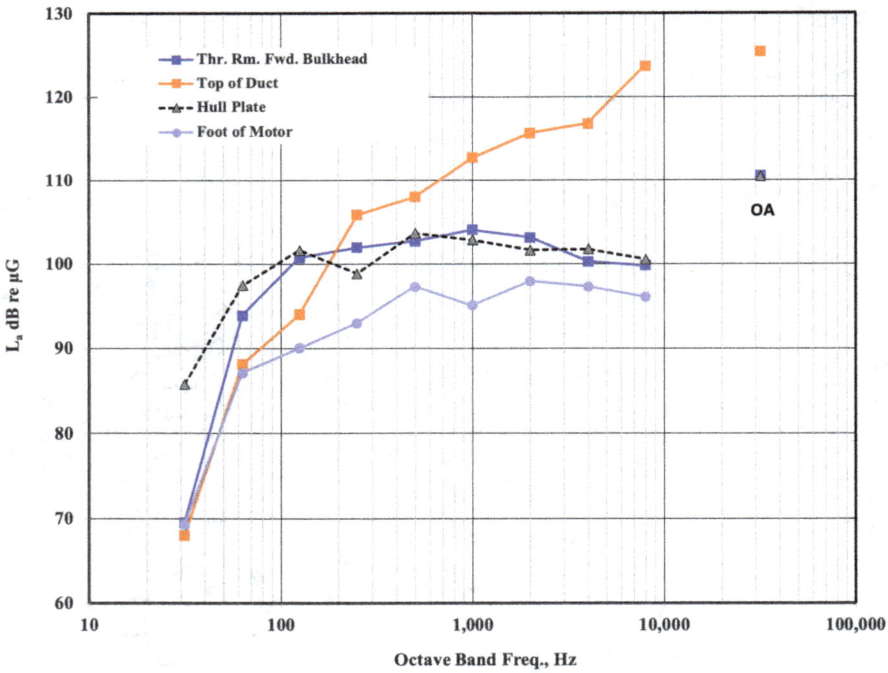

Fig. 5.12 Measured vibration levels induced by bow thruster on ship structures

5.3 Flow Induced Noise

Flow, whether in air or fluid, over structure or systems such as a sonar dome, can excite noise and/or vibration. Unless the speed of flow is fairly high in water, the excitation is not usually too significant. The following subsections describe some empirical approaches to predicting these responses.

5.3.1 *Wave Impact*

For speeds higher than 18 knots, noise levels in the forward compartments close to the water line may have excessive noise levels. Flow and waves will excite vibrations on the hull around the waterline. Measurements show that the maximum vibration level occurs on the side structure adjacent to the forepeak bulkhead (Spence et al. 2016). One can consider the part of hull between decks above and below waterline and between forepeak and next transverse bulkhead as a source of first structureborne sound.

Figure 5.13 presents the measured data at the bow of the catamaran at 28 knots (Spence et al. 2016). Note that the levels are on par with that of diesel and that the

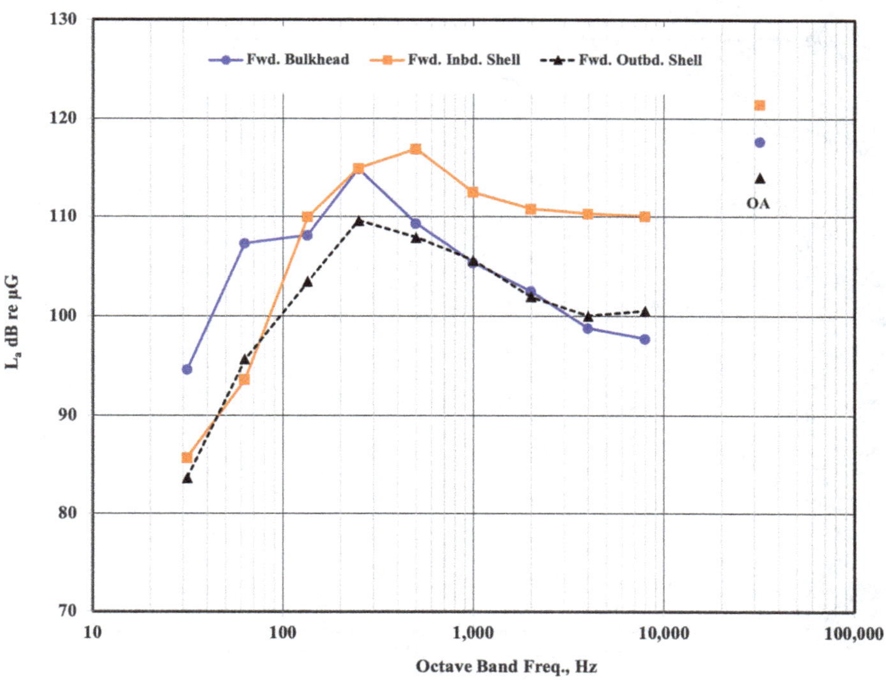

Fig. 5.13 Wave-induced vibration on the forward hull of the catamaran

inboard shell has levels approximately 10 dB higher than the outboard shell. This may be due to wave interactions between the port and starboard hulls. Limited data show a 40 Log speed dependence for noise and vibration due to wave impacts. Empirical prediction for hull response to wave impact is provided in Fischer and Boroditsky 2001, Section 6.4.

Even moderate sea states, SS-1 to SS-2, lead to a remarkable noise increase. For example, if the vessel speed is 12 m/sec (24 knots), the noise level will be higher by 6 dB at a wave height of 0.5 m relative to calm water conditions. The vessel direction relative to the wind also has an influence. Multiple acoustical measurements on high-speed ships show this phenomenon as relatively stable in noise and vibration spectrum shape. Wide band levels (octave or third octave) decrease or increase with changes in conditions, but the spectrum shape remains similar (Spence et al. 2016).

5.3.2 *Hull Response to Flow*

The measured versus calculated flow-induced vibration on the hull for operation at 40 kn is shown in Fig. 5.14. The prediction method is fairly accurate in the mid-frequencies controlling the A-weighted induced noise levels.

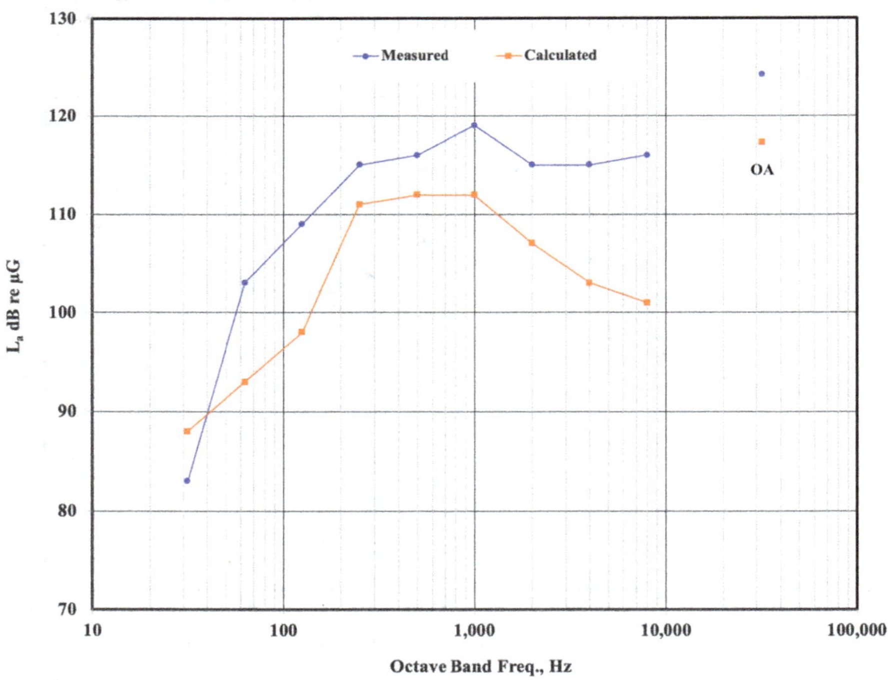

Fig. 5.14 Measured versus calculated hull response to flow noise

Table 5.18 Octave band adjustments for baseline flow-induced sonar self-noise, dB

	Octave Band Center Frequency (Hz)								
Description	31.5	63	125	250	500	1000	2000	4000	8000
Adjustments for Baseline Sonar Platform Noise, dB	39	28	19	16	13	10	7	4	3

5.3.3 *Sonar Self-Noise Due to Flow*

Turbulence is generated as the vessel travels through the water. Fluctuating pressures caused by this turbulence can radiate noise directly into the water and can excite hull vibrations. However, the direct radiation of turbulence is not usually a source of ship radiated noise. Except for fast craft, this phenomenon rarely contributes to the internal noise on the vessel. Turbulence can impact sonar self-noise, even at low speeds, thereby limiting the performance of sonar (Rao 1987).

The following can be used to estimate the sound pressure levels seen on the sonar elements. These would be the 'platform noise levels', $L_{PN\ Base}$, seen prior to being processed by the sonar system.

$$L_{PN\ Base}\ \text{dB re } 1\ \mu\text{Pa} = 34 + 45\ \text{Log}(V) - 20\ \text{Log}(t) + 10\ \text{Log}\ (A) \tag{5.25}$$

where V is the ship speed in knots, t is the thickness of the sonar dome in cm, and A is the surface area of the sonar dome, m^2. The adjustments in Table 5.18 should be added to the base level. Note that the controlling factor is the ship speed, where the noise will increase as the 4.5th power.[11] Thus, going from 10 knots to 20 knots will increase the platform noise by 14 dB.

The predicted levels are just outside any near-field, non-radiating noise created on the sonar array.

5.3.4 *Piping Flow Noise*

Due to extensive pipe runs on ships, noise induced by hydraulic systems can be significant (Nelson 1999). Piping-induced noise is transmitted over both airborne and structureborne paths. The structureborne noise is transmitted into the ship structure by the numerous pipe attachments. This component can contribute not only to the internal noise but also to the ship's radiated noise and sonar self-noise.

An estimate of the in-fluid pressure level L_p caused by pump pulsation frequency is as follows:[12]

---

[11] Many flow related phenomena vary as speed to the sixth power; thus, going from 10 to 20 kts would increase noise by 18 dB.

[12] For this particular component, the pressure is measured in dB re 20μPa not dB re 1μPa

$$L_p \text{ dB re } 20\ \mu\text{Pa} = 136 + 20 \text{ Log } (Q) + 20 \text{ Log}(\Delta p) - 20 \text{ Log } (S) \tag{5.26}$$

where Q is the volume flow rate in liters per minute, Δp is the pressure rise across the input/output in atmospheres, and S is the cross-sectional area of the outlet pipe, cm^2. This equation is only valid for rotating vane, gear, and swash-plate pumps. For screw pumps, the internal sound pressure is:

$$L_p \text{ dB re } 20\ \mu\text{Pa} = 20 \text{ Log}(\Delta p) + 164 \tag{5.27}$$

For the discharge side of a centrifugal pump, the internal pressure is:

$$\begin{aligned} L_p \text{ dB re } 20\ \mu\text{Pa} = {} & 296 + 10 \text{ Log } (Q) + 35 \text{ Log}(\Delta p) \\ & - 40 \text{ Log } (a) - 40 \text{ Log } (RPM) - 20 \text{ Log}(w) \end{aligned} \tag{5.28}$$

where a is the impeller radius in cm and w is the width of the impeller blade tip in cm.

Note that the narrow band spectrum of most pumps contains multiple harmonics of the fundamental frequency, f_0, calculated as:

$$f_0 = z \times RPM/60$$

where z is the number of vanes or part volumes delivered per shaft rotation. There can be anywhere from 10 to 50 harmonics of the fundamental appearing in the spectrum.

The vibration on the pipe wall due to the internal fluid pressure is computed as:

$$L_a\ dB\ re\ 1\ \mu G = L_p - 75 + 15 \text{ Log}(f) - 5 \text{ Log}(d_i) - 10 \text{ Log}\left(1 + \left(32\ {}^{t}/_{d_i}\right)\right) \tag{5.29}$$

where L_p is the internal pipe pressure, f is the octave band center frequency, d_i is the internal diameter of the pipe, cm, and t is the pipe wall thickness. The predicted pipe wall vibration only applies to pipe lengths over 3 m.

5.3.5 Vortex Shedding/Strouhal Number

As with any bluff body subject to cross-flow, small diameter structural components on a ship may respond with high vibration (Roshko 1954). Even relatively low flow rates in air can cause a resonant response. Moreover, especially where the response is resonant, as in the case of 'singing', the excitation can be excessive and structurally damaging. Items generally of concern are rails and masts due to air flow and propeller blades, wings, bilge keels, struts, and rudders due to fluid flow underwater. The real problem develops when the Strouhal shedding frequency matches a structural natural frequency. Then, the system can have a 'locked-in' resonant response. In some cases, this has led to cracked propeller tips or structural fatigue.

5.3.5.1 Structural Response to Air Flow over a Bluff Body

The vortex shedding frequency, f_{st}, for flow over a bluff body is computed as:

$$f_{st} = 0.2\, D/U \tag{5.30}$$

where D is the diameter of the body and U is the flow speed over the body, where D and U are consistent dimensional units. For example, for a speed of 15 kn with flow over a 10 cm diameter pole, the Strouhal frequency would be 15 Hz. If this pole happened to have a structural resonance at 15 Hz, the flow would excite a high vibration at its base. This type of moment coupling can excite excessive vibration in the ship structure. If the pole were located over a noise-sensitive space, habitability criteria might be exceeded. This would be in addition to possibly excessive vibration on the pole. In either case, the result is most likely noise and/or vibration excesses.

Possible vortex shedding problems can easily be checked during the design over the expected flow rate speeds. There is little to be gained by matching the shedding frequency to all the slender components on a vessel unless an FEA has been carried out for structural reasons. These problems can be fixed either by changing the natural frequency of the structure, changing the diameter of the structure, or spiral wrapping the structure to break up the vortices, making the excitation incoherent.

5.3.5.2 Singing Propeller

A "singing propeller" has a tone that is generated by the interaction between a Karman vortex shedding mechanism and the blade's natural mode of vibration (Reed 1979). The shedding is usually from the trailing edge of the propeller blade. However, it may also occur on the leading edge or tip. The lower blade modes – torsional, cantilever from the hub, and/or chordwise – may be excited. Wide and/or thin blades are more likely to sing.

'Singing' can result in a high sound radiated into the water – see Fig. 5.15. The difference in radiated noise between singing and not singing is 20 dB. The tone at 2000 Hz originated at the 0.55 nondimensional radius of the propeller with a 1 mm thick trailing edge. The excitation was the two-nodded cantilever mode. This type of radiated sound can interfere with the ship's sonar systems or even cause high radiated noise affecting mammals.

It is often coupled via the waterborne path to a high noise level inside the vessel, which can be quite objectionable to any crew and passengers. Tones, in general, are found to be more objectionable than broadband noise. In one case, the hull vibration levels over the propeller were reduced by 50 dB by changing the trailing edge to an anti-singing geometry, as shown in Fig. 5.16. A rounded trailing edge is expected to be at least a factor of 3 worse in vibration than a chiseled edge. Furthermore, a singing propeller can cause stress cracks or catastrophic failure of the blade.

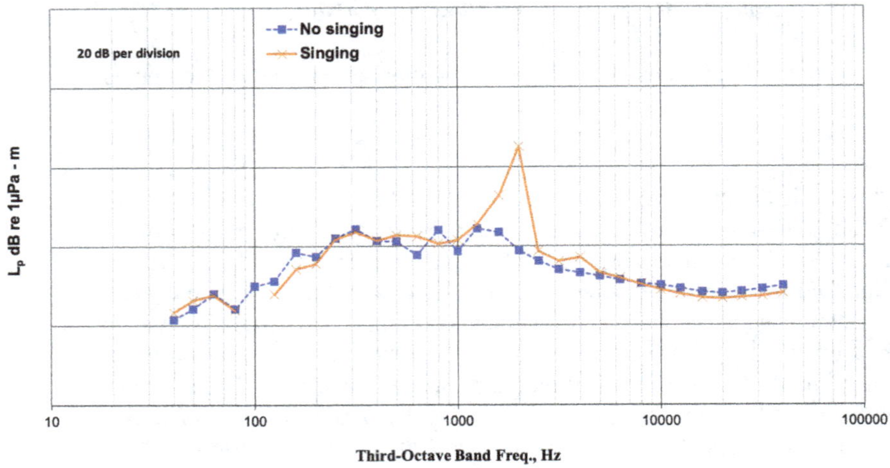

Fig. 5.15 Example of radiated noise due to a singing propeller

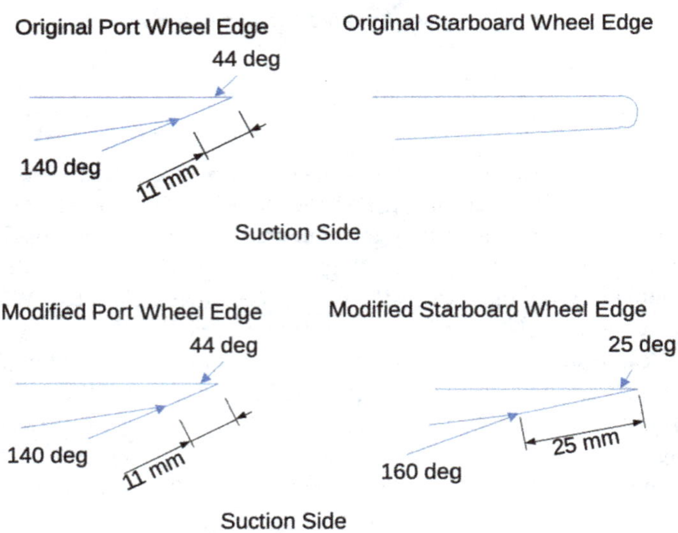

Fig. 5.16 Trailing edge shapes before and after modification

Empirical methods can be used to gauge the possibility of a singing response. Furthermore, Computational Fluid Dynamics (CFD) combined with Finite Element Analysis (FEA) programs can be used for a more thorough investigation. As noted above and in Chap. 10, the most straightforward way to avoid a singing propeller is to place an 'anti-singing' geometry on the trailing edge of the blades. This should be part of a Purchase Specification to the propeller vendor. Once a propeller is found to sing, it takes a rather intense effort to cure it. This may involve divers to assist in underwater testing for blade resonances, divers to modify the trailing edge of the propeller, or dry docking in lieu of the preceding.

If available, a spare propeller can be tested in air to determine its dry resonances. Diagnostic testing can be undertaken to determine blade natural frequencies to see if they match the measured singing frequency (Fischer 2008). The difference between the natural frequency in air and water loaded needs to be accounted for empirically.

For the propeller case, *D* in Eq. (5.30) is replaced with the trailing edge thickness, *d,* and *U* is replaced by U_∞, the flow velocity at infinity. "*d*" is somewhat of an indefinite thickness – somewhere between the edge physical thickness and boundary layer thickness. The physical dimension of the trailing edge seems to provide sufficient accuracy. Depending on the definition of *d*, S_t can vary from 0.15 to 0.25. A S_t of 0.18 seems to correlate well to the measured Strouhal frequencies when the actual thickness of the trailing edge is used.

A propeller can sing through a fairly wide range of inflow velocities or rotation rates. As the propeller rotation rate increases, the point of excitation will move toward the hub. Likewise, as the propeller velocity increases, the singing can move to other modes. To determine the vortex shedding frequency, the free flow velocity needs to be determined as:

$$U_\infty = \pi n D\sqrt{(r/R)^2 + (J/\pi)^2} \tag{5.31}$$

where *n* is the propeller rotation rate; rotations per second; *D* is the diameter; r/R is the nondimensional radius of interest; and *J* is the advance coefficient.

If an FEA of the wetted blade is not available, rough estimates can be obtained by the following methods. Blade natural frequencies, f_b, depend on the stiffness, shape, and mass of the blade. The blade's mass per unit area and its stiffness vary along its span. The lowest natural frequency is typically that of a blade acting as a cantilever supported at the root. A singing propeller usually involves either bending of the blade (fundamental or one-nodded cantilever mode) or the lowest torsional mode (twisting of the blade about the root). Depending on the blade's shape and size, it is also possible to excite the blade in the chordwise direction as a free-free beam. The blade frequency for the fundamental cantilever mode in air, $f_{o\ air}$, is (Carlton 1994):

$$f_{o\ air} = \frac{0.305}{(R - r_h)^2}\left[\left(\frac{E}{\rho}\right)\left(\frac{t_m}{c_m}\right) c_r t_r\right]^{0.5} \tag{5.32}$$

where *R* is the tip radius; r_h is the hub radius; *E* is Young's modulus; ρ is the propeller blade mass density; t_m is the mean blade thickness; t_r is the root blade thickness; c_m is the mean chord length; and c_r is the root chord length. The natural frequency in water can be reduced by a factor of 0.6.

Baker (1940) also provides an estimate of the lowest blade torsional frequency as:

$$f_{t\ air} = \left[0.92/(R - r_n)\right]\ t_{0.5}/c_{0.5}\left(c_r/c_m\right)\sqrt{G/\rho} \tag{5.33}$$

where $t_{0.5}$ and $c_{0.5}$ are the thickness and chord length at $0.5R$, respectively, and G is the shear modulus (or modulus of rigidity) of the material. In water, the torsional mode would be lower by approximately a factor of 0.7.

It has also been found that singing associated with the propeller tip can vary with water qualities, such as air content and nuclei density in the water (Peng et al. 2015). Additionally, there are cases where only one or several blades on a single propeller begin to sing. According to Ross 1983, these variations have led to confusion regarding singing, their detection, and cure.

Databases

Aircraft:

Krebs, W., Butikofer, R., Pluss, S., Thomann, G.: Sound Source Data for Aircraft Noise Simulation. Acta Acust. **90**, 91–100 (2004)

Industrial/Construction Sources:

Imagine, "Description of Source Database", Project Co-ordinator: AEA Technology Rail BV, Document: IMA07TR-050418-DGME02, 6-2-07

NIOSH Health Hazard Evaluation Program Noise Measurement Database: NIOSH Dataset RD-1005-2014-0; https://www.cdc.gov/niosh/data/datasets/rd-1005-2014-0/default.html

Noise Navigator™ Sound Level Database:

https://multimedia.3m.com/mws/media/888553O/noise-navigator-sound-level-hearing-protection-database.pdf

Impulsive noise:

Hobbs, C., Downing, M., Ikelheimer, B., James, M.: *Air Weaponry Noise Source Characterization Protocol*. WR 08–06, Nov. 2008.

Military Sources

Source: US Army Environmental Hygiene Agency Technical Guide 167A dated 02/1991
HEAR NOISE DATA LISTING 11

References

Aatola, S.: The Measurements of Vibrational Power Transmission from Ship Main Engine, Research Report No. BVAL35-021224, VTT Technical Research Center of Finland, Espoo Finland, 2002

Ashcroft, R.: Design Guide for Shipboard Airborne Noise Control. SNAME Ship's Machinery Committee, Alexandria, VA (2018)

Baker, G.S.: Vibration patterns of propeller blades. Trans. NECIES. **57**, 1040–1041 (1940)

Bies, D., Hansen, C.: Engineering Noise Control, Theory and Practice, 4th edn. Unwin Hyman, London (2009)

Bowes, M.A.: Development and Application of an Analytical Method for Predicting Helicopter Transmission Noise. In: Paper C3 Symposium on Internal Noise in Helicopters, U. of Southampton, July, 1979

Brown, N.A.: Cavitation noise problems and solutions. In: Buiten, J. (ed.), Shipboard Acoustics: Proceeding of the 2nd International Symposium on Acoustics (ISSA), The Hague, The Netherlands, Oct. 1986

Buiten, J., de Regt, M.J.A.: Handbook on Shipboard Noise Control, Institute of Applied Physics. TNO-TH, Delft, Holland (n.d.)

Carlton, J.S.: Marine Propellers & Propulsion. Butterworth Heinemann, Oxford (1994)

Carlton, J.: Marine Propeller and Propulsion, 2nd Ed., Elsevier (2007).

EU Machinery Directive 2006/42/EC; https://osha.europa.eu/en/legislation/directives/directive-2006-42-ec-of-the-european-parliament-and-of-the-council

Fischer, R.: Singing propellers – solutions and case histories. Mar. Technol. **45**(4) (2008) SNAME

Fischer, R., Boroditsky, L.: Technical and Research (T&R) Bulletin 3-37 (Supplement), Supplement to the Design Guide for Shipboard Airborne Noise Control. The Society of Naval Architects and Marine Engineers (SNAME), Alexandria VA (2001)

Fischer, R, Courtney, B., Nelson, D.: Design Guide for Shipboard Airborne Noise Control, Society of Naval Architects and Marine Engineers (SNAME), Technical & Research Bulletin 3-37, NJ, 1983

Heckl, M.: Bericht Nr. 193/1988, Entwicklung von Methoden zur Schallpegelprognose fur den Unterkunfts- und Maschinenraumbereich von Schiffen in German), Technische University Berlin, Hamburg, Germany, 1988

Ishii, Y., Miyawaki, K., et al.: Underwater noise prediction and development of special purpose ships. Mitsubishi Heavy Ind. Tech. Rev. **57**(1), 1 (2020)

Jalkanen, J.-P., Johansson, L., et al.: Modelling of ships as a source of underwater noise. Ocean Sci. **14**(6), 1373–1383 (2018)

Nayak, P.R.: Helical and Spur Gear Dynamics, BBN TIR No. 86, Jan. 1973

Nelson, D.A.: NASA Glen Research Center, Reduced-Noise Gas Flow Design Guide, Elgin TX, April 1999

Nilsson, A.C., Tyvand, N.P. (eds.): Noise Sources in Ships: I: Propellers. Nordforsk, Miljovardsserien (1981)

Nilsson, A.C., Tyvand, N.P. (eds.): Noise Sources in Ships: II: Diesel Engines. Nordforsk, Miljovardsserien (1983)

Optiz, H.: Dynamic behavior of spur and helical gears. In: Proceedings of JSME Semi-International Symposium Papers, pp. 199–209, Tokyo, Japan, September 1967

Peng, X., Xu, L., Song, M., Cao, Y.: Vortex singing in tip vortex cavitation under the effect of water quality. J. Phys. Conf. Ser. **656**, 012185 (2015)
Plunt, J., Odegaard, J. (eds.): Noise Sources in Ships, II: Diesel Engines. Nordforsk, Stockholm, Swedem (1983)
Rao, V.B.: On the flow induced structural noise of a Ship's sonar dome. Mar. Technol. **24**(4), 321–331 (1987)
Reed, F.E.: Propeller singing. In: Proceedings, Symposium on Practical Experiences with Flow Induced Vibrations. Springer (1979)
Roshko, A.: On the Drag and Shedding Frequency of Two-Dimensional Bluff Bodies. NACA Tech. Note 3969, 1954
Ross, D.: Mechanics of Underwater Noise, 3rd edn. Peninsula Press, Los Altos Hills, CA (1983)
Spence, J., Favini, E., Boroditsky, L.: Wave Noise on High Speed Ships – Source Characterization and Modeling. In: Noise-Con 2016, Providence, RI, June 2016
Thumann, A., Miller, R.: Fundamentals of Noise Control Engineering, 2nd edn. The Fairmont Press, Inc, Liburn GA (1990)
Tupov, V., Semin, S. A.: Gas Turbine Noise. In: 15th International Congress on Sound and Vibration, Daejeon, Korea, July 2008
VDI-Richtlinien 2159: Emissionskennwerte technischer Schallquellen; Getriebegeräusche; Verein Deutscher Ingenieure, July 1985

Chapter 6
Introduction to Source/Path/Receiver Modeling

This chapter discusses the parameters and variables associated with how noise and vibration couple to the response of the ship structure.

The airborne path typically controls how much noise gets from one compartment to the next. This transmitted noise is controlled by the area of the interface, material properties, and treatments on this interface. Other factors, such as openings in the interface for pipe and duct transits, will be addressed.

The structureborne noise is controlled primarily by the attachment method (bolted or isolation mounted), the construction of the hull foundation, and the projected area of the foundation. Once the vibration enters the ship's structure near a particular source, it becomes structureborne noise. Attenuation is provided by the intervening ship structure between the foundation and a receiver compartment. These attenuations consist mainly of cylindrical spreading, losses at intersections of decks and bulkheads, and inherent damping within the ship's hull. Another path – secondary structureborne noise – is caused by airborne noise from the source incident on the ship's structure, which will then excite compartment boundaries. This secondary structureborne noise is then transmitted as discussed above. These paths are illustrated in Fig. 6.1.

The acoustic power arriving from airborne, structureborne, and secondary structureborne paths at any compartment is then radiated as 'received' noise. The magnitude of the received level depends on the acoustic characteristics of the room. Parameters that control the received level include the total area of the surfaces and the room constant or acoustic absorption of the finish surfaces of the room. If there is an acoustic source within the room, the received noise will depend on both its 'direct' and 'reverberant' path contributions. The overall received noise level is the combination of all these sources/paths.

R. Fischer, L. Boroditsky, *Noise and Vibration Control on Ships*,
https://doi.org/10.1007/978-3-031-55170-3_6

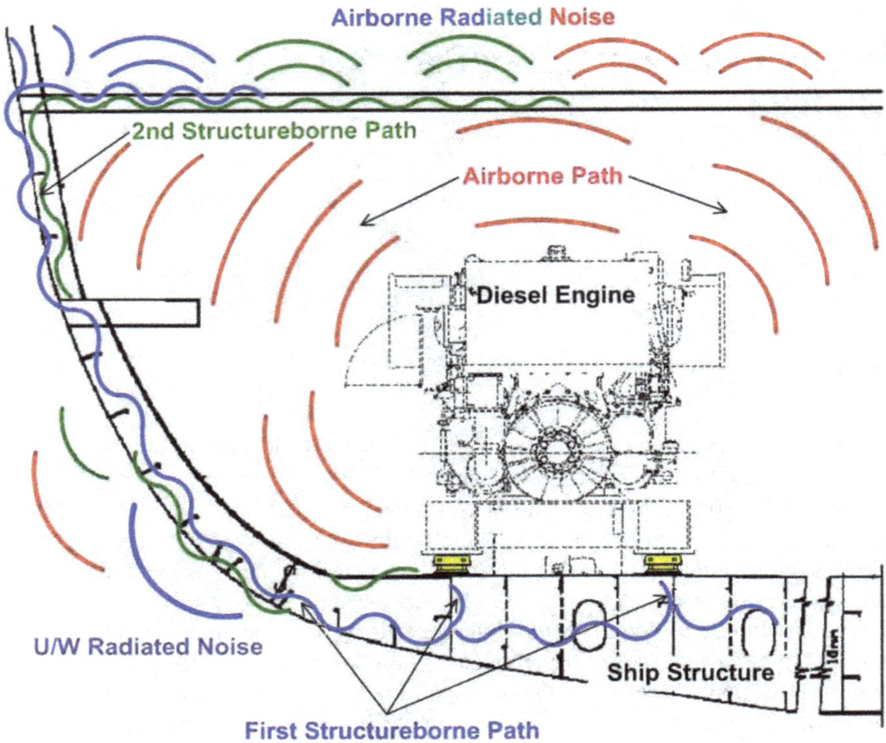

Fig. 6.1 Noise transmission paths

Noise can also be radiated into the free field, either from machinery, intake/exhaust of machinery systems, or fan inlet/exhaust. This noise can propagate freely[1] to an on-deck work station or a location dockside.

All of these parameters are quite frequency dependent; some are more important at low frequencies, some are important at high frequencies, and some are in between.

An effective acoustical design requires detailed and quantitative analysis of the ship's physical details and parameters controlling the resulting acoustic (airborne, fluidborne, and structureborne) noise environment. Modern noise prediction algorithms can help solve this problem. However, a high level of input data and understanding of the output are still needed for a successful project. The noise prediction process is a labor-intensive effort. This includes consideration of general arrangements, hull and foundation structural features, acoustical and non-acoustical coatings/insulation, distributed systems, equipment noise and vibration characteristics, and other relevant information, as discussed in this chapter.

[1] Assuming there are no barriers to block line-of-sight transmission. The insertion loss of 1-D and 2-D barriers is provided in Beranek and Ver 1992.

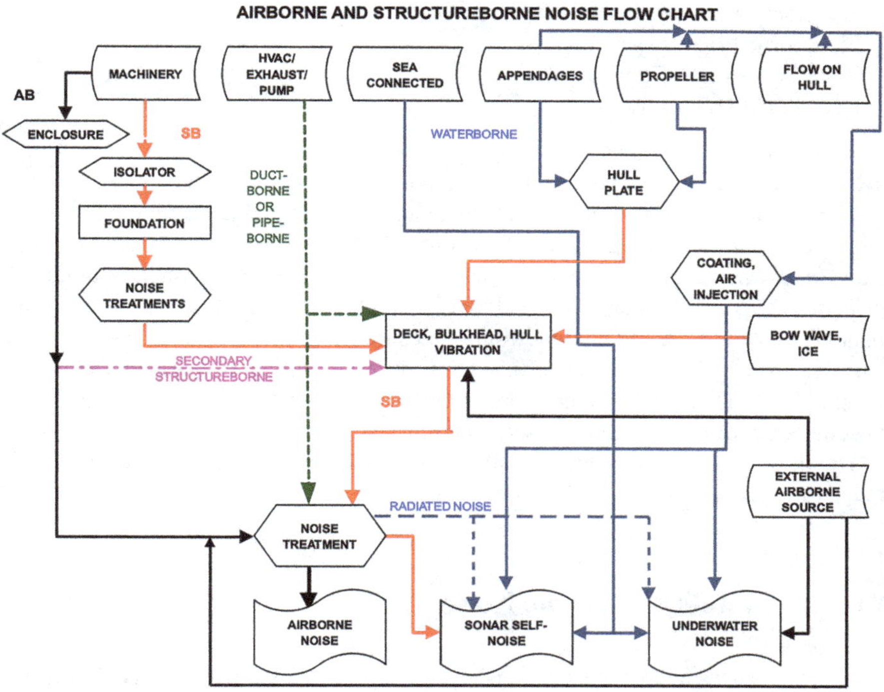

Fig. 6.2 Airborne and structureborne noise flow chart. (Notes: AB Airborne Path, SB Structureborne Path)

As shown in Fig. 6.2, the primary 'sources' with respect to habitability noise are (Fischer et al. 2008):

- Machinery and equipment internal to the ship
- HVAC, machinery intake/exhaust, and distributed piping systems internal to the ship
- Propeller
- Bow wave or operation in ice
- External airborne noise source – fan, propulsion and auxiliary system intake/ exhaust openings, aircraft operations

The additional sources with respect to underwater and sonar self-noise are:

- Sea-connected systems
- Flow over control surfaces/appendages
- Flow over the hull

The 'acoustic paths' are:

- Airborne
- Structureborne

- Secondary Structureborne
- Fluidborne (piping systems)
- Waterborne (e.g., from propeller)
- Duct-borne (HVAC)

As shown in Fig. 6.2, control treatments can be utilized at multiple points and there are interdependences and commonalities among sources, paths, and receivers.

This is where there is a significant advantage of a full 3-D acoustic model of the ship. A computer model can consider all sources, all transmission paths, and receiver acoustic characteristics. The model can also cover the entire frequency range of concern. Thus, 3-D models can develop a complete picture of the acoustic environment. They would present an opportunity to examine, in minute detail, paths losses, coupling loss factors, radiation efficiencies, and controlling source/paths for each compartment. This can be done more completely and more expeditiously than the piecemeal building of a model based on empirical methods.

The three receivers of interest – habitability airborne noise, sonar self-noise, and underwater radiated noise – are discussed in the following sections.

6.1 Habitability Airborne Noise

Machinery spaces would be the noisiest spaces onboard most vessels. Types of internal sources and their acoustic levels are summarized in Chap. 5. Some machinery is unique depending on the purpose of the vessel. For fishery vessels, trawl winches may be significant sources. The piping on dredges can be fairly extensive, so the piping becomes a distributed source, affecting multiple compartments. Propellers, stern and bow thrusters, and water jets often control noise levels at a specific location in the ship.

In these spaces, direct and reverberant noise fields exist. Airborne transmission paths from the source to the receiver within the machinery space may be complicated by barriers effects from large machinery items. The received noise at any location depends primarily on the source of airborne noise radiated from each machinery item and the amount of acoustic absorption (room constant). Rarely does the structureborne noise in a machinery space become a factor.

For compartments adjacent to a machinery space, the airborne noise transmitted through the common interface controls the received noise level. The thickness of the interface plating, the insulation, and any finish surface on the common interface control the received noise level. Compartments adjacent to the main and auxiliary machinery spaces are likely to have high transmitted airborne noise. This is why spaces adjacent to machinery spaces should be those with no or high allowed noise criteria, such as storage areas. An exception may be airborne noise from a gas turbine uptake into adjacent compartments on multiple decks. Again, placing non-sensitive compartments or passageways around intakes/uptakes would minimize the need for acoustic treatments.

Non-machinery spaces may also have internal noise and vibration sources. These can consist of tools, transformers, HVAC systems, etc. The process for the computation of the received noise level is identical to that for machinery spaces. However, these spaces may also be dominated by other sources, such as structureborne and/or secondary structureborne transmitted noise. Furthermore, depending on the physical location within the vessel, these compartments may be impacted by wave-, bow thruster-, or propeller-induced noise.

For rooms without an internal noise source[2] that are not adjacent to a machinery space, there is only one significant noise component. Noise resulting from radiation from the room's boundaries contributes to the received noise within that space. In this case, two factors are important. The first is the amplitude and frequency of structureborne vibration levels from the deck, bulkheads, and deckhead. Structureborne noise can arrive from numerous sources, such as machinery, propulsors, wave impact, ice breaking, and flow excitation of bluff bodies. The second is the ability of these structures to radiate noise. This ability depends on the plate type and thickness as well as the spacing of the framing. Each of these factors should be quantified in the noise modeling process of maritime vessels. The noise radiated from the ship's structure may be attenuated by finish surfaces such as joiner bulkheads or drop ceiling.

Those components that are generated by sound energy radiated to the surrounding air are called *airborne noise*. The notation L_{ab} is used for the airborne noise level throughout this book. Noise radiated by a vibrating hull structure is termed structureborne noise. It is convenient to subdivide this component into *first* and *second* structureborne noise. The first structureborne noise is generated by vibration induced by forces acting on hull structure through connections or support from sources (engines, propellers, flow). Notation, L_{fsb}, is used for this particular noise. The second structureborne noise is generated when airborne noise in a source room excites vibration of the hull structures forming the source room. The noise level generated by this mechanism is denoted as L_{ssb}. The subscripts *fsb* and *ssb* will be used not only for first/second structureborne noise but also for first/second structureborne vibration levels when one needs to describe the different "genesis" of these components.

Separation of first and second structureborne noise, as a part of noise control prediction for ships, has often been used in books and guides (Fischer et al. 1983; Fischer and Boroditsky 2001, and Fischer and Bahtiarian 2017; Plunt 1980). This separation is necessary from a noise control point of view. The acoustical treatments have different amounts of noise reduction achievable relative to each component. Different approaches to control these components will be discussed in Chap. 8 and below in connection with noise and vibration paths through hull structures.

Some sources are truly a combination of several sources that are distant from each other. For example, internal combustion engines have exhaust and intake systems, which include pipe, ducts, and feet attachment points. The termination of a pipe or

[2]Contributions from HVAC and piping systems to these compartments are considered separately.

duct can be considered an independent source of airborne noise. Some of the sources are located interior to the hull; some are outside (propulsors, vent louvers, hydrofoils, landing/taking off aircraft). In the acoustical design process, all significant sources need to be considered.

The idea of the separation of airborne and structureborne noise supposes that each source is considered an independent source. Such division is conditional. However, for engineering practice, the assumption about the independence of these sources is acceptable.

Another convenience is to consider the acoustic power of individual sources. Let the airborne source power be W_{ab} and the structureborne source power be W_s. The values of W_{ab} and W_s and their ratio, strictly speaking, depend on how the source is supported. However, for all practical purposes, it is possible to consider the W_{ab}/W_s *ratio* as a characteristic feature of a source. Understanding this ratio is critical in considering which type of treatment to apply. The influence of structureborne noise depends on this ratio, especially for rooms in the vicinity of the source.

The significance of structureborne noise is obvious from the figures shown below. Figure 6.3 shows the noise levels from a diesel generator measured near a transverse bulkhead in the engine room. Also shown is the noise in an adjacent cabin. Similarly, Fig. 6.4 shows the noise in the bow thruster room and the adjacent State Room. For

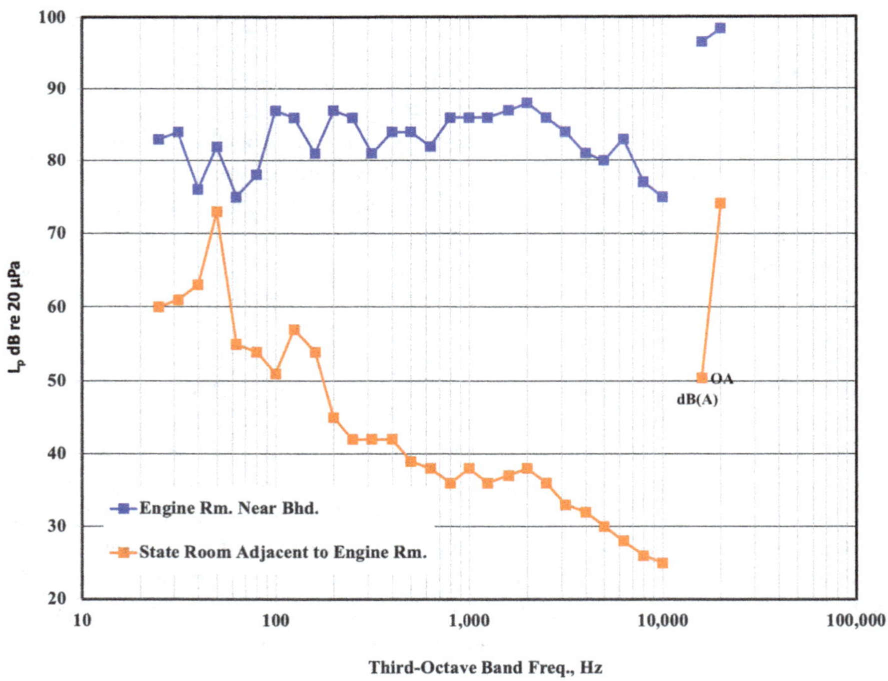

Fig. 6.3 Noise levels from a double-stage resiliently mounted diesel generator

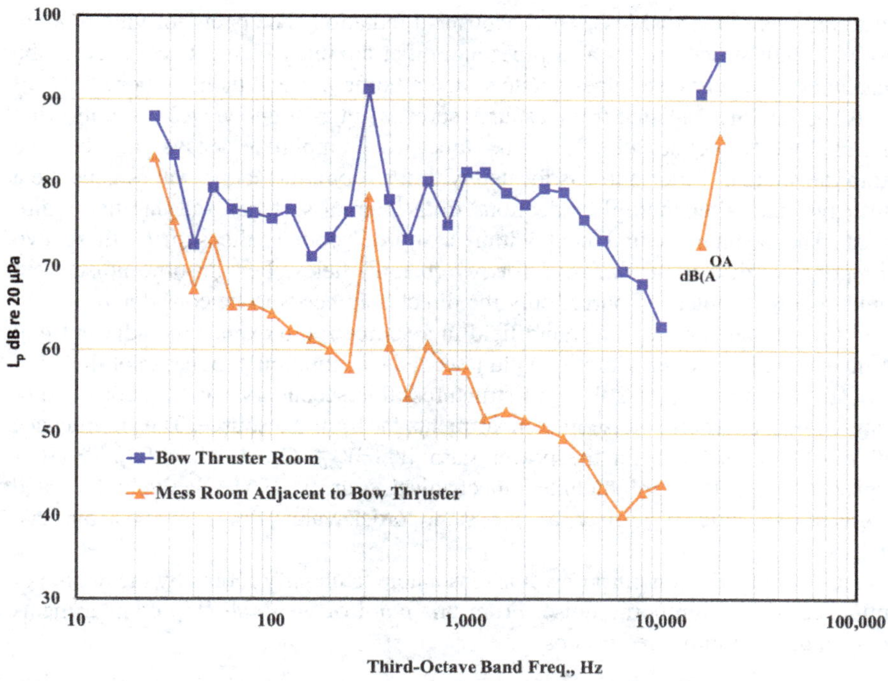

Fig. 6.4 Noise levels from the tunnel bow thruster

this case, the physical parameters of all the partitions between source rooms and adjacent rooms have approximately the same physical and acoustical characteristics. However, the change in noise decay between rooms is quite different: 40–50 dB for the diesel generator and 20–30 dB for the bow thruster. This phenomenon is easy to explain. For the resiliently mounted diesel generator, W_{ab}/W_s is greater than the same ratio for the tunnel thruster. This is due to the resilient mounts that effectively reduce structureborne noise from the diesel. A tunnel thruster is a classic example of a structureborne noise source. Here, cavitation bubbles collapse and excite vibration of the tunnel and adjacent structure. One should keep in mind that ship structures easily transmit structureborne noise, whereas it effectively insulates against airborne noise in the mid- to high-frequencies.

Mechanical sources act on foundations, usually via unbalanced (moment) forces, and through various attachments, such as piping. The result of this excitation is structureborne noise transmitted into surrounding hull structures: foundations, framing and plates. Structureborne noise spreads in all directions away from the source along the hull structure. The process of spreading is accompanied by reflection from many non-uniformities in the hull. Intersections of hull plating with ribs and deck-to-bulkhead connections all contribute to these reflections. The magnitude depends on the thickness ratios making up the intersection. Furthermore, vibration energy is scattered due to thermal losses and acoustical radiation. The paths for vibration are not only hull structures but also pipes and other auxiliary elements connecting the machinery. For instance, pipes excite structureborne noise at their supports.

Airborne noise radiated from a source spreads out as sound waves, forming what is called the "direct sound field". This direct wave amplitude decays with distance from the source. If the source is located within a room, the sound waves generate a diffusive field after multiple reflections from internal surfaces and internal equipment. The resultant sound field within a source room is the sum of these two components: direct and diffusive (or reverberant) fields. For airborne noise from an unenclosed 'outside' source, only the direct field needs to be considered.

As mentioned above, the sound field in a source room excites the adjacent hull structures and induces a vibration field ('secondary vibration'). Because of the good acoustical conductivity of the hull structures, this secondary structureborne noise may influence airborne noise in compartments far from the source. The significance of this factor depends on the power ratio W_{fsb}/W_{ssb}. Cases when $W_{fss} >> W_{ssb}$ typically occur for hard mounted mechanical sources. For effectively mounted engines, W_{fsb} and W_{ssb} may be the same order values, or it is possible that $W_{fss} << W_{ssb}$.

For noise prediction purposes, it is reasonable to classify rooms by the degree of influence of structureborne noise. From this point of view, all ship compartments may be divided into three groups:

- Group 1. Compartments that contain noise sources. Typically, in this room, $L_{ab} >> L_{fsb}$ and always $L_{ab} >> L_{ssb}$.
- Group 2. Compartments adjacent to a source room. In these compartments, all three components should be considered.
- Group 3: Compartments, which are separated from the source compartment by one or several compartments. For these rooms, noise may be only of structureborne nature: L_{fsb} and L_{ssb} are of primary interest.

Such classification is related for any specified couple of "source room-receiver rooms". One room may belong to different groups for sources located in the different rooms. This grouping makes the development of noise prediction algorithms much easier.

Such classification is not valid for all sources. For example, the noise level in a steering room may be controlled by sound radiation from a bottom plate, which is excited by a propeller. This means that only the first structureborne noise is realistically important in this case. The same situation occurs when noise is induced by ice impact or wave impact. Noise is pure first structureborne in such cases.

Another source/path/receiver model considers the impact of an external airborne source. Examples are impact of aircraft operation on compartments below, gas turbine exhaust stack contribution to an on-deck station, and engine room intake fan to a work station. In these cases, the source power level, source directivity, distance between source and receiver, and any blockage of sight-line between source and receiver need consideration. If there is a blockage of line-of-sight, then the effectiveness of this barrier or partial barrier needs to be considered.

6.2 Underwater Radiated Noise

As shown in Fig. 6.2, the radiated noise signature is a composite of noise from machinery, propulsors, sea-connected fluid systems, and flow over the hull and appendages. Therefore, acoustic sources for all these components must be analyzed over the entire frequency range of interest for the radiated signature. In many cases, this goes well beyond the 8000 Hz octave band used for airborne habitability noise considerations. Some underwater noise criteria extend to 200,000 Hz.

Both sonar self-noise and underwater radiated noise involve a completely different acoustic environment compared to the above discussion on habitability noise. Obviously, the fluid parameter and the sound reference pressures are different. The changes in density and sound speed are significant – $c_{0\ air}$ is 332 m/s, while in seawater, $c_{0\ water}$, it is 1500 m/s. In air, the only occasion where temperature is a consideration for the speed of sound is for high-temperature exhausts. In sea water, the speed varies with temperature – 4.5 m/s per 1 °C increase, with pressure – 1.7 m/s per 100 m depth, and with salinity and acidity.[3] While sound absorption in air is frequency dependent, it is only of concern over long distances from the ship, as with respect to aural detectability or noise dock-side. For transmission through water, this frequency-dependent absorption should not be ignored (Fisher and Simmons 1977). High-frequency sound absorption can significantly reduce the received levels over distances normally encountered with underwater noise – particularly of the noise levels impacting ocean life. Another factor is that most of the underwater sources – propellers especially – are directive, so it matters whether the receiver is ahead, aft, abeam, or below the ship/source (Trevorrow and Vasiliev 2008).

For the underwater cases, only direct or freely propagating noise needs to be considered; reverberant noise is neglected. However, the depth of water has a significant impact on the far-field transmission and cannot be neglected. The waterborne path can become very complicated, depending on varying depths, temperature layers, and surface and bottom reflections (Ross 1976). If of interest, refer to Udovydchenkov 2007. In this book, the authors assume the transmission in water can be modeled as either:

- Spherical – for deep water, usually deeper than 200 m, where spreading is spherical (decreases as 20 Log distance, *r*) and absorption losses are ignored for distances less than 1 km
- Cylindrical – for relatively shallow water of constant depth, where spreading decreases as 10 Log *r* and there is no absorption for distances less than 1 km and modal transmission losses are ignored
- In-between cases, where spreading losses can be modeled as 15 Log *r* and modal transmission losses may come into play

[3] http://resource.npl.co.uk/acoustics/techguides/seaabsorption/

There is a fourth waterborne path of interest with respect to sonar self-noise: that of "hull-grazing". Grazing accounts for the attenuation from an external source, such as the propellers, as it moves forward along the hull to the sonar array. This attenuation is on the order of 10 Log r below 500 Hz to 30 Log r above 8000 Hz (Fischer and Collier 2007).

Given its direct coupling to water, propellers radiate efficiently into the far field. Some radiation is considered monopole[4] or dipole[5] with respect to the type of acoustic source and with respect to directivity. Radiation models for non-cavitating and cavitating propellers are addressed in Chap. 9. With respect to transmission to a remote location, Lloyd mirror correction is needed given the 'free-surface' above the propeller causing a reflection out-of-phase with the direct ray (Carey 2009). These corrections depend on the depth of the propeller and the surface roughness or waviness. The propeller acoustic source level generally depends on the diameter, rotation rate, depth of the propeller hub, and blade characteristics such as chord length and thickness. Another significant factor is the wake field coming into the propeller plane. The more uniform this wake is, the lower the expected propeller source level. Obviously, the hull can cause a large wake deficit around the top dead center. Even a tunnel thruster usually has a very poor wake into the propeller. The signature is highly dependent on whether cavitation is present or not, and if present, what is the area of blade covered by this cavitation. As a rule of thumb, cavitation noise is on the order of at least 12 dB higher than noise from a non-cavitating propeller. There is also a difference in the spectrum shape and level between the tip, suction side, pressure side and hub cavitation.

A major nonstarter with respect to underwater noise[6] is a propeller that has a blade or blades that are said to be 'singing' (Fischer 2008). The physics behind a singing propeller is described in Chap. 9. An 'anti-singing' edge is usually sufficient to prevent the blade from resonating due to a vortex created on the trailing edge.

The radiated noise and hull pressures from a non-cavitation propeller are basically low and would only be a concern for a vessel with a very low radiated signature requirement. Large Eddy Simulation (LES) and infinite element methods have been used successfully to address noise from non-cavitating propellers (Sun et al. 2017).

The next order of concern is the underwater noise radiated by the propulsion machinery. This level is controlled by the amount of vibration produced on the hull plating by the machinery and the radiation efficiency of the wetted hull plate.[7] The radiation efficiency is affected by the hull plate thickness, Young's modulus, surface density, frame spacing, and frame moment of inertia. The radiating area must also be factored into consideration. Again, the aspect of concern can change the outcome.

[4]Mono-pole considered a 'thickness' noise source due to unsteady sheet volume fluctuations

[5]Di-pole radiation associated with fluctuating loads on the propeller blade.

[6]And sometimes airborne noise in compartments over the propellers.

[7]The difference between radiation efficiency of dry plate and wetted plate is primarily due to the fluid loading, which is frequency dependent.

The keel aspect radiation is likely to be higher than the beam aspect since much of the machinery is located directly over the keel. The transmission path from machinery to the wetted hull plate is shorter, creating higher vibration on the hull plate.

Sea-connected systems can contribute to underwater radiated noise, particularly if the pump cavitates. The amount of radiated noise depends on the acoustic level in the pump fluid, the piping path from the pump to the overboard discharge (including flex hose), and the pipe diameter. This noise is directive and, as with the propeller, subject to Lloyd mirror effects (Purshouse 1986). Some vessels also have underwater diesel exhaust, as discussed in Sect. 9.2.2.6.

To reduce an underwater signature, it is important to avoid cavitation on any appendage up to the highest speed covered by any criteria. While not likely, flow-induced noise re-radiated from the hull (Fureby et al. 2002) is sometimes of concern for high-speed vessels such as catamarans and trimarans. Prediction of the flow noise generally requires sophisticated fluid flow computations and a thorough understanding of the acoustic coupling (Choi et al. 2020). In Choi's study, the radiated noise from a trimaran at 30 knots peaked at 630 Hz at a level of approximately 140 dB re 1 μPa at 5 m. Assuming cylindrical spreading, this would amount to 147 dB at 1 m. From the appearance of the plots, these data are in one-third octave bands. Above 630 Hz, the radiated noise decreases at approximately −9 dB/octave. Furthermore, the difference between 20 knots and 30 knots was found to be approximately 5 dB.[8]

If there is an airborne noise source above the water interface, there is a chance that it can contribute to the underwater signature. This could happen for an exhaust pointed downward between SWATH[9] hulls. This is also a concern for air-cushion vehicles due to lift fans. When transmitted into the water at near perpendicular incidence, the pressure in the water will be 6 dB above that in the air (Ross 1976). To correct to dB re 1 μPa would add another 26 dB. A more complete model of airborne transmission into water for static and moving sources, such as helicopters, is provided in Zhao-Hui and Ling-Shan 2016.

6.3 Sonar Self-Noise

There are two separate categories of sonar self-noise. The first is called Platform Noise. This is the noise that would be measured by an omni-directional hydrophone located at the sonar system of interest. The second is the appropriately labeled 'self-noise'. This component takes into account the array gain, directivity, and signal-to-noise allowances for the respective sonar system. When specifying a new vessel's allowed noise induced at the sonar, it is usually the allowed platform noise since the actual sonar system's operational parameters may not be known. If the system is

[8] Scaling by the 6th power of velocity would impute a 10 dB change; however, at 20 knots machinery induced noise may dominate and hold up the measured signature.

[9] SWATH – small waterplane area, twin hull

identified, the sonar vendor will often set the allowed platform noise knowing the processing characteristic of their system. The following only considers platform noise aspects since there are quite a few different types of sonar systems. The primary components – sources and paths – of Platform Noise are shown in Fig. 3.8.

The acoustic limits for sonar systems are sometimes specified as an equivalent Sea State (Urick 1984) at a set vessel speed. Otherwise, platform limits are provided as a limit in dB re 1 μPa within a certain frequency range and bandwidth. It is possible given the sonar physical make-up and operating parameters, such as the needed signal-to-noise ratio, operating depth, and bottom reflection to determine the allowed platform or even self-noise for both active and passive sonars.[10] This process utilizes the 'sonar equations' (Loeser 1992).

As with habitability airborne and underwater radiated noise, one of the most significant sources is the propeller. Since most acoustic surveying is performed at relatively low speed, it is often assumed that the propeller is non-cavitating. If it is cavitating, the sonar system will likely be blanked. For most vessels with bow-thrusters, it is assumed that they will not be operational, as they cavitate even at low speed due to the required thrust at low speed. Along the same lines, it is assumed that the appendages are not cavitating within the operating speed for sonar surveying.

Following the propeller contribution, machinery structureborne-induced underwater noise may be of concern. This is particularly true for hard mounted propulsion systems. For this source/path, one needs to determine the coupling of hull vibration-to-underwater radiated noise (radiation for fluid loaded plate is covered in Sect. 9.2). The directivity of radiation from framed plating (Zhang et al. 2022; Berry et al. 1990) along a hull grazing path is one additional factor not usually considered for the underwater radiated signature since the sonar receiver is usually at a tangent to the hull. Abeam, the radiation directivity is usually maximum, whereas at hull grazing, it is significantly lower.

There is another transmission path internal to the vessel. Machinery vibration can be propagated along the tank top and wetted hull plate forward to a bow sonar. As a rule-of-thumb, machinery more than 20 meters from the sonar can be neglected. Furthermore, a baffle may be placed aft of the sonar to block both water-borne and structureborne propulsion machinery contributions to a bow-mounted sonar system.

The impact of any underwater noise source will then depend on its acoustic level, the length of the path from the source to the sonar (whether hull mounted or towed) and the directivity of the source with respect to the sonar. Whereas habitability noise is usually modeled in octave bands, sonar self-noise is modeled in the bandwidth and operational frequency of the sonar system.

Flow over the sonar dome can create measurable sonar platform noise. While prediction techniques exist (Magionesi 2016), this is usually the domain of the sonar vendor. The vendor usually selects the hydrophone cover or dome to minimize this component up to the maximum speed allowed for operation of the sonar.

---

[10] https://man.fas.org/dod-101/navy/docs/fun/part08.htm

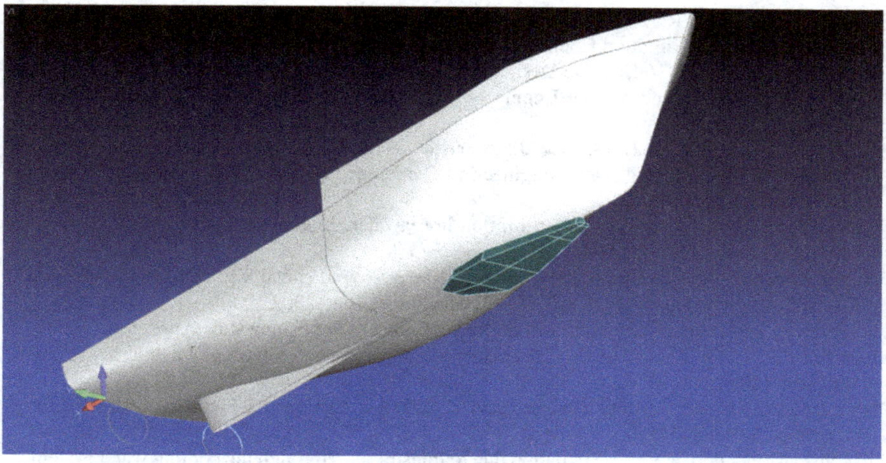

Fig. 6.5 Cow Catcher Fairing for Multi-beam sonar located near bow (Rolland et al. 2010)

Bubble sweep down, whether from the bow or appendages, will most likely decouple any sonar, making it ineffective. In general, computational fluid dynamics (CFD) modeling is needed to ensure that the flow lines from these sources do not cross the sonar dome (Rolland et al. 2010). In some cases, especially for multi-beam sonar located mid-length of the ship, a barrier – baffle or fairing – is used forward of the sonar to prevent bubbles from passing over the dome. One such option is shown in Fig. 6.5. Another option is to have a retractable multi-beam sonar so the system is below any bubble stream and is 'stowed' when not in use.

References

Beranek, L.L., Ver, I. (eds.): Noise and Vibration Control Engineering, Principles and Applications. Wiley, NY (1992)

Berry, A., Guyader, J.L., Nicolas, J.: A general formulation for the sound radiation from baffled rectangular plate with arbitrary boundary conditions. J. Acoust. Soc. Am. **88**(6), 2792–2802 (1990)

Carey, W.M.: Lloyd's mirror – image interference effects. Acoust. Today. **5**(2), 14–20 (2009)

Choi, W.-S., Hong, S.-Y., et al.: Estimation of Turbulent boundary layer induced noise using energy flow analysis for chip Hull Designs. J. Mar. Environ. **234**(1), 196–208 (2020)

Fischer, R.W.: Singing propellers – solutions and case histories. Society of Naval Architects and Marine Engineers (SNAME) Mar. Technol. SNAME News. **45**(4), 221–227 (2008)

Fischer, R., Bahtiarian, M.: Ship noise control. In: Finch, D. (ed.) Encyclopedia of Maritime and Offshore Engineering. Wiley (2017) Online ISBN: 9781118476406

Fischer, R., Boroditsky, L.: Technical and Research (T&R) Bulletin 3–37 (Supplement), Supplement to the Design Guide for Shipboard Airborne Noise Control. The Society of Naval Architects and Marine Engineers (SNAME), Alexandria VA (2001)

Fischer, R.W., Collier, R.: Noise prediction and prevention on ships. In: Crocker, M. (ed.) Handbook of Noise and Vibration Control. Wiley, New York (2007)

Fischer, R, Courtney, B., Nelson, D.: Design Guide for Shipboard Airborne Noise Control. Society of Naval Architects and Marine Engineers (SNAME), Technical & Research Bulletin 3–37, Alexandria, VA, 1983

Fischer, R.W., Boroditsky, L., Spence, J.: Acoustic design tools – present and future integrated software. In: International Naval Engineering Conference (INEC) 2008, Hamburg, Germany, April 2008

Fisher, F.H., Simmons, V.P.: Sound absorption in seawater. Journal of the Acoustical Society of America (JASA). **62**, 558–564 (1977)

Fureby, C., et al.: A First Study of Flow Induced Noise Around Naval Vehicles, Swedish Defence Research Agency, Report FOI-R-0519-SE, Stockholm June 2002

Loeser, H.T. (ed.): Sonar Engineering Handbook. Peninsula Publishing, Los Altos, CA (1992)

Magionesi, F.: Flow induced structural noise on a sonar dome of a ship. In: InterNoise 2016, Hamburg (2016)

Plunt, J.: Methods for Predicting Noise Levels in Ships. Chalmers University of Technology, Goteborg, Sweden (1980)

Purshouse, M.: Underwater noise radiation due to transmission through the cooling water system of a marine diesel engine. In: Buiten, J. (ed.) Ship Acoustics. Martinus Nijhoff Publishers, Dordrecht (1986)

Rolland, D., Forgach, K., Clark, P., Mackes, J.: Hull and sonar mount design for reducing bubble sweepdown on oceanographic research vessels. In: NAVSEA Distribution Statement A: Approved for Public Release: Distribution Unlimited, 2010

Ross, D.: Mechanics of Underwater Noise. Pergamon Press, NY (1976)

Sun, S., Chang, X., Zhang, H., Ye, L.: Non-cavitation noise numerical simulation of propeller behind a ship hull. In: 27th International Ocean and Polar Engineering Conference, S.F., CA June 2017

Trevorrow, M.V., Vasiliev, B.: Directionality and maneruvering effects on a surface ship underwater acoustic signature. Journal of the Acoustical Society of America (JASA). **124**(2), 767–778 (2008)

Udovydchenkov, I.A.: Sound Transmission Through a Fluctuating Ocean: A Modal Approach, Dissertation – University of Miami, Dec. 2007

Urick, R.J.: Ambient Noise in the Sea, Undersea Warfare Technology Office, DTIC ADA460546 (1984)

Zhang, K., Lin, T., et al.: The effects of ribs on the sound radiation efficiency of rectangular plates. Phys. Fluids. **34** (2022)

Zhao-Hui, P., Ling-Shan, Z.: A review of research progress in air-to-water sound transmission. Chinese Physics B. **25**(12), 124306 (2016)

Chapter 7
Prediction Modeling Methods

Any mechanical system used on a ship induces noise and vibration on that vessel. Other equipment, such as transformers, UPSs, and switchgears, need consideration when located in a noise-sensitive compartment. It is essential to understand the overall impact of machinery and equipment on the acoustic environment of the vessel. How it is attached or coupled to the structure will also impact the resulting acoustic environment. In addition to rotating or reciprocating machinery, most vessels have extensive, distributed systems such as piping or HVAC. As discussed in Chap. 6, these systems can easily transmit noise to remote, sensitive sections of the vessel.

The equations and tools that implement the techniques needed to predict the overall impact of *all* the machinery transmitted over *all* the paths to *all* the receiver locations of interest are well documented. These tools are iterated to find an optimal approach – whether it is to reduce cost or weight impacts to achieve the acoustic requirements. In many cases, worst-case scenarios will be assessed, first to determine if there is a potential problem, second to assess the possible magnitude and frequency range of the problem, and third to gain a feel for when more accurate modeling or expertise might be needed. Assumptions and simplifications for the models presented will be discussed.

Acoustical modeling for vessels usually follows the 'source-path-receiver' process discussed in the previous chapters. The following parameters and ship design information all need consideration:

- Variety of sources – mechanical and non-mechanical (equipment), flow-induced, and hydro-acoustic, e.g., propeller
- Propagation paths – airborne, structureborne, fluidborne, distributed systems – piping, HVAC, cabling, etc.
- Coupling factors – airborne-to-airborne, airborne-to-structureborne and inverse, structureborne-to-structureborne, fluidborne-to-structureborne and inverse, piping/HVAC internal noise-to-external noise, and piping wall vibration-to-external noise

R. Fischer, L. Boroditsky, *Noise and Vibration Control on Ships*,
https://doi.org/10.1007/978-3-031-55170-3_7

- Room (or compartment) acoustic characteristics, direct or reverberant field, for enclosed spaces and propagation to external shipboard or far-field locations of interest

Prediction algorithms can employ empirical methods, be based on analyses of measured data from other vessels, or explicit methodologies such as Statistical Energy Analysis (SEA) or Finite Element Analysis (FEA). Many assumptions are made in any modeling process. An exact model of a ship with all its structure, machinery, joiner work, insulation, etc., is not feasible. For example, the joint between the tanktop and a curved hull plate is typically modeled as a 90-degree joint, not an angled joint, with little loss of accuracy. The major concerns about the applicability of any model from FEA to SEA to BEM involve accurate values for frequency-dependent loss factors, radiation efficiency from wet and dry plates, and inclusion of all wave types – bending, longitudinal, and shear. Most models have the greatest difficulty accounting for low frequencies. There was a 60 dB difference in the expected power level at 12.5 Hz between eight different modeling methods (Tanttari 2014). Once created, models are generally updated and verified over time based on real-world testing and feedback.

There are two principal methods for noise prediction. The first is to use measured data on another ship (prototype) to predict noise and vibration on the ship of interest. If the prototype's acoustic sources and general arrangement are similar to the ship under consideration, noise levels for the new ship may be found by 'scaling' the appropriate physical size and dimension and acoustic data. However, this method cannot be considered a comprehensive approach. Only for some very restricted cases can reliable results be reached when the general arrangement, sources, and hull construction for the prototype and design of interest are very similar. The great advantage of this method is its simplicity. The down fall is inaccuracy, hopelessness of determining the physical reasons for any predicted excessive noise levels, and, as a result, over/under selection of control approaches.

The second method involves calculations based on analytically derived or empirical algorithms. This approach has to address a wide variety of fundamental parameters: source noise and vibration characteristics, geometry of ship structures and compartments, material parameters, types of equipment and propulsor, and others, as described in the remaining sections of this chapter. Well-established and reliable methods exist for airborne noise components based on 'room acoustic' algorithms (Beranek and Ver 1992). Measurements on ships show that algorithms based on room acoustics are sufficient to calculate the expected noise levels in the engine room or in any room containing a noise source. Methods go so far as to consider sound scattering due to large objects in a machinery compartment (Nilsson 2004).

A significant step ahead in practical calculation of structureborne noise on ships was provided in the SNAME Design Guide developed by Bolt, Beranek, and Newman, Inc. (BBN) team in 1983 (Fischer et al. 1983; Fischer and Boroditsky 2001). This guide and its supplement developed transfer functions relating the "source vibration level-to-receiver space vibration level". The length of the path and the number of obstacles (joints) between the source and receiver were considered in this guide. Maidanik (1963) algorithms for vibration-to-radiated noise transfer functions were used in the Design Guides. Spreadsheet programs were used for a practical realization of the Guide. Furthermore, as part of the SNAME Supplement, additional material databases were provided covering absorption coefficients, transmission loss (TL), cladding effectiveness, and duct attenuation losses.

A greater opportunity for noise analysis occurred when energy balance methods became known to the engineering community and computer ability for swift calculations became available. Energy balance methods helped to develop 3-D acoustical modeling of a ship for habitability noise analysis. Averaged vibration levels anywhere in the ship could be determined as a result of solving a set of linear algebraic equations. The matrix of a system contains the coupling and damping loss factors, which are derived from geometry and material information (Cremer et al. 1988; Lyon et al. 1995; Westphal 1957). The development of algorithms for the radiation of sound by finite vibration plates (Maidanik 1974; Smith and Lyon 1965) assisted in the prediction of structureborne noise.

The preeminent advancement was the developed and useful realization of energy methods such as Statistical Energy Analysis (SEA). This methodology became available for engineering applications early in the 1960s. SEA may be applied to a grouping of connected structural elements, which model decks, bulkheads and other structural parts of the hull. Statistical energy analysis provides an opportunity to calculate averaged vibration levels of hull structures remote from a source in a format that is convenient for noise prediction. For example, the vibration level may be calculated in octave bands. Once the averaged vibration is known, one can apply formulas for the radiation of vibrating plates to calculate airborne noise levels radiated from any structural elements of interest. The combination of the radiated noise levels from all the surfaces of a compartment provides the expected noise level received by remote compartments. Currently, commercially available SEA computer software exists.

In these, the actual ship structure is an assemblage of rib-stiffened plates, which form the hull. Side and bottom shells, decks, transverse, and longitudinal bulkheads make up this hull. Often, these basic structures are covered with insulation and joiner panels, which can affect the radiated noise.

All the different machinery/equipment items are also attached to this hull structure. Pipes, ducts, cables and other "connectors" are practically everywhere inside a ship. Such diversity and great non-uniformity make a ship a very inconvenient subject for acoustical modeling. However, experience has shown that successful modeling is possible if only the key factors influencing noise generation and spreading are considered. The three-dimensional nature of a ship's structure "provides" multiple paths for acoustical energy spreading from noise/vibration sources. Any acoustical model should therefore address all actual combinations of structures that transmit and radiate structureborne noise.

Other analysis methods include Boundary Element Methods (BEMs), Seybert et al. 1994, and coupled BEMs and Finite Element Methods (FEMs) (Salio 2013). For combined habitability and radiated noise, computer analysis programs are being developed. This includes coupled FE/SEA, which represent the vessel, interior cabins, fluid tanks, underwater fluid loading and noise radiation for frequencies from 0 to 10,000 Hz (Blanchet and Caillet 2014). Vibro-Acoustic Simulation Software, with its VA One, from the ESI Group utilizes the above approaches (https://myesi.esi-group.com/downloads/software-downloads/va-one-2020.0). *DesignerNOISE®* (2002) software is also being used for combined habitability and underwater radiated noise analyses.

BEM is not a general-purpose modeling approach (Carlton et al. 2018). It is often used for specific structural/acoustic modeling, for instance, the interaction between diesel excitation and ship-hull vibration (Zhen et al. 2001). In some cases, the finite element method (FEM) has been combined with the BEM to couple the predicted vibration levels to the acoustic response of the structure (Seybert et al. 1994; Seng and Khoon 1999). Either of these methods is more computationally intensive than SEA. It also requires a detailed understanding of the acoustical properties, especially boundary conditions, to derive an accurate prediction of the noise. Furthermore, they have limited capability to predict the higher-frequency noise levels and are much more computationally intensive.

A summary of potential analysis tools is provided in Table 7.1 (Fischer et al. 2000, 2008).

A comparison of some of these modeling approaches – FE, BEM, SEA and hybrids – is provided in Tanttari, 2014.

Table 7.1 Potential analysis tools

Functional capabilities	Existing analysis tools	Potential analysis tools
1.0 **Noise Model**: Habitability Noise; Airborne and Underwater Radiated Noise	Empirical Database; SNAME	Physics-based structural analysis
1.1 Low- to Mid-Frequency Analysis	CAD 3-D Noise Model; BEM	FEA/IE
1.2 High Frequency Analysis	CAD 3-D Noise Model	SEA/EFEA
1.3 Propeller Underwater Noise Sources	Turbulence Ingestion Strip Analysis; Hydro-acoustic codes (CFD/RANS)	Improved CFD/RANS
1.4 Piping/HVAC	Noise Model; FE models	Wave Transmission Line
2.0 Sonar Self Noise	CAD Noise Model	SEA & TBL analysis
3.0 Vibration Habitability	Low Frequency Ship FEA Model	FEA/SEA
4.0 Optimization	Not fully developed	Logic trees

SNAME (T&R 3-38), FEA finite element analysis, SEA statistical energy analysis, IE infinite elements, EFEA energy finite element analysis, CFD computational fluid dynamics, RANS Reynolds-averaged Navier-Stokes, TBL turbulent boundary layer

7.1 Airborne Noise in a Source Room

As a rule, the noisiest room inside a ship is an engine room. Airborne noise is radiated by machinery into the air volume of the compartment. This component usually dominates the overall noise level. In principle, it is possible that structureborne noise influences or is equal to the airborne level in a source room. For example, an effective enclosure around sources together with ineffective resilient mounts may create such a situation.

Many guidelines recommend using the following formula to calculate the noise level around a source in a room (Beranek and Ver 1992; Bies and Hansen 2009):

$$L_p(r) \text{ dB re } 20\ \mu\text{Pa} = L_w + 10 \text{ Log}\left[\left(Q/4\pi r^2\right) + (4/R)\right] \quad (7.1)$$

where $L_p(r)$ – sound pressure level in a position of interest; L_w – sound power level of the source; r – source center to receiver distance, m; R – room constant, m^2 (quality of absorption in room, see following paragraphs); and Q – directivity factor.

The expression in brackets contains two terms. The first describes the so-called direct field of spherical spreading, where the decay of sound pressure squared is inversely proportional to distance squared. The second quantifies acoustical energy reflected from the room's boundaries.

Some comments need to conveyed about using this equation in ship applications. The formula is derived for a point source or for a "small" source, where the physical size is less than the wavelength of interest. Many noise sources onboard do not meet this requirement. Another comment concerns the room constant.

The room constant, R, in m^2 is equal to:

$$R = S\,\alpha_{sab} \quad (7.2)$$

where α_{sab} is the average absorption coefficient in the room of interest and S is the full surface area of the room of interest, m^2. The absorption coefficients for materials typically used on ships, ranging from 0 to a maximum of 1, are provided in the SNAME Design Guide (Fischer et al. 1983). For some 'generic' shipboard compartments, frequency-dependent absorption coefficients are shown in Table 7.2 (Spiridonov 1978). The values in the table will be used further for the calculation of airborne and structureborne noise in ship compartments.

Table 7.2 Average absorption coefficients, α, for specialized ship compartments

	Octave Band Freq., Hz							
Compartment type	63	125	250	500	1000	2000	4000	8000
Steering gear and refrigerator rooms	0.04	0.06	0.08	0.10	0.13	0.15	0.18	0.20
Engine rooms, boilers	0.04	0.08	0.11	0.14	0.17	0.20	0.24	0.27
Mess, lounge, offices, workstations	0.12	0.17	0.19	0.20	0.21	0.21	0.22	0.22
State rooms	0.12	0.21	0.25	0.27	0.28	0.29	0.29	0.29

Instead of sound power levels, equipment manufacturers usually provide sound pressure levels, $L_{p,\ surface}$, measured at a preset distance from a source (typically 1 m). In this case, formula Eq. (7.1) may be rewritten:

$$L_{w\ machy}\ \text{dB re } 20\ \mu\text{Pa} = 10 * \text{Log} \sum\nolimits_n \left[10^{\left(L_{p,surface\ n}/10\right)} \times 10^{\left(+10\ \text{Log} S_n/10\right)}\right] \quad (7.3a)$$

$$L_p(r)\ \text{dB re } 20\ \mu\text{Pa} = L_{w\ machy} + 10\ \text{Log}\left[\left(Q/4\pi r^2\right) + (4/R)\right] \quad (7.3b)$$

where $L_p(r)$ is the resulting noise in the engine room, $L_{w\ machy}$ is the power level of the averaged noise level on each measured surface, $L_{p\ surface}$, and S_n is the surface area for each side[1] of the source, m^2. The total noise level at a particular position in the machinery space would be the combination of all the individual $L_p(r)$. For this equation to be valid, at least 5 sides need to be measured for $L_{p\ machy}$.

A comparison of measured and calculated data with this formula shows that this approach does not create significant problems in the majority of cases. However, if the larger size of the source is on the order of meters and one side is much larger than the other side, a non-directional sound source may be inappropriate for direct field calculation. Different approaches exist to address this issue (Boroditsky and Fischer 2004; Janecek, 1989). If one considers the real source as an assembly of point sources with equal strength, it is possible to assume that the front of the resultant wave will have a shape similar to the source shape. From this known analytical surface, an ellipsoid shape may be a convenient model.

Airborne noise in an engine room is calculated not only for comparison with noise criteria but also because noise levels near engine room boundaries are input data for the second structureborne noise calculation. This means that noise levels on structures forming an engine room should be determined.

Very often, an engine room does not have a simple shape. The main section, where sources are located, may have openings to another section (usually an opening to a smaller compartment above or an uptake casing). Different methods may be used to determine noise levels in this additional volume to an engine room. Two of them may be useful for a simple engineering approach.

It is possible to formulate an energy balance equation for two (or more) coupled rooms (Xiang and Summers 2007; Gomzikov and Izak 1981). First, assume that only the reverberant part of the energy in room 1 is transmitted through the opening without losses. Then, assume that there is a noise source power, W, in Room 1. In this case, the energy balance equation may be written:

$$W + (cA_0 w_2/4) = cw_1(\alpha_1 A_1 + A_0)/4 \quad (7.4a)$$

$$(cA_0 w_1/4) = cw_2(\alpha_2 A_2 + A_0)/4 \quad (7.4b)$$

[1] Typically, the four sides and top surfaces are measured.

where A_0 is the opening area; A_1, A_2, α_1, α_2 are the surface areas of rooms 1 and 2 and the average absorption coefficient in rooms 1 and 2; w_1, w_2 are the energy densities in rooms 1 and 2; and c is the speed of sound in air. The left-hand sides in the equations are the energy coming into the room; the right-hand sides are the energy leaving the room. The first equation describes the balance of energy for Room 1, and the second equation is the same for Room 2. The absorption coefficient for opening is assumed to be equal to one.

Another approach is possible if one considers aperture as a source for adjacent Volume 2. The power of this source (Beranek 1986), noted as W_{12}, is equal to:

$$W_{12} = \left(p_{12}^2/\rho c\right)(A_0/k) \tag{7.5}$$

where p_{12}^2 is the sound pressure in the aperture plane and k is the coefficient depending on the kind of sound field. For the diffusive field $k = 4$, for a direct field normal incident $k = 1$. Assume $k = 4$. Using Eq. (7.5) as a source power in Eq. (7.1), one can calculate the noise level at distance r_2 from the aperture center:

$$L_p(r) \text{ dB re } 20\ \mu\text{PA} = L_{p12} + 10\,\text{Log}\left[\left(A_0/16\pi r^2\right) + (A_0/R_2)\right] \tag{7.6}$$

where L_{p12} is the noise level in the aperture plane; r_2 is the distance from the center of the opening to the position of interest; and R_2 is the room constant for Room 2. When R_2 is calculated, one should use an absorption coefficient equal to 1 for the opening. A third volume coupled with the second volume may also be addressed in the same manner.

7.2 Noise Levels Transmitted through a Partition Between Source and Adjacent Rooms

In the noise prediction process, transmitted noise is calculated only for rooms adjacent to a source room and separated by a deck or a bulkhead. The basic equation for this noise component (Beranek 1986):

$$L_{pab} \text{ dB re } 20\ \mu\text{Pa} = L_{ps} - TL + \text{Log}\left(A_p/R'\right) \tag{7.7}$$

where L_{pab} – level of airborne noise transmitted through partition from source room to adjacent room; L_{ps} – level of airborne noise in the source room near the partition; *TL* - transmission loss of a partition; A_p - area of a partition; and R'- room constant of a receiver room.

The last term of this formula is easy to determine if an averaged absorption coefficient is known (this is usually the case). *TL* is expressed in decibel ratio of energy incident on a partition-to-energy transmitted through partition into a perfectly absorbing space. Many papers are devoted to the analytical definition of *TL*.

Table 7.3 Transmission loss, dB, and break points based on critical frequency

Material	Frequency, Hz			
	$0.25\,f_{cr.}$	$0.5\,f_{cr.}$	$f_{cr.}$	$2\,f_{cr.}$
Steel, TL dB	35	37	30	39
Titanium, TL dB	31	33	26	35
Aluminum Alloy, TL dB	29	31	23	31
Fiberglass, TL dB	28	31	28	34
Plywood, TL dB	26	29	26	32
Glass, TL dB	32	34	27	34
Plexiglass, TL dB	34	36	30	38

No theoretical results are perfect for ship partitions. Keep in mind that a ship partition is a finite size metal panel with ribs or a composite multi-layer panel or a double wall panel. Practice shows that the best results are empirical methods based in part on measured *TL* data in a lab or better yet on a representative ship. The authors have successfully used these empirical methods for ship noise predictions (Bogolepov 1978).

A single wall TL calculation is based on the classical "mass law" with a coincidence effect correction. The ribs on the hull structure also influence the transmission loss. The theoretical model was corrected in accordance with the experimental data. The practical procedure, consists of the following steps:

1. Critical frequency, f_{cr}, determination.
2. Assignment of transmission loss corresponding to frequencies: $0.25f_{cr.}$, $0.50f_{cr.}$, $f_{cr.}$ and $2f_{cr.}$ in accordance with the empirical values shown in Table 7.3.
3. Determination of the transmission loss between these points by linear interpolation.
4. Change in transmission loss values for frequencies below $0.25f_{cr}$ by −4 dB/octave
5. Adjust the transmission loss at frequencies higher than $2.0f_{cr}$ by +8 dB/octave

The critical frequency is one of the key parameters for noise prediction for ships. The critical frequency is equal to (Beranek and Ver 1992):

$$f_{cr} = \left(c^2/1.8t\right)\sqrt{\rho_m/E}\;\;Hz \tag{7.8}$$

where c – speed of sound in air, m/s; ρ_m – density of plate, kg/m^3; t is the plate thickness, m; and E – Young's modulus of the plate, N/m^2.

The above equations apply to a flat plate. A corrugated plate, sometimes used in marine constructions, has some different acoustic properties. First, there are two different moments of inertia and bending stiffness. These differences affect the first panel resonance and the coincidence frequency. In general, the first panel resonance may be several octaves above that of a flat plate. Second, there are two coincidence frequencies. Thus, the 'plateau region of transmission' will extend over a larger frequency range than that typical of a flat panel. Measured data show the corrugated

Table 7.4 Acoustic parameters for flat and corrugated panels

	Bending stiffness, N-m	First panel resonance, Hz	Coincidence freq., Hz
Panel A, corrugated	240,000 and 241	30	163 and 5115
Panel B, corrugated	90,000 and 263	18	264 and 4897
Flat panel	296	29	4852

panel has a lower transmission loss in these mid-frequencies than a flat plate of the same thickness. A comparison of the panel resonance and coincidence frequency for a 2.5 mm thick flat plate and two different corrugated panels is provided in Table 7.4 (Soares 2008).

If necessary, transmission loss may be increased by placing a lightweight, sound absorptive material on the plate surface. Measurements show that this absorption layer effect on *TL* is additive. In other words, the transmission loss of a panel with an absorptive layer, TL_{al}, may be determined as:

$$TL_{al}\ dB = TL_0 + \Delta TL_{al} \tag{7.9}$$

where TL_0 is the transmission loss of the untreated partition and ΔTL_{al} is the additional transmission loss due to the absorptive layer.

Mineral wool and fiberglass with densities between 48 and 144 kg/m^3 (3–9 lb./ft^3) are usually used in the shipbuilding industry for thermal or acoustical insulation. The authors have measured transmission loss of a marine bulkhead with different absorption materials attached and with a variety of material thickness. The measurements show that the material density does not significantly influence the absorption layer effectiveness. However, the thickness of the absorptive material is a key parameter for ΔTL_{al} – used in Eq. 7.9. Measurements show that at low frequency $\Delta TL_{al} = 0$ or negative, positive effects start from some "cut off" frequency, which depends on layer thickness. Spiridonov (1978) suggested the following empirical formula:

$$\Delta TL_{al} = (A\ \mathrm{Log}(H) + 3)\ F\left({}^{f}\!/\!_{f_0}\right)\ dB \tag{7.10}$$

where *A* is an empirical constant; *H* is the thickness of an absorptive layer, cm; and $F(f/fo)$ is an empirical function, which is equal to zero when $f = fo$. The authors found that this equation approximates the experimental data when $A = 20$ and the function $F(f/fo)$ is determined by the curve in Fig. 7.1.

In this function, $f_O = 400/(1 + H^{1/2})$, where H is the thickness of an absorptive layer in cm. Figure 7.2 shows the measured 'Delta' in transmission loss for a 5 cm thick fiberglass versus that predicted by Eq. 7.10, demonstrating the validity of the equation. Lab measurements have shown that the *TL* is independent of the base plate mass and stiffness.

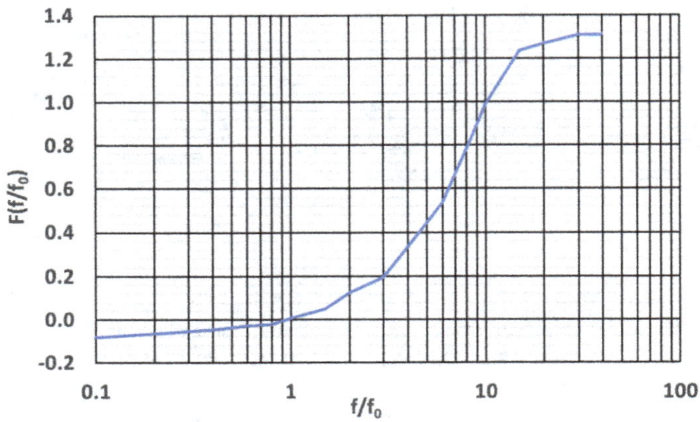

Fig. 7.1 Plot of function $F(f/f_0)$

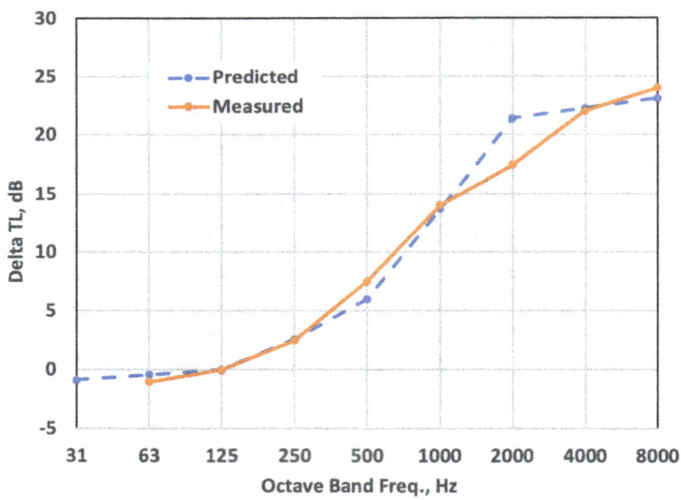

Fig. 7.2 Delta *TL*, 5 cm thick fiberglass

Facings, often referred to as joiner panels, predominantly act to reduce radiation efficiency (or acceptance, where applicable). Facings can either be hard or resiliently mounted, with appropriate differences in effectiveness. A 'hard' mounted joiner refers to whether there is a direct mechanical connection to the structure that the panel is covering. A joiner panel that is bolted directly to a bulkhead would be an example of a hard connection. Joiner panels that are mounted to the ship structure via rubber lined channels are considered 'resilient.' The presence of absorptive materials between the facing and structure should also be taken into account. Algorithms for joiner panel effectiveness are based on experimental data. Note that, based on experimental results, the changes in radiation efficiency, transmission loss, and acceptance are all considered to be equivalent when joiner panels are applied to a bare structure. Therefore, a single 'delta' is computed when analyzing the effects of joiner panels.

For example, a common joiner panel used in the marine industry is a sandwich of thin steel with a mineral wool core. The 'delta' of this resiliently mounted panel is calculated as follows. The cut-on frequency, f_{on}, for effectiveness is calculated based on a simple lumped parameter mass-spring-mass system using the formula (Boroditsky et al. 2007)

$$f_{on} = 2\pi\sqrt{\left(\rho_0 c_0^2\right)\ \left(\left(\rho_b + \rho_{jp}\right)/\left(\rho_b \rho_{jp} AG\right)\right)} \tag{7.11}$$

where ρ_b and ρ_{jp} are the surface densities of the base structure and joiner panel, respectively (kg/m^2); ρ_O is the density of air; c_O is the speed of sound in air; and AG is the 'air gap' thickness or distance between the joiner and structure (ignoring insulation) in meters. The delta (in dB) is then computed using three equations, which are valid for low, middle, and high frequencies. These frequency ranges are determined using the parameter:

$$f_{mid} = \rho_b c_0 / \rho_{jp} AG$$

Low frequencies are defined when f/f_{mid} is less than 0.6, where f is the octave band center frequency. Middle frequencies are defined when f/f_{mid} is between 0.6 and 1.2, and high frequencies are defined when f/f_{mid} is greater than 1.2.

The low-frequency delta is given by the following equation:

$$\Delta_{Low} = 11\ \mathrm{Log}\left(f_{c,b}/f_{on}\right) + 7\ \mathrm{Log}(AG) + 5 \tag{7.12}$$

where $f_{c,b}$ is the critical frequency of the base structure (without a joiner panel). Note that the values provided by the above equation are always forced to be greater than – 2 dB. The equation used to determine the delta at middle frequencies is:

$$\Delta_{Mid} = 10\ \mathrm{Log}(R1) + \Delta_{Low} \tag{7.13}$$

$$R1 = c_O \rho_{jp} / \left(2\ AG\ \rho_b\ f_{c,b}\right) \tag{7.14}$$

Delta at high frequencies is given by

$$\Delta_{High} = \Delta(f_{-1}) + 2 \tag{7.15}$$

where $\Delta(f_{-1})$ is the delta value for the previous octave band.

If absorption is present between the joiner panel and structure, a correction factor is added to the above calculations in each octave band. Note that it is assumed that the absorption thickness is at least ½ of the total air gap thickness, and the absorption is assumed to be 50 kg/m^3 density fiberglass. The correction, C, is calculated as follows:

$$C = 20\,\mathrm{Log}(f/f_{on}) \tag{7.16}$$

Note that C is forced to always be greater than or equal to zero. In addition, an upper limit is applied to this correction, given by

$$C_{Limit} = 4\,\mathrm{Log}\left(\rho_b/\rho_{jp}\right) + 6\,\mathrm{Log}(AG) + 11 \tag{7.17}$$

If the air gap thickness is greater than 100 mm, an additional correction is applied in the octave band containing the frequency f_{mid} and the octave band above f_{mid}:

$$C_{large\ gap} = 26\,\mathrm{Log}(AG) + 26 \tag{7.18}$$

The final joiner panel correction with insulation is calculated by adding the delta to the various corrections listed above. An example of the accuracy of this method is provided in Fig. 7.3.

This figure shows a range of joiner panel options covering different base materials both with and without absorption. The accuracy is generally good, within a few dB. It is noted again that this approach is used for the most common joiner panel types used in modern shipbuilding.

In terms of the depth of the gap between the joiner panel and the base structure, an increase from a 5 cm gap to a 10 cm gap is only a few dB below 2000 Hz and

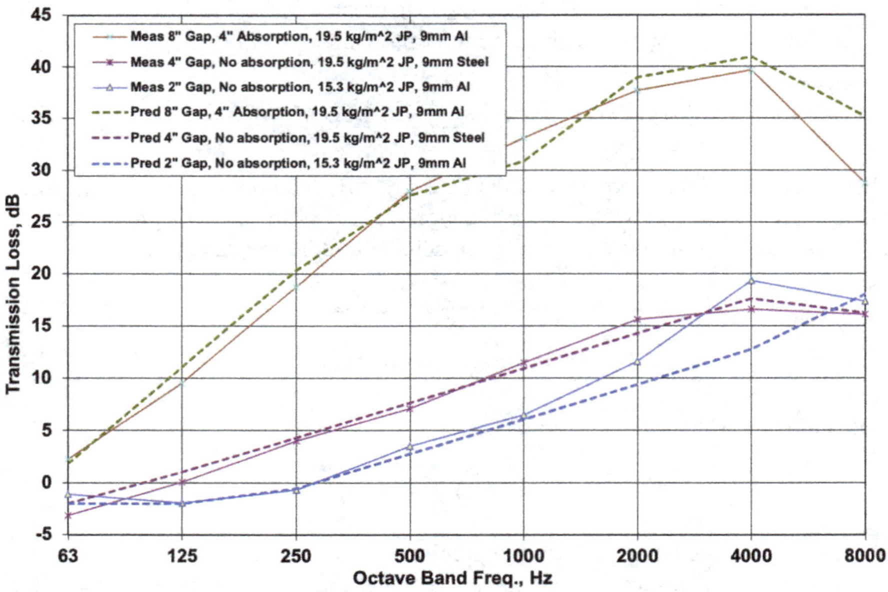

Fig. 7.3 Joiner panel, JP, additional transmission loss

minimal above that frequency. The most effective treatment is to include an insulation material in the gap that does not bridge the gap.[2]

The transmission loss of multiple partitions can be computed analytically (Brekke 1981). Resonant and non-resonant (forced) transmission loss for single, double, and triple partitions are considered. Resonant transmission is important above the panel's critical frequency. The resonant transmission loss, TL_r, is computed as:

$$TL_r \text{ dB} = 20 \text{ Log}(m) + 30 \text{ Log}(f) - 20 \text{ Log}(\sigma) + 10 \text{ Log}(\eta_{2t}) - 10 \text{ Log}(f_c) - 44 \tag{7.19}$$

where m is the surface mass; σ is the radiation efficiency; f_c is the plate critical frequency; and η_{2t} is the total loss factor, computed as:

$$\eta_{2t} = 2.2/fT_2$$

where T_2 is the reverberation time, sec, of the panel.

The non-resonant transmission loss, TL_{nr}, is:

$$TL_{nr} \text{ dB} = 20 \text{ Log}(f\rho_s) - 10 \text{ Log}\left(\text{Ln}\left(k\sqrt{A}\right)\right) + 20 \text{ Log}\left(1 - (f/f_c)^2\right) - 42 \tag{7.20}$$

where ρ_s is the surface mass, kg/m^2, A is the panel area m^2, and $k = 2\pi f/c$, m^{-1}. Valid for $f \geq 100$ Hz and $A > 1\text{m}^2$.

7.3 Structureborne Noise Calculation

The following subchapters consider the energy flow balance, energy transmissibility, and loss factors with respect to structureborne noise.

7.3.1 Energy Flow Balance for the Ship Structural Element

Empirical structureborne noise prediction methods are outlined in Petersson and Plunt 1980 and Plunt 1980. This approach provides an analysis of the vibration transmission of complex structures using mobility[3] (inverse of impedance) methods. The results of full-scale measurements and predictions are presented.

[2]Laboratory tests conducted by Noise Control Engineering, LTD in support of SBIR N98–092—Shipboard Airborne Noise Control Design and Diagnostic Tool, NAVSEA and NSWC-CD program.

[3]Structural impedance, Z, is equal to F/v; or velocity for a given force acting on the structure; mobility, M, is 1/Z.

Other algorithms for structureborne noise are based on stastical energy analysis (Lyon 1975; Lyon and DeJong 1995; Smith 1979). However, algorithms, which the authors propose to use in practice, have some difference from established SEA (Burroughs et al. 1997) for some key issues.

A whole ship structure or part of a ship is treated as a system containing subsystems (elements). An element is part of a hull restricted by adjacent intersecting elements. For example, part of a transverse bulkhead between the nearest decks and between the nearest longitudinal bulkheads is an element. In general, the order of the size of an element includes several frames (several meters).

In developing an 'energy flow balance' approach, testing was undertaken on a ship. Multiple noise and vibration measurements were taken with a tapping machine (pure L_{fsb} source), speakers (pure L_{ab} source), and diesel generator as realistic noise sources. These tests confirm an overwhelming contribution of structureborne noise for the majority of compartments of interest (Boroditsky et al. 1974). Broad band vibration measurements on all structural elements (excluding the mechanical source element) show a very small dependence on the location of the sensor. Given the test results, the assumption regarding a diffusive vibration field looks reasonable. Similar to the reverberation field in room acoustics, the vibration level of an element may be described with one figure in the frequency band of interest.

For a steady-state source, the vibrational energy going into a structural element is equal to the energy leaving. When the input power is known, it is possible to compile the energy balance equation for each structural element forming the compartments of interest.

Based on the similarity with the room acoustic calculation of reverberant field density and known mean free path between boundary collisions for 2-D media, it is possible to compile the following system of equations (Westphal 1957):

$$\left(\sum_{j=1}^{N} \alpha_{ij} + \beta_i\right) u_i - \sum_{j=1}^{N} \alpha_{ji} u_j = \sum_{k=1}^{k=K_i} W_{ik} \tag{7.21}$$

where $\alpha_{ij} = 2\,\tau_{ij} l_{ij} c_{ib}/\pi$ and $\alpha_{ji} = 2\,\tau_{ji} l_{ji} c_j/\pi$ are coupling loss factors; $\beta_i = 2\,\pi\,A_i\,\eta_i f$ is a damping loss; u_i and u_j are unknown energy densities for structural elements; K_i is the number of sources in the i element and W_{ik} *is the* source power.

In these formulas:

τ_{ij} and τ_{ji} – energy transmissibility coefficients calculated with formula Eq. (7.26)
l_{ij}, l_{ji} – the length of the common part of contour i-th and j-th structural element
c_i and c_j – phase bending wave speeds in the i-th and j-th elements
A_i – structural element area
η_i – loss factor of the i -th element
N – number of elements in the global system.

The speed of bending waves c_i and c_j is easily calculated if the mechanical parameters of plating are known (Cremer et al. 1988). For metal plates $c_i = 10^2 (t_i f)^{1/2}$, where t_i – plate thickness, m; f – frequency, Hz.

Bear in mind, these equations require sound power or vibration energy inputs for sources. This can often be a subject of frustration among engineers because sound pressure and acceleration levels are more commonly available and measured. Conversion to power is necessary via an impedance calculation or measurements. Experimental methods to determine impedances with a hammer or shaker can provide drive point impedance. In reality, one has to deal with distributed force excitation. Additionally, this conversion to power can be another potential source of inaccuracies. This problem may be overcome if reliable algorithms exist to convert vibration levels to power levels.

However, there is a more serious problem with the "source element." This problem is the assumption of a diffusive field for the source element. For example, a relatively small source is located at a random position on an element. Experimentally, it has been shown that vibration levels near different boundaries of the source element will be different, depending on this source location. The same is true for elements adjacent to the source elements. The difference may be on the order of tens of decibels. To overcome this, the authors used an approach developed by Spiridonov 1978.

Vibration levels on top of a foundation on a source element are used as input data. Furthermore, it is assumed that the vibration level on the base element boundary does not depend on adjacent element vibration levels. Based on experimental data, the vibration level decay from the foundation to the boundary of the base element can be calculated. This calculation takes into consideration the following main factors and assumptions:

- vibration level on the machinery foundation top plate is equal to the base element framing vibration near foundation;
- difference between plate and framing vibration depends only on the frequency ratio to the lowest resonance frequency of the plate between frames.
- decay of vibration is determined by the expanding direct wave front, and this front has a shape of co-focused ellipses where the eccentricity of ellipses depends on the foundation length/width ratio.

With such an approach, it is possible to find a vibration level on any side of the base element boundary. The vibration level (or density of vibration energy in terms of the SEA approach) may be different for different sides of the base element. As with airborne noise from a large source, there is a dependence on the shape of the foundation and its location on the base element. In any case, the density of energy on any side of the base element can be directly calculated. The energy density at each connection on the source element boundary, multiplied by the appropriate coupling loss factors, forms the forced side of the global energy matrix equation for elements adjacent to the source element. The energy balance equation for a 'source element' will be excluded from the global balance system of equations.

A similar approach is possible for secondary structureborne (SSB) noise calculation. Noise levels near all boundary elements forming an engine room may be calculated in accordance with the above-described algorithms. Noise spreading through openings is also counted. Then, the well-known transfer function L_p-L_a

(see Sect. 2.8) is used to determine vibration levels excited by airborne noise. The vibrational energy density of each engine room element is multiplied by an appropriate coupling loss factor to contribute to the forced side of adjacent elements. Again, it is assumed that there is no return energy flow from adjacent elements to engine room elements. Engine room elements are excluded from the energy global matrix.

In this case, the right-hand part exists only for elements adjacent to the source elements and is equal to:

$$W_{ik} = \propto_{ki} u_k \tag{7.22}$$

where u_k is the density of the energy on the contour of the source structural element. Solving this system of equations provides the energy density on each structural element. The vibration velocity of interest v^2_i is equal to u_i/m_i, where m_i is the surface mass for the *i-th* structural element.

The energy balance idea is used in statistical energy analysis (Lyon 1975). The reciprocity principle and mode energy exchange for finite size structures are the basis for compiling the energy balance system of equations. The 'source element' problem discussed above is the same as for the system of equations in Eq. 7.21.

The following empirical procedure was used to determine u_k. The analytical expression that the program uses to calculate vibration decay from the foundation to any side of the base element boundary is as follows:

$$\Delta L \text{ dB} = 10 \text{ Log}\left(\left(\frac{r_1 + r_2}{a + 2r_0}\right)\left(\frac{E(e_r)}{E(e_0)}\right)\right) \tag{7.23}$$

where r_1 and r_2 are the distances from the ends of the foundation to the middle of the joints forming the base element (see Fig. 7.4); a is the length of the foundation; r_0 is half of the minimum frame spacing; and E is the function shown in Fig. 7.5.

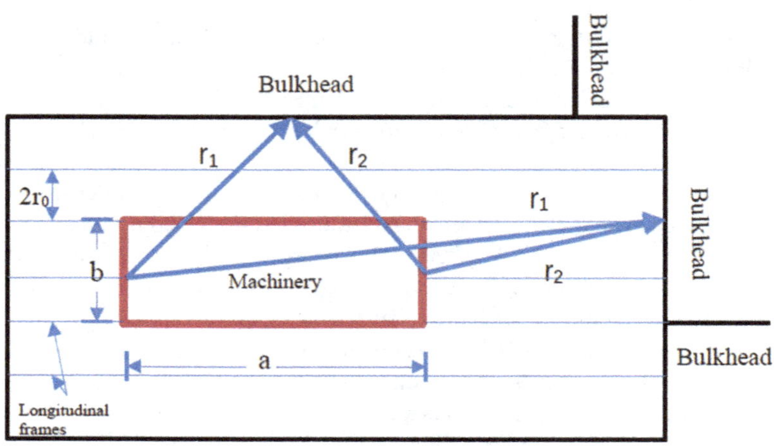

Fig. 7.4 Geometry for Eq. (7.23)

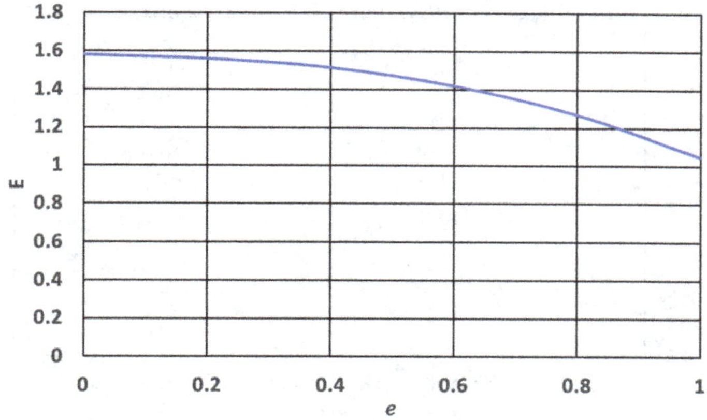

Fig. 7.5 Plot of the E function for Eq. (7.23)

The arguments of function, E, are defined by the following formulas.

$$e_0 = (a - b)(a + b + 4r_0)/(a + 2r_0)^2 \tag{7.24}$$

$$e_r = (a - b)(a + b + 4r_0)/(r_1 + r_2)^2 \tag{7.25}$$

These formulas show a dependence of vibration decay on the shape of the foundation and its location on the base element. In any case, the density of energy on any side of the base element can be directly calculated, and the equation for the base element will be excluded from the global system.

7.3.2 Energy Transmissibility Coefficient for Ship Structural Connections

Transmissibility coefficients are calculated analytically for L, T and cross shape structural junctions. The following equation is used for practical calculation of transmissibility coefficients:

$$\tau_{ij} = \tau_{ij0}\psi\left(t_j/t_i\right) \tag{7.26}$$

where τ_{ij0} – transmissibility coefficient for normal incident; $\Psi(t_j/t_i)$ – a function of thickness ratio j-element to i-element.

Coefficient τ_{ij} characterizes an energy transmitted from the i-th element to the j-th element through the connection; τ_{ji} describes an energy flow in the opposite direction. Note that $\tau_{ijO} = \tau_{jiO}$. Function $\Psi(t_j/t_i)$ takes into consideration that diffuse

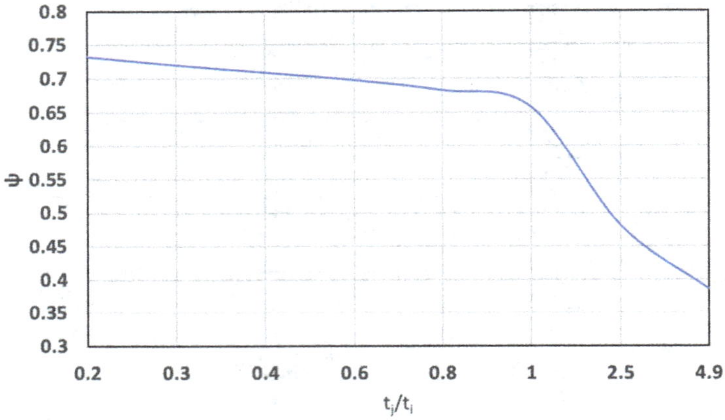

Fig. 7.6 Plot of function $\Psi(t_j/t_i)$

transmission from a thick element to a thin element is higher than that from a thin element to a thick element because of full reflection past the critical angle. As an approximation: $\Psi(t_j/t_i) \times \sqrt{t_j} = \Psi(t_i/t_j)\sqrt{\ t_i}$.

With an error not more than 1%, the function $\Psi(t_j/t_i)$ may be described by the following integral (Spiridonov 1978):

$$\psi\left(t_j/t_i\right) = \int_0^1 \sqrt{\left(1 - x^2\right)\left(1 - \left(x^2/b_1^2\right)\right)}\, dx \tag{7.27}$$

where $b_1^2 = t_j^2/t_i^2$

This function is shown graphically in Fig. 7.6.

Transmissibility for a normal incident depends on the thickness and physical constant of the element material (Cremer et al. 1988). For L, T, and X (or cross connections):

$$\tau_{ij0} = 2\beta_{ij} \Big/ \left[1 + \sum_{q=1}^{q=M} \left(\beta_{ij}\right)_q\right]^2 \tag{7.28}$$

where

$$\beta_{ij} = \left(\rho_j t_j/\rho_i t_i\right)\left(f_{cri}/f_{crj}\right)^{1.5}$$

and ρ, t, and f_{cr} are the material density, thickness and plate critical frequency of the i-th and j-th elements, respectively.

The sum in the denominator is for all structural elements adjacent to the i-th element from which the energy is flowing. The number of terms in this sum depends on the angle connection type: for cross connection, M = 3; for T-connection, M = 2; and for L-connection, M = 1.

One should keep in mind that the transmissibility coefficients discussed above were derived for plates that do not have framing (stiffeners and girders). This is not representative of a hull structure. Realistically, any deck-bulkhead connection has a plating connection; however, there is also a stiffener connection. The influence of framing on the transmissibility coefficient was not studied. Measurement shows that the plating vibration level is higher than the framing vibration by 3–7 dB, depending on the framing cross section size and shape and frequency. The mass of framing is 2–4 times lower than the mass of plating for most structural elements. This means that framing vibrational energy is approximately an order less than that for plating energy. This gives optimism that the plating connection transmissibility is applicable for vibration spreading analysis.

Note that for low frequencies, it is best to smear the beam-plate properties into an equivalent plate, having a higher surface mass, thickness, moment of inertia, and bending stiffness.

7.3.3 Loss Factor for Hull Structure

The material loss factor, η_i, is recognized from measurements (Beranek 1986; Fahy and Thompson 2015). For metals such as aluminum and steel, this value is on the order of thousands; for composites, it is hundreds. If the metal plate is covered with special damping materials, the loss factor may be on the order of one tenth.

The loss factor, η_t, for a single layer of unconstrained damping is (Beranek and Ver 1992):

$$\eta_t = \eta_1 \left(E_2/E_1 \right) h_2 \left(3 + 6h_2 + 4h_2^2\right) \tag{7.29}$$

where η_1 is the loss factor of the damping material, E is the Young's modulus of each layer – 1 for base plate and 2 for damping layer, and $h_2 = t_2/t_1$. Note that damping will change the bending stiffness and critical frequency for the composite. The bending stiffness, B_t, becomes:

$$B_t = \frac{E_1 t_1}{12(1-\nu_1^2)}\left[t_1^2 + \left(\frac{E_2(t_1+t_2)}{2(E_1+E_2)}\right)^2\right] + \frac{E_2 t_2}{12(1-\nu_2^2)}\left[t_2^2 + \left(\frac{E_1(t_1+t_2)}{2(E_1+E_2)}\right)^2\right] \tag{7.30}$$

where ν_ι is the Poisson's ratio for each layer.

The composite surface mass, m_t, is $m_t = \rho_1 t_1 + \rho_2 t_2$, and the critical frequency, $f_{c\ composite}$ Hz, is:

$$f_{c\ composite} = \left(c^2/2\pi\right)\sqrt{m_t/B_t} \tag{7.31}$$

The accuracy of the loss factor value is significant for the noise prediction process. Different references give significant discrepancies for a steel ship hull loss factor – from approximately 0.0001 to 0.02. For a diffusive vibration field on a structural element, the density of energy is inversely proportional to the loss factor.

The diversity of measured loss factors may be a result of different completion stages for the hull and methods of measurements. The section of the hull may be just plating and stiffeners or may have attached joiner panels, coatings, cables, piping, etc. Any additional items attached to the hull usually increase the measured loss factor (Favini and Spence 2014).

Two methods can be used to measure the loss factor of a ship structure. One is a reverberation method, in which a structure of interest is excited by a shaker. When excitation is cut off, the difference in vibration levels at two moments in time shows a slope of the vibration decay. Loss factors may be derived from these data (Beranek and Ver 1992). An advantage of this method is that it can be used over a wide range of frequencies, typically from 60 to 8000 Hz.

The most popular and more precise method is to derive the loss factor from the width of the resonance curve (Beranek and Ver 1992). This method is only possible to use for frequencies where a resonant response can be clearly determined, e.g., when the density of natural frequencies is relatively small. For ship structures, this limit is approximately 300 to 400 Hz.

The loss factor measured on ship structures the sum of three components: internal losses; *losses on the contour,* where energy is spreading through the boundary of the excited structure; and radiation losses. Below the critical frequency for the plating structure, the radiation loss is relatively small *and* may be neglected. Energy spreading through the boundary of the excited part of hull may be evaluated and separated from internal loss.

Vibrational energy decays exponentially with time. If the energy density at the "cut off" moment is u_O, the density at time t is $u = u_O\ e^{-\eta_m \omega t}$, where η_m is the measured loss factor. Over a unit of time, the structural element will lose energy $dU/dt = \eta_m \omega A u$, where U is the vibrational energy *of the structure and A is the* area of the structural element. Additionally, energy loss may be expressed as a sum of losses due to internal loss (subject of interest) and energy loss due to spreading through the boundaries $dU/dt = u(\eta_{in} A\omega + c_g\ \tau\ P/\pi)$, where η_{in} is the internal loss factor, c_g is the group speed of energy spreading, and τ *is the* energy transmissibility coefficient. It is assumed that a diffusive field was created on the measurement structure before the "shut off" moment.

From the equality of the right side of the equations in the previous paragraphs, the internal loss factor may be derived: $\eta_{in} = \eta_m - c_g\ \tau\ P/\pi\ A\ \omega$. The second term in this equation may be considered a "spreading loss factor".

Table 7.5 lists the mean values of the measured loss factor on an untreated steel vessel in addition to spreading and internal losses. These data show that the untreated steel structure internal loss factor is less dependent of frequency than that from field measurements and close to lab measurements.

Table 7.5 Measured damping coefficients

Octave band freq., Hz	Measured	Spreading loss	Internal loss
63	0.020	0.015	0.006
125	0.015	0.010	0.005
250	0.012	0.007	0.005
500	0.010	0.005	0.005
1000	0.009	0.004	0.005
2000	0.007	0.003	0.004
4000	0.005	0.002	0.003
8000	0.004	0.001	0.003

Table 7.6 Loss factors for various ship structures and subsystems

Structure/Substructures	Loss factor
Dry hull plate	0.004
Wet hull plate	0.015
Plate supporting machinery	0.01
Deck, covered by floating floors	0.03
Deck, covered by levelling compound	0.012
Bulkheads, lined with joiner panels	0.008

The above-described procedure to calculate the vibration level on the boundary of the "source" element is used for mechanical sources of noise and vibration. However, there are several cases when one can determine the average vibration levels of the source elements. For example, when part of the bottom plate is above the propulsor, it is considered a "source" element (when propulsor noise is of interest). Algorithms for cavitating propeller-induced hull vibration were developed (see Sect. 5.2.1) and used in noise prediction software. To calculate flow noise or ice impact noise, one can use part of the side shell crossing a water line in the bow section of a hull as a source element.

Solving the system in Eq. (7.21) provides vibration levels on each element. The first structureborne vibration levels are combined with the second structureborne vibration levels. Once the combined structureborne vibration level L_a is known, it is possible to use the transfer function L_a–L_p (see Sect. 7.5) to calculate radiated noise levels from each element in the rooms of interest. Combining noise levels from all elements forming the room of interest provides the overall received noise levels.

The transfer function mentioned above does not address the effectiveness of joiner panels and absorptive material layers. The influence of these on radiation and excitation by sound is treated in the same manner as transmission loss.

Furthermore, the above approach only considers bending wave transmission. Longitudinal waves may play an important role, as found by Kim et al. 1996. This component may increase noise levels by 10 dB at high frequencies and for compartments remote from the source.

Loss factors for various ship structures and subsystems are included in Table 7.6 (Plunt 1980).

7.4 Second Structureborne Noise

Structural vibration in the audio range may be excited not only through the engine connection to the foundation but also induced by airborne noise from a loud source. Calculations and measurements show that this phenomenon is important, especially when a source is effectively resiliently mounted. The isolation in this case reduces the first structureborne noise.

Plating response to sound excitation was developed in the 1960s (Smith 1962; Maidanik 1962) for finite size plates in reverberant sound fields. This issue was discussed in Chap. 2 as the transfer function $L_p - L_a$. For a noise prediction program, it is more practical to calculate the broadband vibration velocity level, L_{vi}, with the following equation:

$$L_{vi}\ \text{dB re } 5\times10^{-8}\,\frac{\text{m}}{\text{s}} = L_{pi} - 10\,\text{Log}\left(2\pi m_i f_{cri}/\rho c\right) + 10\,\text{Log}(\mu_i) - 20\,\text{Log}\left(f/f_{cri}\right) - \Delta_1 L_i - \Delta_2 L_i \tag{7.32}$$

where L_{pi} is the noise level near element i center; m_i is the surface mass of the plate; f_{cr} is the critical frequency of element i; μ_i is the ratio of power loss by acoustical radiation to total power loss, which is a sum of power dissipated in the plate and radiated sound power, $\mu_i = 2\sigma_i\rho c/(2\sigma_i\,\rho c + \eta_i\,\omega m_i)$; $\Delta_1 L_i$ is the effectiveness of noise absorption materials located on the source side of the element; and $\Delta_2 L_i$ *is* the effectiveness of the joiner panel on the source side of element i.

Each plating structure forming the source room may be considered a source element, whose vibration near the boundary is calculated with Eq. (7.32). In this case, Eqs. (7.21 and 7.22) can be used to calculate the secondary vibration levels of any element in the system.

7.5 Sound Radiation from Vibrating Ship Structures

Based on the algorithms described above, one can calculate the vibration levels of any deck or bulkhead forming a room of interest. The next step is to calculate the noise radiated from these structures into the compartment. The relationship between vibration, v, and radiated sound power, W, is controlled with the following equation:

$$W = \rho c \sigma A v^2 \tag{7.33}$$

where v^2 – vibration velocity squared, m^2/sec^4; A-area of a plate, m^2; σ – radiation efficiency; ρc- air characteristic impedance, kg/sec-m^2. If v^2 is known from previously discussed algorithms, then only σ needs to be determined. A great deal of effort was performed in 1960s to find a realistic value for σ for the ribbed ship structure. As noted by (Cremer et al. 1988), a flat plate is not an efficient noise radiator below the coincidence frequency, which contradicts measured data.

Maidanik (1963) analytically determined that a finite size ribbed plate can radiate noise and quantified the radiation efficiency.

When the vibration level of the plates of interest and radiation efficiency are known, the noise radiated by the i-th panel is calculated with the following equation:

$$L_i \text{ dB } 1\ \mu\text{Pa} = L_{vi} + 10\,\text{Log}(A_i/R) + 10\,\text{Log}(\sigma_i) - \Delta_1 L_i - \Delta_2 L_i \tag{7.34}$$

where L_{vi} is the velocity vibration level relative to $v_O = 5 \times 10^{-8}$ m/sec; A_i is the panel area, m^2; R is the full absorption in the compartment of interest, m^2; σ_i is the radiation efficiency; and $\Delta_1 L_i$ and $\Delta_2 L_i$ are the radiated noise reduction due to the absorptive layer and joiner panel, respectively.

The first structureborne vibration levels are combined with the second structureborne vibration levels, and this combined level is used for noise prediction.

The authors conducted experiments to validate Eq. (7.34). Figures 7.7, 7.8, and 7.9 show the calculated and measured transfer functions $L_a - L_p$. Maidanik's (1963) algorithm was used for the radiation efficiency with a correction considering the influence of the room's finite size influence (Spiridonov 1978). In Figs. 7.7, 7.8, and 7.9, the calculated data are the prediction values determined by a program developed by NCE (Fischer 2001).

The sound pressure level in the near-, mid-, and far-field from a panel can be related to the radiated power, W, as follows (Kaminsky et al. 1998):

$$\begin{aligned} &L_{p\,near} \text{ dB } 20\ \mu\text{Pa}@d = L_w \text{ dB re } 1\text{ pW} - 10\,\text{Log}(lw) - 1 \textit{ if } d < w/\pi \\ &L_{p\,mid} \text{ dB } 20\ \mu\text{Pa}@d = L_w\, dB\, re\, 1\, pW - 10\,\text{Log}(l) - 10\,\text{Log}(d) - 6 \textit{ if } w/\pi < d < l/\pi \\ &L_{p\,far} \text{ dB } 20\ \mu\text{Pa}@d = L_w\, dB\, re\, 1\, pW - 20\,\text{Log}(d) - 11 \textit{ if } d \geq l/\pi \end{aligned} \tag{7.35}$$

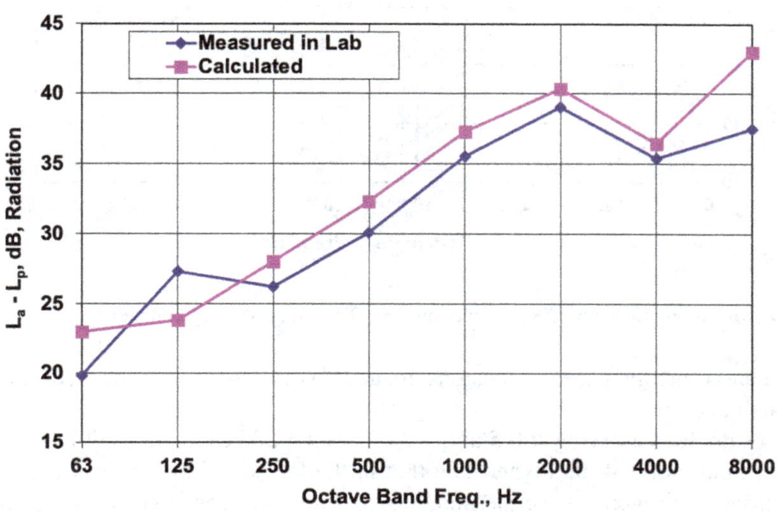

Fig. 7.7 Radiation efficiency transfer function for a 3 mm (1/8″) aluminum bulkhead

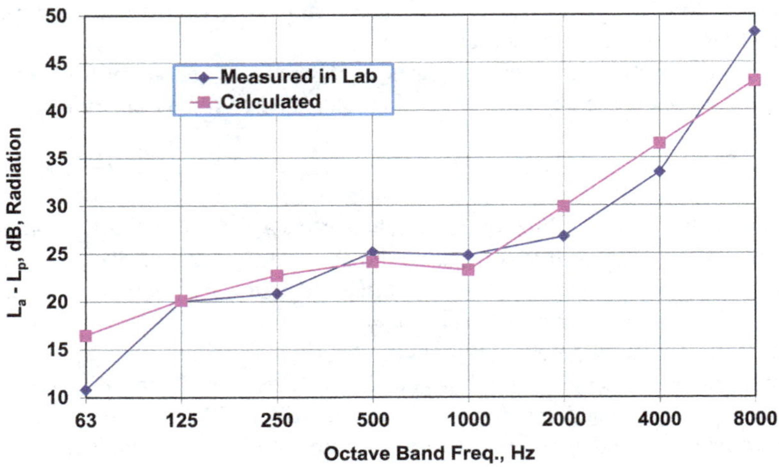

Fig. 7.8 Radiation efficiency transfer function for a 9.5 mm (3/8″) aluminum bulkhead

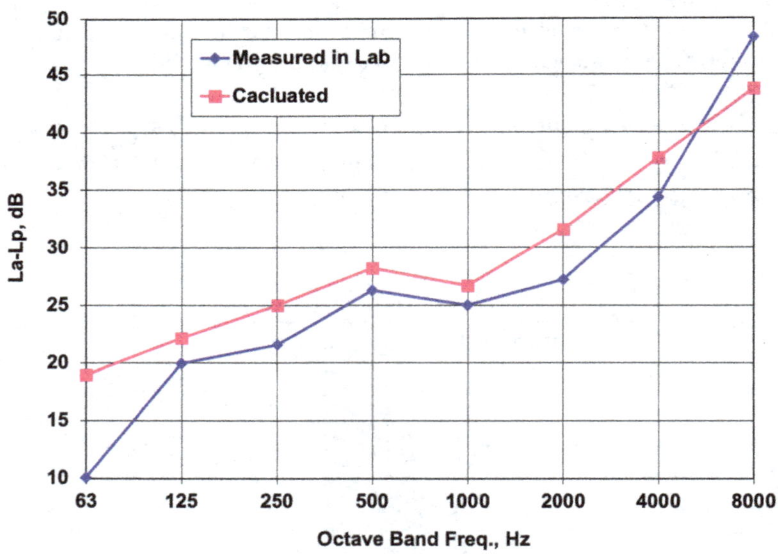

Fig. 7.9 Radiation efficiency transfer function for a 9.5 mm (3/8″) steel bulkhead

where l and w are the panel dimensions, m, and d is the distance of the measurement from the panel, m.

For example, measuring at 0.5 m from a 2 × 3 m bulkhead will result in a level in the near field that is 4 dB higher than that in the far field. Thus, measuring near a solid surface will increase the measured level. If the noise criteria specify measurement at the head of a bunk, this increase due to the near field should be considered in the modeling. Furthermore, if the head of the bunk is in a corner, the increase will be another 6 dB.

Table 7.7 Increase in noise by measuring 1 m from the surface or corner

	Freq., Hz.						
Increase in noise, dB	10	12	16	20	25	31.5	40
1 m from corner	11	11	11	10	10	9	8
1 m from surface	6	6	6	5	5	5	4

Measuring noise 1 m from a large surface is predicted to increase the low-frequency free field noise (Agerkvist and Jacobsen 1993) by the values shown in Table 7.7. Thus, measuring at pillow positions in bunks or at desk positions less than one meter will have higher noise levels in the lower bands than those measured in the center of the room. Again, this consideration should be discussed with the generator of the acoustic requirements. The noise environment seen by the crew in these locations is 'realistic', and therefore, the treatment should be directed at ensuring very low noise at the bunk head and desks, with the attendant penalty in terms of space and weight. If bunks are located perpendicular to the bulkhead, the pillow end should always be away from the bulkhead. One potential answer may be active noise cancellation, which is known to work at low frequencies. Note that the low-frequency increase near a surface will unlikely increase the A-weighted noise level; however, if the overall noise is dominated by low frequencies, it may.

7.6 HVAC Noise

Heating, Ventilation, and Air Conditioning (HVAC) noise prediction methods include both empirical (spreadsheet type) and 3-D computer-aided modeling. Empirical methods are provided in the SNAME Design Guide (Fischer et al. 1983) and the American Society of Heating, Refrigerating and Air-Conditioning Engineers (ASHRAE) (ASHRAE 2010; Reynolds and Bledsoe 1991). CAD programs are available from fan and HVAC system vendors (Kandekar et al. 2019), Siemens, AUTOCAD, and AcousticCalc to name a few.

The noise prediction methodology is generally one wherein each element from the fan through the ducted HVAC system to the vent terminal produces and/or attenuates a certain sound power level. From there, one simply adds the sequential set of sound power or sound attenuation levels to come up with a resulting sound power level emanating from the diffuser (Bahtiarian 2014).

The one component not normally included in general software is the 'break in' and 'break out' noise. Ducting passing through a noisy machinery space may have sufficient 'break in' noise to flank the noise produced by the fan and regeneration elements downstream. Ducting passing through noise-sensitive compartments may add to the received level in the compartment by 'break out' noise. Both of these components are addressed in the *DesignerNOISE*® software covered in the following chapter.

7.7 3-D Noise Prediction and Analysis Software

Various 3-D acoustic models are available commercially. Like FEA, many SEA codes make an attempt to be a completely universal tool that is applicable for cars, sea-going vessels, rockets, etc. Like any universal tool, there are inherit tradeoffs that must be made, particularly when fundamental parameters such as necessary element size are still debated in the SEA community. The authors are not familiar with the applicability of any commercial codes, other than *DesignerNOISE®*, to shipboard noise.

Prediction software is available from regulatory agencies such as the Indian Registry (IRClass) and commercial vendors of Multiphysics Software such as COMSOL, ANSYS, LMS-SEADS (SEA software), Vibro-Acoustics from ESI Group, Siemens Acoustic Simulation Software, SEAM, Adaptive Corporation – Wave6, and CDH Vibro-Acoustic Optimization Program (Third-Party Product MathWorks). Some shipyards, such as Mitsubishi, have developed their own SEA codes. Other published papers on SEA modeling include DeJong 2003; Tso and Hansen 1997; Jensen 1976; Elmallawany 1985; and Hynna et al. 1995. An example of a design study via a hybrid SEA approach is given by Vladimir et al., n. d.

The authors pioneered a specialized tool at the behest of the U.S. Navy. This program is discussed in the following chapter. *DesignerNOISE®* software[4] became commercially available in 2005. The algorithms used in this software are based on a 3-dimensional acoustical modeling process. A Graphical User Interface (GUI) helps the user enter all necessary input data to develop the model. During one run session, the program calculates the octave band, overall and A-weighted noise levels in all rooms of interest; vibration levels of all structural elements of the model (decks, bulkheads); and airborne and structureborne noise contribution for every room of interest. The contribution from every source over every path is evaluated. All of this information gives the user the opportunity to determine the physical reasons for excessive noise levels and to find the optimal method for noise reduction.

DesignerNOISE® algorithms make use of the sound pressure and acceleration inputs through the implementation of the following procedures. For secondary structureborne noise, the airborne noise level at each element is computed using spreading and reverberation methods. The calculated spreading losses are due to source size and take into account the near field effects of the source level measurement (if any) and elliptical spreading that would occur from a 3-D source. The computed L_p at each element is then converted into a vibration level on the element through the use of an acceptance transfer function ($L_p - L_a$), which is automatically calculated by the program based on the element particulars. Once the vibration levels are known for each element of the source room, their energy densities can be computed. These energy densities are multiplied with the appropriate coupling loss factors and are used as the forcing side of the global energy matrix equation for SSB calculations.

[4]Developed under a Small Business Innovative Research (SBIR) for the U.S. Navy and available commercially from Noise Control Engineering, LLC, www.noise-control.com

As noted above, the levels at each intersection with the "source element" (i.e., the element on which the source imparts vibration) are calculated by applying the appropriate spreading losses to the source levels.

Use of these algorithms not only allows the user to input convenient source data resulting from direct measurements but can also account for positional differences of the source in the machinery space. These factors enhance the usability and accuracy of the program.

7.7.1 Noise Prediction with a Focus on Ships

DesignerNOISE® was created specifically to analyze noise on surface ships. Many assumptions have been made that fit with this narrow focus: thin stiffened plate construction, large element size (determined by intersecting bulkheads), dominant transmission paths, source entry methods, etc. However, because of these assumptions, the code is very efficient, and the solution processor can be run in a matter of seconds even for large models. This facilitates rapid treatment trade-off studies that can be used to find the most efficient noise control design for any particular situation.

In addition, because *DesignerNOISE®* is ship specific. Treatment and construction materials, as well as the acoustical effects related to these materials, are focused on those important and common to the marine industry. This increases the ease with which the designer can create models and apply treatments, without spending time investigating treatments that may not be applicable.

7.7.2 Noise Prediction Software for Ships

Together, the above-discussed equations may be used for the noise prediction procedure. However, the process would only be practical if the entry process were straightforward. The software has a Graphical User Interface (GUI), which can depict a ship model and all relevant components needed for the analysis. Two companies, Proteus Engineering and Noise Control Engineering, have developed software called *DesignerNOISE®* under a US Navy grant.[5]

Figure 7.10 demonstrates an example of a ship cut-away view of a *DesignerNOISE®* model.

Model creation is typically started with the export of hull information from a 3-D naval architecture program. The hull form is usually available at the early stage of a project as a general arrangement drawing. The user can enter horizontal and vertical planes, whose coordinates correspond to decks, transverse and longitudinal

[5] SBIR Phase II, Topic No. 98–02, Advanced Ship Acoustics, Office of Naval Research and NAVEA

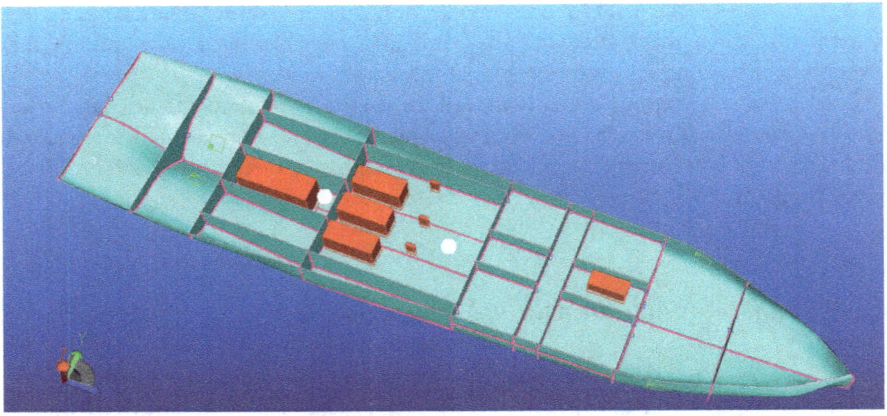

Fig. 7.10 Three-dimensional (3-D) ship model from *DesignerNOISE®*

bulkheads taken from the general arrangement drawings. The model development is a two-step procedure. The first is the compilation of a 'topology' model. The topology model gives only geometric parameters. The next step is a project model, where a user can enter the physical parameters of hull and joiner structures. In this way, the user creates the following entities, which are necessary to run noise prediction programs:

– whole model ("the object")
– closed volumes ("compartments")
– plates, which forms compartments ("elements").
– connections, which are boundaries between each element

All necessary geometric parameters become known for noise prediction: length of connections, area and shape of elements, and volume and shape of compartments. A user can enter element material, element thickness, parameters of coating, insulation materials, joiner panel parameters and other information influencing the expected noise level. Each entity can be selected on a computer screen to receive all necessary information. Figure 7.11 demonstrates a selected element highlighted in yellow. A user can see and change particulars of this element, e.g., wet[6] or dry loading. Similarly, a compartment can be selected, and its acoustic properties, such as absorptive treatments, changed (see Fig. 7.12).

The fifth entity is the noise and vibration source levels. The selected mechanical source is shown in yellow in Fig. 7.13. In accordance with the above-discussed algorithms, a user enters coordinates and sizes of foundation, source sizes (length, width and height) elevation about source element and noise and vibration levels. The program calculates noise levels near all elements in the source room to obtain input data for the secondary structureborne noise calculation.

[6] If 'wet' loaded, the radiation efficiency includes the impact of water loading.

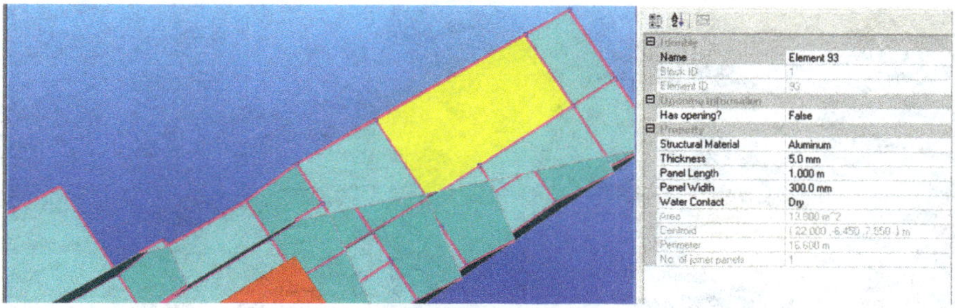

Fig. 7.11 Graphic showing 'selected' structural element (in yellow)

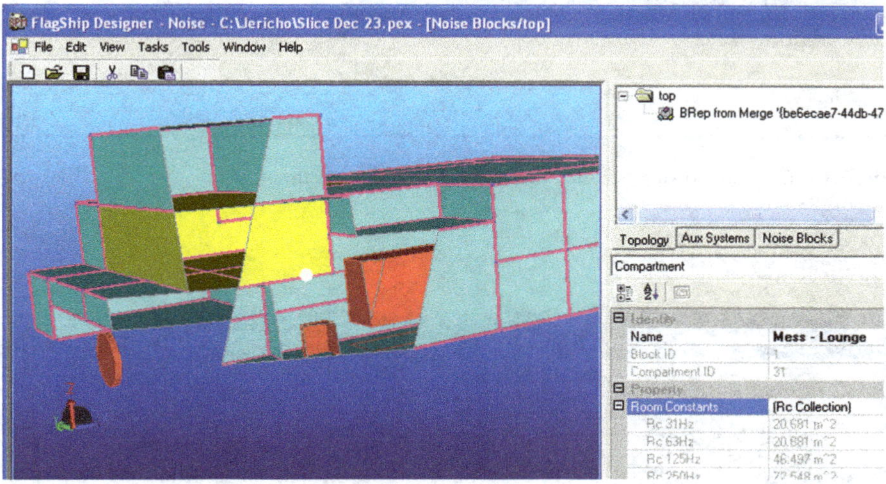

Fig. 7.12 Graphic showing 'selected' compartment (in yellow)

When the model is completed, the operator can run the program. For a several thousand ton ship, it takes seconds. The results may be presented in several formats.

Figure 7.14 depicts octave band received noise levels from one source in one selected compartment – Mess-Lounge. Three components from this particular source are shown: airborne noise, structureborne noise, and second structure borne noise. These are compared to the criteria to determine if there is an excess and over what frequency range it might occur.

Figure 7.15 shows the calculated octave band vibration levels of three adjacent selected elements from one source. The source, the starboard propulsion engine, and the selected elements of interest are shown in Fig. 7.16. One can see how the vibration level decreases farther away from a source. A 5–10 dB decay at each connection is in line with the experimental data.

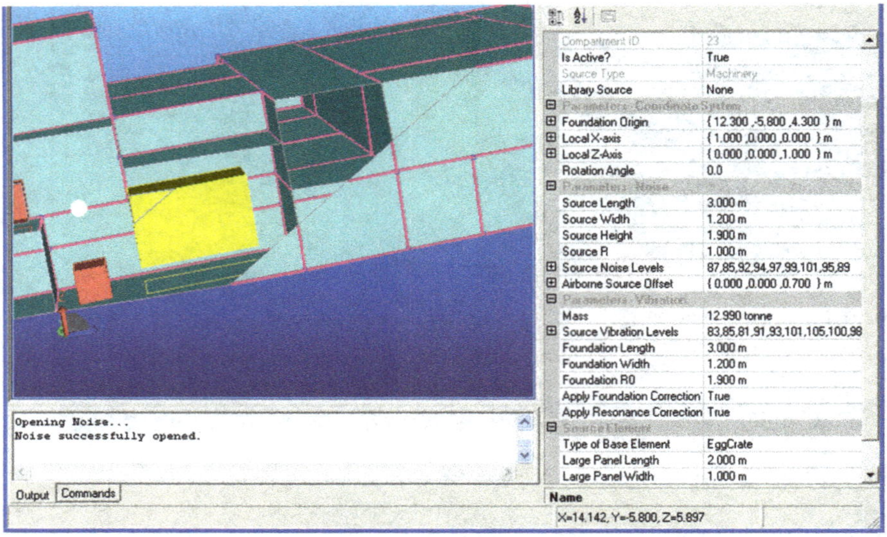

Fig. 7.13 Graphic showing 'selected' mechanical source (in yellow)

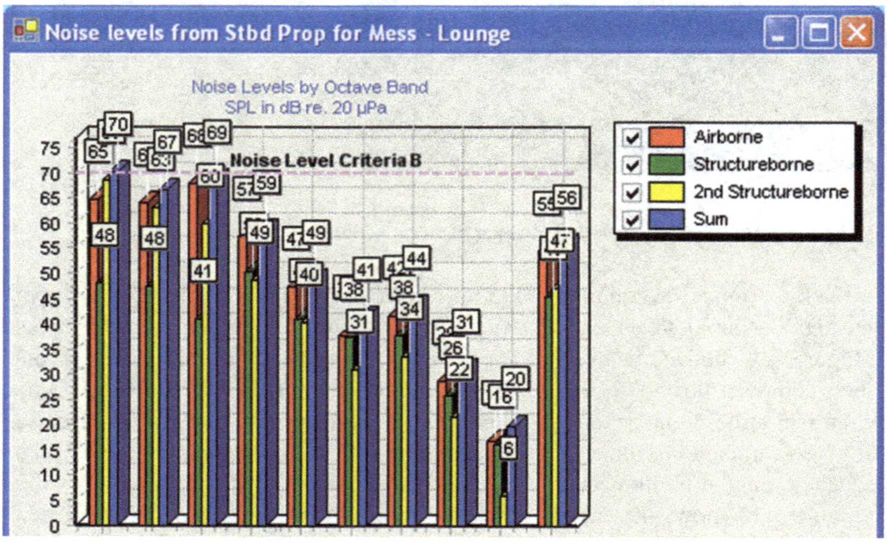

Fig. 7.14 Calculated noise levels in compartment

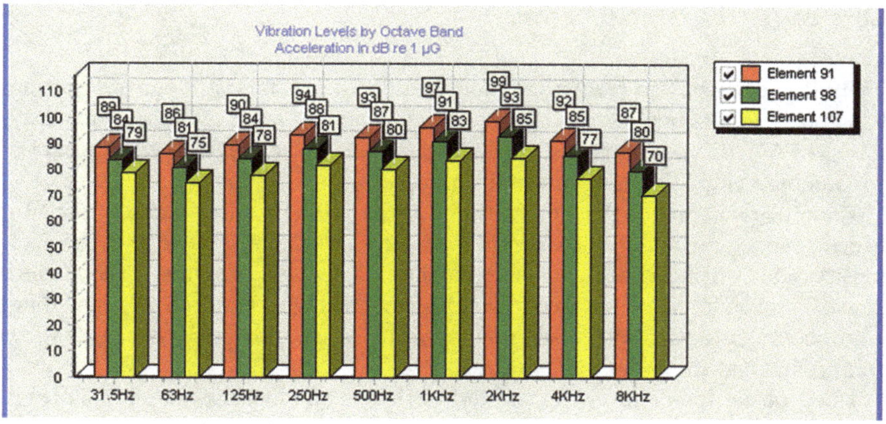

Fig. 7.15 Calculated vibration level of the selected element (re 1 μG)

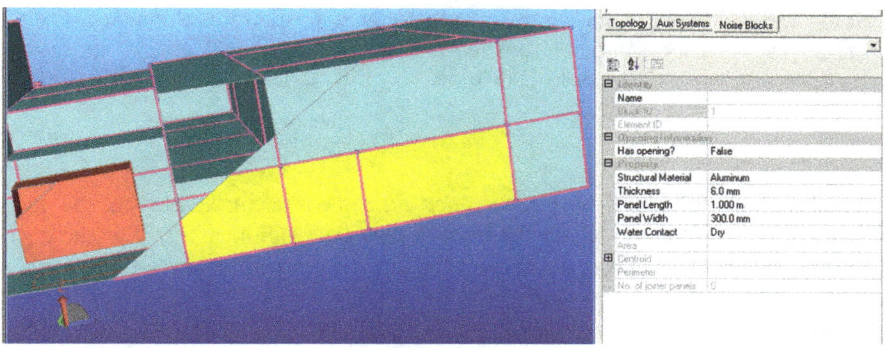

Fig. 7.16 Selected structural multiple elements (in yellow)

The detailed results of all calculations can be shown in an Excel report format. The report shows the overall dB(A) levels in each compartment of interest from each source and the contribution of each physical component: airborne, structureborne and second structureborne noise.

Other related information provided includes radiation efficiency of elements, damping loss factors for all elements, and transmission loss for partitions between engine rooms and compartments of interest. This information, along with noise and vibration octave band levels, helps to find the most noteworthy source, to determine critical components (airborne or structureborne), and to find the most acoustically active element in any compartment of interest. Based on this information, one can choose the optimal noise treatment. When the model is completed, many iterations of noise treatment may be analyzed over a short period of time. This facilitates what-if studies or investigation of optimal treatments.

7.8 In-Air Far-Field Noise

Often, it is necessary to predict the in-air noise while a ship is in port or cruising along a river (Badino et al. 2015). Various criteria are used for community noise (EPA n.d.), consideration of annoyance – particularly to low frequencies (Moorehouse et al. 2005), and sleep disturbance.

The process to predict the airborne noise in the far field utilizes the source-path-receiver model already discussed in this chapter. The major difference is the path transmission, which is open, so the reverberant component is ignored. The attenuation of sound outdoors is covered in ISO Standard 9613-2 – Part 2 (2024), including atmospheric absorption losses and environmental excess losses from transmission over ground and snow and through trees.

The acoustic source levels for gas turbine intake and exhaust, diesel intake and exhaust, and ventilation fans, which are the primary sources affecting far-field noise, are covered by the SNAME Design Guide and its Supplement (Fischer et al. 1983; Fischer and Boroditsky 2001).

Another potential contributor is the noise radiated by a gas turbine exhaust duct wall itself, rather than the noise transmitted into the open after it is attenuated by any silencer. The radiated far field from the GT exhaust duct wall due to the inside exhaust power level (Giampaolo 2006)

$$L_w = L_{w_i} - (TL + 6) + 10\,\mathrm{Log}(S/A) \tag{7.33}$$

where L_{wi} is the power level inside the duct; *TL* is the transmission loss of the duct wall (assume that the non-silenced section will radiate significantly); *S* is the duct exterior surface area, m^2; *A* is the duct cross-sectional area, m^2; and + 6 dB accounts for the free field.

Once the source level is known, the received level at some distance can be computed as shown in Eq. 7.32 with the reverberant term deleted and a directivity term, D_c, added, as shown below:

$$L_p(r)\ \mathrm{dB\ re\ 20\ \mu Pa} = L_w + 10\,\mathrm{Log}\left(\left(1/4\pi r^2\right)\right) + D_c - D_a \tag{7.34}$$

where D_c is a directivity term, which is equal to 0 for an omnidirectional source radiating into 4π steradians (or less if directional for part of the 4π steradians; i.e., +3 dB if radiating into a hemisphere or + 6 dB if radiating into a quarter of a sphere, and D_a is the atmospheric attenuation. The above equations can also be used to predict the airborne noise impacting deck work stations on board the ship.

Equation 7.34 assumes spherical spreading, which is accurate enough for most of the sources encountered. Atmospheric attenuation covers atmospheric absorption, ground effects, and barrier attenuation. ISO standards provide frequency-dependent values for these conditions. There is another condition encountered in the marine field. This is the effect of propagation over water (Van Renterghem et al. 2014 and Bolin and Boue 2009) and in fog (Wiener 1961). These papers discuss the positive and negative effects of transmission over ocean waters. The propagation losses for

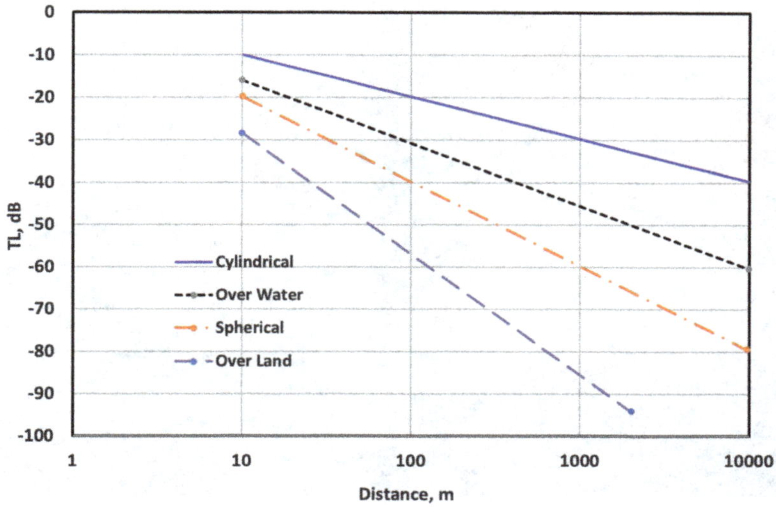

Fig. 7.17 Propagation losses for in-air transmission

spherical and cylindrical and those expected over land and water are shown in Fig. 7.17 (Pettit and Fischer 2007).

Propagation over water is not well understood. ISO 9613 Acoustics -Attenuation of sound during propagation outdoors, Part 2: General method of calculation does not address propagation over water. One significant factor, given the temperature change with elevation over water, is the existence of inversion layers. These layers can trap noise and increase the variability of noise transmission.

The US military has a standard used to determine aural non-detectability - Mil-Std-1474. Using this standard, a power boat with a source level of 103 dB re 20 μPa at 1 m is non-detectable at 5 n-mi over land.

Badino et al. 2015 has more information on a hybrid experimental-computational approach to predicting the impact of ship noise emissions dockside and a possible new measurement procedure for "an effective characterization of the ship source." One other factor may be the received noise indoors due to a vessel operating dockside. The noise reduction of a typical house wall (windows closed) is on the order of 8 dB at low frequencies to 16 dB at higher frequencies.

CAD software can be used successfully to predict the contribution of external sources to on-deck noise or internal compartments aboard a ship. One such program, already discussed, *DesignerNOISE*®, uses 2- and 3-dimensional sources, architectural algorithms and SEA to follow acoustical paths from external sources to on-deck receivers or compartments within the vessel (Favini and Spence 2014). Three-dimensional sources have both airborne and structureborne components. One study is illustrated in Fig. 7.18 for an air cushion vessel. Here, the sources are shown in red, including propulsion propellers, gas turbine intake and exhaust, and lift fans. The receiver is the Pilothouse (forward compartments). The measured and

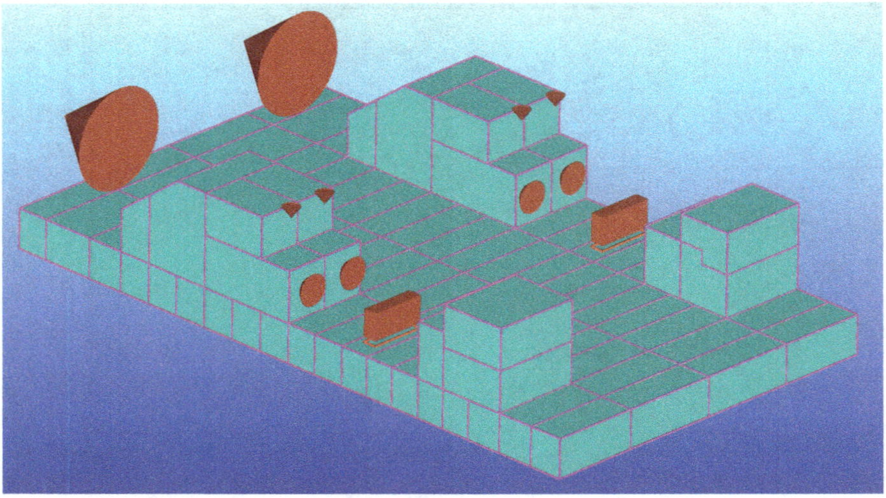

Fig. 7.18 Air cushion vessel model from *DesignerNOISE®*

predicted levels in the Pilothouse are within 1 dB at 91 dB(A).The measured and predicted vibration levels on the aft bulkhead of the Pilothouse from all these airborne and structureborne sources are within +3 to −7 dB.

Another example is show in Fig. 7.19, which shows predicted versus measured noise in Chaplain's compartment, which is below the launch area for F-18 aircraft. The noise levels are just over 101 dB(A).

7.9 Reverse Engineering

Sometimes it may be necessary to conduct the noise/vibration analyses 'in-reverse'. For instance, this would apply if one is trying to determine the allowable airborne and structureborne source level given an airborne noise limit. This method can also be used to derive a 'budget' for various machinery and systems.

This 'reverse engineering' uses an acoustic model of the entire ship. Then, a transfer function is developed from the 'budget' or 'assumed' airborne and structureborne levels and the resulting noise in a compartment of concern. If the resulting noise is over the criteria, the assumed or budget levels must be reduced accordingly. If the resulting noise levels are well under the criteria, the assumed or budget levels can be increased accordingly.

Basically, this entails working 3-D CAD acoustic models of the ship's structure 'backward' from the specified noise or vibration hull vibration to determine the 'allowable' source levels for the machinery in question. It may be necessary to use both FEA and SEA models depending on the frequency range of concern. The FEA

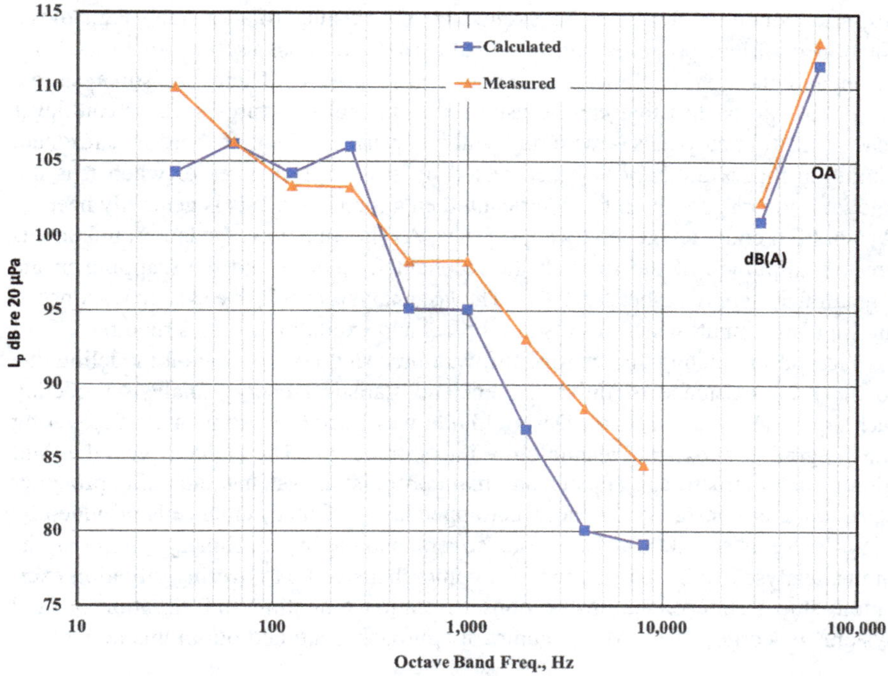

Fig. 7.19 Measured v predicted noise for aircraft launch impact on internal compartment

model is typically used for frequencies below 250 Hz, and the SEA model is typically used for higher frequencies. In this process, transfer functions are developed from the receiver back to the source, and the results should be acceptable w levels.

Another method of reverse engineering, if the radiated power, W_{spec}, is specified, is to compute the allowed hull vibration level, v_{max}:

$$v_{max}^2(f) = W_{spec}(f) / (\sigma_{rad}\ \rho c\ A_{tot}) \tag{7.35}$$

where A_{tot} is the total radiating area of the hull.

7.10 Vibration Modeling

The vibration levels of the ship's structure depend on the magnitude of the excitation forces, the flexibility of the ship's structure – locally or globally, the dynamic response at different frequencies, and the damping loss factors.

Whether habitability, machinery, or structural vibration is of interest, there are generally two controlling factors of concern. The first to address is whether there is a resonant condition that may be excited by an existing rotation rate, e.g., diesel rotation or firing, propeller shaft or blade rate. The most useful tool to access a resonant condition is finite element analysis software. There are some simple empirical approaches that can be used for elemental structures such as cantilever masts and framed plates – with and without water loading. If there is a resonant condition that is unlikely to be corrected by structural changes or when it is not known if the response will be significant, then a forced analysis is generally needed. Again, FEA codes are generally appropriate. As shown below, the largest unknowns are the amplitude of the force at the resonant frequency and the damping in the structural system (Grushetsky 2007). The last analysis checks the system response to an impulsive input, which results in a broadband excitation of the structure.

Vibration modeling and control software preceded noise vibration modeling due to the prior existence of finite element and statistical energy analysis modeling techniques. A Ship Vibration Design Guide was issued by Noonan 1989 covering the integration of existing technologies into ship design. The Guide covers the 'hull girder, major structural elements, and main propulsion systems, including propeller selection, stern design and underwater appendages.' Other guidance is provided by Lloyd's Register (2006). This note covers habitability/machinery vibration; its limits, analyses, and measurements. A comprehensive text covering vibration excitation, ship response, shafting response, isolation mounting and vibration in fluid systems is Veritec 1985. Many empirical approaches are laid out in this text.

7.10.1 Habitability Vibration Modeling

Typically, habitability vibration modeling includes two steps: analysis of dynamic forces exciting the hull and response of the hull and its substructures. The most basic forces of interest are those generated by propulsion machinery, by propulsors, and by shafts. Diesel engines and other reciprocating machinery may develop forces and moments, which are sufficient to create excessive ship structural or habitability vibration. These forces/moments may excite the machinery itself, the ship foundation on which it is attached, and the hull girder or other structure within the vessel. The force amplitude can be derived if the dynamic stiffness of the foundation and vibration amplitude are known.

Propulsors generate two types of fluctuating forces. These are hull pressure forces and bearing forces. The hull pressure originates from hydro-acoustic pressure variations caused by the passage of a propeller through non-uniform flow or wake. These hull pressure forces are affected by propeller-hull clearance in the vertical and horizontal directions, by blade loading (thrust), and by changes in the local pressure field around the blades. The occurrence of cavitation significantly increases the hull pressure forces. The frequencies of propeller impact are shaft rate, blade rate and 2x and 3x multiples of the blade rate. Hull pressures may be calculated during the design stage based on parameters such as propeller rpm, clearance, blade form, hull thickness and framing. The state-of-the-art propeller pressure forces should be below 2 kPa (van Terwisga et al. 2007).[7]

The analysis of the hull response can be segregated into two groups: global response and local response. For the global response, one can consider the hull as a beam (of varying stiffness) floating in water and vibrating in a 'free-free' mode (Veritec 1985). Propulsors and engine excitation create moments and forces that can induce vertical, horizontal transverse, longitudinal and torsional vibration of a hull, acting as a whole beam. If the excitation frequency from a particular force is close to or coincides with a natural frequency of any of these modes, then the whole-body vibration may be significant. For vertical and horizontal vibration, the first 2 to 3 modes of vibration are of greatest interest. For hull torsional vibration, only the first mode (one-node) is important. The torsional mode may be coupled with bending modes. The lowest natural frequency of the whole body for commercial ships is typically below 10 Hz.

Local structures may exhibit a vibration response independent of the 'whole' body response. These substructures, such as decks, machinery platforms, or the superstructure, of sufficient mass and flexibility need also be considered in the vibration analysis process. These substructures have their own natural frequencies, which again should not coincide with primary excitation frequencies. The natural frequencies of these substructures are usually higher than that of the whole hull. The frequency depends on the structural stiffness, mass distribution, and method of attachment of the substructure to the main structure (boundary condition).

The normal practice is to calculate a few natural frequencies of the hull once information about the hull structure and weight distribution becomes available (including mass loading by water – DND 1978). The main goal of this calculation is to check how close the excitation frequencies are to the computed natural hull frequencies. As a design rule of thumb, forcing frequencies with the machinery and propulsor at maximum continuous speed should differ by 20–30% from the hull mode natural frequencies. This ratio depends on the sophistication and accuracy of the hull modeling process.

The amplitude of the response is critically dependent on using the correct damping coefficients (Klinge 1995). Furthermore, if the damping loss factor is off by a factor of 5, the error at a resonant response can be wrong by 14 dB (Spence and Dempsey 2010). Different areas of the vessel will be subject to different damping coefficients. The total damping is a combination of structural damping, hydrodynamic damping, and internal components such as cargo, insulation and joiner treatments (Favini and Spence 2014), which generally cannot be ignored.

Furthermore, the response of the hull will vary with the load/ballast condition and weight distribution along the hull. In a ballast condition, the vibration amplitude may

[7] Lloyd's Guidance Notes suggest a reasonable probability of acceptable vibration for the following 'typical' hull surface pressures: Cruise ship: 1–2 kPa; RoRo Ferry: 2–4 kPa; Container and fast cargo ships: 3–6 kPA; and Slow bulk carriers: 4–7 kPa.

be higher or lower than when lightly loaded. An increase in vibration in shallow water has also been observed.

There are three approaches to analyze the response of a hull or substructures. The most accurate and time-consuming method is to build a finite element model (FEM).[8] Within FEM, one can further choose to perform a 'coarse' mesh model or a 'fine' mesh model. The former is more appropriate for a 'whole' body analysis; the latter is more appropriate for vibration analysis of a 'local' structure. The second approach would involve an empirical analysis. Simplified methods exist to determine the vibrations of free-free beams, framed panels, cantilever structures, etc. These methods are obviously less accurate than those using modern FEA techniques. The final method is to use historical data from a close prototype vessel and account for differences in excitation frequencies and machinery. This type of scaling is also risky but can be used as a first cut as to likely problem areas.

Analytical or empirical methods should be used to predict the lowest natural frequencies of the superstructure and mast. These analyses may be performed using FEA of the stack, mast, and exhaust systems (intakes/uptakes). The predicted system natural frequencies should be compared to propulsion excitation frequencies and hull natural frequencies to ensure resonance avoidance for all operating conditions.

In addition to resonance avoidance with hull modes, potential foundation resonances should be avoided. Free edges of foundation top plates have been known to cause excessive structureborne noise (Fischer 1988).

7.10.2 Machinery and Equipment Vibration Modeling

Tracing the internal forces and moments caused by operating machinery to the vibration levels on the feet of the equipment is best left to the manufacturers or lab measurements. However, it is important to understand, at some general level, the importance of these components. This helps with the determination of excitation frequencies and whether longitudinal, transverse, or vertical excitation is significant. Another component is the torsional vibration in shafting. Jenze 1997, provides 'some general comments regarding the vibration characteristics of modern ship machinery installations.' Understanding how machinery vibrates due to internal sources also assists with diagnostic type testing and/or understanding malfunctions.

In regard to specialized equipment, such as the need for propulsion machinery in icebreakers, it is important to know if these changes may have a positive or negative influence on the machine's acoustic source level (DNV 2010). In this case, the coupling of torsional vibrations to the ship has a significant influence.

[8] FEM can also be utilized for low frequency airborne noise predictions (Xu, 2017)

The US Coast Guard has a Design Standard for Evaluation of Ship Propulsion Machinery Vibration, 1999. This document has some useful factors to be applied to machinery systems under different ship operating conditions, such as torsional and longitudinal calculations in hard turns with multiple shafts open stern (3× to the expected broadband 1–1000 Hz). It allows broadband vibration limits for thrusters of 18 mm/s rms with intermittent maneuvering and 10 mm/s for continuous use. Cardan shafts, which have a large 2× rotation rate component, are allowed a lateral vibration limit of 10 mm/s, 1–1000 Hz.

It is always useful to be conservative in estimating equipment vibration levels – error on the high side. One reason for this is the variability in vibration levels. One study showed that the standard deviation for non-propulsion machinery ranged from 4 to 8 dB (McIntosh 2014).

7.10.3 Piping System Vibration

DNV provides a comprehensive reference for the structural analysis of piping systems (DNV 2008). DNV also has vibration class rules for allowable pipe vibration (2011). The 2008 document addresses other codes (including ASME, DNV Offshore Standards, etc.), flexibility analysis, thermal expansion, wall thickness, and FEA methods. It also includes examples. It is estimated that between 10 and 15% of piping failures are due to vibration-induced fatigue or excessive vibrations. As with other mechanical systems, the excitation is due to connections to the pump equipment – excitation at reciprocating or rotating vane pass frequency or pulsation; flow-induced vibration; and high pressure drops at valves. Cavitation may be a concern for a system that is not correctly designed and installed. More than once, a pump has been installed backward to reverse inflow and outflow.

Also note that if a fire main is to be run dead-heading, the entire system should be carefully isolation mounted.

There are a multitude of software programs (CAD or CAE) to assist in the prediction of piping vibration. CAESAR Pipe Stress Analysis is often used on US Navy projects. Others include FEA software or software from CAEPIPE, AUTOPIPE, SIEMENS, MSC, EDI, and Solidworks.

References

Agerkvist, F.T., Jacobsen, F.: Sound power determination in reverberation rooms at low frequencies. J. Sound Vib. **166**(1), 179–190 (1993)

ASHRAE Handbook, HVAC Applications, Chapter 48: Noise and Vibration Control, Atlanta, GA (2010)

Badino, A., Borelli, D., et al.: Airborne noise emissions from ships: experimental characterization of the source and propagation over land. Appl. Acoust. **104**, 158–171 (2015)

Bahtiarian, M.: HVAC noise prediction and treatments for ships. Noise-Con 2014, Ft. Lauderdale FL, Sept. 2014

Beranek, L.: Acoustics. American Institute of Physics, NY (1986)

Beranek, L.L., Ver, I.L. (eds.): Noise and Vibration Control Engineering – Principles and Applications. Wiley (1992)

Bies, D.A., Hansen, C.H.: Engineering Noise Control, Theory and Practice, 4th edn. CRC Press, Boca Raton (2009)

Blanchet, D., Caillet, A.: Full frequency noise and vibration control onboards ships. In: 21st International Congress on Sound and Vibration, Beijing, China, July 2014

Bogolepov, I.: Sound insulation on ships, Ch.10 in Handbook "Ship acoustics", Sudostroenie, (1978) (in Russian)

Bolin, K., Boue, M.: Long range sound propagation over a sea surface. J. Acoust. Soc. Am. **126**(5), 2191–2197 (2009)

Boroditsky, L., Fischer, R.: "Direct field of a finite size source," Inter-Noise 04, Prague, Czech Republic, Aud. (2004)

Boroditsky, L., Spiridonov, V.: Structure borne Noise Control in Ship Compartments, Sudostroenie, Leningrad, 1974 (in Russian)

Boroditsky, L., Spence, J., Fischer, R.: Predicting shipboard noise using 3-D acoustic modeling. Noise Control Eng. J. **55**(2), 246–256 (2007)

Brekke, A.: Calculation methods for the transmission loss of single, double, and triple partitions. Appl. Acoust. **14** (1981)

Burroughs, C.B., Fischer, R., Kern, F.R.: An introduction to statistical energy analysis. J. Acoust. Soc. Am. (JASA). **101**(4), 1779–1789 (1997)

Carlton, J., Jukes, P., Choo, Y.-S. (Eds.) Encyclopedia of Maritime and Offshore Engineering, Vol. 3 (016), R. Fischer & M. Bahtiarian, Ship noise control, Wiley, 2018

Cremer, L., Heckle, M., Ungar, E.: Structureborne Sound: Structural Vibrations, and Sound Radiation at Audio Frequencies, 2nd edn. Springer, NY (1988)

DeJong, R.: Statistical Energy Analysis of Ship Structureborne Sound. InterNoise 2003, Korea, 2003

DesignerNOISE®: Software program created under SBIR Phase II, Topic No. 98–02, Advanced Ship Acoustics, Office of Naval Research and NAVEA, 2002

Det Norske Veritas (DNV), *Recommended Practice, DNV-RP-D101, Structural Analysis of Piping Systems,* Oct. 2008

Det Norske Veritas (DNV): *Ice Strengthening of Propulsion Machinery*, Classification Notes No. 51.1, Sept. 2010

Det Norske Veritas (DNV): Rules for Classification of Ships, Newbuildings, Special Equipment and Systems, Additional Class, Part 6, Chapter 15, Vibration Class, 2011

DND (Department of National Defence – Canada): Hull vibration standard for Canadian warships," Design Standard DMEM II, Issue I, September 1978

Elmallawany, A.: Calculation of sound insulation of ribbed panels using statistical energy analysis. Appl. Acoust. **18**, 271–281 (1985)

EPA: The Environmental Protection Agency's Model Community Noise Control Ordinance, n.d.

Fahy, F., Thompson, D. (eds.): Fundamentals of Sound and Vibration. CRC Press, Boca Raton (2015)

Favini, E, Spence, J.: Impacts of insulation and joiner treatments on low frequency vibration in marine structures with finite element applications. In: SNAME Annual Meeting and Expo., Houston TX, Oct. 2014

Fischer, R.: Case history: noise and vibration induced by cantilever foundations on ships. Noise Control Engineering Journal. **31**(3), 159–165 (1988)

Fischer, R.: Shipboard airborne noise prediction algorithms. 142nd Acoustical Society of America Meeting, Fort Lauderdale, 2001

Fischer, R., Boroditsky, L.: Technical and Research (T&R) Bulletin 3-37 (Supplement), Supplement to the Design Guide for Shipboard Airborne Noise Control. The Society of Naval Architects and Marine Engineers (SNAME), Alexandria VA (2001)

Fischer, R., Courtney, B., Nelson, D.: Design Guide for Shipboard Airborne Noise Control. Society of Naval Architects and Marine Engineers (SNAME), Technical & Research Bulletin 3-37, Alexandria, VA (1983)

Fischer, R., Boroditsky, L., Spence, J.: Acoustic Design Tools – Present and Future Integrated Software. International Cooperation on Marine Engineering Systems (ICMES), SNAME, New York (May 2000)

Fischer, R., L. Boroditsky, and J. Spence, "Acoustic design tools – present and future integrated software," International Naval Engineering Conference, IMAREST, Hamburg, Germany, 2008

Giampaolo, T.: *Gas Turbine Handbook: Principles and Practices*. The Fairmont Press, Lilburn, GA (2006)

Gomzikov, E.A. and G. D. Izak, Sudostroenie, 180 p., Noise Control Design for Ships (Proektirovanie protivoshumovogo kompleksa sudov), (in Russian), 1981

Grushetsky, I.: Internal loss factors of ship structures and its application for computing vibration and noise. Tech. Acoust. **4**, 1–10 (2007)

Hynna, A., Klinge, P., Vuoksinen, J.: Prediction of Structureborne sound transmission in large welded ship structures using statistical energy analysis. J. Sound Vib. **180**, 583–607 (1995)

International Standard (ISO) 9613-2: Acoustics-Attenuation of sound during propagation outdoors – Part 2 General method of calculation, 2024

Janecek, P.: "A rectangular plane sound source", Applied Acoustics, 27, (1989)

Jensen, J. O.: Calculation of Structureborne Noise Transmission in Ships Using the "Statistical Energy Analysis" Approach, Computer Science (1976)

Jenze, J.: Some vibration aspects of modern ship installations. Wartsil NSD, Oct. 1997

Kaminsky, C., Van Buskirk, J., Forzelski, T., Qian, Y.: "Statistical Energy Analysis of a Vehicle Interior Equipped with New Lightweight, High-absorption Trim Material," Noise-Con 98, Institute of Noise Control Engineering, Ypsilanti, MI (1998)

Kandekar, A., Nagarhalli, P., et al.: HVAC System Noise Prediction through CFD Simulation SAE Technical Paper 2019-26-0210. SAE International (2019)

Kim, J.S., Kim, H.S., et al.: Effects of inplane modes in SEA on structure-borne noise transmission in ship structures. J. Hydrosp. Technol. **2**(1), 18–26 (1996)

Klinge, P.: Prediction of Structureborne transmission in large welded ship structures using SEA. J. Sound Vib. **180**(4) (1995)

Lloyd's Register, Ship Vibration and Noise, Guidance Notes, Revision 2.1, July 2006

Lyon, R.H.: Statistical Energy Analysis of Dynamical Systems. MIT Press, Cambridge, MA (1975)

Lyon, R.H., DeJong, R.G.: Theory and Application of Statistical Energy Analysis. Buttersworths-Heimann, Boston, MA (1995)

Maidanik, G.: Response of Ribbed Panels to Reverberant Acoustic Fields. J. Acoust. Soc. Am. **34**, (1962)

Maidanik, G.: Radiation efficiency of panels. J. Acoust. Soc. Am. (JASA). **35**(1), 115 (1963)

Maidanik, G.: Vibrational and radiative classifications of modes of a baffled finite panel. J. Acoust. Soc. Am. (JASA). **34**(115), 447–455 (1974)

McIntosh, D.J.: Real variability in ship systems' noise and vibration, design and through-life management implicatinos for underwater noise and habitability Inter-Noise 2014, Melbourne, Australia, Nov. 2014

Moorehouse, A., Waddington, D., Adams, M.: Procedure for the assessment of low frequency noise complaints, University of Salford, Contract No. NANR45, Feb. 2005

Nilsson, E.: Decay processes in rooms with non-diffuse sound fields. Part II: Effects of irregularities. Build. Acoust. **11**(2), 133–143 (2004)

Noonan, E.F.: Ship Vibration Design Guide, Ship Structure Committee, SSC-350, USCG Headquarters, Washington DC, July 1989

Petersson, B., Plunt, J.: Structureborne Sound Transmission from Machinery to Foundations, Chalmers University of Technology, Report 80–19, Gothenburg, Oct. 1980,
Pettit, L., Fischer, R.: How Can Acoustic Detectability of Riverine Craft Be Predicted and Controlled? In: Spec Ops East 2007 Symposium and Expo, Fayetteville, NC, 2007
Plunt, J.: Methods for Predicting Noise Levels in Ships, Experiences from Empirical and SEA Calculation Methods, Part 1: Noise level prediction methods for ships, based on empirical data, Dept. of Building Acoustics, Report 80-07, Gothenburg, April, 1980
Reynolds, D., Bledsoe, J.: Algorithms for HVAC Acoustics. ASHRAE, Atlanta, GA (1991)
Salio, M.P.: Numerical assessment of underwater noise radiated by a cruise ship. In: Ships and Offshore Structures. Taylor & Francis (2013)
Seng, J., Khoon, L.: Prediction of Underwater Radiated Noise of a Ship Section Using SEA and Coupled FEM/BEM Approach, Naval Platform Technology Seminar, DSO National Laboratories, 1999
Seybert, A.F., Wu, T.W., Wu, X.F.: Experimental Validation of Finite Element and Boundary Element Methods for Predicting Structural Vibration and Radiated Noise, NASA Contractor Report 4561, Report ARL-CR-109, January 1994
Smith, P.W.: Response and radiation of structural modes excited by sound. J. Acoust. Soc. Am. (JASA). **34**(640), 640–647 (1962)
Smith, P.W.: Statistical models of coupled dynamical systems and the transition from weak to strong coupling. J. Acoust. Soc. Am. (JASA). **65**(115), 695–698 (1979)
Smith, P.W., Lyon, R.H., *Sound and Structural Vibration*, NASA CR-160, Washington, DC, 1965
Soares, C.: Chapter: 6 inlets, exhausts, and noise suppression. In: Gas Turbines: A Handbook of Air, Land, and Sea Applications. Elsevier, Inc (2008)
Spence, J., Dempsey, R.: Finite Element Vibration Analysis and Measurement of a 255 Foot Vehicle Ferry. Noise-Con 2010, Baltimore, MD, April 2010
Spiridonov, V.: Chapter 15: Shipboard noise prediction. In: Handbook "Ship Acoustics". Sudostroenie (1978) (In Russian)
Tanttari, J.: Simulation of Underwater Sound Radiation from Ship Structures in Full Frequency Range, VTT-R-oo724-12, Tampere Finland, 2014
Tso, Y.K., Hansen, C.H.: Transmission in ships using statistical energy analysis. Acoust. Aust. **25**(3), 97–102 (1997)
Van Renterghem, T., Botteldooren, D., Dekoninck, L.: Airborne sound propagation over sea during offshore wind farm piling. J. Acoust. Soc. Am. **135**(2), 599–609 (2014)
van Terwisga, T., van Wijngaarden, E., Bosschers, J., Kuiper, G.: Achievements and challenges in cavitation research on ship propellers. Int. Shipbuild. Prog. **54**(2–3), 165–187 (2007)
Veritec: Vibration Control in Ships, A.S. Hovik, Norway, 1985
Vladimir, N., Loncar, I., Ancic, I., Senjanovic, I.: Prediction of Noise Performance of Ro-Ro Passenger Ship by the Hybrid Statistical Energy Analysis. ISSN 0554-6397, Pomorski zbornik, n.d..
Westphal, W.: Ausbreitung von Korperschall in Gebauden, (in German). Akustisch Beihefte, Heft. **1**, 335–348 (1957)
Wiener, F.M.: Sound propagation over ocean waters in fog. J. Acoust. Soc. Am. **33**(9), 1200–1205 (1961)
Xiang, N., Summers, J.: Acoustics in coupled rooms: Modeling and data analysis. In: International Symposium on Room Acoustics, Seville, Sept. 2007
Xu, Y.-l., Huang, X., et al.: Numerical analysis and optimization of noise of ship cabins in the low frequency. J. Vibroeng. **19**(3), 2234–2246 (2017)
Zhen, H., Liu, G., Tao, J., Lan, K.: FEM/BEM analysis of diesel piston-slap induced ship hull vibration and underwater noise. Appl. Acoust. **62**, 341–358 (2001)

Chapter 8
Airborne Noise and Vibration Control Treatments

Noise and vibration control for most land-based systems is typically straightforward. Effective controls usually exist. An important factor and difference from land-based mechanical systems is the 'coupling' of the noise and vibration to the ship structure and then from this structure to its environment – whether internal or external to the vessel. This coupling is quite different across both air or fluid interfaces. Not all treatments used for land-based facilities are appropriate for use in marine fields.

In selecting treatments for machinery/equipment-induced noise and vibration, it is necessary to understand:

- the frequency range over which the various treatments are effective
- the magnitude of their impact on noise or vibration or the combination of noise and vibration
- whether the treatment is most effective against the source, the transmission path, and/or receiver
- limitations and assumptions on the treatment's effectiveness, as installed
- impact of the treatment on the system(s) to which it is applied, and
- whether marine regulatory requirements are met, e.g., captive mounts, fire, smoke, toxicity, and outgassing[1]

Another factor is consideration of the propulsor for the vessel. Cavitation is a critical factor in assessing and controlling noise produced by a vessel – both above and below the waterline. Propagation of noise and vibration throughout the vessel requires a great degree of understanding before appropriate noise and vibration treatments can be selected. There is cross-coupling of noise and vibration energy that needs consideration before the entire system of treatments can be optimized.

[1] e.g., SOLAS Class A60, A30, A15, A0 and B15, B0 Structural Fire Requirements

R. Fischer, L. Boroditsky, *Noise and Vibration Control on Ships*,
https://doi.org/10.1007/978-3-031-55170-3_8

Six approaches to noise control can be classified as follows:

1. Use of low noise and/or vibration machinery and equipment ("Buy Quiet" approach – see Sect. 8.1)
2. Choice of general arrangement, which is rational from an acoustic control point of view of keeping sources and sensitive spaces far apart
3. Noise and vibration controls at the sources, e.g., isolation mounts, cladding of machinery surfaces, acoustic insulation in machinery spaces, and full or partial enclosures and barriers
4. Noise and vibration reduction along the path, e.g., damping, impedance mismatches, stiff foundations
5. Noise and vibration reduction at the receiver, e.g., High Transmission Loss (HTL) treatment, absorption, floating deck/room, and
6. Special noise control treatment, e.g., silencers/mufflers, barriers, tuned absorbers and 'active' controls, innovative propeller design, isolation mounted superstructure, etc.

Most treatments used today are considered passive. Active control of noise and vibration is being investigated and should prove useful, particularly at controlling low-frequency tones. Active controls have been very effective in reducing low-frequency HVAC system noise (Gelin 1997). Any active system usually consists of an actuator – speaker or shaker – and an error measuring sensor. This sensor is usually a microphone for an active noise control system and an accelerometer for a vibration control system. Active noise and vibration control treatments usually have a feedback loop of some type to actively control the source or response. "Tuned treatments" can be active or passive and usually target a tone or multiple tones. Details on active systems are beyond the scope of this text. Useful references are Berkman and Bender 1997; Daley et al. 2004; Winberg et al. 2001; and Lung 1997.

The first group of controls listed above is fully in the hands of machinery and propulsor designers and manufacturers. Ship designers also have a choice to select more or less noise and vibro-active machinery. Quieter machinery may be more expensive but requires less noise control effort. Here, an economic trade-off may be needed to determine which may be more feasible and have a positive impact on any Return On Investment (ROI). Many organizations such as NASA and the mining industry have moved to a 'Buy Quiet' approach.

The second group of controls should be executed early in the ship's design phase. A common-sense general arrangement requires the compartment with the most stringent criteria to be located farthest from noise sources. Rooms such as lockers, storage areas, passageways, or machinery rooms with low noise equipment may be used as a buffer between noisy and quiet rooms. It is useful to concentrate the noise equipment in one or a few locations. In this case, it is easier to provide a significant distance from sources to rooms with stringent noise requirements. It also provides an opportunity to concentrate noise control treatments around the sources. As a rule, this is a more productive approach to noise control than to disperse treatments throughout the ship.

With respect to the third group, methods for airborne noise reduction (L_{ab}) are well known from architectural acoustics (Rettinger 1977). Approaches to increase the transmission loss of partitions or sound absorption of surfaces, in principle, are well covered in the literature. However, the "naval" edition of these treatments has some restrictions. Specifics will be discussed in the following sections.

Treatments of the structureborne noise components (L_{fsb} and L_{ssb}) are less known in shipbuilding practice. This is in spite of the circumstances where structureborne noise dominates the ship's acoustic environment in the majority of cases. Numerous treatments exist to control structureborne noise: resilient mounts, floating floors, damping tiles, special foundations, additional insulation in hull element connections and others. Some of them, such as floating floors and damping tiles, can also reduce the transmission of airborne noise. These treatments will be discussed in the following sections.

The fourth approach is to place noise and vibration treatments along the structural path from the machinery source to the receiver location of interest. For machinery sources, this includes treatments such as damping and structural impedance mis-matches. For HVAC systems, this would typically include lined ducts, radiused turns, etc.

Per the fifth element, the acoustic energy reaching a receiver location can be reduced by increased absorption, acoustic ceilings, tuned absorbers, and/or the use of floating decks or bulkheads. High Transmission Loss (HTL) materials can be applied to the bulkhead or underside of the compartment deck if the adjacent space is noisy. HTL can also be used as a cladding on the interior surfaces to reduce radiation into the compartment.

Finally, there are all types of advanced techniques being developed as the acoustic criteria become more stringent and prevalent. For machinery, this could include active isolation systems. For the propeller, this may include optimized hydro-acoustic design, such as skewed and wake-adapted strategies.

The extent of noise prediction output data discussed in the previous chapter provides an opportunity to make optimal decisions regarding noise and vibration control approaches. The key factors for this decision are the frequency range of concern, amount of excess relative to the criteria, number of rooms with excesses, source contribution, component contribution (*AB, FSB, SSB*), and acoustic characteristics of the receiver space. Depending on these factors, the designer may choose treatments that have an effect on all rooms of interest or just a few of them. Some treatments may influence only one component, some all of them. Furthermore, special attention should be given to whether the excess and treatment effectiveness cover the appropriate frequency range.

Additional 'design guidance' for the treatments discussed below is provided in Chap. 10.

8.1 Low Level Noise and Low Vibration Machinery and Equipment – Buy Quiet

"The U.S. Navy has a 'Buy Quiet' policy for equipment aboard ships" using the tactic that selecting quiet equipment is the 'best way to reduce noise at the source' (Avery and Gray 2009). Using this approach would require those involved with the noise control process to obtain vendor acoustic data before purchasing to choose the quietest equipment available. Source-level data from multiple vendors can be compared. These data can also be used in preliminary or baseline noise analyses. While quieter equipment is often more expensive than non-quieted equipment, it often saves with respect to longevity due to the use of better materials, such as bearings. It may also save more than this cost differential due to the need for fewer treatments.

The National Institute for Occupational Safety and Health (NIOSH) and National Aeronautics and Space Administration (NASA) have both successfully implemented Buy Quiet programs (Franks et al. 1996; Maling Jr. 2012; Cooper and Nelson 1997). NIOSH has stated that one of the benefits with respect to hearing loss compensation from this type of program is an 'estimated \$100 per dBA of saving' (Nelson 2011). Buy Quiet is a key element of Technology for a Quieter America 2010.

Another way to 'buy quiet' is to select machinery and equipment that is inherently quiet. This includes the use of rotating rather than reciprocating machinery and the use of drip-proof in lieu of totally enclosed fan-cooled electric motors. Spur gears are noisier than double helical gears. Enclosed equipment, where the vendor has accounted for not only the acoustic performance of the enclosure but also cooling and fire control, is better than using custom enclosures. Access for maintenance is an important factor in utilizing any type of enclosure. Rotary screw or gear pumps are more desirable than piston pumps, which generally require both vibration isolation and enclosures.

The shipyard's Purchasing Department should develop, establish, and implement procedures to integrate a Buy Quiet program. The Purchasing Department should maintain a list of vendors who can supply quiet equipment. Furthermore, suppliers of machinery and equipment should provide acoustic data and information on the feasibility of meeting any proposed guarantee levels and related confidence levels.

8.2 Optimized Arrangements

8.2.1 Habitability Noise

High airborne noise, typical in machinery spaces, is an important noise contributor to adjacent spaces. It is of less importance to non-adjacent spaces, especially for any Diesel-Electric (D-E) vessel with isolation mounted gensets. In some cases, for D-E vessels, it is helpful to place a high transmission loss treatment on the bulkhead, side hulls, and overhead to reduce secondary structureborne noise. Another less common method is to use enclosures on noisy machinery. Furthermore, lockers, storage areas, passageways, and similar nominally unoccupied spaces should be located between machinery spaces and nearby noise-sensitive compartments.

In contrast, structureborne noise can be transmitted throughout the vessel. Thus, it is important to locate noise-sensitive spaces as far as possible, both longitudinally and vertically, from critical sources such as propulsion machinery and propulsors. Structureborne noise losses are typically 3 to 5 dB per divisional bulkhead and per deck. If excessive noise is expected, the appropriate treatments are path damping, the use of blocking masses on stanchions or foundations, or significant changes in plate thickness at deck/bulkhead intersections.

Stanchions can provide a vibration short-circuit from the tank top to upper decks (Burroughs et al. 1979). Table 8.1 shows the increase in the vibration at the center of a room with and without a stanchion, measured at a ship mock-up test site. An increase of 5 dB in the structureborne noise is generally expected due to the presence of a stanchion. This increase will be significantly greater if the bottom of the stanchion is located on the foundation next to a substantial vibration source, such as diesel.

This path can be attenuated if layers of Distributed Isolation Material (DIM) are applied to the top of the stanchion. Figure 8.1 shows 4 layers of DIM on top of a 61 × 46 cm platform used to attenuate the stanchion path.

Table 8.1 Increased structureborne noise due to stanchion path

Freq., Hz	125	160	200	250	315	400	500	630
Increase, dB	5	3	4	4	6	1	0	13

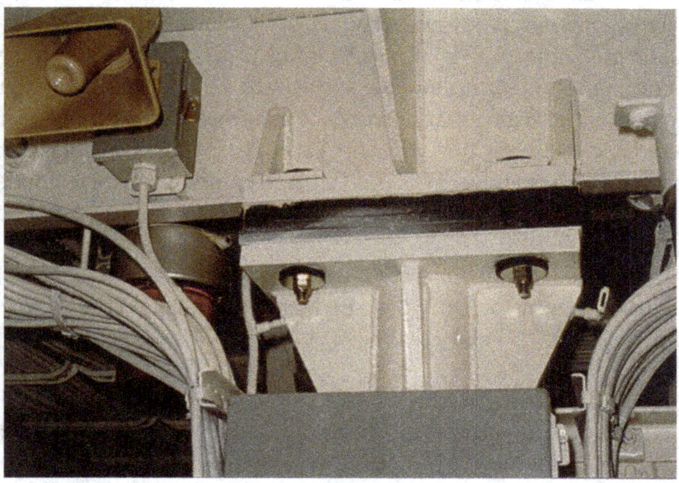

Fig. 8.1 Distributed Isolation Material (DIM) isolation at top of the stanchion

8.2.2 Underwater Radiated Noise

In contrast, the higher the vessel's machinery is located, the easier it is to meet underwater noise criteria. Some approaches have utilized a small gas turbine generator above the 01 Deck for use in 'quiet operation' (Bonetti and Calcagno 1986). Minesweepers and submarines have used U-shaped cradles attached to the Main Deck to support isolation mounted diesels (Fusil 2004). Double hulls offer significant advantages over single shell plating, effectively reducing the airborne transmission path.

8.3 Source Noise Controls

Unbalanced or misaligned machinery is a major cause of excessive noise or vibration throughout the ship. Therefore, a fairly stringent balance and alignment specification should be a first line of 'source control.' As stated above, the mantra for noise control is to specify or select machinery or equipment with inherently low noise and/or vibration source levels. This type of equipment will likely reduce or eliminate the need for additional treatments

8.3.1 Resilient Mounts for Marine Machinery

Isolation is the best method for reducing the structureborne noise path from a piece of equipment. Isolation disconnects the ship's structure from the vibrating machinery. There are numerous types and vendors of isolation mounts. These isolation systems can be as simple as a set of metal springs or a rubber pad; all the way up to complex mounting solutions that employ specialized visco-elastic compounds mated to metal structure. The octave band frequency range of interest from a habitability noise standpoint is between 31.5 Hz and 8000 Hz. Modern resilient mounts are effective for frequencies above 31.5 Hz. Generally, more complex mounting solutions are more expensive. These systems also require extra engineering time for their design.

Low-frequency isolation mounts are the most effective passive treatment used for controlling structureborne noise from shipboard machinery (US Navy Resilient Mount Handbook 2005) – see Figure 8.2 a, b, and c for examples. Low-frequency mounting systems are designed so that the natural frequency of the mounted equipment falls well below the lowest operating, rotational frequency. This mounting method is extremely effective. Most low-frequency isolation systems have natural frequencies between 5–10 Hz. These mounts generally consist of some type of rubber along with a metal top and bottom to allow mounting between the

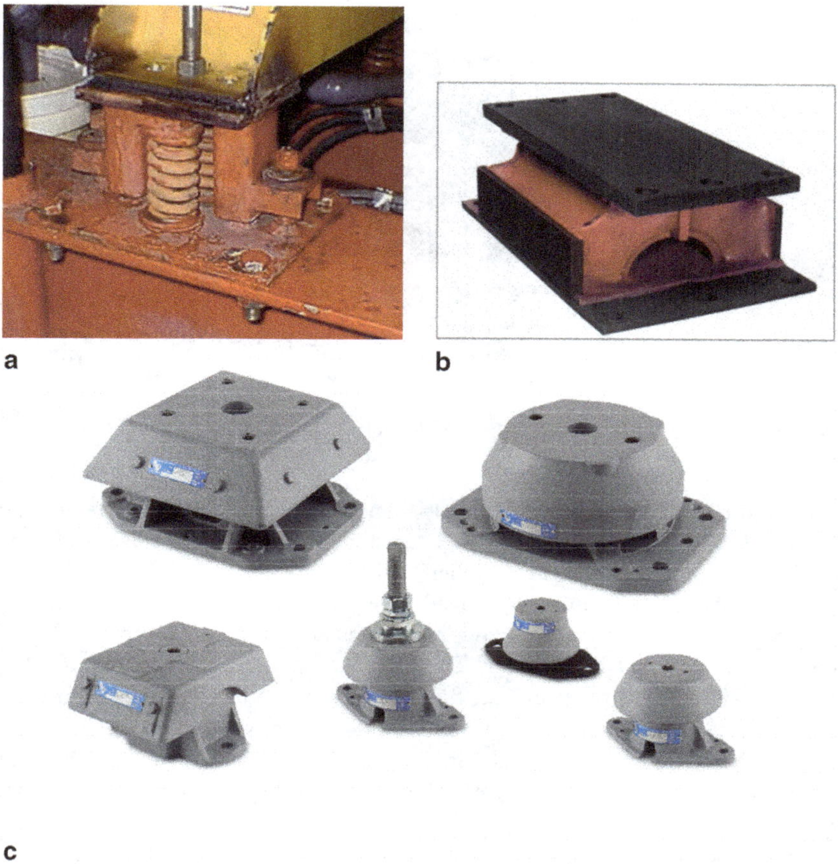

Fig. 8.2 Examples of low-frequency isolation mounts. (a) Coil Spring Mount. (b) Elastomeric Mount. (Photo from https://www.trelleborg.com/en). (c) Line of Marine Isolators. (www.christiegrey.com)

equipment feet and the foundation. A well-designed low-frequency mounting system can attenuate the vibration levels (structureborne noise), and thus both habitability and underwater radiated noise, by as much as 20 dB.

A built-in captive mechanism should be used for all mounted systems. A captive mount will 'fail safe' if the rubber is sheared away or melted. In a non-captive mount, rubber is the only thing that connects the equipment to the foundation. An alternative with respect to fire, assuming that the transmission loss is sufficient, is the use of an all metallic mount. If a non-captive mount is used, a snubbing system that will limit the motion of the equipment in extreme situations is generally needed.

The preferred, in almost all cases, is the enclosed marine elastomeric mount. Coil springs, most often used in industrial applications, generally short out when loaded laterally, as occurs in a seaway. An exception to this is an isolator designed to take

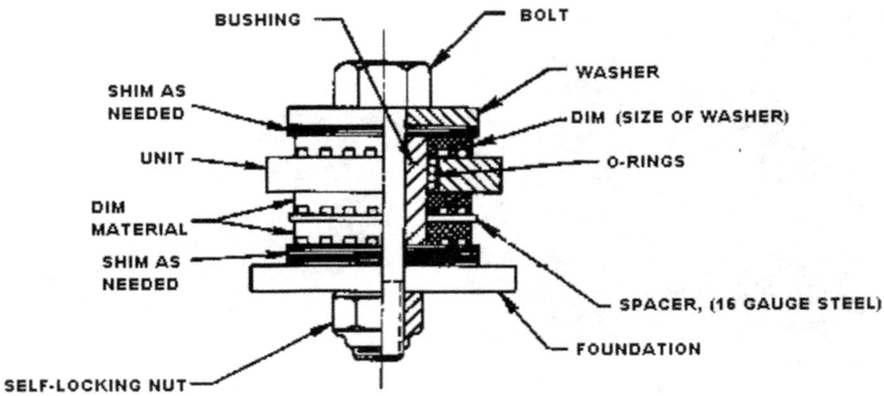

Fig. 8.3 Cross-section of Distributed Isolation Mount (DIM) type mount

forward and reverse thrusts. This type of isolator is generally used on small, direct drive vessels without any thrust bearing. A relatively soft flexible shaft coupling is also needed for this style system (Fischer 1988).

Rubber bushings are 'exposed' to adverse shipboard environment conditions such as high temperatures and oil and thus are usually not a great option. There is a fourth type of mount to consider. This is Distributed Isolation Material (DIM) (NAVSEA 1982), illustrated in Fig. 8.3. DIM pads are dimensioned to meet an allowed load per area, N/mm^2, that varies from vendor to vendor.[2] DIM mounts are meant to provide isolation at 'high' frequencies, generally above 120 Hz. High-frequency mounting effectiveness generally attenuates the vibration and underwater radiated noise by 3–7 dB.

Furthermore, many vendors of marine rubber mounts can control the amount of damping to reduce the height of the amplitude response at the resonant frequency. Figure 8.4 shows a system with a natural frequency of 7 Hz and a damping ratio of 0.05. Shifting to a lower resonant frequency is accomplished by increasing the supported mass (keeping the mount stiffness the same) and adding damping. The result of changing the damping to 0.25 and increasing the mass to result in a natural frequency of 5 Hz, which reduces the amplitude of the response, as shown in the "Shifted" curve.

Metal mesh mounts may be appropriate for high temperature environments. They usually have high damping and, often, nonlinear characteristics. A new generation of mounts combines rubber elements with fluid dampers, providing tunable, highly damped mounts. Wire rope and coiled mounts are often used for the dual purpose of isolation and shock. Again, these mounts have more inherent damping than rubber or spring mounts due to the sliding friction between the strands that make up the wire rope. New mount types are continually being developed, especially for naval use.

[2]Typically, varying from a low of 172,000 Pa (25 psi) to a high of 689,500 Pa (100 psi)

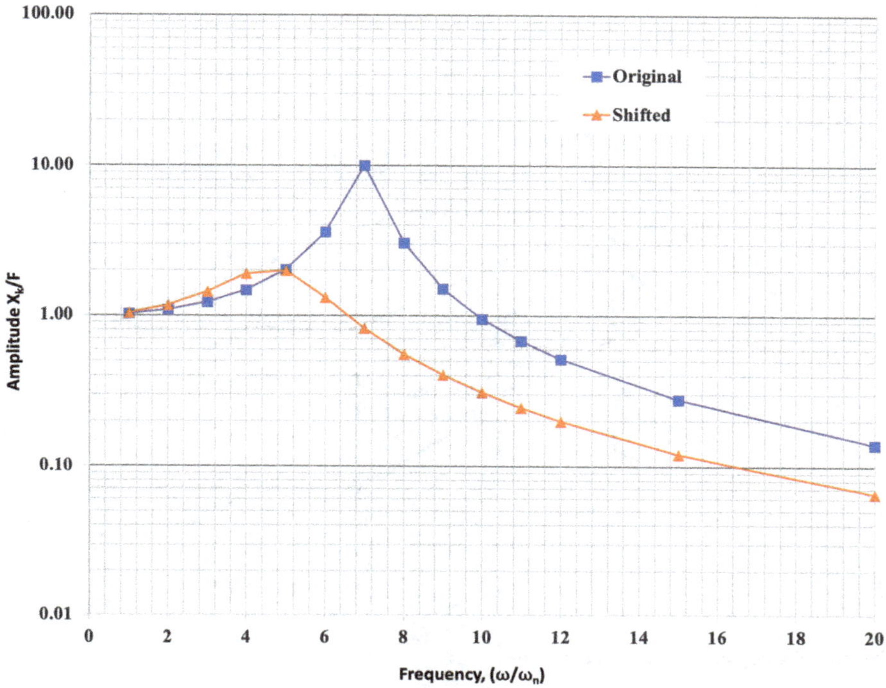

Fig. 8.4 Impact of damping and mass on isolation system performance
Amplitude X_k/F is the normalized response
Frequency (ω/ω_n) is the normalized frequency response

Examples are XM mounts using "two or more leaf springs in opposing pairs to form an elliptical leaf spring assembly. The spaces between the leaves are filled knitted mesh cushions to provide damping (Caparo patent)."

The practical design of a single-stage resilient mount system should consider motions in six degrees of freedom – see Fig. 8.5 and Sect. 10.1.4. Keep in mind that the mount is usually softest in the vertical direction; therefore, mount system resonances in the longitudinal and athwartships directions are usually higher. The natural frequencies in all six Degrees of Freedom (DOFs) should be computed and compared to the machinery drive frequencies – whether constant or variable speed. These natural frequencies can easily be computed with any finite element software. Rotational stiffnesses are almost always ignored. In many cases, these stiffnesses are not measured by vendors. Two-stage isolation systems typically have twelve degrees of freedom and two vertical natural frequencies with which to be concerned.

The goal of the 6-DOF calculation is to determine the natural frequencies of this system and to avoid overlap with (1) external resonances – e.g., hull modes and propeller blade rate – and (2) machinery internal excitation frequencies. This approach is useful to avoid excessive movement of an engine in harsh marine sea states and to have a sufficient Transmission Loss (TL) at low frequency. Designing

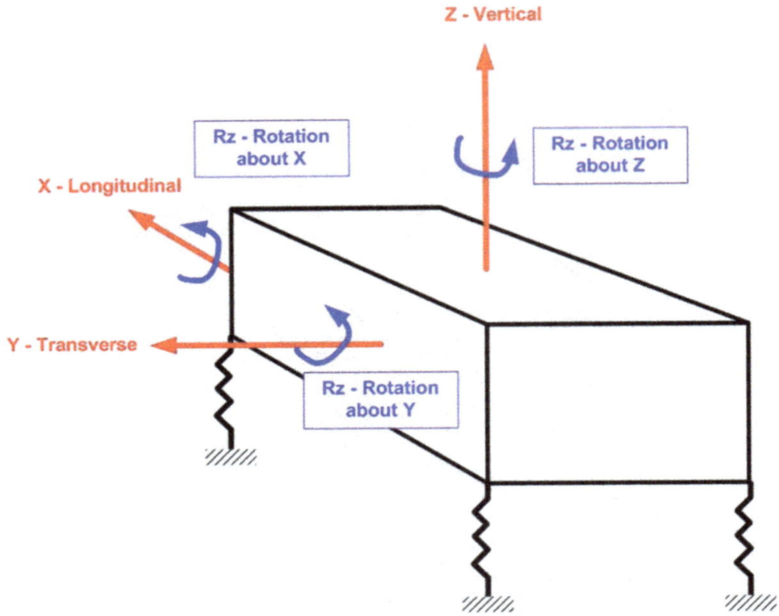

Fig. 8.5 Six degrees of freedom diagram

for a "low-frequency mount" and resonance avoidance does not provide practical information needed to predict mount effectiveness or transmission loss over the frequency range between 63 and 8000 Hz. The authors tend to rely on a lower envelope of measured transmission loss[3] data on similar types of mounts.

In general, the lower the natural frequency of the mounts (less than 10 Hz), the greater the effectiveness in the middle- and high-frequency range.[4] There are no established algorithms applicable for mount analysis to determine the mount effectiveness in the middle to high frequency range for practical third octave and octave bands. Advances in Computer Aided Engineering (CAE) may soon cover this gap in technology. Because of this, it is useful to base performance on vendor-provided data taken in a factory or data collected on-board representative ships. This is especially true for more sophisticated two-stage and active vibration mount systems.

As a final approach, data provided in design guides may be utilized. The performance values listed in the SNAME Design Guide 3-37 (Fischer et al. 1983) are considered 'conservative' and are on the lower end of the expected performance. One factor in determining the effectiveness of a mount system is to consider whether

[3]Transmission loss is the difference in vibration measured top and bottom of the mount; insertion loss is data measured on the foundation top with and without the mount, which is more difficult to measure in practice.

[4]Theoretically, a pure spring-mass-damper system will quickly approach unrealistic performance values and should never be used to predict mount performance.

Table 8.2 The measured vibration levels on the marine foundation for gensets on single-stage mounts, dB re 1 μG

Octave bands, Hz	31	63	125	250	500	1000	2000	4000	8000
Above mounts, La dB re 1 μG	94	109	117	114	117	126	127	119	113
Below mounts, La dB re 1 μG	77	88	101	94	92	89	88	87	79
Transmission Loss, dB	17	21	16	20	24	37	39	33	34

Table 8.3 Transmission loss (TL) for DIM mounts, dB

Octave bands, Hz	31	63	125	250	500	1000	2000	4000	8000
Genset 1; DIM *TL*	−3	0	4	7	11	14	11	5	11
Genset 2; DIM *TL*	7	11	5	6	19	17	14	24	27
Genset 3;DIM *TL*	4	6	12	4	5	16	22	21	18

the equipment is a 'force' source or a 'velocity' source. A velocity source would be a large piece of equipment whose velocity would not change significantly even when attached to the ship structure. In the SNAME Design Guide (Fischer et al. 1983), these are considered machinery of medium and heavy weight. Similarly, a force source is a piece of machinery whose force is not significantly changed upon attachment to a ship structure. As noted by Kim and Singh 2002, a velocity source is a good approximation from mid- to high frequencies. Again, the SNAME Design Guide considers this to be equipment in the 'light weight class.' This would be pumps and fans. The difference in the effectiveness of a typical isolator between a heavy weight and a lightweight piece of equipment on a high-impedance ship foundation can be as high as 17 dB at lower frequencies, with better performance obtained for the lightweight system.

When an engineer evaluates the acoustical effectiveness of resilient mounts, the difference between vibration levels above and below mounts is the parameter typically used for assessment. Table 8.2 shows an example of this procedure. Vibration levels of the diesel generator were measured on a research vessel – above and below the mounts. The diesel generator was installed on US Navy standard 5 Hz mounts (US Navy S9073-A2-HBK-010, 2005). Table 8.3 shows the measured transmission loss for gensets installed on DIM.

If one considers a resiliently mounted engine as a "mass-spring" system, transmission loss should theoretically increase by 12 dB per octave for frequencies above the resonance frequency. Such growth is not achievable for actual systems. Many reasons can be asserted for this relatively modest effectiveness: wave process in the body of the mount, relatively small stiffness of contacting surfaces, flanking airborne excitation, and others such as poor impedances[5] above and below the mount. Examples of a high-impedance support above and below a resilient mount are shown in Figs. 8.6 and 8.10.

[5]Impedance, in this sense, is the dynamic stiffness of the foot supporting the machine (above mount) and the stiffness of the foundation (below the mount). It is defined as the input force over the output velocity.

Fig. 8.6 Example of a 'Stiff' foundation

Table 8.4 Transmission loss for the double-stage mounting system, dB

Octave bands, Hz	31	63	125	250	500	1000	2000	4000	8000
Research Vessel 1	35	47	57	47	37	31	30	33	32
Research Vessel 2	38	48	64	53	45	42	41	42	40

In principle, softer mounts have better effectiveness. However, when the isolator is too soft, the displacement may be too high. Additional snubbers may be needed to reduce vertical, lateral, or longitudinal impacts.

When a significant reduction in structureborne noise is needed, double stage mounts may be needed. Table 8.4 shows the measured *TL* for Gensets on two different research vessels. Typically, a two-stage system has upper mounts located between the base used to support the equipment and an intermediate raft and a lower set of mounts between this raft and the ship's foundation – see Fig. 8.7. The rule of thumb is that the intermediate raft weighs as much as the equipment, but in many cases, a lower value is used, providing less than optimal performance.

Often, double-stage isolators and sometimes single-stage isolators can be 'flanked' by airborne noise (Fischer 2006a, 2006b). High airborne noise impinging on the tank top and other plating below the machinery can easily excite this structure and create significant secondary vibration levels (the definition of second structureborne noise). These are effectively higher than the structureborne noise transmitted directly through the mounting system. Prediction methods and treatment methods to minimize this flanking are provided by Fischer 2006a, 2006b. One approach is to place a solid barrier between the bottom of the diesel and the tanktop to reduce the airborne noise levels exciting the tanktop. The effectiveness of this approach is shown in the acceleration measurements graphed in Fig. 8.8. Note that

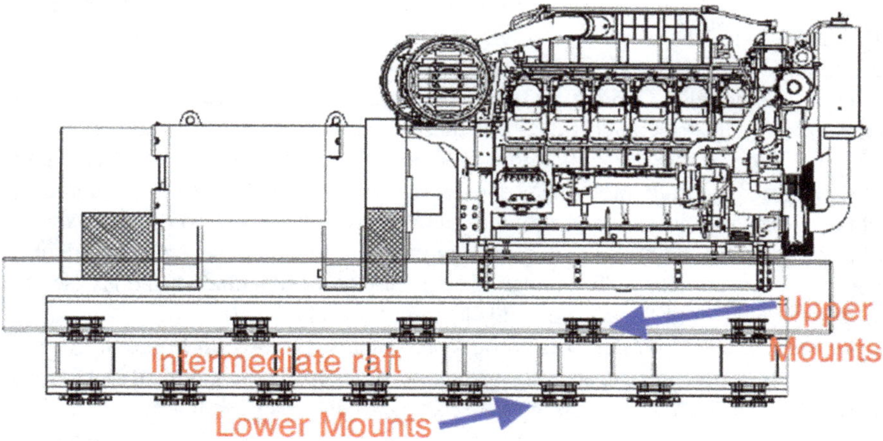

Fig. 8.7 Two-stage diesel genset isolation system for the Quiet Research Vessel

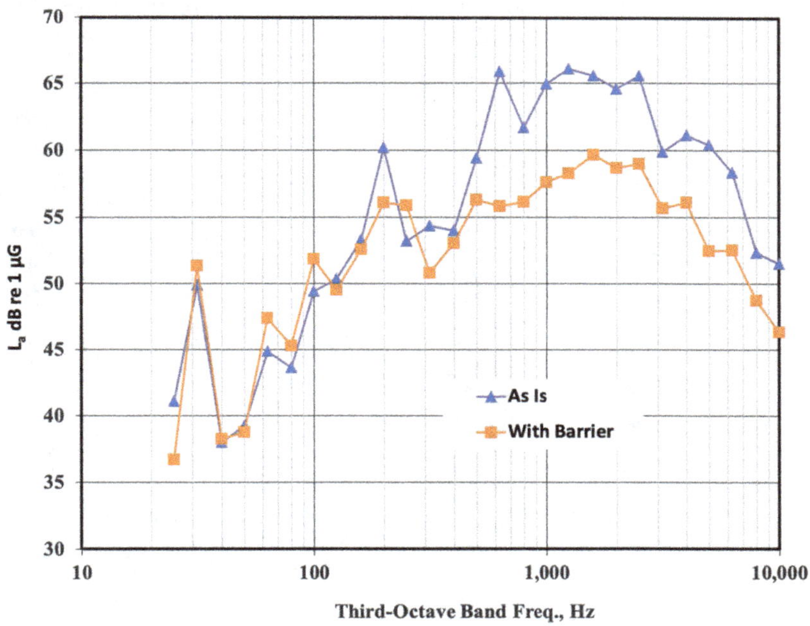

Fig. 8.8 Impact of airborne flanking on vibration isolation system

utilizing a solid intermediate raft would not only reduce the first structureborne noise but could also reduce second structureborne noise. This approach is very effective in reducing the underwater radiated noise of the ship.

The same model of Christie & Grey mounts was used for the following case study on two ships. The principal difference is that Vessel 1 does not have a solid platform raft as the intermediate mass; therefore, it is flanked by airborne noise. Vessel 2 has a solid intermediate raft, blocking airborne transmission to the tanktop; therefore, better performance is achieved at higher frequencies. These data confirm the significance of airborne noise excitation. The highest *TL* will be observed when double stage mounts accompany a full or partial enclosure of the machinery. In this case, the *TL* over a majority of the octave bands is approximately 50 dB.

Another way to optimize D-E propulsion systems is shown in Fig. 8.9. For this design, only the centerline gensets were two-stage isolation mounted on a solid intermediate raft. These gensets were used when the vessel needed to operate in a quiet mode. The outboard gensets were only single-stage isolation mounted to reduce habitability noise. It is also possible, with more detailed analyses, to optimize a two-stage system with respect to the natural frequencies (Zhiqiang et al. 2016) or placement of mounts (Yu et al. 2011). Figure 8.10 show the installation of a two-stage isolation system. Note that gussets are used on the foot of the diesel and on the intermediate raft. Thick plates are used on the construction of the raft. An additional thick 'doubler' (billet) plate is installed on top of the foundation to provide additional transmission loss.

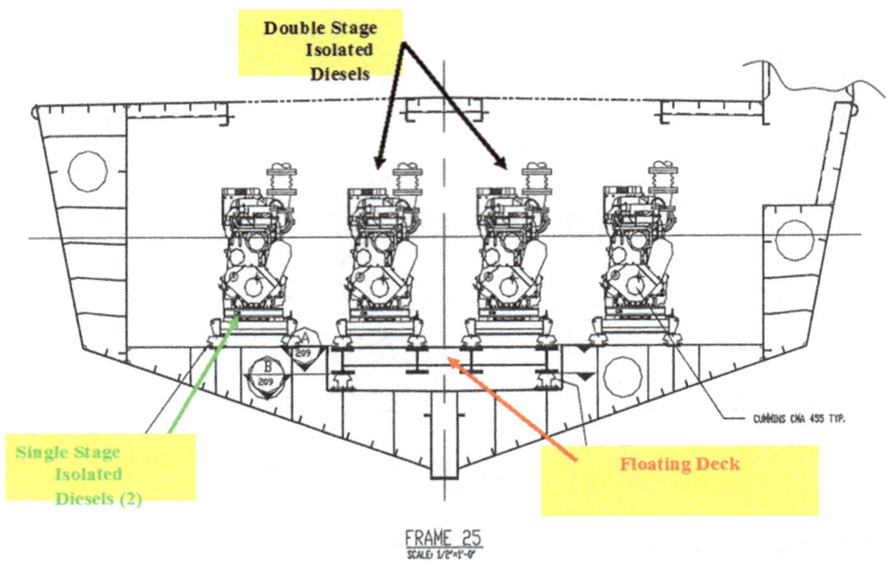

Fig. 8.9 Optimized diesel genset isolation system for a Quiet Research Vessel

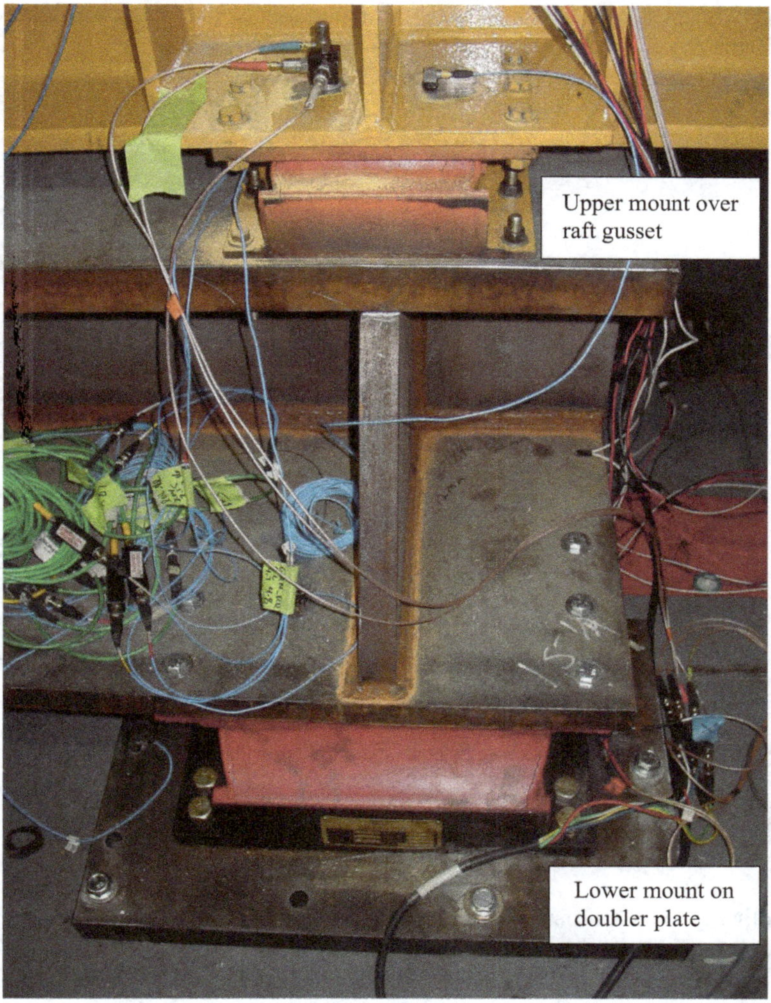

Fig. 8.10 High impedance construction above and below 2-Stage isolation system

Active vibration isolation usually consists of standard low-frequency isolation mounts coupled with electromechanical actuators that can impart reactive forces into the mounted system (see Fig. 8.11). In theory, these systems would monitor the vibration of the system and then impart opposite forces that cancel out the vibrations, particularly at multiple rotational tones. Active mount systems are still in development, and their laboratory results look promising. In actual installations, however, there have been few publicized positive results over a broad frequency range. Most active systems will be employed on top of the intermediate raft of a two-stage mounting system (Niu et al. 2005). Active systems typically require extensive instrumentation, tuning, and maintenance to achieve reasonable results (Yang and Suani 2012).

Fig. 8.11 Actuator for the active control system

Table 8.5 Transmission loss for a single-stage mount system with a billet

Freq., Hz	31	63	125	250	500	1000	2000	4000	8000
TL, dB	48	52	49	36	33	38	31	42	33

Another type of optimization can provide a significant gain in transmission loss. The use of a billet on top of the foundation and directly below the mount will significantly increase the TL, even for a single-stage system. In this case, the impedance mismatch between the mounts and foundation will be increased. The authors measured the vibration level above and below mounts located on a 100 mm thick billet. The system was used on a supply vessel under a genset operating at 1700 kW at 1800 rpm. Table 8.5 shows the measured TL for this generator. Another method to optimize a mounting system is to use DIM for the upper stage and low-frequency mounts below the intermediate raft – lowering the height of the entire mount system.

Comparing data between Tables 8.2 and 8.5 shows a significant increase in transmission loss in the frequency range between 31 and 500 Hz. Comparing Table 8.4 and 8.5 shows that the transmission loss is comparable to that of two-stage mounts in the frequency bands of 31 Hz and 63 Hz.

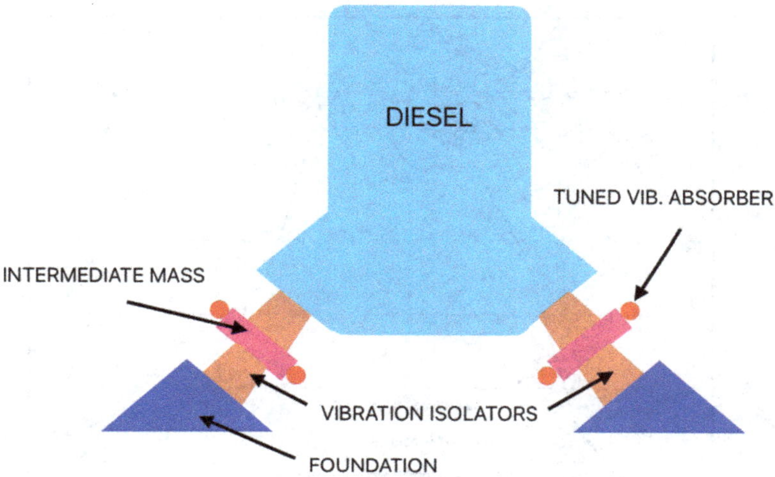

Fig. 8.12 Isolation system used on diesel submarines

Canted or 'focused' mounting systems decouple the vibration modes without the plane of the isolators at the center of gravity of the system.[6] These systems have requirements on the axial-to-radial stiffness ratio and angle of the isolator axis. Diesel submarines have low acoustic signatures, with a great deal of attention given to the isolation of the diesel propulsion system. The isolation system developed for Collin's submarine is shown in Fig. 8.12 (Li et al. 2004). The diesel is a two-stage isolation mounted with mounts canted or focused toward the center of gravity of the diesel. This reduces coupling to roll modes. In addition, there is an active vibration control system, and the intermediate mass has a 'tuned absorber', as discussed below.

A mass-spring system can be added on top of each isolator to reduce the response to a specific troublesome tone for a constant-speed device. Given that there may be space constraints above the mount, a cantilever/mass system could also be used. The measured data show that this approach can reduce the vibration by 5–9 dB at the troublesome tone. Tunable vibration absorbers are a novel approach that allow "a smart variable stiffness element . . . capable of retuning itself in response to a time-varying excitation frequency" (Bonello et al. 2004).

Based on experience, it has been shown that the main response is dominated by the first structureborne noise for mechanical sources on a single-stage system. The first structureborne noise exceeds the second structureborne noise by 10–15 dB. This ratio will likely be reduced when a double stage mount or billets are used. Figure 8.13 shows the calculated and measured vibration levels on a longitudinal bulkhead located near the genset on a supply vessel. These data are for a system isolation

[6] https://hutchinsonai.com/wp-content/uploads/2021/09/Focused-MTG-.pdf

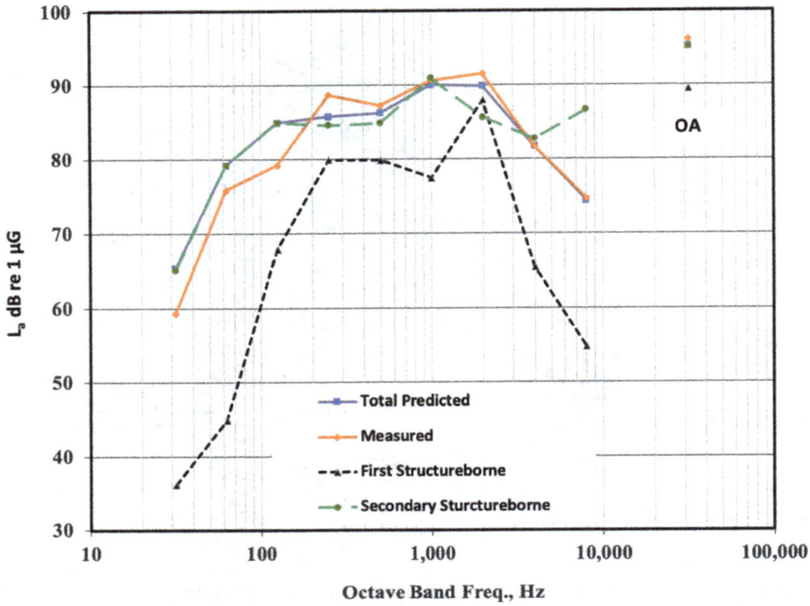

Fig. 8.13 Calculated and measured vibration levels on the bulkhead near the generator

mounted with a billet. Measured noise and vibration levels around the generator were used as source data for this calculation. *DesignerNOISE*® software was used for this calculation. One can see from this chart a 10 to 15 dB prevalence of second structureborne noise. This fact would significantly change the noise control preference.

Another very effective isolator is an air mount. With a very low stiffness, these mounts can have very low natural frequencies, 1 to 2 Hz (He et al. 2014). Furthermore, changing the air pressure to the mounts allows a self-levelling advantage. The disadvantage is that lateral movements can be significant if limit stops are not incorporated. A captive system is also needed should mounts fail.

8.3.2 *Isolation Mounted Superstructures*

Superstructures on ships, fast ferries,[7] river boats, dredges and quarters modules on rigs have been successfully placed on isolation mounts (Payer 1979; Van der Meer 2016). This is known as base isolation. The ultimate floating room is isolation of all the rooms of the superstructure. This treatment is one of the few that provides both

[7] Balearia's Eleanor Roosevelt

airborne and structureborne noise reduction. The large gap between the ship's structural deck and the house lower deck provides significant airborne transmission loss. The isolation mounts provide meaningful vibration reduction if properly selected and installed, without any vibration shorts through the multiple connections needed. Noise reductions on the order of 30 to 40 dB have been achieved (Ort 1976).

Their design is essentially an extension of an isolation mounted system. The exciting force comes from below instead of from above. There are additional factors in the design, such as wind loads and dynamic sea loads. Fairly large isolators are typically used for this treatment (Payer 1979).

The mounts are located on frame intersections on the deck and frame intersections on the house structure. A careful finite element model is needed to account for all mount system and house modes.

Careful flex connections are needed to bridge the gap from the ship's deck to the base of the floating superstructure. A captive system must also be designed and built into the system.

8.3.3 Flexible Shaft Connections

Flexible shaft couplings are usually needed at the connection between an isolation mounted machinery item and any connection to its drive shaft. There are a variety of coupling types made of various materials and sizes. As much as anything else the discussion above for isolation mounts also applies to resilient shaft couplings. They have differing static and dynamic stiffnesses. Any resilient coupling elements may also have standing waves. The elements will also be affected by preload and temperature.

The selection of proper coupling is determined by the torque carried by the shaft, by the angular and parallel misalignment that must be anticipated, and by the variation in the axial separation between the two coupled shaft ends. Reasonable standards for shaft balance, alignment, and runout should be invoked. The engine/gearbox manufacturer should supply guidance on these limits.

Torsional, radial, and longitudinal rigidity of the coupling are factors to be considered in any torsional analysis. Couplings should be able to absorb radial shaft displacements without causing a self-excited torsional vibration. Flexible couplings also add a complication with respect to providing an understanding of the response to transient conditions, such as propeller pulsations (Murawski 2004).

The coupling also takes up significant axial space that must be factored into the system design. Many manufacturers lack complete data needed to fully evaluate the isolation performance of the coupling. Most can provide torque transmission characteristics, torsional vibration control, and mis-alignment capability. A coupling needs to be factored into alignment procedures and any sag expected over its lifetime.

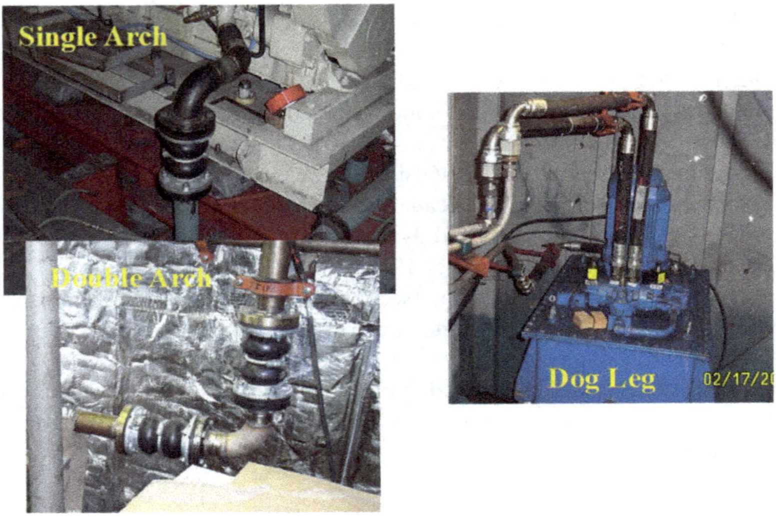

Fig. 8.14 Typical flexible hose connections

8.3.4 Flexible Hose Connections

Examples of flexible (flex) hose connections are shown in Fig. 8.14. The figure on the right-hand side is the standard 'dog-leg' configuration. The two figures on the left-hand side show single- and double-arch flex hoses. The flex shown in the bottom left-hand picture usually provides the greatest amount of noise control – reducing both the structureborne and fluidborne transmission from the machinery side to the ship side.

Several theoretical methods exist to predict the isolation performance of flex hoses – Kosinska and Nogalski 1993; Munjal, 1995, 1996, and 1997; Chapnik, 1996 and 2000. The measured vibration transmission loss for a double-arch flex hose is shown in Fig. 8.15.

8.4 Enclosure

An enclosure is an effective treatment to reduce the airborne noise component. Currently, manufacturers of small diesel generators can supply acoustical enclosures. This is a particularly good approach for gensets that are used on the harbor side to provide power. Figure 8.16 shows **A-weighted** one-third octave noise levels measured around a 154 kW diesel generator – with and without an enclosure.

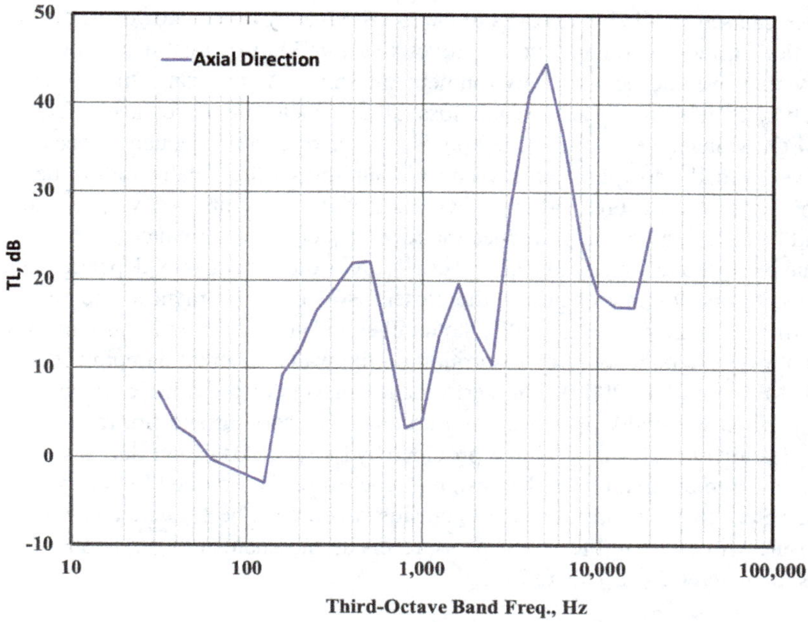

Fig. 8.15 Measured flex hose vibration transmission loss, dB

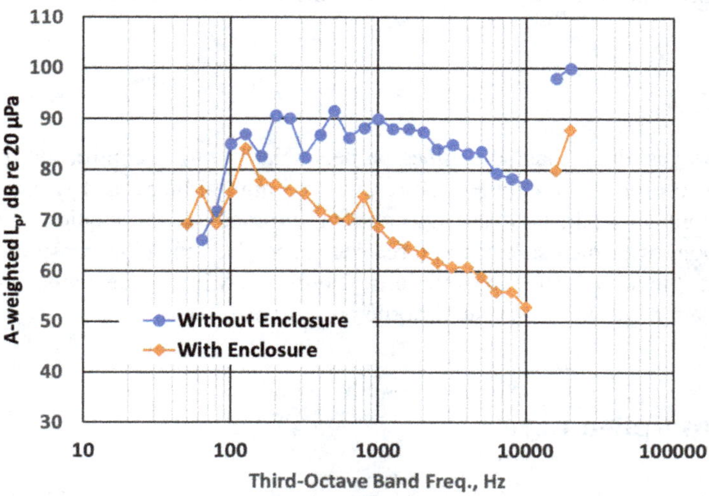

Fig. 8.16 Noise level around gensets with and without an enclosure

One can see small effectiveness in the low-frequency range. However, the 18 dB reduction in the A-weighted noise proves that this kind of treatment may significantly improve the acoustic environment on ships. At the same time, the noise reduction is less than transmission loss of the enclosure panel itself. The zero (or slightly negative) noise reduction is expected at low frequency because of relatively small absorption effectiveness, coupling of the small volume between machinery and enclosure, and usual existence of short circuits. Active noise control would be an effective way to reduce the low-frequency performance.

The performance of any machinery enclosure is quite frequency dependent and is affected by the volume enclosed, the closeness of the equipment to the enclosure wall, transmission loss of the walls, the number and size of penetrations – especially ventilation louvers, amount of absorption on the walls, inherent damping, etc. (Ver 1973; Ming and Pan 2004). The performance of vendor-provided enclosures with their product is usually optimized and accounts for cooling and fire requirements. Optimization of enclosure design is provided by Lee et al. 2003 and Beranek and Ver 1992. A complete treatise on the design of enclosures is provided by Ver 1973.

One item to be aware of when enclosing a source. The limited volume of the enclosure will increase the sound pressure inside the enclosure, p_{in}, relative to the free source level, p_{free}. This ratio is:

$$p_{in}/p_{free} = \left(c^2/(2\pi f)^2\right)\left(4\pi r/V_0\right) \tag{8.1}$$

where r is the distance from the source in the free field, m, and V_O is the enclosed volume, m^3. The frequency $f_{in\ =\ free}$, where the inside and free pressure levels are equal, is:

$$f_{in=free} = 2.35 \times 10^5 r/V_0$$

On marine vessels, it may sometimes be beneficial to turn a control room into an enclosure to protect the engine room crew from high airborne and structureborne engine room noise. In this case, the optimal approach is to isolation mount the control room as an enclosure. Careful seals are needed around high transmission loss doors and the usual cable penetrations. Double or triple pane explosion proof windows are required to prevent airborne noise transmission.

8.5 Absorption Layers

Mineral wool or fiberglass layers are very prevalent among shipbuilder for noise control.[8] The marine insulation industry has a variety of materials differing in density, stiffness, porosity and resistivity, all of which affect its acoustical

[8] Foams are not generally approved by regulatory agencies due to fire, smoke, and toxicity concerns.

absorption. Often, the same material is used for noise control and for thermal and/or fire insulation, thus providing overall weight savings.

Absorption layers influence noise levels through (1) increasing absorption in the room, thereby reducing the reverberant component of airborne noise, and (2) influencing airborne transmission loss (TL_{ab}), radiation efficiency and acceptance.[9] Therefore, accounting for only one of these parameters and ignoring the other can actually lead to a higher than expected received noise level or an overtreatment of any noise environment.

The reverberant component of the noise level in a room, as covered in Chap. 7, depends on the room surface area and material absorption coefficient. These are the controlling parameters of the room constant. Reverberant noise levels are inversely proportional to room constant (see Sect. 7.1). Actual ship compartments without special noise absorptive treatments have some natural absorption offered due to normal fittings such as furniture, bedding, and pipe insulation. Statistically determined averaged absorption coefficients for different types of rooms are shown in Table 7.2.

For example, consider a room with a size of 6 × 6 × 2.5 m. A typical Mess is approximately this size. The full surface area is 132 m^2. In accordance with Table 7.2 at a frequency of 500 Hz, the average absorption coefficient is 0.2. The average room constant at this frequency is $0.2 \times 132 = 26$ m^2. Assume that the whole ceiling (6 by 6 m) is covered with material having an absorption coefficient of 0.8 at 500 Hz. In this case, the average absorption coefficient for this room is $(0.8 \times 36 + 0.2 \times (132–36))/132 = 0.36$. The average room constant in this case is $0.36 \times 132 = 48$ m^2. One can see that the expected reverberant noise reduction is $10 \log (48/26) = 2.5$ dB; therefore, treatment of one surface leads to a noticeable but not significant[10] noise reduction.

Most material vendors can provide frequency-dependent absorption coefficients. As with most vendor information, these data should have been collected to a standard test procedure that is quoted with their information. For absorption measurements, ASTM C423 or ISO 354:2003 are often utilized. Note that due to the test method and edge effects, the absorption coefficient is often listed as greater than 1 in the mid- to high-frequencies. Since absorption cannot physically be greater than 1, these values should always be rounded down to 1. Often, the vendor-provided data only range from 125 Hz to 4000 Hz. Sometimes using a similar material that already covers the 31.5 to 8000 Hz range will allow one to determine the possible slope of the data at the low and high ends. The absorption coefficients for hundreds of acoustic materials are provided in http://www.bobgolds.com/AbsorptionCoefficients.htm.

Another factor evident in ship constructions is that thin metal bulkheads easily 'transmit' noise into the adjacent space. This provides an 'apparent' absorption

[9] Acceptance is the vibration response to an incident airborne noise levels, an important consideration with respect to secondary structureborne noise.

[10] Authors consider 5 dB a significant noise or vibration reduction.

Table 8.6 Apparent absorption coefficient of thin steel bulkhead

TOB Freq, Hz.	16	20	25	31.5	40	50	63	100
Absorption Coef., α	0.71	0.51	0.36	0.26	0.18	0.13	0.1	0.06

coefficient – especially below 100 Hz (Fahy and Thompson 2015; Rosenhouse 1984). At higher frequencies, the bulkhead effectively reflects incident noise, so there is no apparent absorption factor.

For example, a steel bulkhead with a thickness of 0.635 mm will exhibit apparent absorption coefficients, as shown in Table 8.6. This is one reason the SNAME Design Guide in Table 7.7 recommends that as a minimum, the following room constants be used: 61, 15, 4, and 1 m^2 for frequencies between 31.5 and 250 Hz.

For most shipboard applications, the exposed surface of the absorptive material should be faced with a vapor barrier,[11] as a minimum, and covered with a *perforated* metal. The perforations should be a minimum of 25 to 35% open. This cover protects the material from damage. Both of these covers generally reduce the absorption coefficient, especially at low frequencies (Sugie et al. 2006). However, there are also methods to tune the absorptive material to a band of frequencies to reduce a troublesome tone (Sgard et al. 2005).

8.5.1 Absorption Material Influence on TL, *Radiation Efficiency and Acceptance*

Absorption layers will significantly change key noise control parameters: transmission loss, radiation ability, and excitation by incident airborne noise ('acceptance'). Any sound wave penetrates through a loose fibrous material with a significant loss of energy. Thus, if the absorptive layer is attached to a metal surface (bulkhead or deck), the acoustic energy reaching the surface will be reduced. Noise Control Engineering, Ltd. (NCE) tested how fibrous absorption material influences these three parameters. The influence is a function of absorption material features and parameters. The controlling parameters are thickness, density, and flow resistivity of the absorptive materials (Beranek and Ver 1992).

As an example, Fig. 8.17 shows an increase in the transmission loss of the metal panel with different densities and thicknesses of absorption layers.

One can see that a significant increase in *TL* can be reached with relatively light absorptive layers. The positive effects of the absorptive layer start for these specimens in the 125 Hz band and steadily increase with a slope of 5 to 6 dB per octave. Remarkably, this effect does not depend on the metal plate parameters: the same effect is found for the 3.2 mm aluminum and 9.5 mm thick steel panels. Thickness is more of an influencing factor than the density of the material. Figure 8.18 demonstrates the difference between material densities of 48 kg/m^3 (3 lb./cu.ft.) and

[11] The intent is to avoid absorption or wicking of oil, diesel, or other harmful vapors.

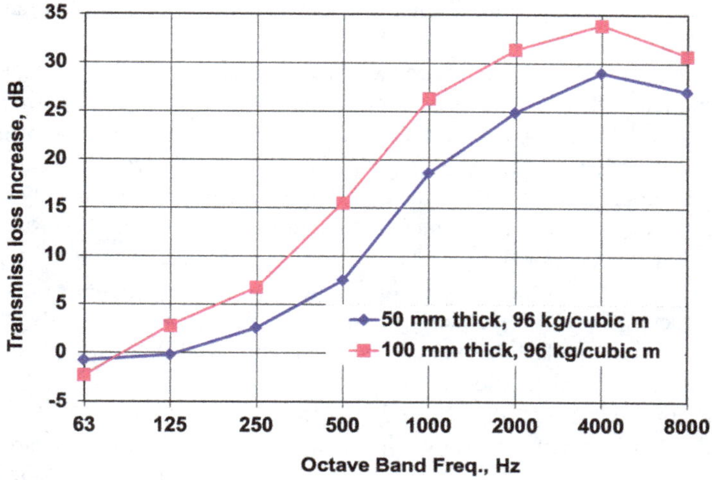

Fig. 8.17 Transmission loss increase with 96 kg/m^3 (6 lb./cu. ft.) mineral wool. (Note 2″ = 50 mm; 6 lb./cu.ft. = 96 kg/m^3; 4″ = 100 mm)

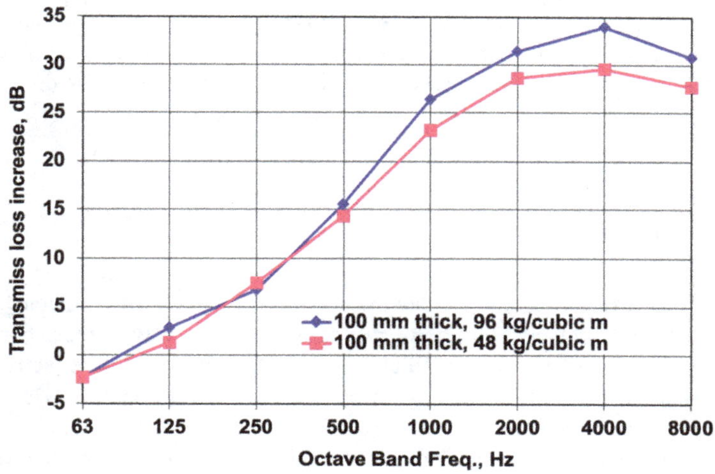

Fig. 8.18 Transmission loss increase with materials of different density. (Note 4″ = 100 mm; 3 lb./cu.ft. = 48 kg/m^3; 6 lb./cu.ft. = 96 kg/m^3)

96 kg/m^3 (6lb/cu.ft.). The material that is twice as dense has a slightly better performance in the high-frequency range. Obviously, thickness is more of an influencing factor.

The higher density insulations typically increase the effective *TL* at frequencies above 1000 Hz. This increase is in the 3 to 6 dB range. Thicker insulation also increases the *TL*. Going from 5 cm to 10 cm can provide an additional TL of 5 to 9 dB.

As stated above, the total noise reduction due to absorptive layers depends on several parameters. This fact gives an opportunity to make the additive correction for TL, radiation efficiency and/or acceptance based on only the absorptive material properties. Insulation placed on a base plate will also effectively decrease the composite structure's radiation efficiency. This decrease is approximately equal to the increase associated with the change in transmission loss. Similarly, insulation on a base plate will decrease the acceptance of the composite structure. Again, in proportion to the effect of the change in TL, however, there is a weak dependence on the characteristics of the underlying structure. For a baseline study, one can assume that any particular treatment equally influences all three key parameters – TL, acceptance, and radiation efficiency.

In many practical cases, the absorptive layer is spaced some distance from the metal surface. Measurements show that this air gap influences the absorption coefficient but does not influence transmission loss, radiation efficiency, or acceptance.

Different materials can be 'layered' together to optimize and improve or achieve the desired transmission loss and/or absorption (Guy 1989; Congyun and Qibai 2005). Guy provides equations to account for different types of materials, air gaps, and base structures. Bear in mind that most theoretical models can predict unattainable values given their basic ideal assumptions. At best, these models do allow one to determine general trends and resonance conditions for comparative purposes.

8.5.2 Tuned Absorber

Low-frequency absorption can be obtained with commercially available passive acoustic absorbers such as Helmholtz resonators, quarter wave tubes, perforated acoustic liners, or passive acoustic radiators. Novel treatments involve 'active tuned acoustic absorbers', which are being developed by vendors such as Deicon and Kinetics.

A perforated facing, with a cavity backing, can be tuned to maximum absorption at a frequency (Bies 2009), f_{max}

$$f_{max} = \frac{5500P^{0.5}}{[L(t + 0.85d(1 - 0.22d/a))]^{0.5}} \text{ Hz} \tag{8.2}$$

where P is the percent open area of the perforated facing, L is the depth of the cavity, mm, t is the facing thickness, mm, d is the diameter of the perforations, mm, and a is the spacing between the holes, mm.

For two solid walls separated by an air gap, d, the resonant frequency, f_{gap}, of the two panels as masses against the springiness of the air is approximated by:

$$f_{gap} = \frac{1}{2\pi}\sqrt{\frac{2\sqrt{2s'}}{\rho_d}}\ \text{Hz} \tag{8.3}$$

where ρ_d is the mass density of either panel, and

$$s' = \begin{cases} \dfrac{\rho_0 c_0^2}{d} & for\, f < f_d = \frac{c_0}{2\pi d} \quad \text{N/m}^2 \\ 2\pi f \rho_0 c_0 & for\, f > f_d \quad \text{N/m}^2 \end{cases}$$

New materials involving meta-layers are being developed to tune absorption to lower frequencies. Examples are provided by Sergeev et al. 2023 and Sui et al. 2015. These materials would need regulatory approval for fire, smoke, and toxicity. Metamaterials are also being investigated with respect to their decoupling properties with respect to underwater noise (Huang et al. 2016).

8.6 Joiner Panels, Floating Floors, and High Transmission Loss (HTL) Treatments

Joiner panels are often a part of a finish surface in accommodations. The acoustic effectiveness of joiner panels was discussed in Chap. 7. It is a common situation when noise radiation from uncovered structural elements controls the noise levels. In

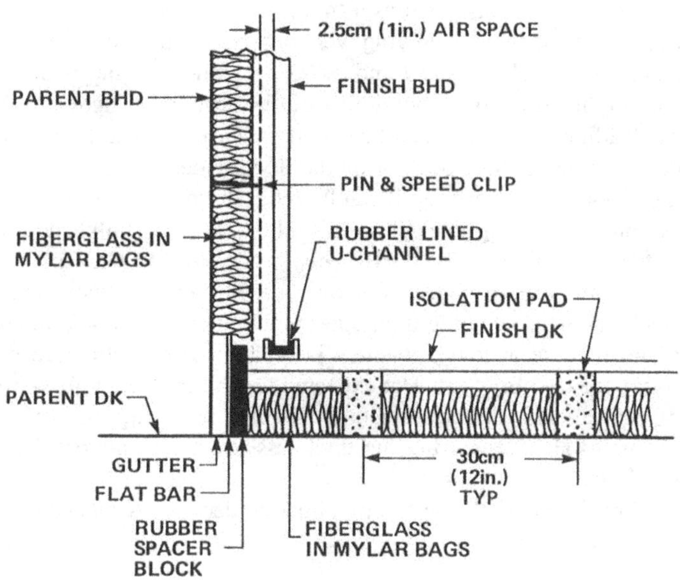

Fig. 8.19 Typical floating room details

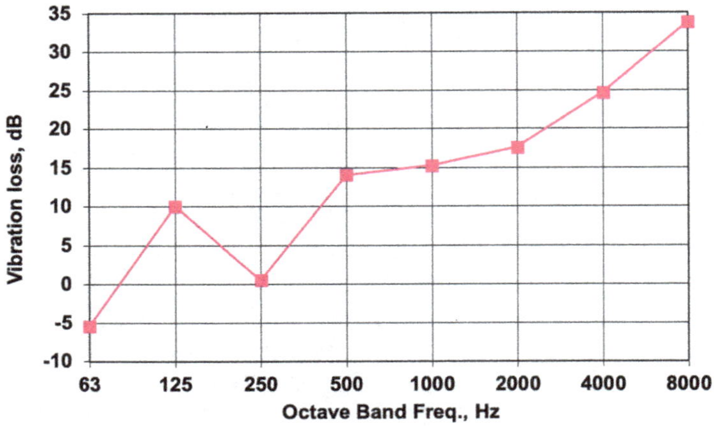

Fig. 8.20 Vibration level difference between the structural deck and floating floor

this case, adding a joiner panel with some gap from the hull structure will significantly reduce the received noise levels.

One very effective noise treatment is a floating floor. Figure 8.19 shows details for a floating floor and a floating bulkhead, both making up a 'floating room'. The gap on the deck is on the order of 50 to 100 mm. In this case, isolation pads, typically on 30 cm spacing, support the floating deck. A bagged fiberglass less than the thickness of the gap is used on the void to improve airborne transmission loss from below. A smaller gap, at least 2.5 cm, is recommended for use on the bulkhead.

As an alternative, the floating floor can be supported directly by fiberglass batting. There are additional variations, such as a 'multi-layered' floating floor (Song et al. 2017). Multiple vendors are supplying versions of floating deck materials and composites that meet SOLAS A-30 and A-60 fire codes. Some of the trade-offs are thickness, weight, curing time, possibility of rust developing under the treatment, and allowable loading on the finish surface. Performance is generally increased with the use of damping on the base deck or on the floating deck.

A floating floor is able to significantly reduce radiation from the deck by 15–20 dB. If only a floating deck is used without floating bulkheads, the noise reduction in the cabin is less, 7–10 dB. This is because the deck is not the sole surface radiating into a room. Fig. 8.20 shows the difference between the vibration levels of a deck below the floating floor and the vibration levels on a floating floor. Some vibration increase at low frequency is typical for this structure due to the air-cavity resonance (see Eq. 8.3). The expected natural frequency for a 5 mm steel deck on a 50 mm thick/96 kg/m3 mineral wool batt, including the stiffness of the entrapped air, is 40 Hz. Decreasing the thickness to 25 mm increases the natural frequency to 55 Hz.

The controlling factors in the design of a floating deck are (Cha and Chun 2007):

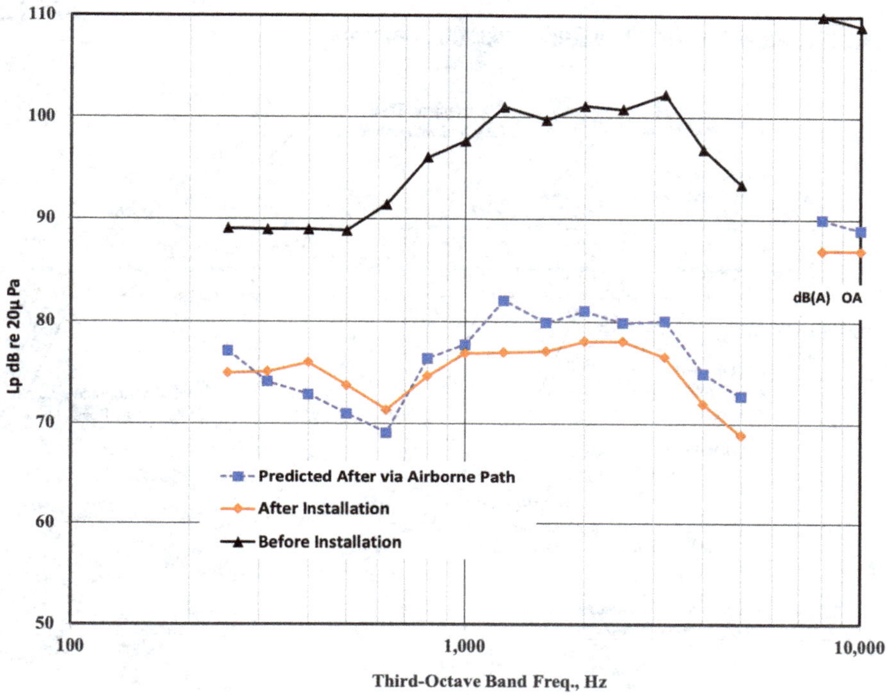

Fig. 8.21 Floating room performance – Laboratory testing

- the height of the gap – bigger is better
- the spring constant of the isolation system supporting the deck – whether distributed isolators or a continuous layer of fiberglass – which is limited by the stiffness of any air trapped between the structural deck and floating deck
- the surface mass of the floating deck
- finish surface on the floating deck – especially any applied damping treatment

Figure 8.21 shows measured and predicted data taken at a land-based mock-up of a ship structure. In laboratory conditions, the noise reduction only starts above 125 Hz, and the reduction is on the order of 20–25 dB.

There are several vendors acquainted with this type of installation onboard ships. Insertion loss (noise with and without the floating room installed) data are typically available. Their data should be taken in accordance with a standard test procedure, the materials should be acceptable to regulatory agencies, and provisions for penetrations and prevention of water egress should be taken into consideration.

Another high transmission loss treatment utilized on ships is layers of fiberglass-limp mass layer material-fiberglass. If this composite covers the surface of the machinery, it is considered 'cladding.' If installed over structural bulkheads or the underside of decks, it is a High-Transmission Loss (HTL) treatment. This treatment

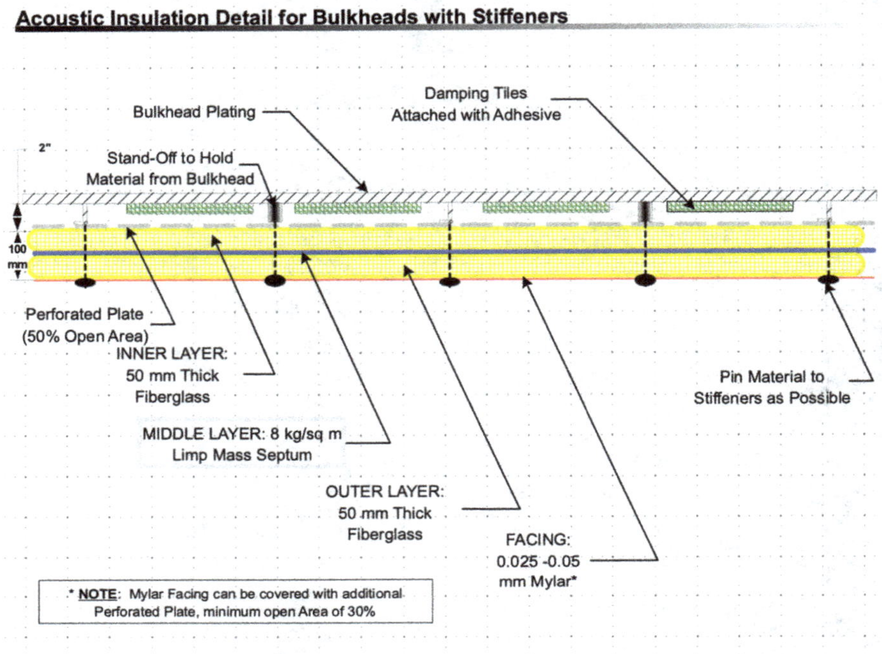

Fig. 8.22 High Transmission Loss (HTL) treatment cross-section

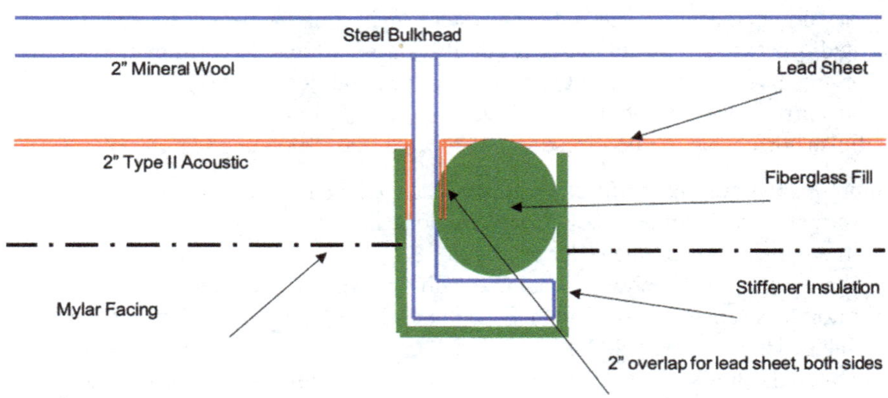

Fig. 8.23 Stiffener Wrap Details

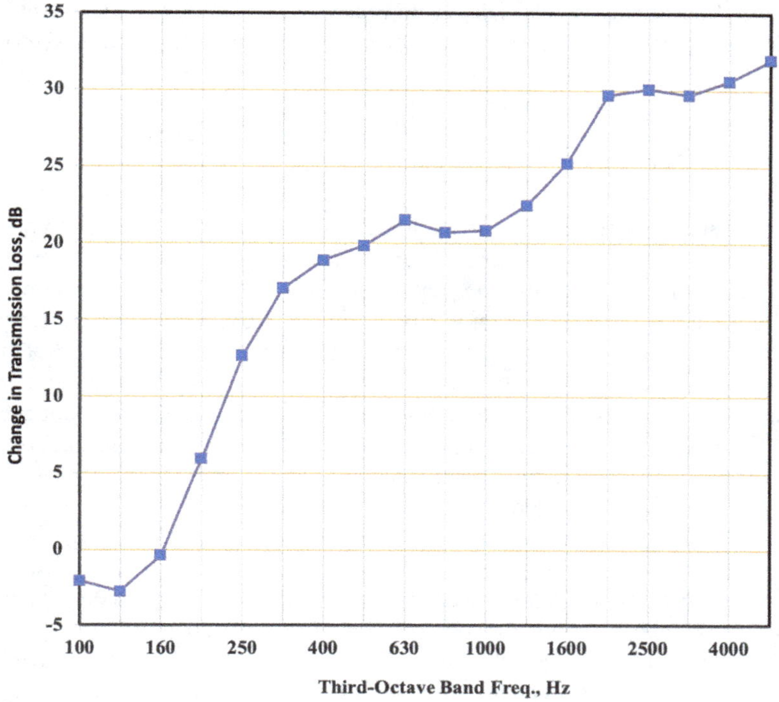

Fig. 8.24 Added transmission loss of HTL treatment

cross-section is shown in Fig. 8.22. The figure also shows the installation of a damping tile on the bulkhead. The composition of the intermediate layer must meet regulatory requirements. Typically, this involves a limp lead layer, which has its own handling drawbacks. The recommended wrap around the stiffener is shown in Fig. 8.23.

The measured average increase in transmission loss over five HTL treatments is provided in Fig. 8.24. The limp surface mass was between 5 and 9 kg/m^2.

Composite bulkheads are commonly used. Based on various tests, it has been determined that the core stiffness controls the acoustic properties of composites – at least those with skins on both sides. Heron, 1979 found that "as the core stiffness is changed from high values through to low, the acoustic properties go from medium to bad to good. If possible low core shear plates should be used, but if this is not possible, then high core shear plates should be used." Furthermore, the airborne TL of composite honeycomb panels can be improved by 10–15 dB above 800 Hz by changing the bonding techniques. Feng 2003, found that "laminates are better to be glued in meshes instead of continuously bonding." Composites have also been utilized for high-speed vessels. They have been selected for their 'built-in noise and vibration damping, which gives a low hydro-acoustic signature' (Vallbo 2005).

8.7 Silencers/Mufflers

Silencers[12] and mufflers are treatments inserted in duct or piping systems to prevent source noise from reaching the atmosphere or other components in the duct or piping system. Sources can be gas turbines and diesel intakes/exhausts, fans, and pumps. There are a large variety of silencers or mufflers – dry, wet, tuned, absorptive, dissipative, active, etc.. The muffler presents an impedance mismatch to the acoustic energy from the source side to the receiver side (Munjal 2014). Note that an internally lined acoustic duct is not considered a silencer since it does not present much of a path impedance. It is just an attenuation that depends on the size and length of the duct. As with most treatments, not just any muffler or silencer can be selected based on its expected acoustic performance. The attendant pressure loss needs to be factored into the selection process. The physical size of the muffler or silencer can also be a limiting factor if not considered early in the design. As a matter of good design practices, the silencer or muffler should be supported on isolation mounts. These mounts should be able to withstand high temperatures.

Gas turbine silencers are generally 'custom designed' based on the available exhaust cross section and length, temperature, flow rates and allowable pressure drop, physical space for the number and types of turns, and exhaust source level. Most often, parallel baffles are used. These need to be aerodynamically shaped, especially on the inlet side. Given the high temperatures, the selection of an acoustic lining is critical.

Table 8.7 provides general uses and properties for acoustic materials (Elden Ray Jr. 2006):

The absorption characteristics of rock wool (60 kg/m^3) and ceramic fiber 80 k (80 kg/m^3) are similar (Lee and Liu 2006). The transmission loss of these two materials is similar up to 2000 Hz. Above this frequency, there is a dip of approximately 8 dB between 2000 and 4000 Hz. All of the materials in Table 8.7 have approximately the same NRC value.

Numerous vendors can provide effective mufflers and silencers. Again, some vendor-provided insertion losses are very optimistic, particularly at low frequencies. The common thread is the data are based on flawed or biased testing. Existing computer aided design software (CAD), based on computational fluid dynamics

[12] Silencers are typically 'absorptive' devices, whereas, mufflers typically refer to impedance mis-match devices.

Table 8.7 Gas turbine silencer material (Ray 2006)

Silencer Application	Acoustic Material	Retainer Material	Temperature Range	Thermal Cycling	Durability	Resistance to Friability[1]
Inlet	Mineral Wool	Cloth Mesh[2]	N/A	N/A	Good	Good
	Fibrous Glass Blanket/Batt	Cloth Mesh[2]	N/A	N/A	Good	Good
Exhaust	Mineral Wool	Wire Mesh	~425°C 800 °F	Good	Moderate	Moderate
	Ceramic	Wire Mesh	~1100 °C 2000 °F	Excellent	Poor[3]	Low[3]
	Basalt	Wire Mesh	~760 °C 1400 °F	Poor[4]	Good[4]	Moderate
	Fibrous Glass Blanket/Batt	Wire Mesh	~760 °C 1000 °F	Good	Good	Moderate

[1] Friability – the propensity to breakdown or disintegrate
[2] A very low flow resistance is needed to maintain performance
[3] Packing of material and low vibration critical for longevity
[4] Protective thermal shock blanked needed for longevity
Perforated sheet should have an open area ratio greater than 35% on inlet and 25% on exhaust

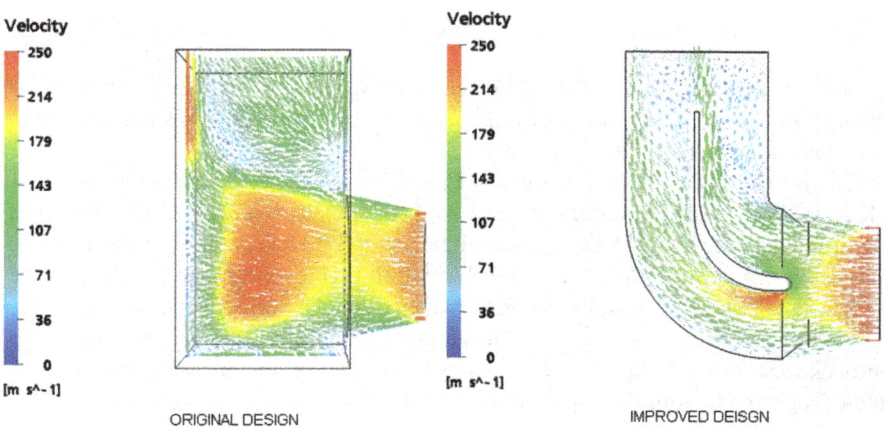

Fig. 8.25 Improved flow with turning vane (Permission M&I Power Technology (CDN) Inc.)

(CFD), can be used to design custom or unusual silencers (Jones and Kessissoglou 2009; Ramakrishnan 2010). Use these programs with caution below 125 Hz. Conservative insertion losses for various sized silencers and mufflers are provided in the SNAME Design Guide (Fischer et al. 1983).

Fig. 8.26 Exhaust Stack – Before added barrier & after added barrier

One factor that needs to be addressed, particularly for high-speed flows over silencer surfaces, is flow-induced noise. This component can limit the acoustical effectiveness of a well-designed system.

One optimization method, shown in Fig. 8.25, effectively reduces the need for silencers. Well-designed turning vanes (lined or unlined) can be used to ensure that the flow path segments do not cause turbulence and flow separation. Both of these can produce high noise levels that are difficult to predict and control. If flow noise occurs, it can effectively short circuit the silencer/muffler effectiveness.

Another noise control treatment for diesel or gas turbine exhaust contribution to on-deck locations is to build a 'stack barrier.' This barrier enhances the loss along a line-of-sight path from the top of the exhaust stack to a receiver location, e.g., an on-deck work station or the pilothouse. This approach is particularly effective at aft work stations, which often can look directly into the exhaust pipe opening. An example is given in Fig. 8.26.

8.8 Damping Tiles and Sprays

Starting in the 1960s, naval shipbuilding industries started to use special damping coatings on hull plating to reduce structureborne noise. This effect is based on a high level of energy loss during plate vibrations due to the high viscosity of the damping material bonded to the structure. Some initial results were not promising (Buiten 1972; Fahey 1976). One reason for this is that all major structureborne paths need treatment or the damping is 'flanked' by vibration taking the easiest path to the receiver. The waveguide behavior of stiffened plates can also limit the effectiveness of damping. Basic concepts, theory, and details of damping are provided in Nashif et al. 1985. An early and experimental study of ship damping was carried out by Chang and Carroll 1981.

There are two main types of damping applied in the marine industry. They include constrained layer damping and unconstrained layer damping. The damping material is typically applied only to plate elements between beams and stiffeners, as the latter do not typically benefit from damping. Thickness is the most important parameter for damping effectiveness. The rule-of-thumb is that the damping thickness should equal the plating thickness. In many cases, damping as a tile material generally has the best performance. Spray-on treatments are being developed that combine damping effectiveness with other properties, such as anti-sweat, thermal and/or fire properties. The spray-on materials are typically easier to apply than the damping tile, which needs to be held in place while the glue dries. Dry time versus humidity, cure time, surface preparation, and application thickness without sagging are some concerns when contemplating spray-on damping treatments. It is important that damping material has regulatory approval, especially for use in accommodation areas.

Tiles are applied with adhesive to the ship's structure. The adhesive for unconstrained damping can either be applied by hand with a trowel or brush or sprayed on.

Vendors of damping material often provide reduced frequency curves or nomograms that allow one to determine the loss modulus and loss factor for a particular frequency of interest and temperature. An example of such a curve is provided in Fig. 8.27 from Pyrotek. The procedure taken to determine the dynamic modulus and loss factor for a particular frequency and temperature is provided with the figure.

There are optimization techniques for both constrained and unconstrained damping treatments. For instance, the top constraining plate can be segmented to tune the damping effectiveness to a particular frequency (Plunkett and Lee 1969). Another approach is to add a spacer between the base plate and the constraining plate (Subramanian et al. 2004).

The application of damping tiles to the foundation floors for a diesel generator is shown in Fig. 8.28. These are unconstrained tiles – black in color.

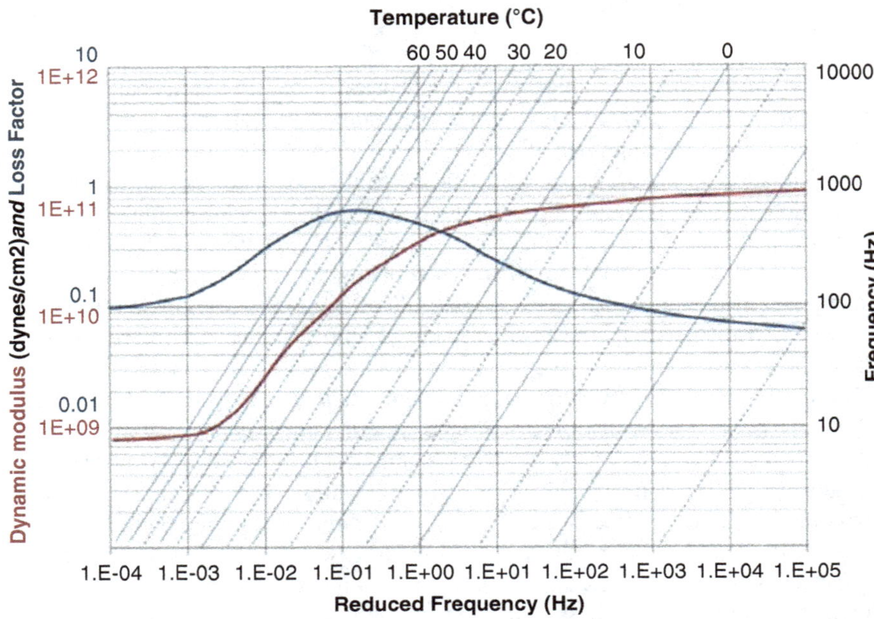

Fig. 8.27 Reduced-frequency damping curve
Reduced Frequency Nomogram (Report No. 19313-PH); (https://www.pyroteknc.com/products/)
How to read a reduced frequency nomogram:
1. Start by selecting the frequency (Hz) on the right-hand vertical axis.
2. This value is followed horizontally to the left, where the diagonal temperature isotherm intersects.
3. Draw a vertical line through the frequency and isotherm intersection, find the point where this line intersects the modulus and loss factor curves.
4. Draw horizontal lines from these points to the left-hand vertical axis to read the values.

8.8.1 Constrained Layer Damping

Constrained layer damping is one of the most effective types of damping treatments. The optimum approach is that the damping material and constraining cover plate be on the order of thickness of the base plate. The damping tile is installed with the pieces trimmed-to-fit. A steel or aluminum plate is then glued to the top of the damping tile. In some cases, the metal plate is secured by threaded studs to the ship's plating (this is particularly useful for slightly curved surfaces and/or to meet shock requirements). These tiles are nominally 0.3 × 0.3 m.

As an alternative, the factory can supply a viscoelastic layer that is permanently bonded with epoxy to a thin steel or aluminum plate. This bonded plate will be the finish side of the damping tile. For most marine-based cases, these vendor-supplied constrained layer damping tiles with thin sheets are not very effective. Thus, a custom-designed system with relatively thick constraining plates works best. To

Fig. 8.28 Un-constrained damping tiles on foundation floors (Floors on machinery foundations are vertical surfaces) (below isolation mounts)

save weight, it has been found that aluminum rather than steel constraining plates work relatively well (Fischer 2008). A detailed study showed that "For damping materials with the appropriate Young's Modulus, there is relatively little difference in the performance of aluminum v. steel constraining layer." Therefore, an aluminum constraining plate is recommended for any location requiring a constraining cover plate" (Fischer 2005).

Damping converts vibration energy to heat because of the imperfect elasticity of the material (hysteresis). This constrained layer damping is very effective because shear is developed between the ship plate and the viscoelastic layer. This shearing also exists between the viscoelastic layer and the constraining metal on the top. Although effective, this damping material is expensive, tricky to install, adds significant weight, and cannot be installed in some spaces because the epoxy used is toxic if burned. Constrained layer treatments are not typically installed inside potable water or fuel tanks. If the constrained layer is on the underside of a deck, it needs to be held in place by a physical restraint until the epoxy sets. Another approach is to weld several threaded studs in place that penetrate and secure the restraining plate.

8.8.2 *Unconstrained Layer Damping*

Unconstrained layer damping is not as effective as the constrained layer variety. This is because there is less shearing of the damping layer. This type of treatment is most effective at reducing the resonant response of structures. Unconstrained layer damping can be applied either as glued on tiles or as a spray-on treatment. For all damping, the thicker the treatment relative to the base plate is, the greater the effectiveness. The performance of most damping treatments is highly temperature dependent. Thus, a ship that operates in warm waters may have a different effectiveness of hull damping when operating in cold waters.

Noise prediction algorithms show that energy dissipation in ship structures is a significant factor influencing noise and vibration levels throughout the ship. The loss factor η is used to quantify energy dissipation. The loss factor is equal to part of the energy that dissipates without returning during time T/2π, where T is a period of vibration. However, the loss factor is not the only parameter that controls energy loss. As shown (Cremer and Heckl 1973), for a unit volume of material with a modulus of elasticity, *E,* the energy lost during 1 sec of longitudinal vibration is equal to ½ $E\eta\omega\varepsilon^2$, where ε is the deformation amplitude.

This equation shows that energy loss is proportional not only to the loss factor, which is expected but also to the modulus of elasticity and deformation squared. This equation clarifies the major parameters that control energy loss in a vibrating structure. It can be used as guidance for damping material developers who should consider stiffness, deformation, and loss factors. These parameters are "instructions" for damping material developers. The production "ηE" is sometimes called the "loss modulus" to underline the importance of stiffness in the energy loss process. Damping systems differ by the type of deformation, which provides energy loss. Stiff damping addresses longitudinal deformation. Soft damping coatings and constrained damping tiles affect shear deformation. Additionally, damping material with a high stiffness leads to a lower "critical frequency" and increased radiation efficiency. Thus, damping materials potentially have both positive and negative effects. This has led to the checkered history of many damping treatments (Fahey 1976).

The stiffness, viscosity and deformation of the damping material are used in practical damping treatments in different ways. As noted above, there are stiff and soft damping materials. Stiff materials are usually composites, for which the modulus of elasticity is less than that of metal by a factor of three. The density of popular stiff damping materials varies between 1200 and 2000 kg/m^3. These materials may be used in tile shapes and as rolled materials. They should be firmly bonded with structural plates. Optimally, the thickness of the stiff damping material should be close to the thickness of the base plating or higher. Recently, damping "spray on" has become popular (Spence 2011; Rao 2003) due to technological advantages. Commonly, "spray on" materials are less efficient than the same mass of a layered composite. Table 8.8 shows the vibration reduction measured for a standard damping tile and several vendors and thicknesses of spray-on damping.

Sometimes stiff damping materials are used with a spacer layer, which may be thicker than the basic plate and damping layer. As a "spacer", the designer can use a stiff and light foam or a metal honeycomb layer. A spacer acts as a lever to increase the deformation of the damping material to increase the loss factor, as mentioned above. Typically, longitudinal deformation of damping layers creates a damping effect.

Table 8.8 Vibration Reduction for Sample Damping Materials

Freq, Hz		60	170	340	560	Avg
Material	Thickness, mm	Vibration Reduction, dB				
Damping tile	3.175	7	11	12	14	12
Vendor						
1	5.5	5	9	9	12	9
2	4.2	3	7	8	11	8
2	3.8	4	8	8	10	8
2	2.54	2	7	8	10	8
3	1.78	1	6	7	10	7
3	1	3	5	5	8	6

Soft damping materials are usually used in combination with a thin metal "constraining" layer atop the damping layer. Sometimes multilayer combinations of viscoelastic and constraining layers are used in practice. During bending vibration, the constrained layer "slides" along the soft viscoelastic layer, and the shear deformation of the damping layer provides a damping effect (Kihlman 1976).

From noise prediction and noise control points of view, it is critically important to acquire accurate loss factor values for untreated and treated panels. Many methods exist for loss factor measurement (ASTM E756-05 and SAE J1637 Feb93). These are discussed in Koruk and Sanliturk 2010. The most popular and practically available method is the half-power bandwidth of the resonance curve method (Treszkai et al. 2020; Gu and Sheng 2015). ASTM E756-05 is the standard procedure for this test method. If one excites a multi-resonance system (strip or plate) with a swept sine signal (pure tone) and measures a response, a loss factor on each resonance frequency is equal to:

$$\eta = \Delta f / f \tag{8.4}$$

where f is the resonance frequency and Δf is the width of the resonance curve at a vibration level, which is 3 dB lower than the maximum resonance level at frequency f.

The authors found that the loss factors of a similarly coated structure measured on a beam and on a plate differ. Typically, the damping coating actually used on a ship structure (plate and stiffeners) has a slightly lower loss factor than the same configuration on a beam. The main reason is that the stiffeners on the ship are not damped and that the damping effectiveness on a beam would be minimal.

Even if the loss factor is measured correctly, the effectiveness of damping for a real structure is less than $10lg(\eta/\eta_0)$. The chief reason is non-resonant modes, which are not as sensitive to damping as the resonant mode. The contribution of non-resonant modes to the overall wide band vibration level may be high enough to reduce the apparent damping effectiveness. Another reason is that the radiation

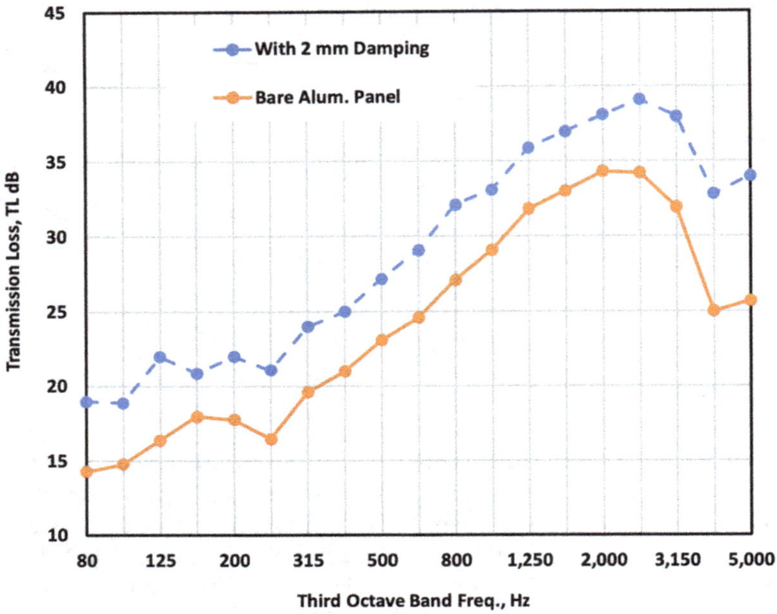

Fig. 8.29 Transmission loss before and after application of damping

Table 8.9 Loss factors of the 3 mm thick aluminum ribbed bulkhead

Bulkhead Material	Coating Type	Plate thick., mm	Coating thick., mm	Octave Band Damping Loss Factor				
				63	125	250	500	1000
Aluminum	No coating	3	0	0.007	0.006	0.007	0.0047	0.0032
Aluminum	Damping	3	2	0.063	0.055	0.057		

efficiency increases, causing typically stiff coatings and light coatings to decrease the critical frequency of plating. This leads to higher radiation efficiency (Kou et al. 2015).

An accurate value of loss factors is necessary for accurate noise prediction during the design stage. From a noise control point of view, it is important to determine how the increased loss factor practically affects the critical acoustical parameters of a structure. For example, Fig. 8.29 shows how one damping material influences the airborne transmission loss of a bulkhead, as tested in a reverberant chamber. The measured loss factors of this bulkhead are shown in Table 8.9.

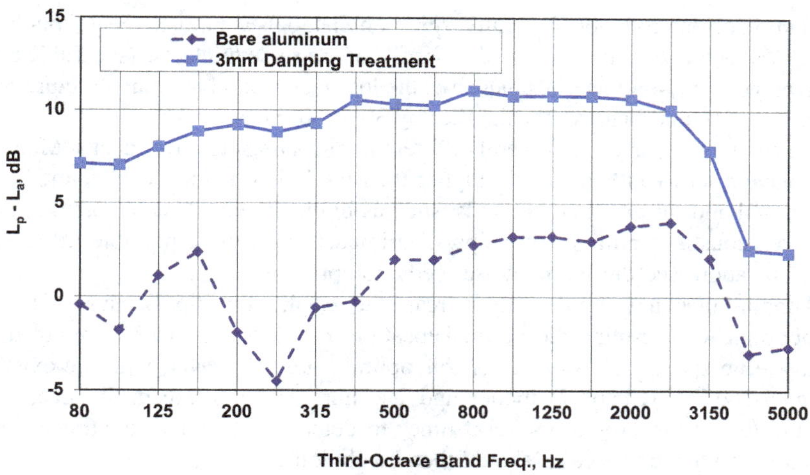

Fig. 8.30 Acceptance of the aluminum bulkhead before and after damping

Below the critical frequency (4000 Hz), the transmission loss is increased by approximately 5 dB, and near the critical frequency, it is increased by 9 dB. However, the expected improvement at 250 Hz is 10 log (0.057/0.007) = 9 dB, which only corresponds to the measurement around the critical frequency. Similar "reduced" values in transmission loss were found for any treated bulkhead.

For many applications, it is necessary to know the response of a structure to incident airborne noise excitation (acceptance). This parameter may be quantified as the difference between the sound pressure level (L_p) and the plate acceleration level (L_a). Figure 8.30 shows the measured acceptance of an aluminum bulkhead. In this case, the vibration reduction in many third-octave bands is close to the expected 10 Log (η/η_O), where η_O is the loss factor of a structure before damping treatment and η is the same after damping treatment.

The difference between the effectiveness for transmission loss and acceptance is an indication that the ability of the structure to radiate noise increases after damping. This is possible if the system stiffness increase exceeds the mass increase. The opposite effect is also possible, for example, with "soft" (low Young's Modulus) damping materials, when the mass increase is higher than that for bending stiffness. In this case, the "decoupling" effect may help to further reduce noise.

Currently, there are only a few effective damping products that are commercially available that also may comply with marine fire, smoke, and toxicity regulations. Known examples are EAR Isodamp C-1002 (roll), Mascoat Delta dB (spray-on) https://mascoat.com/sound-control/sound-control-db-marine-sound-proofing, and Pyrotek Decidamp SP150 (spray-on) https://www.pyroteknc.com/products/

decidamp/decidamp-dc30/. Typically, spray-on materials are more recent[13] but have lower loss factors than composite tiles or rolls of the same thickness (see Table 8.8). On average, spray-on materials increases the loss factor of a bare hull structure by a factor of 7–10 times and composite tiles by 30 to 50 times.

The marine industry uses several different hard coatings to cover decks. Many of them have a loss factor markedly higher than the hull structure loss factor. Some latex-based and oil-based compounds, such as bitumen, have loss factors that may improve acoustic conditions on ships. The impact of all these possible treatments should be addressed during the noise prediction process.

Damping tiles may significantly increase the weight of a ship. Because of this, it is important to determine the optimal location and distribution of damping tiles. There are no specific algorithms for this optimization. However, with the existing *DesignerNOISE®* (Chap. 7) model and its short program execution time, it is possible to run the program several times to determine damping treatment noise reduction. Two extreme cases should then be obvious.

The greatest noise reduction is expected in the room where the damping tiles are applied. Similarly, the greatest vibration level reduction will be on the structure where damping tiles are applied. If a small number of rooms need to be treated, it is best to cover these room surfaces. Of greatest importance is the treatment of structures with the highest radiation activity. If a large number of rooms need to be treated, it is reasonable to install damping tiles around the sources controlling the excessive noise levels. The noise reduction in this case is expected to be lower than that in the "local" application because of the variety of flanking paths for structureborne noise.

8.9 Ventilation Treatments

As a distributed system, the HVAC can be an important noise contributor to compartments – manned and with machinery/equipment. HVAC systems also impact work spaces and on-deck stations. Ventilation noise is produced by a combination of aerodynamic (fan and flow regenerated) and mechanical noise. In addition to any other noisy system, HVAC-induced noise can be abated or reduced by following the good acoustic design practices discussed in Chap. 10. As a general objective, HVAC noise should be designed to be at least 5 dB below the compartment's noise limit. This is to ensure that when other airborne and structureborne contributions are considered, the total received level is below the criteria. As with distributed systems, fan coil units located within a compartment should be purchased from the vendor to be 5 dB below the compartment's criteria. These units should be tested to the latest American National Standards Institute (ANSI)/Air Movement and Control Association (AMCA) test standard.

[13]Developed to be easily applied on ship structures

The treatments for ventilation noise are as follows:

- silencers
- internally lined duct
- externally lined duct

These frequency-dependent treatments range in effectiveness from 3 to as high as 40 dB.[14] Extensive tables for the Noise Reduction (NR) afforded by the treatments listed above are provided in the SNAME Design Guide (Fischer et al. 1983) and will not be repeated herein. Heating, Ventilation, and Air Conditioning (HVAC) vendors can often supply ventilation designs with treatments that are familiar to them and work with their systems. The major hurdle is designing and actually constructing a ventilation system that does not violate innate assumptions or create sound shorts. Some of the innate assumptions, such as a flow speed and radiused turns, are described in Chap. 10.

The airborne transmission path in straight duct runs is often flanked by structureborne paths. The airborne noise in the duct excites vibrations of the duct walls, which then propagate along the duct wall with little attenuation. This second structureborne noise is transmitted along the duct wall and radiated downstream as airborne noise in the duct. Unless the structureborne path is broken with flexible duct couplings, the maximum attenuation of in-duct airborne noise that can be expected is generally approximately 25 dB for the mid-frequencies. Therefore, without flexible duct couplings, the length of the lined duct should not exceed 6 m (20 ft.). Nine meters (30 ft.) of lined duct would not generally be more effective than 6 m (20 ft.) unless structureborne flanking in the duct wall is treated. In general, most HVAC systems should have at least 3 m (10 ft.) of internally lined duct upstream and downstream of all fans to reduce fan in-duct noise transmission. Most ventilation systems with flows greater than 5 m^3/sec (10,000 cfm) should have ducts with internal lining with a thickness of at least 50 mm (2 in.) for a distance of approximately 4.5 m (15 ft.) to reduce fan airborne noise transmission in ducts.

If additional attenuation in the duct is necessary, duct turns should be lined for a distance of at least three duct diameters before and after the turn. This should provide an additional 3 to 6 dB of attenuation at high frequencies, depending on the size of the duct. The larger the diameter of the duct is, the more effective the line turn. Noise in high-velocity HVAC systems can generally be controlled with the use of acoustically lined, flexible ducts and a two-stage expansion prior to entering the cabin (Neale 2006).

When the pressure drop can be tolerated and space permits, HVAC silencers can be used to provide significant noise attenuation for systems with noise excesses. Another noise control treatment of in-duct airborne noise transmissions is to line the plena with acoustic absorption material. Lining the plena will create a "silencer" in

[14]The authors recommend limiting any treatment to 40 dB or less since under realistic ship conditions multiple flanking paths such as structureborne flanking and/or break-in noise almost always exist.

the system. The advantage of lined plena is that the attenuation characteristics are high at low to mid-frequencies, whereas lined duct attenuation is low. Careful design of the plena and the placement of baffles, if necessary, can lead to the removal of discrete or narrowband tones caused by many fans. The plena chamber(s) can be designed to have a damped acoustic resonance at the offending tone-frequency of the fan. The attenuation offered by a lined plenum chamber can be as high as 30 dB with a strong frequency dependence. A three-pass plenum (two baffles in the flow path) can provide up to 50 dB of attenuation. The chief disadvantage of a lined plenum is its high-pressure loss. If the plena already exist in a system, lining the chamber walls can provide significant noise attenuation (Beranek, 1971).

Airborne noise inside the duct, if transmitted through the duct wall (breakout noise) into a noise-sensitive space, can be as important as the noise transmitted through the diffuser or terminal into the space. High noise levels at low frequencies that exist in most HVAC systems are usually difficult to prevent from being transmitted through duct walls. To counter this, it has been found that round or oval ducts provide higher losses for breakout noise transmitted through the duct walls than those found with rectangular or square ducts. This is particularly true at the low to mid frequencies. For this reason, round or oval ducts should be used wherever possible in HVAC systems for ducts passing through noise-sensitive space.

Lagging treatments can be used to reduce the noise radiated from the surface of vibrating duct systems (breakout noise). Most lagging treatments consist of a layer of fiberglass covered by a layer of impervious material such as a sheet of aluminum or, better yet, a limp-mass layer. The limp mass layer would need to meet regulatory standards. The effectiveness of such a cladding treatment is small at low frequencies but can reach 20 dB at high frequencies.

Other guidance and considerations are provided in Gungor 2003. HAVC prediction and control software programs are available (Kandekar et al. 2019, AcousticCalc and Vibro-acoustic). *DesignerNOISE*® has a prediction package that maps noise due to any HVAC distributed system to the room containing a pass-through duct (break-out noise) and/or a terminal or terminals.

Add-on controls for tonal duct borne noise include the side-branch resonator (Hansen 2005), Helmholtz resonator (Fahy and Thompson 2015), Herschel-Quincke Tube (Karlsson and Glav 2008), and Quarter-Wave tube resonator (Hansen 2005). These are obviously used to control the tonal noise in a duct or pipe (including diesel exhaust). These types of passive systems can reduce tonal noise by 10 dB or more (Munjal 2014). Duct- or pipe-borne noise can effectively be controlled by active systems. The technology is very mature for this type of control. Active control is generally accomplished with a downstream sensor and speaker that generates 'anti-noise' as controlled by the active system. There are limits, such as hot gases and a corrosive atmosphere, with their use in exhaust systems.

Exhaust hoods in galleys can be very noisy due to their high velocities. Due to cleaning requirements, fibrous materials cannot be used as silencers or to line ducts.

There are 'no-media' silencers that can often be used in these systems. These can be steam-cleaned to meet health requirements.

ASHRAE has complete documentation on HVAC noise and its treatment (Schaffer 2005). Active noise control has proven effective in some cases (Gelin 1997).

8.10 Bow Thruster Treatments

Bow thrusters are problematic since accommodations are generally located next to the thruster room. As shown in Fig. 8.31, there are multiple paths that need treatment. Treatment approaches are illustrated in Fig. 8.32 (Fischer, 2006a, b and

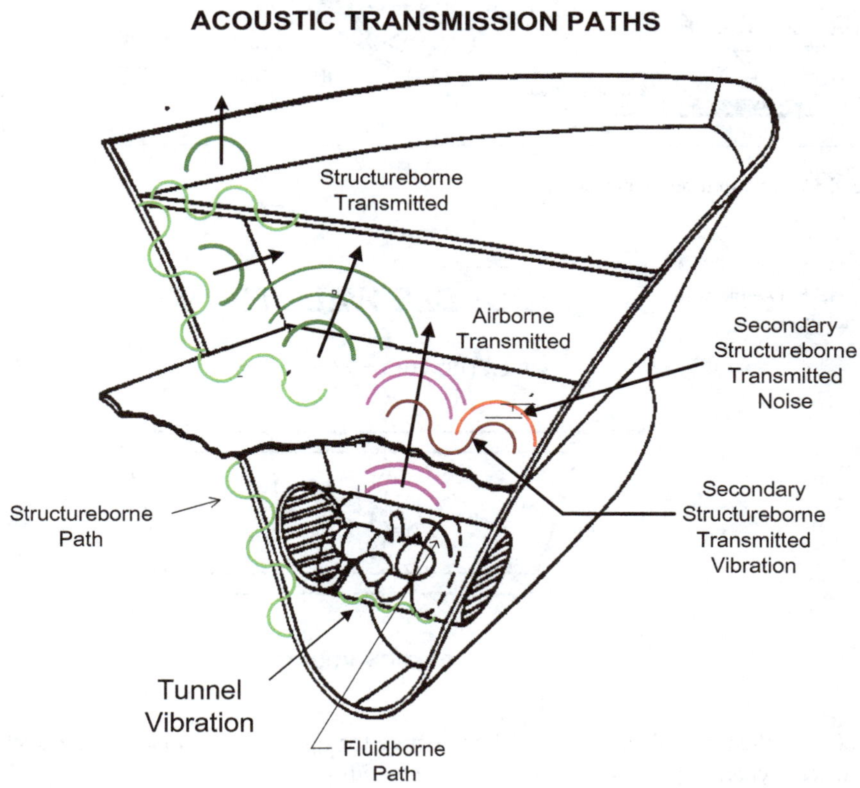

Fig. 8.31 Bow thruster acoustic paths

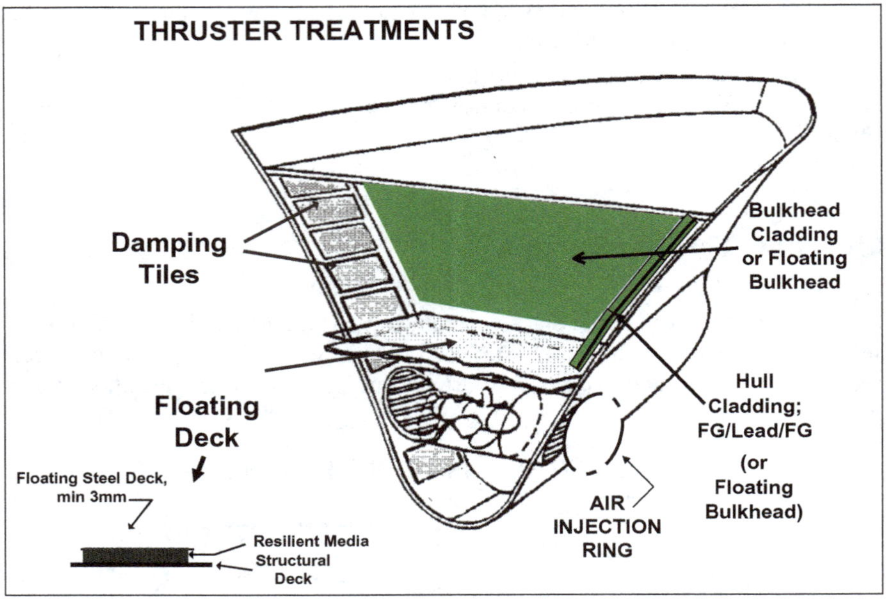

Fig. 8.32 Bow thruster treatments

Fig. 8.33 Double-wall tunnel thruster

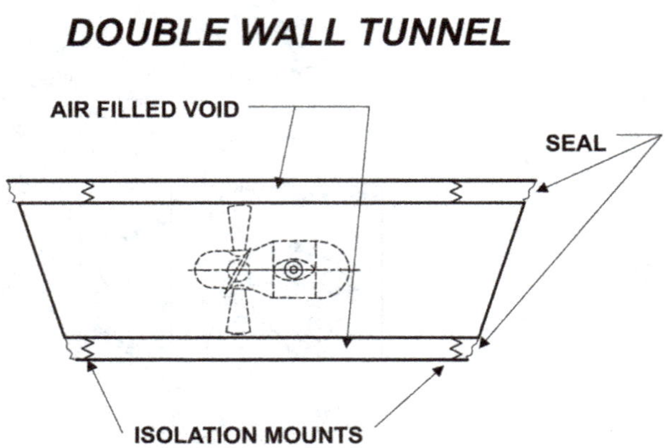

2000). Furthermore, these systems often operate for significant periods of time, such as when dynamic positioning, that interferes with crew rest (Irving and Richards 1998).

The first order of treatment, as with many noisy systems, is to reduce the hydro-acoustic noise from the thruster's impeller (Brown et al. 2001). To avoid or at least reduce the amount of treatment, design guidance is provided in Chap. 10. One treatment offering an approximately 15 dB reduction in the A-weighted noise is to

Table 8.10 Effectiveness of bow thruster treatments

Treatment	Noise and/or vibration reduction
Rounded tunnel inlets	Reduced vibration and just noticeable noise reduction, <3 dB(A)
Air injection	Increased vibration[a] and limited noise reduction, <3 dB(A)
Resiliently mounted tunnel within structural tunnel	Minor vibration change, maximum noise reduction 5–9 dB (A) – See Fig. 8.33.
Above with air mixture in-between inner and outer walls	Minor effect at low frequencies, significant noise reduction of 11–15 dB(A)
Damping treatment of shell plating and/or bulkheads	Minor noise impact <1–2 dB(A), noticeable improvement with damping treatment of all paths
Damping treatment of deck in receiver space	Reduced resonant vibration, maximum noise reduction 2–4 dB(A)
Floating floor in receiver space	Negligible impact on vibration, normally in the range of 5–7 dB(A)
Floating room in receiver space	Negligible impact on vibration, normally in the range of 8–10 dB(A)

[a]Assumed to be low frequency resonance caused by bubble layer

isolation mount the tunnel from the ship's structure – see Fig. 8.33 (Nilsson 1986). Thruster vendors are now offering such systems.

Andresen n.d., provides the following estimates of the noise or vibration reduction afforded by various treatments in Table 8.10.

Some vendors will supply air injection systems with their thruster. These "Prairie" type masker systems decouple the wall from the high hydro-dynamic noise induced by the impeller. Air can be injected near the tunnel inlet and/or near the impeller. This generally takes the form of perforated air emitter pipes located on the perimeter of the duct. This air would be pulled into the thruster and mixed with the cavitation, resulting in a gaseous cavitation-ventilation mixture. This would cushion the bubble collapse and thereby reduce the induced vibration and structureborne noise. The additional air would also provide some absorption in the tunnel.

These air systems have provided significant interior noise and underwater noise reductions. The measured noise reduction on one system is 10 dB in the frequency range above 500 Hz. However, these systems also tend to increase the low-frequency noise in the range below 100 Hz. This is generally acceptable since the ear is less sensitive at these low frequencies. For instance, a person's perception of a noise level of 92 dB at 80 Hz is the same as a level of 70 dB at 1000 Hz. The amount of reduction achieved by air injection is limited by the potential for other flanking paths for hydro-acoustic-induced noise.

Passive treatments can be employed to reduce thruster-induced noise and vibration. Selection of the optimum approach depends on many non-acoustic factors, such as available space, weight constraints, cost and impact on ship operation. In addition to "source" treatments, two other types are generally classified as treatment of the acoustic path and treatment of the receiver space. Source treatments are those directed at the hydro-acoustic environment. Active noise or vibration cancellation treatments are not yet a proven abatement approach.

Only one type of path treatment is generally considered: damping. Noise is carried to adjacent compartments primarily over the structureborne path. Although the airborne noise level in the bow thruster compartment itself is usually on the order of 100 dB(A) or greater, receiver spaces are rarely directly adjacent to the thruster compartment. Thus, airborne transmitted noise is usually not a concern. However, if there is a receiver space with a common interface to the thruster room, this interface should be treated with a high transmission loss material, e.g., a limp mass-loaded material sandwiched between compliant layers such as fiberglass – the thicker the more effective.

The effectiveness of the damping depends on the total path length from the thruster to the compartment of interest. Damping treatments are typically most effective over the mid- to high-frequencies, the same range over which the source levels are highest. Given the proper amount of damping treatment, that is, sufficient path length to the receiver, and treatment of all potential structureborne paths by a well-designed damping treatment, the received A-weighted noise level may be reduced by 5 to 7 dB, at best. Constrained damping has the potential to reduce noise by 8 to 12 dB but at a significant weight penalty.

Some thruster systems can be inherently quieter than tunnel thrusters. This includes systems that are sometimes installed on mine-sweepers. By the very nature of their operation, thrusters have to be acoustically quiet. One such vendor is the Omni-Thruster. Another approach is RIM thrusters, a relatively new technology. A RIM thruster has the tip of the blades attached to the inner ring of the rotor with the stator inside the ship. The blade is often a carbon fiber-reinforced plastic (CFRP). They should have better flow fields, offering higher cavitation inception speeds. Verifiable measured data on their performance are limited. Another option is drop-down thrusters. These have a lower impact on habitability noise since the path length through the water is greater and there is much less of a structureborne path.

8.11 Hydraulic Noise Control Treatments

As with HVAC systems discussed in Sect. 8.9, the major hurdle is designing and actually constructing a hydraulic system that does not violate innate assumptions or create sound shorts. Some of the innate assumptions, such as flow speed, size of radiused turns, and piping resilient supports, are described in Chap. 10.

When treatment is needed, there are limited choices. Various hydraulic silencers exist that reduce the pump pressure pulsations (tones) entering the fluid. From there, the pipe wall is excited and eventually the vessel. As with exhaust silencers, these can either be tuned silencers or broadband silencers. A flexible hose will attenuate both the fluidborne and structureborne noise from the pump entering the piping system (Purshouse 1986). Finally, for piping systems with high vibrations, resilient attachment points should be used on all pipe supports. Rubber grommets of 30 durometer or less should be used.

8.12 Propeller Noise Control Approaches

The most effective way to minimize propulsor noise is through careful hydro-acoustic design. The designer's goal should be to increase the cavitation inception speed (vessel speed at which cavitation begins). The following measures help to increase the cavitation inception speed:

- Uniform flow into the propulsor disk, especially near the top-dead center
- Skewed blades of the propulsor
- Increasing propeller diameter thereby reducing the needed rpm
- Increasing number of blades
- Incorporating stern shape into the propeller design (Bodger et al. 2014)
- Adding flow path devices to achieve a more uniform inflow to the propeller plane

To reduce structureborne induced noise, the rule-of-thumb for tip-hull and leading edge-strut/rudder clearance is nominally 20% of the propeller diameter. All other things being equal, the hull vibration response will decrease with increasing clearances. Vibration of hull structure around propulsors may be reduced by damping or by increasing hull plate mass and stiffness.

It is a non-trivial process to obtain a 'quiet' propulsor, as the design is a balance of many factors. It requires an advanced understanding of hydro-acoustics and the use of advanced design tools.

Model propellers are often tested in special test channels. Acoustic data from these tests are often used to improve propeller performance. Again, scaling (Helma 2014) and inferring the acoustic properties (Aktas and Turkmen 2014) from these tests should be done with care. Even determining cavitation from these test facilities or even at full scale should be carefully chosen, as visual and acoustic cavitation inception can be quite different, as noted by (Johannsen et al. 2014): "Purely visual cavitation investigations, however, suffer from the fact that cavitation might be hearable already before it can be detected by eye. On the other hand, scaling effects involved in vortex cavitation inception can spoil the result of a purely acoustic inception test at model scale." The difference can sometimes amount to 2 knots.

8.12.1 Propeller-Induced Habitability Noise

Compartments above or on aft-lower decks can have high propeller-induced noise, especially once the propeller cavitates. Retro-fit treatments, other than hull damping, are usually ineffective and costly. During the design, the impact of increasing the plate thickness and stiffener impedance can be evaluated. Treatments such as adding sand bags (see Fig. 8.34) to provide a small amount of damping and increased mass have been tried with limited success – especially at the low frequencies corresponding to the blade rate and its multiples. Increasing the tip clearance from

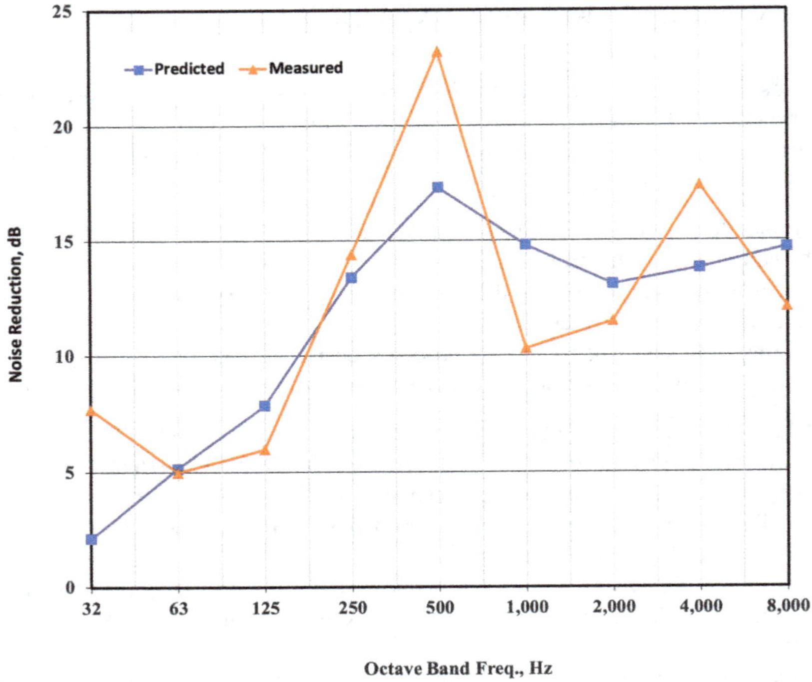

Fig. 8.34 Predicted v. Measured noise improvement with mass loading of hull over propellers

10% of diameter to 20% can decrease the hull vibration at a blade rate by approximately 5 dB (Veritec 1985; Noonan and Feldman 1978).

Prairie/Masker[15] systems on naval vessels have reduced the A-weighted airborne noise in aft, lower deck compartments by up to 14 dB. These systems are only effective above 125 Hz. Increased noise below this frequency occurs at the bubble layer resonant frequency, which may actually increase the blade rate noise components.

8.13 Barriers

Barriers, in the sense of highway noise treatments, are not applicable shipboard. Highway barriers are meant to treat a 'line source' such as a line of cars/trucks. Barriers used on ships are generally small and near the source or the receiver. Due to their physical size, the noise not only goes over the top with respect to line-of-sight

[15] Injection of air layers on the propeller leading edge and through hull air emission belts forward of the main engine room.

but also tends to travel around the sides, creating a flanking path. Per the SNAME Design Guide, if the barrier is extensive, the noise reduction can be from a low of 5 to 9 dB from 31.5 Hz to 8000 Hz to a high of 11 to 24 dB for a relatively tall barrier with no flanking (Fischer et al. 1983). If there are flanking paths (finite sized barrier), then the effectiveness can be reduced by 3 to 8 dB over the 31.5 to 8000 Hz range. A complete discussion predicting barrier effectiveness is given in Attenborough 2014.

COMSOL and other multi-physics software can easily be used to accurately predict the performance of any type, size, and style barrier (Li and Wong 2005).

Barriers are not effective in reverberant spaces. The only case where barriers are effective is for communication stations, where basically 5 sides form the barrier. The transmission loss of the barrier construction material itself must be significantly higher than the barrier insertion loss.

Conversely, if a source is placed near a reflecting surface by distance d and the free field is measured at some distance D, then the source level will increase by 3 dB for a d/D ratio of less than 0.1 and by 2 dB if the d/D ratio is between 0.1 and 0.28. If d/D is greater than 7, there is no impact. Along the same vein, a deck can provide a hard surface that enhances the noise at a receiver location (or cause interference, as described for Lloyd's mirror in Chap. 9).

8.14 Treatment Summary

The effectiveness of the primary treatments against airborne (AB), structureborne (SB) and secondary structureborne (SSB) noise is summarized in Table 8.11 (Fischer 2014). In terms of the overall noise control approach and optimizing it, one needs an estimate of the cost of the treatments. Table 8.11 presents a very rough order of magnitude of the material costs for several treatments as of 2014. Another factor is the installation cost. These vary widely from shipyard to shipyard. Other factors are impact on schedule and integration between the various trades. For instance, if the design calls for a deep frame and it is installed before the HVAC duct design is completed, the duct may get installed with very sharp bends to go around the frame. This may lead to an unanticipated noisy system.

Table 8.12 provides an example of a recommended treatment matrix for a quiet research vessel as a result of a detailed noise analysis, including airborne habitability noise and underwater radiated noise:

Other system type treatments to be considered for 'quiet vessel' design include:

- use of filters on the electric drive system
- hydro-acoustic design of hull form – bow to stern
- treatment of all sea-connected systems with flexible hoses and decoupling material near hull openings
- design and installation of a low-speed hydraulic system with a hydro-acoustic muffler resiliently isolated from a ship structure
- development of a list of the minimum and quietest equipment to be operating when the ship needs to be in a 'quiet mode'.

Table 8.11 Approximate noise reduction effectiveness and cost estimates by treatment type

			Approximate Cost		Noise reduction, dB	
Name of treatment	Treatment location	Effect location	Material alone	Path Affected[a]	Between 63 to 500 Hz	Between 500 to 8000 Hz
Low Noise Equipment	Source rooms	All rooms	1× to 5× cost of standard equipment	AB, FSB, SSB	26	10–15
Engine enclosure	Source rooms	All rooms	\$100–\$200 sq. meter	AB, SSB	5–10	10–20
Magnetic bearings	Machinery	All rooms	Not known	FSB	10–30	?
Absorptive layers	Source room, cabin	Source room, cabins	\$10–\$50 sq. meter	AB, FSB, SSB	2–4	4–8
One stage resilient mounts	Source room	Cabins	\$80–\$4000 per unit; 4 mounts	FSB	3–10	10–20
Two stage resilient mounts	Source room	Cabins	\$5000–\$30,000 per unit	FSB	6–15	15–35
Silencers	Source room	Source room	\$500–\$2000	AB	10–20	15–30
Damping Tile	Source room, cabins	Cabins	\$100[1]–\$600[2]/ sq. meter	AB, FSB, SSB	3–5	4–8
Floating floor	Cabins	Cabins	\$150–\$700 sq. meter	FSB, SSB	6–8	7–9
High Transmission Loss Treatment	Source Room	Cabins	\$35 to \$50/ sq. meter	AB, SSB	3–5	9–12
Joiner panel	Source room	Cabins	Insulation costs above	AB, SSB	1–4	3–6
Joiner panel	Cabins	Cabins		AB, FSB, SSB	2–5	4–7

1 – damping tile less than 6 mm thick
2 – damping tile between 6 mm and 13 mm thick
[a]AB – airborne path, FSB – first structureborne path (structural), and SSB – secondary structureborne path

Table 8.12 Example of 'critical equipment & recommended treatments'

Noise critical systems and associated noise control		Resilient mounts	Special sub-base	High impedance foundation	Foundation damping	Flexible connection	Resilient pipe hangers	Low noise motors
Propulsion	Propulsion motor		YES		YES			YES
	Motor cooling pump			YES		YES		
	Motor cooling fan	YES						
	Motor lube oil pump			YES	YES			
	DC Filter	YES						
	Propulsion transformer	YES						
	Propulsion thyristor	YES						
Electrical	Diesel generators	Compound	YES	YES	YES	YES	YES	
	Diesel exhaust piping	YES				YES	YES	
	Diesel exhaust mufflers	YES				YES		
	Transformers	YES				YES		
	Solid state frequency converters	YES				YES		
	Power supply cooling					YES		
Auxiliary	Fuel oil purifier	YES				YES		
	Sea water cooling pumps	YES				YES	YES	
HVAC	A/C supply fans	YES				YES	YES	
	A/C refrigeration plants	Compound		YES	YES	YES	YES	
	A/C chill water pumps	YES				YES	YES	
	Ventilation fans	YES				YES		

(continued)

Table 8.12 (continued)

Noise critical systems and associated noise control		Resilient mounts	Special sub-base	High impedance foundation	Foundation damping	Flexible connection	Resilient pipe hangers	Low noise motors
	Fan coil assemblies	YES				YES		
	Fan coil units	YES				YES		
Domestic fresh water	Fresh water pumps	YES				YES		
	Hot water CIRC pumps	YES				YES		
Sanitary	Sewage treatment plant	YES				YES	YES	
	VCHT system pumps	YES				YES	YES	
	Macerator	YES				YES		
Refrigeration	Ship service refriger-ation plants	Compound		YES	YES	YES	YES	
Compressed air	Air compressors	Compound		YES	YES	YES		
Steering gear	Hydraulic plant	YES		YES	YES	YES	YES	
Bow thruster				YES	YES			
Trawling system	Hydraulic power plant	YES		YES	YES	YES	YES	YES

References

AcousticCalc, http://acousticcalc.com/

Aktas, B., Turkmen, S.: Systematic Cavitation Tunnel Tests of a Propeller in Uniform and Inclined Flow Conditions as Part of a Round Robin Test Campaign. In: Proceeding 1st International Meeting – Propeller Noise & Vibration, Istanbul, Turkey, Nov. 2014

Andresen, K.: Thrusters for auxiliary propulsion and maneuverability. In: Ship Propulsion Systems, n.d.

Attenborough, K.: Sound propagation in the atmosphere. In: Rossing, T. (ed.) Springer Handbook of Acoustics, 2nd edn. Springer, Dordrecht (2014)

Avery, L., Gray, D.: Shipboard Noise Control Lessons Learned. BMT Group Technical Instruction # 043, Sept. 2009

Beranek, L. (ed.), *Noise and Vibration Control,* McGraw-Hill Book Co., New York, 1971

Beranek, L.L., Ver, I.L. (eds.): Noise and Vibration Control Engineering, Principles and Applications. John Wiley & Sons, New York (1992)

Berkman, E.F., Bender, E.K.: Perspectives on active noise and vibration control. Sound and Vibration Magazine, Jan. 1997

Bies D.A., Hansen, C.H.: Engineering Noise Control, Theory and Practice, 4th edn. CRC Press, 1062 Boca Raton (2009)

Bodger, L., Helma, S., Sasaki, N.: Vibration Control by Propeller Design. In: Proceeding 1st International Meeting – Propeller Noise & Vibration, Istanbul, Turkey, Nov. 2014

Bonello, P., Brennan, M.J., Elliott, S.J.: The Design of an :daptive Tuned Vibration Absorber with a Variable Shape Stiffness Element. Institute of Sound & Vibration Research (ISVR) Technical Memo No. 941, Aug. 2004

Bonetti, E., Calcagno, P.: In: Buiten, J. (ed.) Low Noise Levels as the First Task of a Vessel. A Description and Some Remarks about Acoustic Quieting Design Criteria and Features of SACLANT ASW Centre Oceanographic Research Vessel Shipboard Acoustics. Martinus Nijhoff Publishers, The Netherlands (1986)

Brown, N.A., Dai, T.K., et. al.: DP Bow Thruster Noise Remediation in R/V ROGER REVELLE, AGOR-24. In: Marine Technology Society (MTS) Dynamic Positioning Conference, 2001

Buiten, J.: Noise Reduction on a Rhine-Cruise Ship Due to Damping Material, vol. 21, pp. 159–167. Academic Press Incorporated, London (1972)

Burroughs, C., Fischer, R., Shiflet, F.: Structureborne noise transmission through ship stanchions. In: 98th Meeting of the Acoustical Society of America, Salt Lake City, Nov. 1979.

Cha, S.-I., Chun, H.-H.: Insertion loo prediction of floating floors used in ship cabins. Appl. Acoust. **69**, 913–917 (2007)

Chang, P., Carroll, T.: Experimental program for the determination of Hull structural damping coefficients, Ship Structural Committee Report SSC-306, 1981

Chapnik, B., Currie, I.: The effect of finite length flexible segment on acoustic wave propagation in piping systems. In: InterNoise 96. Institute of Noise Control Engineers (INCE) (1996)

Chapnik, B., Currie, I.: Noise reduction using finite-length flexible segments. JVA. **1222**(2), 94–108 (2000)

Congyun, Z., Qibai, H.: A method for calculating the absorption coefficient of a multi-layer absorbent using the electro-acoustic analogy. Appl. Acoust. **66**, 879–887 (2005)

Cooper, B.A., Nelson, D.A.: A "Buy Quiet" program for NASA Lewis Research Center: Specifying low equipment noise emission levels. Proceeding of NOISE-CON 97. The 1997 National Conference on Noise Control Engineering, Univ. Park, PA, June 1997, 1996

Cremer, L., Heckl, M.: In: Unger, E.E. (ed.) Structure-Borne Sound. Springer Verlag, NY (1973)

Daley, S., Johnson, F.A., Pearson, J.B., Dixon, R.: Active vibration control for marine applications. Control. Eng. Pract. **12**, 465–474 (2004)

Fahey, F.J.: A theoretical evaluation of the effectiveness of reducing vibration transmission in the hull-frame of a ship by damping. In: International Symposium on Shipboard Acoustics, The Netherlands, 1976

Fahy, F., Thompson, D. (eds.): Fundamentals of Sound and Vibration. CRC Press, Boca Raton (2015)

Feng, L.: A modified honeycomb panel to increase sound transmission loss. In: Tenth International Congress on Sound and Vibration, Stockholm, Sweden, July 2003

Fischer, R.W.: Patrol boat habitability noise control. Nav. Eng. J. **100**, 73–89 (1988)

Fischer, R.W.: Bow thruster induced noise and vibration. In: Marine Technology Society (MTS) Dynamic Positioning Conference, Houston, Tx., 2000

Fischer, R.F.: Acoustic design, construction, and testing of NOAA's fisheries research vessel. In: Lloyd's Ship Noise and Vibration Conference, London, 2005

Fischer, R.W.: Airborne noise flanking of shipboard vibration isolation systems. Sound and Vibration, Dec. 2006a

Fischer, R.W.: Addressing thruster noise and vibration. In: World Maritime Technology Conference, London, March 2006b

Fischer, R.: Acoustic Design of NOAA's FRV/FSV, Oceantech Expo '08, Providence RI, Feb 2008

Fischer, R.W.: Noise control decision process for marine vessels. In: SNAME Ship Production Symposium, Maritime Convention, Houston, 2014

Fischer, R.W., Burroughs, C. B., Nelson, D.L.: Design Guide for Shipboard Airborne Noise Control, Society of Naval Architects and Marine Engineers (SNAME) Technical and Research Bulletin 3-37, Jersey City, NJ, Jan. 1983

Franks, J.R., Stephenson, M.R., Merry, C.J.: Preventing Occupational Hearing Loss: A Practical Guide. NIOSH, Cincinnati (1996)

Fusil, E.: Submarine Signature and Vulnerability. ASW Asia (2004)

Gelin, L.J.: Active Noise Control: A Tutorial for HVAC Designers. In: ASHRAE J. Online, Aug. 1997

Gu, J., Sheng, M.: A method for damping estimation based on least square fit. Am. J. Eng. Res. (AJER). **4**(7), 205–209 (2015)

Gungor, F.E.: Computer Aided Noise Prediction in Heating, Ventilation, and Air Conditioning Systems, A Thesis submitted to The Middle East Technical University, Sept. 2003

Guy, R.W.: A preliminary study model for the absorption or transmission of sound in multi-layer systems. Noise Control Eng. J. **33**(3), 117 (1989)

Hansen, C.: Noise Control, from Concept to Application. Taylor & Francis, London (2005)

He, L., Xu, W., Bu, W., Shi, L.: Dynamic analysis and design of air spring mounting system for marine propulsion. J. Sound Vib. **333**, 4912–4929 (2014)

Helma, S.: A scaling procedure for modern propeller designs. In: Proceeding 1st International Meeting – Propeller Noise & Vibration, Istanbul, Turkey, Nov. 2014

Heron, K.: Acoustic radiation from honeycomb sandwich plates. Royal Aircraft Establishment, Tech Memo AERO 1813 (1979)

Huang, L.-Z., Xiao, Y., Wen, J.-H.: Analysis of underwater decoupling properties of a locally resonant acoustic metamaterial coating. Chin. Phys. Soc. IOP Sci. **25**(2), 024302 (2016)

Irving, L., Richards, R.L.: Acoustics and Noise Control Handbook for Architects and Builders. Kreiger Publishing Company, Malabar, FL (1998)

Johannsen, C., Zarbock, O., Bretschneidter, J.: Experimental Determination of the Difference in Visual and Acoustic Cavitation Inception. In: Proceeding 1st International Meeting – Propeller Noise & Vibration, Istanbul, Turkey, Nov. 2014

Jones, P., Kessissoglou, N.: An evaluation of current commercial acoustic FEA software for modeling small complex mufflers geometries: prediction vs experiment. In: Proceeding of Acoustics, Adelaide, Australia, Nov. 2009

Kandekar, A., Nagarhalli, P., et.al.: HVAC System Noise Prediction through CFD Simulation. In: SAE Technical Paper 2019-26-0210, 2019

Karlsson, M., Glav, R.: The Herschel-Quincke tube: the attenuation conditions and their sensitivity to mean flow. J. Acoust. Soc. Am. **124**(2), 723–732 (2008)

Kihlman, T.: Noise reduction with constrained visco-elastic layers, a case history. In: International Symposium Ship Acoustic (ISSA), The Netherlands, 1976

Kim, S., Singh, R.: Effect of source characteristics on structure-borne noise in an isolation system over mid and high frequencies. In: Inter-noise 2002, Dearborn, MI, Aug. 2002

Koruk, H., Sanliturk, K.: On measuring dynamic properties of damping materials using Oberst Beam method. In: Proceedings of the ASME 2010 Biennial Conference on Engineering Systems Design and Analysis, ESDA2010, Istanbul, July 2010

Kosinska, N., Nogalski, K.: Structureborne sound diminishing through visoelastic joints in a pipe installation. In: InterNoise 93. Institute of Noise Control Engineers (INCE) (1993)

Kou, Y., Liu, B., Tian, J.: Radiation efficiency of damped plates. J. Acoust. Soc. Am. (JASA). (2), 137, 1032–1035 (2015)

Lee, J.-C., Liu, T.-Y.: Experimental study of the performance of ship fire protection materials. J. Mar. Sci. Technol. **14**(1), 4 (2006)

Lee, Y.Y., Yuen, K.K., Lam, K.C., Lam, H.F.: Noise reduction techniques for close-fitting enclosures at structural/acoustic resonance. JSME Int. J. Series C. **46**(1), 161–167 (2003)

Li, K.M., Wong, H.Y.: A review of commonly used analytical and empirical formulae for predicting sound diffracted by a thin screen. Appl. Acoust. **66**, 45–76 (2005)

Li, X., Howard, C.Q., Hansen, C.H., Winbery, M.: Feasibility of active vibration isolation of diesel engines in Collins class submarine. Navy Eng. Bull.25–26 (March 2004)

Lung, R.C.N.: Active control of machinery noise in a marine environment – lessons learned. In: Fifth International Congress on Sound and Vibration, Adelaide, S. Australia, 1997

Maling Jr., G.: International standard for specifying noise emission. In: Inter-Noise 2012, New York (2012)

Ming, R., Pan, J.: Insertion loss of an acoustic enclosure. J. Acoust. Soc. Am. (JASA). **116**(6), 3453–3459 (2004)

Munjal, M.L.: Acoustics of Ducts and Mufflers. Wiley (2014)

Munjal, M., Thawani, P.: A simple model for wave propagation along and across a hose. In: InterNoise 95. Institute of Noise Control Engineers (INCE) (1995)

Munjal, M., Thawani, P.: Acoustic Performance of Hoses, a Parametric Study, vol. 44, p. 274. Noise Control Eng. J., INCE (1996)

Munjal, M., Thawani, P.: Prediction of the vibro-acoustic transmission loss of planar hose-pipe systems. J. Acoust. Soc. Am. (JASA). **101**(5), 2524–2535 (1997)

Murawski, L.: Axial vibrations of a propulsion system taking into account the couplings and the boundary conditions. J. Mar. Sci. Technol. **9**(4), 171–181 (2004)

Nashif, A., Jones, D., Henderson, J.: Vibration Damping. John Wiley & Sons, New York (1985)

NAVSEA: DIM Mount Design Handbook, NAVSEA S9078-AA-HBK-010-DIM, Sept. 1982

Neale, J.: Experimental and Numerical Investigation of Noise Generation from the Expansion of High Velocity HVAC Flows on Board Ocean Going Fast Ferries, Thesis, School of Mechanical and Manufacturing Engineering, University of New South Wales, June 2006

Nelson, D.: Adaptation of NASA Buy-Quiet Process Roadmap for Construction Firms, p. 212. National Institute for Occupational Safety and Health (2011)

Nilsson, A.C.: Noise Reduction of Bow Thrusters. ISSA (1986)

Niu, J., Song, K., Lim, C.W.: On active vibration isolation of floating raft system. J. Sound Vib. **285**, 391–406 (2005)

Noonan, E.F., Feldman, S.: State of the art for shipboard vibration and noise control. In: Ship Vibration Symposium. Society of Naval Architects and Marine Engineers (SNMAE), NJ (1978)

Ort, A.: Resilient Mounting of Deckhouses for Push-boats and Motor Cargoboats. International Symposium Ship Acoustic (ISSA), The Netherlands (1976)

Payer, H.G.: Elastic Foundation of Deckhouses. SNAME STAR Symposium, Houston, April 1979

Plunkett, R., Lee, C.: Length Optimization for Constrained Viscoelastic Layer Damping. Air Force Materials Laboratory (AFML) TR-68-376, July 1969

Purshouse, M.: Underwater noise radiated due to transmission through the cooling water system of a marine diesel engine. In: Buiten, J. (ed.) Shipboard Acoustics. Martinus Nijhoff Publishers, Dordrecht (1986)

Ramakrishnan, R.: Validation of COMSOL multiphysics and acoustical performance of splitter silencers. Can. Acoust. **38**(3), 178–179 (2010)

Rao, M.D.: Recent applications of viscoelastic damping for noise control in automobiles and commercial airplanes. J. Sound Vib. **262**(3), 457–474 (2003)

Ray Jr., E.: Chapter 10: Combustion Turbine Acoustics And Noise Control. In: Giampaolo, T. (ed.) Gas Turbine Handbook; Principles and Practices. The Fairmont Press, Inc, Liburn, GA (2006)

Rettinger, M.: Acoustic Design and Noise Control. Chemical Publishing Co., New York, NY (1977)

Rosenhouse, G.: A modified formula for absorption influence on sound transmission through partitions. Appl. Acoust. **17**, 379–389 (1984)

Schaffer, M.: Practical Guide to Noise and Vibration Control for HVAC Systems, 2nd edn. ASHRAE Handbook/Manual/Guide (2005)

Sergeev, S., Fleury, R., Lissek, H.: Ultrabroadband sound control with deep-subwavelength plasmacoustic metalayers. Nat. Commun. **14**, 2874 (2023)

Sgard, F., Olny, X., Atalla, N., Castel, F.: On the use of perforations to improve the sound absorption of porous materials. Appl. Acoust. **66**, 625–651 (2005)

Song, H.-H., Hong, S.-Y., Kwon, H.-W.:Development of multi-layered floating floor for cabin noise reduction. In: ICMEAS, Materials and Science Engineering, Vol. 280 (2017)

Spence, J.: Measurement of spray-on damping effectiveness and application to bow thruster noise on ships. NCEJ. **59**(2), 126–134 (2011)

Subramanian, S., Surampudi, R., Thomson, K., Vallurupalli, S.: Optimization of damping treatments for structureborne noise reduction. Sound Vib. **38**(9), 14–19 (2004)

Sugie, S., Yoshimura, J., Ogawa, H.: Absorption characteristics of fibrous material covered with perforated facing and film. Acoust. Sci. & Tech. **27**(2), 87–96 (2006)

Sui, N., Yan, X., Huang, T.-Y.: A lightweight yet sound-proof honeycomb acoustic metamaterial. Appl. Phys. Lett. **106**, 171905 (2015)

Technology for a Quieter America, National Academy of Engineering, National Academies Press, Washington, DC, 2010

Treszkai, M.F., Sipos, D., Feszty, D.: Damping determination by half-power bandwidth Mehod for a slightly damped rectangular steel plate in the mid-frequency range. Acta Technica Jaurinensis. **13**(3), 177–196 (2020)

US Navy Resilient Mount Handbook, S9073-A2-HBK-010, Rev. 1, Export Controlled, Sept. 2005

Vallbo, S.: Material selection considerations for polymer composite structures in naval applications. J. Sandw. Struct. Mater. **7**, 413–429 (2005)

Van der Meer, L.: Vibration Isolation for Deck-Mounted Superstructures, Thesis, Delft University of Technology, April, 2016

Ver, I.L.: Reduction of noise by acoustic enclosures. In: Isolation of Mechanical Vibration, Impact and Noise, Colloquium by American Society of Mechanical Engineers (ASME) AMD-Vol. 1, NY, NY, 1973

Veritec: Vibration Control in Ships. Marine Technology Consultants, Norway (1985)

Vibro-Acoustics: https://noisecontrol.vibro-acoustics.com/v-a-design/

Winberg, M., Johansson, S., Lag, T.: Active control of engine induced noise in a naval application. In: The 8th International Congress on Sound and Vibration, Hong Kong, China, 2001

Yang, T.J., Suani, Z.J.: Active vibration isolation system for a diesel engine. Noise Control Eng. J. **60**(3), 267–282 (2012)

Yu, W., Xinglin, C., Guangmin, L.: The vibration characteristics analysis of ship foundation based on arrangement of isolators. Appl. Mech. Mater. **55–57**, 2202–2205 (2011)

Zhiqiang, L.V., Lin, H.E., Channggeng, S.: Optimization of natural frequencies of large-scale two-stage raft system. J. Phys. Conf. Ser. **744**, 012180 (2016)

Chapter 9
Underwater Acoustics

9.1 Underwater Acoustics

This chapter provides a basic understanding and overview of the critical factors controlling underwater radiated noise from commercial vessels. It is intended to assist the marine community – shipyards, designers, regulators and owners – in understanding the controlling physical parameters and their impact on underwater noise.[1] As noted in the Introduction, this understanding allows noise problems to be addressed and abated via proper attention to the overall ship systems and acoustic designs. This includes program planning, analysis, treatment selection, construction quality assurance (QA), and verification testing. Understanding and sequencing these steps will lead to a quiet vessel in an optimal manner. The optimization is from the viewpoint of reducing noise and minimizing the usual adverse impacts of treatments on ship weight, space and cost. Cost includes materials, installation, and maintenance.

As discussed in Chap. 5, each vessel has numerous systems – mechanical and hydrodynamic – that induce underwater radiated noise. Acoustic sources are those items on the vessels that are capable of converting part of their mechanical energy into acoustic energy. Even at a very small conversion ratio of 10^{-10}, a 1 kW mechanical source could produce a sound pressure level in the water of 100 dB re 1 μPa at a distance of 1 m (denoted as 1 μPa–m). Quieter sections of the ocean have levels on the order of 70–90 dB re 1 μPa, so ship mechanical sources could significantly impact the environment.

The most prominent acoustic sources, in order, are propulsors, propulsion-drive equipment, compressors, pumps, and piping systems. Next operating conditions need to be considered – low or high speed, sea state, equipment line-up, and physical

[1] See ISO 18405:2017 Underwater Acoustics – Terminology

R. Fischer, L. Boroditsky, *Noise and Vibration Control on Ships*,
https://doi.org/10.1007/978-3-031-55170-3_9

size and location of the equipment. Finally, for high-speed craft, additional sources are flow/turbulence over the hull and appendages and, possibly, underwater exhaust. For research and naval vessels, sea-connected systems may be of importance.

The possible acoustic transmission paths include airborne; structureborne; the interaction between these two paths – secondary structureborne; and fluidborne. One major difference relative to habitability noise is the coupling of hull vibration and radiation of a water-loaded plate into the ocean.

The ocean, as a 'receiver', has its own unique set of acoustic characteristics that need to be considered in predicting the 'far-field' noise induced by a ship (Ross 1976). In this case, the 'receiver' – the ocean – is much more complicated than for receivers in compartments onboard the vessel. For instance, there is little reverberation (only in shallow water – McDaniel 1993). The ocean bottom, salinity and temperature of the water, and sea surface all strongly influence propagation from the ship to the point of interest. Other major differences between sound in-air and in-water are as follows:

- density – 1000 kg/m^3, speed of sound in water at approximately 1500 m/s, and acoustic impedance, 1.53×10^6 rayl, are all significantly higher than in air[2, 3]
- existence of both upper and lower acoustic boundaries in the ocean (ocean surface and ocean floor) affects propagation differently than in air
- sound propagation losses in the ocean have different frequency dependencies than those in air, allowing low frequencies to propagate long distances in water
- directivity of the source, especially in the near acoustic field, is more important than for noise in reverberant compartments onboard the ship.

Low noise levels from all types of vessels can be achieved with appropriate attention to proper design details, the use of inherently quiet equipment, and the application of state-of-the-art control technologies. Computational fluid dynamics (CFD) for the quiet hydroacoustic design of propellers (Hallander 2013) is becoming a necessity. A quiet ship is not the result of black magic; rather, it is the result of understanding the controlling system parameters and paying attention to the proper details and processes as outlined herein.

Modeling underwater noise follows the same basic 'source, path, receiver' approach as considered in Chap. 6 for airborne and structureborne noise. Again, each of these characteristics is frequency dependent, requiring a basic understanding of the relative importance of low, middle and/or high frequencies.

The impact of various ship types on the radiated noise areas of interest – environment, sonar-self noise, surveying/acoustic research, and naval operations – is highlighted in Table 9.1.

---

[2]In air speed of sound is 343 m/s, density is 1.2 kg/m^2, and acoustic impedance is 415 rayl.

[3]Quoted noise decibel values for objects in air and those in water generally differ by 62 dB, where $dB_{water} = dB_{air} + 62$ dB. This is due to density and sound speed differences and different reference values of 20 μPa and 1 μPa, respectively

Table 9.1 Ship-radiated noise impacts

Type	Ocean environment	Sonar self-noise	Surveying/Acoustic research	Radiated signature
Commercial	X			
Specialized NGO vessels	X	X	X	X
Naval noncombatant		X	X	X
Naval combatant		X		X

Note that ship signatures are often reported with respect to their narrow band (one Hertz) spectra or their broadband -octave or third-octave band spectra. Narrow band signatures were usually denoted L_p dB re 1 μPa//*Hz* @ 1 m, SPL dB re 1 μPa//Hz @ 1 m, L_p dB re 1 μPa^2//Hz @ 1 m, and SPL dB re 1 μPa^2//Hz @ 1 m, all of which are basically equivalent. Broadband signatures are usually denoted ***without*** the "//Hz" label, for example: L_p dB re 1 μPa @ 1 m, SPL dB re 1 μPa @ 1 m, L_p dB re 1 μPa^2 @1 m, SPL dB re 1 μPa^2 @ 1 m, all of which are basically equivalent, and the bandwidth, octave or third-octave, is usually denoted on the x-axis of the graph. Note that with the latest ISO Standard, 18,405(2017), Underwater Acoustic Terminology, μPa @ 1 m has been replaced by μPa–m. To be as consistent as possible, graphs of ship signatures presented herein are in third-octave bands unless otherwise denoted. The authors have used this standard wherever possible. The source level is referenced to a fictious 1 m from the 'acoustic center' of the ship. This acoustic center can be considered to be near the propellers, if cavitating, or the location of the engine room for machinery.

9.1.1 Underwater Noise and Environmental Sensitive Areas

The world's oceans are becoming noisier. The levels show a 3 dB increase in noise every decade (McDonald et al. 2006). As a result, underwater radiated noise is receiving increased worldwide attention from regulatory agencies. This includes IMO, government agencies, particularly Canada and Europe, and nongovernment organizations (NGOs), including research agencies. As an example, the European Achieve *QU*ieter *O*ceans (AQUO) project was started in 2012 to "provide policy makers with practical guidelines and solutions, in order to mitigate underwater noise footprint due to noise radiated from ships" (Audoly et al. 2015). Another program is *S*hip-oriented *I*nnovative so*L*utions to r*E*duce *N*oise and *V*ibration – SILENV. The status of the impact of underwater noise in Europe is summarized in Cruz 2021.

There is an audio database of ship radiated noise maintained by ShipsEar (Santos-Dominguez et al. 2016); http://atlanttic.uvigo.es/underwaternoise/. Another ship noise database – SONIC Ship Underwater Radiated Noise Database – is maintained by the University of Southampton: https://web-archive.southampton.ac.uk/

vesselnoise.soton.ac.uk/. SILENV and AQUO collected radiated noise data. Various labs and institutions are also collecting and providing access to measured radiated noise data. Not all these databases provide signature 'source levels', e.g., noise levels scaled to a distance of 1 m. Kipple and Gabriele 2007 provides a summary of the radiated source levels for vessels ranging from smaller than 6 m to larger than 76 m.

Representative ship radiated signatures are presented throughout this chapter. Vessels of the same class and builder may have similar signatures, but not all vessels of the same type, e.g., super tankers, have similar signatures. In some databases and even some of the signatures used in this book, the operating conditions are not known or identified. Thus, the signature could vary up or down depending on whether noisy equipment was secured or operating while the data were collected. Other important variables not often reported or stated are the distance from the ship to the measurement array and the impact of any ambient noise on the measurements. With the new IMO test procedures (ISO 17208), these variables should be better defined, and the data should be more useful.

Some of the first studies on the subject of vessel noise and its impact on the ocean were undertaken by LGL, Ltd. and Bolt, Beranek, & Newman Inc. (BBN) for Mineral Management Service (MMS)[4] in the mid – 1970s. Underwater noise radiated from commercial vessels is becoming a more of a concern, especially for cruise vessels and workboats operating in environmentally sensitive areas. Environmentally sensitive areas are generally the areas where protected species feed and spawn. In the U.S., these are usually species protected by the Endangered Species Act[5] and Marine Mammal Protection Act (MMPA).[6] Because sound is easily transmitted in the ocean, the vessel does not always need to be physically present near the area of concern. IMO has a definition for 'Particularly Sensitive Sea Areas (PSSA)', adopted in 1991 by the Assembly of the IMO. In 2001, ocean noise was considered a pollutant with respect to a PSSA. A PSSA must have 'ecological, socioeconomic or scientific value' and be 'vulnerable to damage by international shipping.' Member states can propose a PSSA.

The Sakhalin Energy project (Borisov and Gritsenko 2003) is one of the largest developments requiring a thorough understanding of the impact of shipping (and construction) on a sensitive ocean environment around Sakhalin Island. An Environment Impact Statement (EIS) was required, and as a result, vessels operating in and around the project needed a measurement of the ship radiated noise.

Shipping is believed to be the dominant source of noise in the oceans (NRC 2005; Richardson et al. 1995; and Simmonds et al. 2003). Vessels radiate noise effectively from 1 Hz to several hundred thousand Hertz. This frequency range covers the frequency range of hearing for whales, fish, and many mammals. It is necessary to understand which operating conditions and systems control the vessel's signature to

[4]Currently Bureau of Ocean Energy Management, Regulation and Enforcement (BOEMRE)

[5]Endangered Species Act of 1973, Public Law 93–205, (*16 U.S.C. 1531–1544*).

[6]Marine Mammal Protection Act of 1972, Public Law 92–522 (*16 U.S.C. 1361 et seq.*).

optimally choose how to better design and build quieter ships or to abate the noise from existing ships. To effect a change in a vessel's radiated noise, the effectiveness of various treatment types needs to be fully understood (also see Chap. 8).

A guide to the intersection between underwater sound and the marine mammal acoustic environment was provided by Bradley and Stern 2008. Not enough information is available today regarding the threat to marine inhabitants from underwater anthropological noise to definitively assess all impacts to animals in all geographical regions. Marine biologists do not know specific noise characteristics (both amplitude and frequency) that bother and/or harm many species of marine animals. Per Southall and Hatch 2008, the lowest frequency range was 5 Hz for baleen whales, while the highest was over 200,000 Hz for porpoises. Detailed information is typically available only for animals that we can readily communicate with, such as dolphins. While the body of literature on this topic is certainly expanding, there still exist many large gaps in our knowledge of hearing capabilities as well as reaction thresholds for many species. (A complicating factor, which is also present in humans, is the fact that a sound that might bother one animal of a certain species may not bother another of the same species. This can be due to many factors, such as age, sex, hunger, etc.).

Another factor is the possible impact of climate on the underwater noise environment, both with respect to temperature and ocean acidity. Climate is measurably altering the thermal structure of the oceans (Possenti et al. 2023). A change in temperature affects both the speed of sound and the existence of sound channels. Furthermore, ocean acidification decreases sound absorption below 10,000 Hz. Both these factors will increase the distance over which sound effectively propagates.

Various underwater noise criteria in terms of sound pressure are covered in Sect. 3.4. However, underwater noise has another variable considered with respect to its impact on marine wildlife. This is called the Sound Exposure Level (SEL) (Southall et al. 2019). This frequency-weighted value factors in the duration of the exposure (somewhat equivalent to Time Weighed Average). The SEL is computed as:

$$SEL \text{ dB re } 1\ \mu\text{Pa}^2 - \text{s} = L_p + 10 \text{ Log } (Duration, sec.) \tag{9.1}$$

where L_p is the steady state sound pressure level, dB re 1 μPa, over the duration.

9.1.2 *Whole-Ship 'Total' Radiated Noise Prediction Methods – Empirical*

Most analytical prediction methods[7] (Ross 1976; Urick 1983; and Coates 1989) are simple at best and only provide a yard stick with which to estimate a ship's total[8]

[7]There are at least six empirical ship source models in addition to Ross, Urick, & Coates – Wales & Heitmeyer (W&H), Lurton, ANRTRA, RANDI-2&3, Wittekind, and AQUO (Yan et al. 2021).

[8]Machinery and propulsor contributions.

radiated noise. These are based on generalized parameters such as ship length, displacement, propeller diameter, and/or propeller rotation rate. Prediction methods based on empirical approaches or simple formulas have limited validity. For a critical design, propeller signatures should be addressed by the appropriate experts and, where possible, should include model testing in cavitation tunnels. Scaling laws may allow one to more accurately predict full-scale propeller performance and optimize the design features of the hull and propeller interaction.

Ross used the following equation to predict a passenger vessel's one-third band (TOB) radiated noise above 100 Hz based on the 'trend' of data collected during WWII and on other vessels:

$$L_p \text{ dB re 1 } \mu\text{Pa–m TOB} = 190 + 53 \text{ Log}\,(\text{V}/10) - 20 \text{ Log}\,(\text{f}_{\text{ctr}}, \text{Hz}) + 10 \text{ Log}\,(.23\text{f}_{\text{ctr}}) \tag{9.2}$$

where V is the ship speed in knots and f_{ctr} is the center frequency of the band of interest in Hz.

Ross also uses the following equation to predict the propeller-induced noise in one-third octave bands, where the slope is based on −20 log (frequency):

$$L_p \text{ dB re 1 } \mu\text{Pa–m TOB} = 195 + 60 \text{ Log}\,(\text{U}_t/25) + 10 \text{ Log}(\text{B}/4) - 20 \text{ Log}\,(\text{f}_{\text{ctr}}, \text{Hz}) + 10 \text{ Log}\,(.23\text{f}_{\text{ctr}}) \tag{9.3}$$

where U_t is the propeller tip speed in m/s and B is the number of blades.

An example of predicted levels for a 10 kn ship with a propeller speed of 200 rpm is provided in Fig. 9.1. Note that the predicted overall[9] (OA) noise level is approximately the same for the Urick and Ross methods. The source level provided by Urick is:

$$L_p \text{ dB re1 } \mu\text{Pa–m TOB} = 51 \text{ Log}\,(\text{U}_t) + 15 \text{ Log}\,(\text{T}) - 20 \text{ Log}\,(\text{f}_{\text{ctr}}) + 10 \text{ Log}\,(.23\text{ f}_{\text{ctr}}) + 73 \tag{9.4}$$

where T is the vessel tonnage. Similarly, Coates 1989 uses the following equation for third octave bands above 1000 Hz:

$$L_p \text{ dB re1 } \mu\text{Pa–m TOB} = 35 + 10 \text{ Log}\,(\text{T}) + 60 \text{ Log}\,(\text{V}) - 20 \text{ Log}\,(\text{f}_{\text{ctr}}) + 10 \text{ Log}\,(.23\text{ f}_{\text{ctr}}) \tag{9.5}$$

Note the typical dependence of the velocity on the sixth power in Eq. 9.5.

[9]The overall level is a single value that represents the summation of all the acoustic energy in the frequency range of concern; for ships. This is typically from a low of 10 Hz to a high of 50,000 Hz. This OA value is presented in each figure's legend, where possible. It can be used as a quick way to review the relative radiated signature from each vessel.

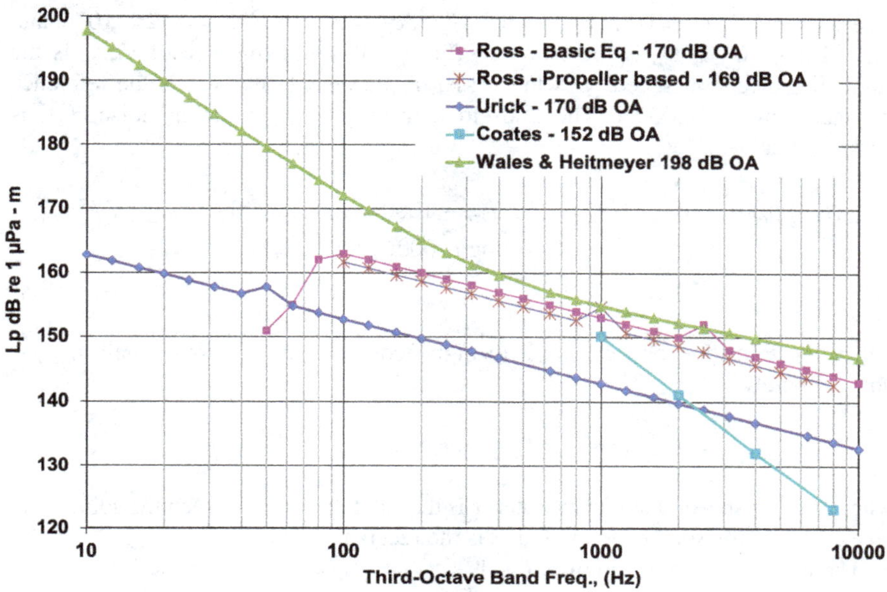

Fig. 9.1 Analytically predicted radiated noise at 10 knots

Heitmeyer et al. 2004 provide the following analytical equation for modeling the mean "merchant" ship noise in one-third octave bands (note that it is independent of any ship parameters):

$$L_p\,\text{dB re1 µPa–m TOB} = 230 - 36\ 10\ \text{Log}\ (\text{f}_{\text{ctr}}) + 10\ \text{Log} \left\{ \left[1 + (\text{f}_{\text{ctr}}/340)^2 \right]^{0.917} \right\} + 10\ \text{Log}\ (.23\ \text{f}_{\text{ctr}}) \tag{9.6}$$

As shown in Fig. 9.1, these approaches are significantly different at low frequencies.

Another empirical prediction approach divides the signature into three parts – low-frequency cavitation noise, mid- to high-frequency cavitation noise, and machinery-induced noise (Wittekind 2014). These equations also account for ship design parameters, such as the block coefficient and displacement.

The low-frequency cavitation noise L_{S1} is computed as:

$$L_{S1}(f_{ctr})\ \text{dB re 1 µPa//Hz} = \sum_{n=0}^{5} c_n f_{ctr}^n + 125 + 80\ \text{Log}\ \left(4C_B V/V_{incep}\right) + 6.67\ \text{Log}\ \left(\nabla/\nabla_{ref}\right) \tag{9.7}$$

where $c_0 = 125$, $c_1 = 0.35$, $c_2 = -8 \times 10^{-3}$, $c_3 = 6 \times 10^{-5}$, $c_4 = -2 \times 10^{-7}$, and $c_5 = 2.2 \times 10^{-10}$; C_B is the block coefficient; V is the ship speed; V_{incep} is the cavitation inception speed; ∇ is the vessel displacement, and ∇_{ref} is the reference displacement = 10,000 t. The mid- to high-frequency cavitation noise, L_{s2}, is computed as:

$$L_{s2}(f_{ctr})\text{ dB re 1 μPa//Hz} = -5\text{Ln}f_{ctr} - 1000/f_{ctr} + 10 + 6.67\text{ Log}\left(\nabla/\nabla_{ref}\right) + 60\text{ Log}\left(1000 C_B V/V_{incep}\right) \tag{9.8}$$

where the transition between mid- to high-frequency and low frequency, f_{tran}, is approximately

$$f_{tran} = cr/8z_s z_r$$

where c is the speed of sound in water; r is the slant distance between the source and receiver; z_s is the source depth; and z_r is the receiver depth.

The machinery radiated noise, L_{s3}, is:

$$L_{S3}(f_{ctr})\text{ dB re 1 μPa//Hz} = 10^{-7}f_{ctr}^2 - 0.01 f_{ctr} + 140 + 15\text{ Log}(m) + 10\text{ Log}(n) + E \tag{9.9}$$

where m is the mass, tonnes; n is the number of operating main engines; and E is an engine mounting parameter where $E = 0$ if the engine is resiliently mounted and $E = 15$ if not.

The total signature in each one-third octave band, L_S *dB re* 1 μPa−m, is the combination of Eqs. [9.7] through [9.9] as:

$$L_S\text{ dB re 1 μPa–m TOB} = 10\text{Log}\left(10^{\frac{L_{S1}}{10}} + 10^{\frac{L_{S2}}{10}} + 10^{\frac{L_{S3}}{10}}\right) + 10\text{Log}\left(.23\,\mathrm{f_{ctr}}\right) \tag{9.10}$$

Per the AQUO project, Audoly et al. 2015, the measured ship narrow band signature can be decomposed into three components. These are low, $L_{p-prop\ low}$, high-frequency propeller noise, L_{p-cav}, and machinery noise $L_{p-machy}$. The equation is as follows:

$$L_S\text{ dB re 1 μPa–m//Hz} = 10\text{ Log}\left(10^{L_{p-prop\ low}/10} + 10^{L_{p-cav}/10} + 10^{L_{machy}}\right) \tag{9.11}$$

These are further subdivided into 'characteristic' patterns that have two frequency divisions for each component; with unknown coefficients, K, as such:

$$L_{p-machy} = K_1 + K_4 \,\mathrm{Log}(V), \quad f < f_{machy,} \tag{9.12}$$

$$L_{p-machy} = K_2 + K_3 \,\mathrm{Log}(f) + K_4 \,\mathrm{Log}(V), \quad f > f_{machy,}$$

$$L_{p-prop\ low} = K_5 + K_6 \,\mathrm{Log}(f) + K_9 \,\mathrm{Log}(V), \quad f < f_{prop\ low,}$$

$$L_{p-prop\ low} = K_7 + K_8 \,\mathrm{Log}(f) + K_9 \,\mathrm{Log}(V), \quad f > f_{prop\ low,}$$

$$L_{p-cav} = K_{10} + K_{11} \,\mathrm{Log}(f) + K_{12} \,\mathrm{Log}(V), \quad f < f_{cav,}$$

$$L_{p-cav} = K_{13} + K_{14} \,\mathrm{Log}(f) + K_{15} \,\mathrm{Log}(V), \quad f > f_{cav,}$$

where V is the ship speed, kn.

Dependance on and typical values of K are presented in Audoly et al. 2015. Peng et al. 2018, shows the application of this approach to a fishing boat.

This model was applied to a fast craft, Audoly 2017. The resulting radiated noise model in dB re 1 μPa–m//Hz (uncorrected for the Lloyd mirror) is:

$$L_{machy}(f, U_a, L_{ref}) = 122 + 15 \,\mathrm{Log}(V) \qquad \text{for f} < 300 \text{ Hz} \tag{9.13}$$

$$L_{machy}(f, U_a, L_{ref}) = 182 - 25 \,\mathrm{Log}(f) + 15 \,\mathrm{Log}(V) \qquad \text{for f} > 300 \text{ Hz}$$

$$L_{cav}(f, \mathrm{U_a}, L_{ref}) = 58 + 10 \,\mathrm{Log}(f) + 60 \,\mathrm{Log}(V) \qquad \text{for f} < 80 \text{ Hz}$$

$$L_{cav}(f, U_a, L_{ref}) = 112 - 20 \,\mathrm{Log}(f) + 60 \,\mathrm{Log}(V) \qquad \text{for f} > 80 \text{ Hz}$$

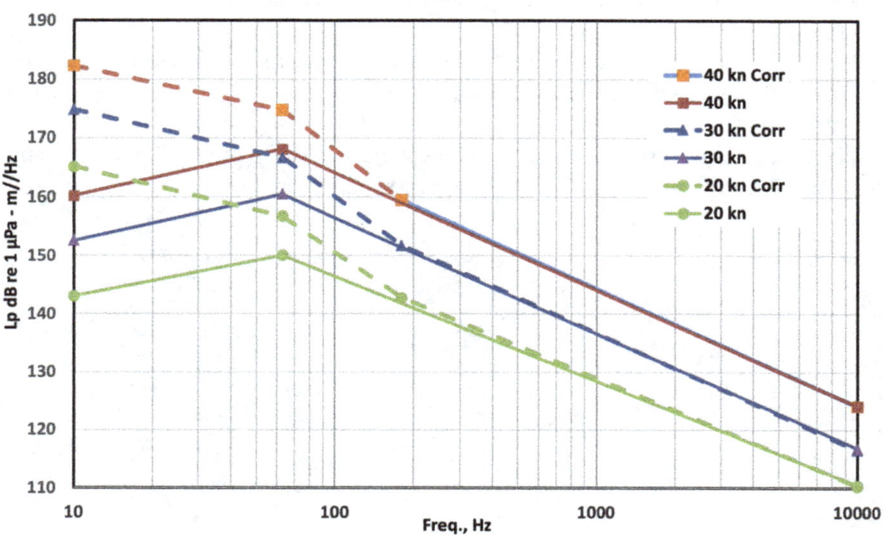

Fig. 9.2 Predicted radiated noise of high-speed craft versus speed (Audoly 2017)

For the propeller, note the significant increase with the sixth power of speed. For machinery contributions, the increase is only the 1.5th power. The predicted levels versus speed are shown in Fig. 9.2.

The 'corr' in Fig. 9.2 is the Lloyd mirror correction (Audoly 2017). This correction is detailed in Sect. 9.7.

A summary of nine empirical approaches, some of which are postulated above, is provided in Yan et al. 2021. More details on semiempirical prediction methods accounting for tip vortex noise are addressed by Bosschers 2018 and Lee et al. 2014. These references also address the hull vibration induced by the tip vortex. Additional empirical models are presented in Yan et al. 2021. Some 'parametric' approaches are provided in Traverso et al. 2016.

9.1.3 Measured/Representative Ship Radiated Noise Levels

In contrast to the prediction methods provided above, representative levels are provided in Tables 9.2 (Ainslie et al. 2009) and 9.3 (Yatsenko 2012). MMO 2015, Table 9.1 has an extensive list of numerous ship noise sources collated from the available literature.

Table 9.4 provides representative source levels for various vessels (adopted from Ainslie et al. 2009). Note that the spectrum shape for all types of vessels is very similar and dominated by high levels below 100 Hz. This region would have blade rate tones and broadband noise, as shown in Fig. 9.15 for typical cavitation noise. The actual 'peak' frequency for any vessel depends strongly on the propeller characteristics, as discussed in Sect. 9.1. Note that the average "tanker", "large tanker" and "super tanker" radiated signatures are above the DNV Silent E – Transit criterion (see Sect. 3.4).

Measured data for craft under 25 m in length are presented in Parsons 2021. The following charts provide measured noise levels for various types of vessels and a discussion of pertinent details.

The first group of cruise ships covers diesel-electric (DE), direct-diesel (DD) and steam turbine (ST) propulsion plants. The vessels are relatively large, ranging from

Table 9.2 Representative ship source levels below 300 Hz, dB re 1 μPa–m *TOB*

Ship type	Length, m	speed, m/s	Third-Octave Band Center Freq., Hz					
			10	25	50	100	300	OA, dB
Supertanker	244–366	7.7–11.3	189	197	196	189	175	200
Large tanker	153–214	7.7–9.3	179	187	187	180	167	190
Tanker	122–153	6.2–8.2	171	179	180	173	161	183
Merchant	84–122	5.1–7.7	165	173	174	168	155	177
Fishing	15–46	3.6–5.1	143	151	152	146	135	155

OA overall, sum over the frequency range measured

Table 9.3 Ship signatures – peak frequency and level, dB re 1 μPa–1 m

Source	Peak frequency or band	Peak source level, dB re 1 μPa–1 m
Icebreaking ship (full power in ice)	10–1000 Hz	193
Large tankers and bulk carriers[a]	10–30 Hz	180–186
Container ship[b]	7–33 Hz	181
64 m rig suppl tender[a]	Broadband	177
Tug towing barge[a]	1000-5000	145–171
20 m fishing vessel[a]	Broadband	168
Trawler[c]	100 Hz	158
25 m SWATH ferry with (2) 950 hp. inboard diesels[b]	315 Hz	166
13 m catamaran with (2) 200 hp. inboard diesels[a]	315/1600 Hz	159/160
Bertram cabin cruiser, (2) 165 hp. outboards[a]	400 Hz	156
8 m RHIB (2) 250 ho outboards[c]	315–5000 Hz.	177–180
Cutter-suction dredge – Working	100 Hz tonal	180
Clamshell dredge – Working	250 Hz pulses	150–162
Pile driving operation	Low tonal pulses	170–180
Seismic survey	0–1000 Hz	200–232
Mobile offshore drilling unit	10–4000 Hz	154–160
Supply vessel	1–500 Hz	182

[a]Recorded at 10–11 knots; [b]Recorded at 15 knots; [c]Unrecorded speed or speed range. Data sourced from Richardson et al. 1995; Dames & Moore 1996; Au and Green 2000, McCauley et al. 2000; University of Rhode Island, undated; and DSTO data for Port of Dampier

23 to 77 thousand gross tons and from 189 m to 260 m in length. Data were collected at various speeds, nominally 10 to 20 knots. No consistent ship characteristics were discernable from the data.

The spectra for the cruise vessels, operating at 10 kn, are typically dominated by low-frequency tones, below 100 Hz, as shown in Fig. 9.3 (Kipple 2002). The minimum, maximum and 'power' averaged data for the cruise vessels are shown. The radiated noise at 10 knots for steam, direct drive, and diesel electric-driven ships is provided in Fig. 9.4 (Kipple 2002). Note that these signatures are just below DNV's Silent E – Transit limit. For a vessel with an Azipod drive, the spectrum content[10] is quite different at 10 kn and 20 kn, see Fig. 9.5. However, the overall levels are comparable (Kipple 2002). However, at 5000 Hz, the noise level for operation at 20 kn is over 11 dB higher than for the signature for operation at 10 kn. The difference above 5000 Hz is basically the difference in cavitation-induced noise for the two operating speeds.

For a particular mammal, the difference at this frequency may be critical, whereas if only the overall level were considered, these two signatures would be considered

[10]There is insufficient data provided to identify the tone in the 10- knot signature at 315 Hz. Tones such as this are not atypical; and may be the gear mesh frequency in the Azipod.

Table 9.4 Representative ship source levels, dB re 1 μPa–1 m

	Supertanker	Large Tanker	Merchant	Tanker	Fishing
Third-octave band Freq, Hz	L_p dB re 1 μPa–m				
10	189	179	164	171	142
12.5	191	181	167	173	144
16	193	183	169	175	146
20	195	184	171	177	148
25	196	186	172	178	150
31.5	197	187	173	180	152
40	197	187	174	180	152
50	196	186	173	179	151
63	194	185	171	177	150
80	191	182	169	175	148
100	188	179	167	173	146
125	185	177	164	170	143
160	183	174	162	167	141
200	180	172	159	165	
250	177	169	157	162	
315	175	167	155	160	
400	172	165	152	158	
500	170	162	150	156	
630	169	161	149	155	
800	168	161	148	154	
1000	167	160	148	153	

to have an equivalent impact on marine life. Once again, the signatures of various vessels operating at 10 kn are compared in Fig. 9.6.[11] For comparative purposes, the DNV Silent E criterion is also plotted in this figure. The shape of the measured spectra varies from broadband (without apparent tones) for the 'Large surface vessel' to tonal (peaks in the third-octave band signature) for the cruise vessels – see Fig. 9.7 (Kipple 2002).

The tones normally seen for a diesel propulsion system are shown in Fig. 9.8. The first twenty-two half harmonics of the diesel rotation rate are easily identified. Blade rate harmonics at $n = 1, 4$, and 8 are significant. Firing rate tones are strong for the first three harmonics.

Representative noise spectra for the cargo ship M.V. OVERSEAS HARRIETTE (Arveson and Vendittis 2000) are adapted for Fig. 9.9. M.V. OVERSEAS HARRIETTE is a bulk cargo vessel (length 173 m and displacement 25,515 tons) powered by a direct-drive low-speed diesel engine. The signature has a strong tone at 25 Hz for all operating speeds except the highest. It is important to note that the tone does not change with the propulsion speed. Thus, this tone is not related to ship

[11] Greater details on ship size, propulsion, etc., are not readily available.

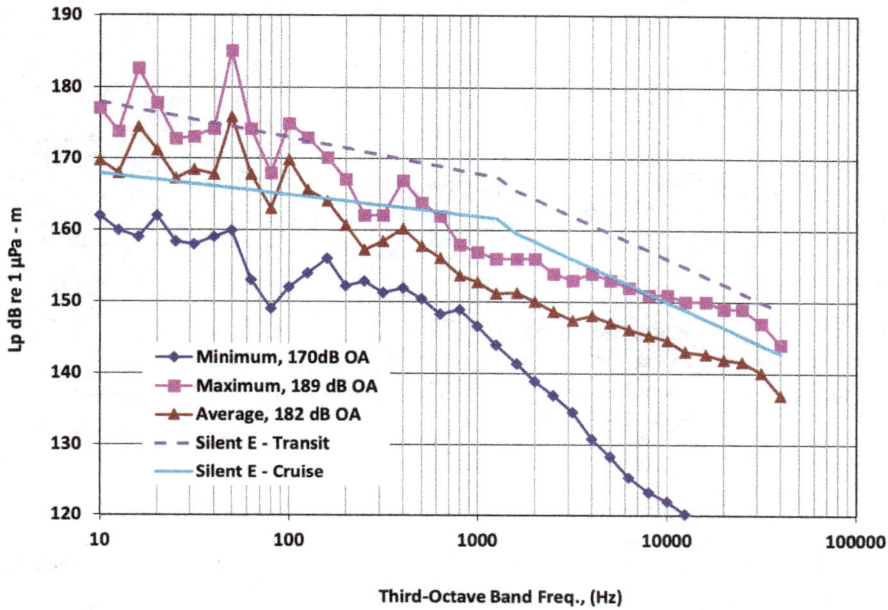

Fig. 9.3 Radiated noise – cruise ships at 10 knots

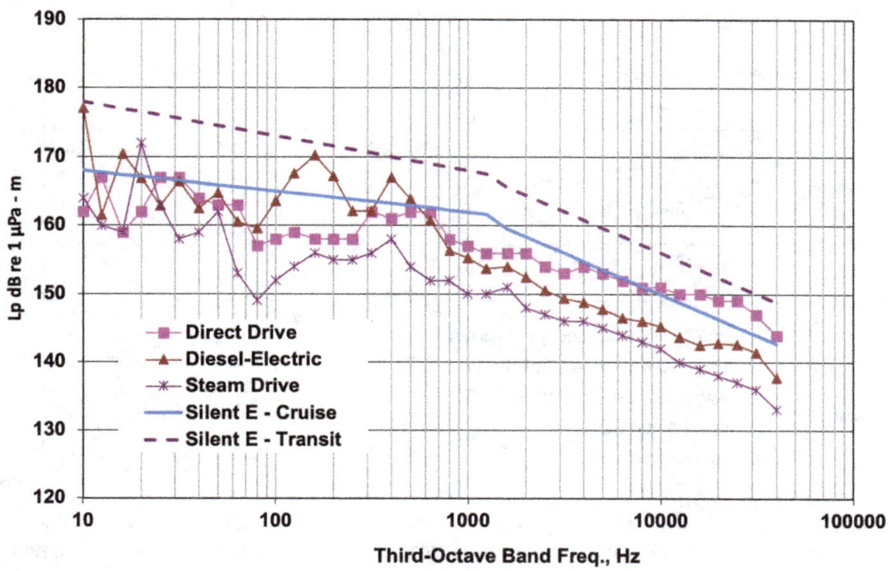

Fig. 9.4 Radiated noise of cruise ships versus Silent Class limits

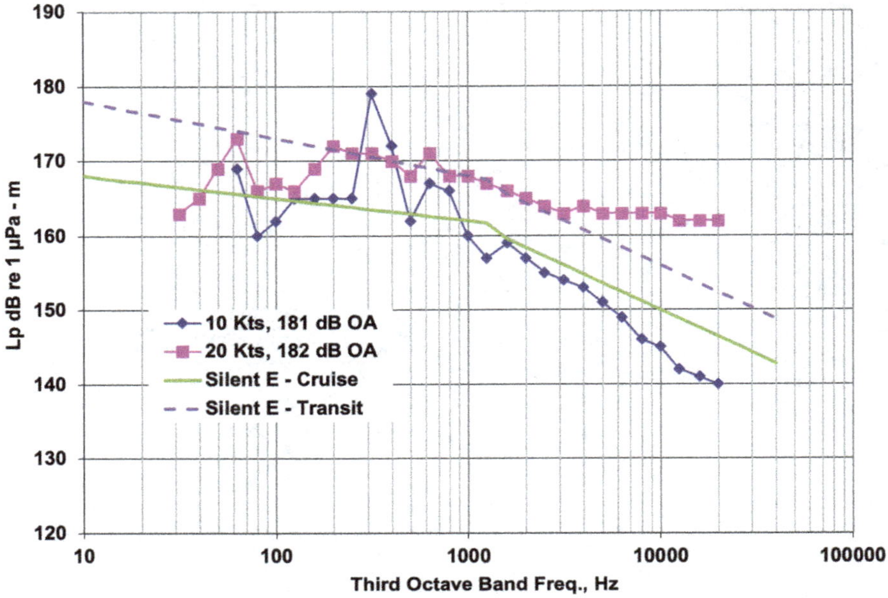

Fig. 9.5 Radiated noise – Azipod propulsion

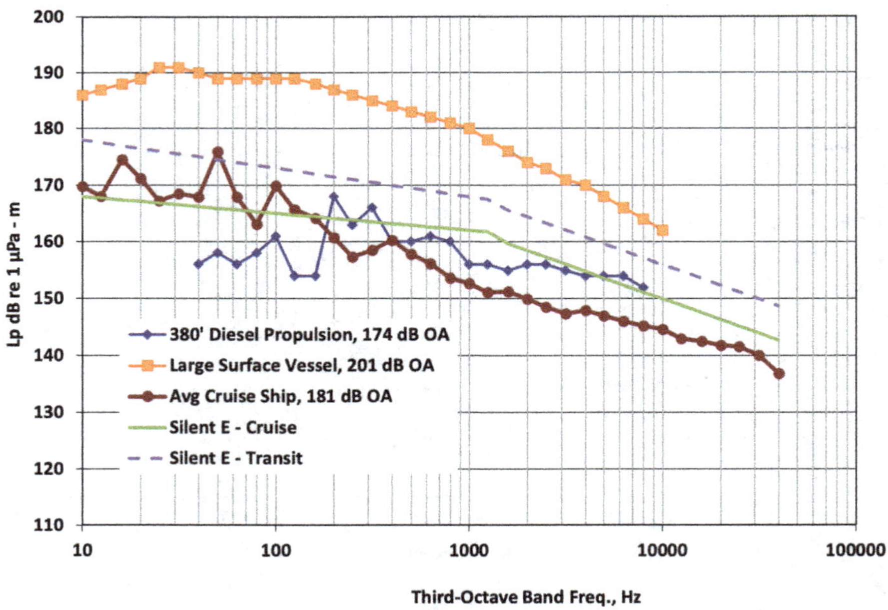

Fig. 9.6 Radiated noise – various types of ships @ 10 knots

Fig. 9.7 Radiated tonal signature – cruise ship

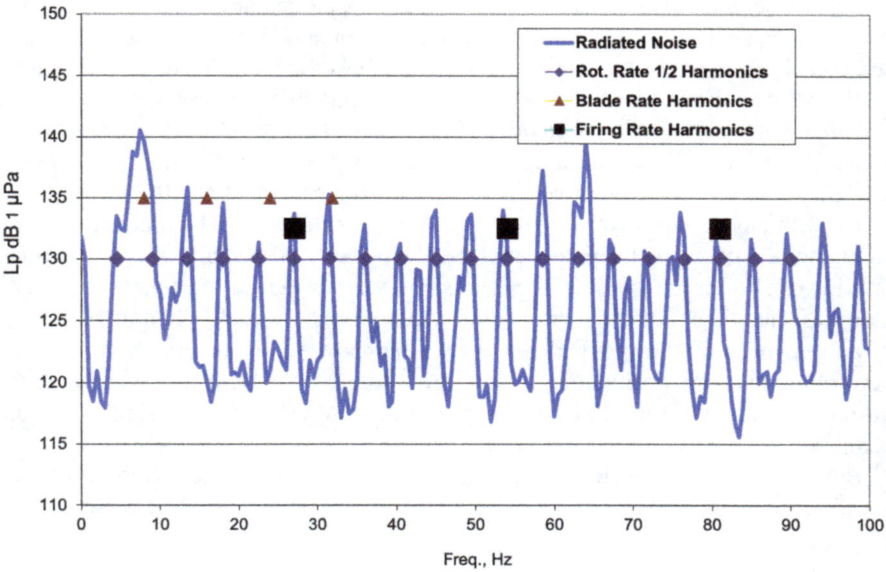

Fig. 9.8 Radiated noise – rotation, blade, and firing rate harmonics

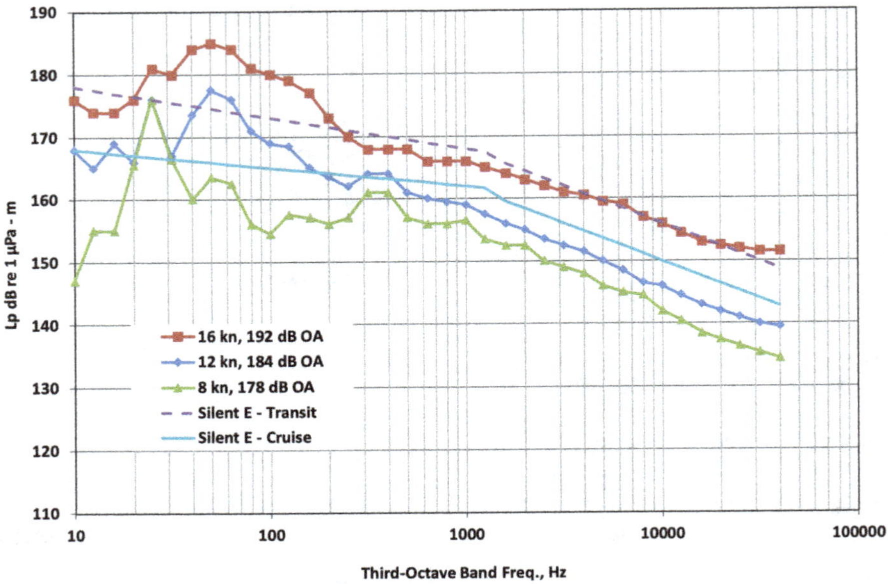

Fig. 9.9 Cargo ship radiated noise; 25,525 tons, 173 m- LOA, 11,200 hp. diesel

speed or machinery that varies in rotation rate. This particular tone is controlled by the diesel generator, which operates at a constant rotation rate, independent of ship speed. This vessel at 8 knots might just meet the DNV Silent E- Transit Speed limit.

Workboat noise levels for several vessels operating in an environmentally sensitive area, Sakhalin Island, are provided in Fig. 9.10 (Hannay 2004). These levels are above DNV Silent E, Cruise or Transit. Table 9.5 presents radiated noise data for dredging and offshore drilling vessels (adopted from Ainslie et al. 2009).

As with any of the spectra shown in this chapter, there may be tones related to strong noise induced by the propeller at the blade rate or the machinery at its rotation or firing rate.[12] It is not possible in most cases to identify the specific tones in any one plot presented herein due to a lack of information on the vessels' equipment.

An example of the radiated noise for a support vessel in transit and while performing dynamic positioning is provided in Fig. 9.11. Note that during dynamic positioning, the noise levels increase by approximately 10 dB. The signatures are compared to the DNV Silent E criteria.

Bear in mind that radiated noise from drill ships, FPSO[13]s, semi-submersibles, and floating platforms will be dominated by their machinery-induced noise unless

[12] Firing rate = rotation rate (rps) × number of cylinders × 2/number of strokes

[13] FPSO – Floating Production Storage and Offloading vessel

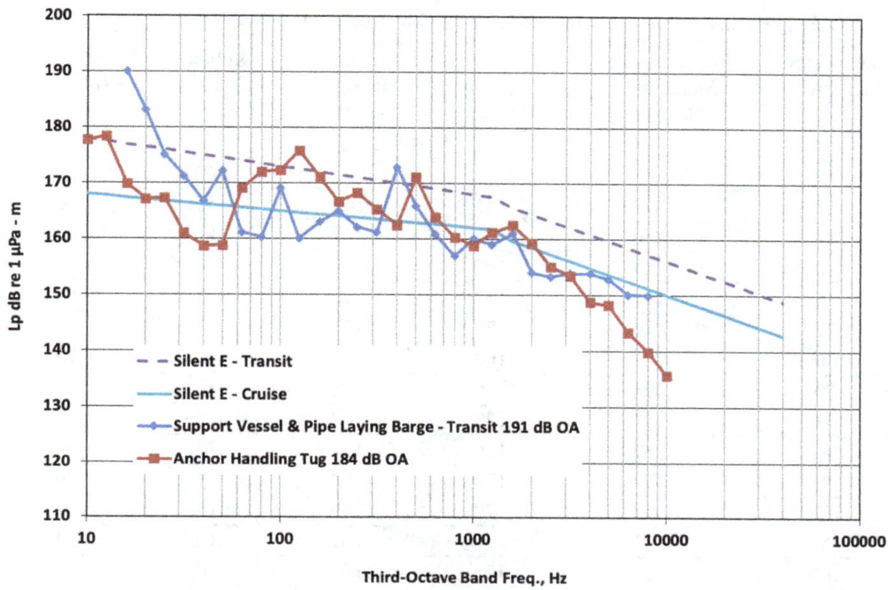

Fig. 9.10 Workboat radiated noise, operating off Sakhalin Island

underway. Machinery-induced noise from these vessels is likely to be below several hundred Hertz. When underway or when using thrusters, propeller cavitation would likely be the dominant radiated noise source. Figure 9.12 shows a representative radiated noise signature (adapted from Erbe et al. 2013).

In summary, the overall noise level from commercial vessels varies from a low of 150 to a high of 230 dB re dB 1 µPa–m. The frequency characteristics will vary widely depending on the operating speed, type of propulsor, and propulsion drive equipment. In general, as vessels grow larger and higher in speed, their radiated signature increases. There are insufficient data currently available to develop general scaling laws for these changes.

9.2 Ship-Radiated Noise Sources

As noted above, at full speed, the propeller-induced radiated noise is most likely the controlling factor over most of the frequency range of interest. At lower operational speeds, the signature may be composed of machinery-induced noise in the low- to mid-frequency range and propeller vortex shedding/turbulence ingestion in the mid- to high-frequency range (Abrahamsen 2012 and de Jong 2009).

Noise is also generated from other fluid/ship interactions, such as water flow over the ship's hull or wave motion against the hull. The noise level that these interactions

Table 9.5 Dredging and drillship radiated noise, dB re 1 μPa–1 m

Third-octave Band Freq., Hz	Gerardus Mercator THSD	Aquarius Dredger	Beaver Mackenzie Dredger	Kulluk Drilling Unit	Explorer II Drillship
	L_p dB re 1 μPa–m				
10	183			151	
13	178	151	147	155	160
17	180	163	147	158	166
20	168	160	151	160	163
25	167	167	155	159	167
32	170	170	154	161	174
40	171	169	155	162	172
50	178	170	162	167	174
63	178	178	167	164	172
80	167	176	161	161	169
100	173	178	160	160	176
125	165	177	159	162	176
160	171	175	161	161	174
200	170	175	160	161	172
250	164	174	162	164	177
315	172	171	158	161	176
400	170	168	157	162	173
500	167	167	158	159	173
630	165		161	158	168
800	165		156	156	167
1000	161		153	152	166
1250	158		152	150	
1600	155		148	148	
2000	152		147	145	
2500	149		145	143	
3150	146		144	140	
4000	144		141	137	
5000	144		141	135	
6300	144		144	134	
8000	146		142		
10,000	142		142		
OA	188	186	172	174	186

produce in the water is usually small when compared to the other sources of noise on typical commercial vessels. These components, excluding mechanical sources inside the hull, are summarized in Fig. 9.13 and will be described in greater detail in the following sections (adopted from Jebsen 2004).

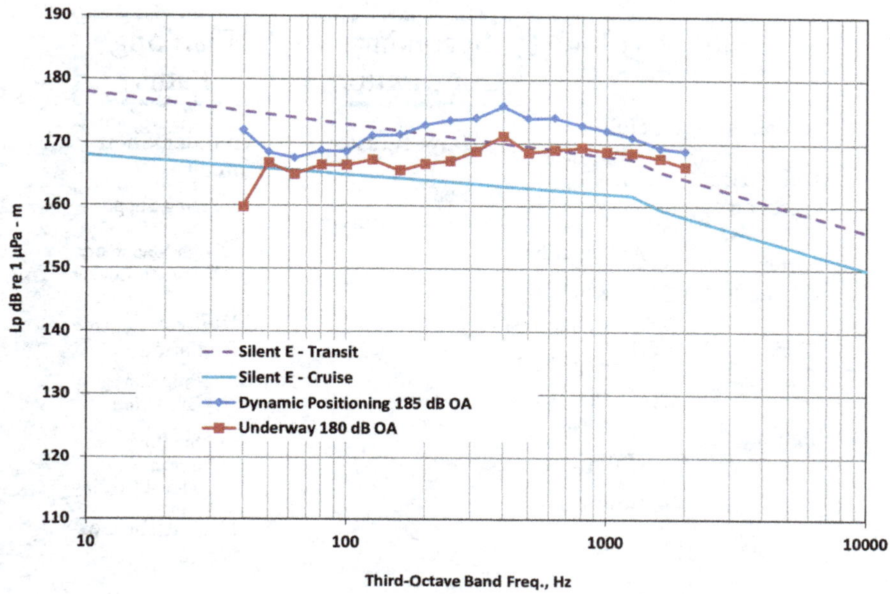

Fig. 9.11 Offshore support vessel, underway and while dynamic positioning

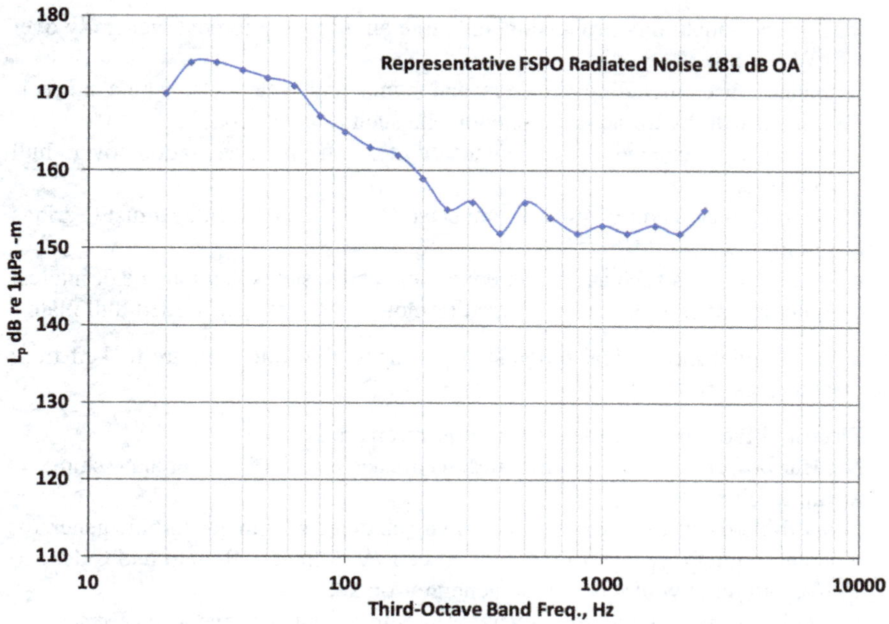

Fig. 9.12 Representative FSPO radiated noise signature

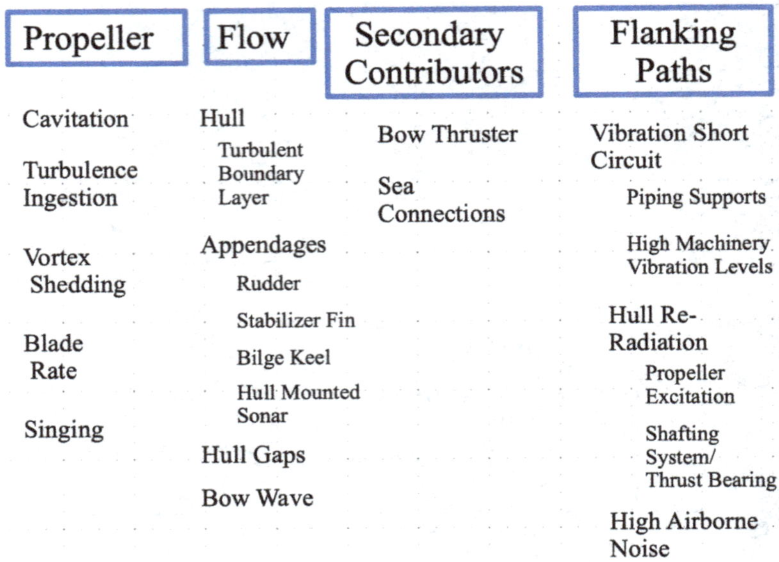

Fig. 9.13 Summary of non-machinery radiated noise sources

As shown in the figure, the major groups of sources are as follows:

- machinery – including propulsion and large auxiliary equipment, generally over 37 kW (50 hp)
- propeller noise, including thrusters and pump and water jets, where noise is associated with hydro-acoustic phenomena such as cavitation
- flow noise – created by turbulence/vortex shedding/cavitation over hull components
- flanking path created by systems connected to machinery such as thrust bearings or piping
- secondary sources such as smaller equipment and systems that may be of interest on quiet research vessels or if located or mounted directly on wetted hull plate.

In terms of quiet mechanical propulsion packages, the following are ranked from quietest to noisiest:

- Electric drive – battery, fuel cell, or non-mechanical drive
- Nuclear power – used on submarines, icebreakers, and a few merchant ships
- Steam turbine
- Gas turbine-electric – used on research vessel, quietest with gas turbine generator mounted high above Main Deck, again see R/V Alliance (Bonetti and Calcagno 1986); still quiet with gas turbine generator on keel
- Wankel-electric – again with generator system mounted above Main Deck
- Diesel – Electric (DE) drive – used on some submarines and quiet research vessels such as the R/V Alliance; motor with quiet sine waveform drive or superconducting electric motor

- Gas turbine-geared
- Diesel-geared
- Diesel-direct drive

9.2.1 *Propeller/Propulsor as an Underwater Acoustic Source*

Propeller cavitation is, by far, the strongest source of ship radiated noise and the greatest underwater threat to marine mammals. This noise is both tonal and broadband in frequency, extending from less than 10 Hertz to well beyond 50,000 Hz, see Fig. 9.14 (Brown 1976). Reduction of ship propeller cavitation will be the most effective means to protect ocean animals.

Cavitation is basically a consequence of blade lift fluctuations on the blade operating in a non-uniform inflow. Cavitation usually occurs on propellers that are heavily loaded and/or subject to high thrust loading. The cavitation 'sound' is basically a 'boiling' of the water wherein bubbles are formed. These collapse violently as they move away from the low-pressure zone near the blade. The impulsive noise generated by the various sized bubbles creates a broadband noise spectrum (Brown 1976). The noise generated from this event is the dominant source of underwater noise once the propeller cavitates. Additionally, cavitation-induced noise often dominates habitability noise in aft compartments.

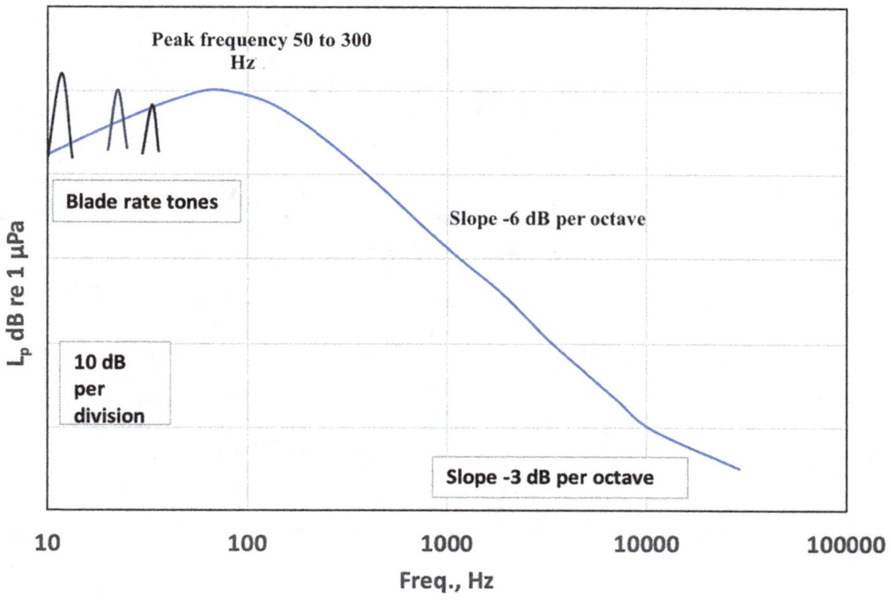

Fig. 9.14 Typical narrow band and broadband propeller cavitation spectra

The cavitation level depends on the type of cavitation – tip vortex, suction side sheet bubble, pressure side sheet bubble, cloud-type, and hub vortex – and the total area of cavitation, as illustrated in Fig. 5.4. Propeller radiated noise can also be directional, especially at low frequencies, depending on the type of propulsor and its location relative to the hull.

As already noted, propellers often exhibit some directionality in their radiated signature. Figures 9.15 and 9.16 show directional patterns depending on the operating conditions, *j*, and the advance coefficient. *j* is determined as:

$$j = U_a / \left(N_p D_p\right) \tag{9.14}$$

where U_a is the ship's advance speed and *Np* and *Dp* are the rotation speed in rpm and the diameter of the propeller, respectively. For a lower j, or higher speeds, Fig. 9.15 shows the directivity peaks fore and aft, whereas for higher j's, or lower speeds, the peak is abeam, Fig. 9.16. The abeam directivity is an indication of the monopole nature of the radiated noise from the propeller. The fore/aft directivity is an indicator of dipole radiation (Mousavi et al. 2014).

As discussed above, the noise radiated by vessels is controlled by many factors, including the vessel size, power, speed, type of propeller or propulsor, and type of propulsion and auxiliary equipment. Below the cavitation inception speed, the signature is dictated by the machinery contribution to the signature.

For a non-cavitating propeller, the signature is controlled at low frequencies by tones associated with the blade rate caused by inflow turbulence (IT). There may also

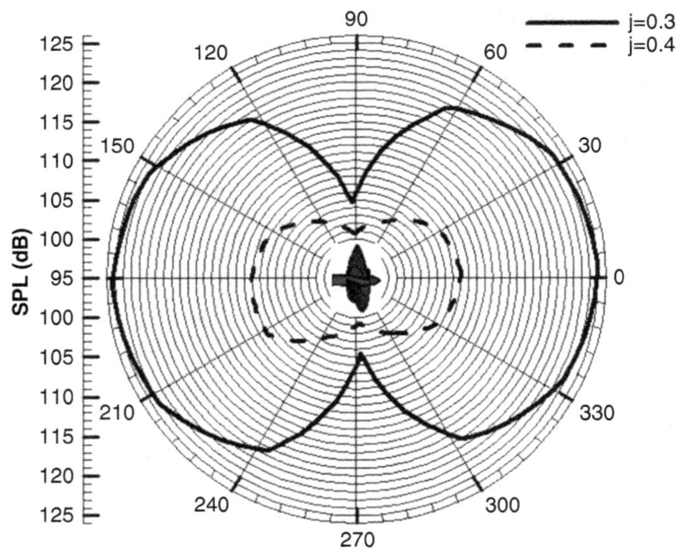

Fig. 9.15 Propeller directivity, low advance coefficient, J (Mousavi et al. 2014)

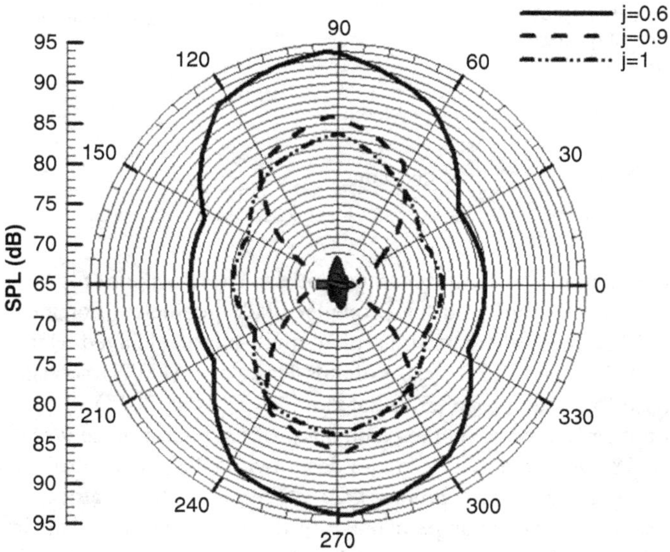

Fig. 9.16 Propeller directivity, high advance coefficient, J (Mousavi et al. 2014)

be some low-frequency tones associated with wake or flow noise. The discrete frequency at the blade rates results from fluctuations in the cavity volume as the propeller blades move through the non-uniform inflow presented by the vessel's hull wake field and by the shaft angle. Due to large-scale turbulence fluctuations (and ship motions), these lines have some bandwidth. Inflow turbulence from the leading edge and vortex shedding off the trailing edge are both significant in the mid- to high-frequency spectrum. While these sources are monopoles, radiation appears dipole, hence directivity, due to the proximity to the free surface (Urick 1983).

Above the cavitation inception speed, the broadband noise is regulated by the cavitation contribution, propeller blade rate contributions, and tones associated with the rotation rate of the prime mover and drive system. Note that the cavitation inception speed discussed herein is that for normal operation in a calm sea (<Sea State (SS) 2) since cavitation nominally increases in turns (Trevorrow and Vasilyeva 2008), stops, and increased sea state. Environmental effects such as vessel heading with respect to wind and sea state can significantly alter the inception speed. Towing or marine growth on blades or vessels will also affect the inception speed and radiated signature. A rudder angle of up to 20 degrees can decrease the cavitation inception speed by 3 to 4 knots. Furthermore, the radiated level may increase by 1 dB for a rudder change of 10 degrees and by 3 to 4 dB for a change of 20 degrees (Verkuyl and van Terwisga 2000). As shown in Table 9.6, the sea state can also affect the inception speed. Measured data for a fixed pitch, single screw vessel shows the following (adapted from Verkuyl and van Terwisga 2000):

Detailed and accurate prediction methods exist to address propeller radiated noise. The more sophisticated propeller prediction methods involve RANS/lifting

Table 9.6 Sea state impact on cavitation inception speed

Sea State	Decrease in inception speed, kn
Calm	0
Following	−1
Stern quartering	+1.5
Beam	+3
Bow quartering	+5
Head	~ + 7

surface and computational fluid dynamic models. These account for the interaction of the flow around the hull and the propeller blade geometry. For the EU project AQUO, "A CFD method, consisting of a multi-phase Delayed Detached Eddy Simulation (DDES) and a Ffowcs Williams-Hawkings (FWH) acoustic analogy" have been used to model propeller radiated noise (Li et al. 2015).

Some of the early reports on the calculation of cavitation noise were carried out by Greeley 1978. He found that the major design variables that control cavitation are "diameter, expanded area ratio, percentage lift by camber, wakefield non-uniformities, ship speed as percentage of design speed, and shaft immersion." Greeley also provided methods to determine sheet cavitation volume, and details of how the volume history affects blade rate harmonics.

Modeling the acoustic signature of a propeller usually requires extensive experience and sophisticated computer software. Details of which are beyond the scope of this book. The reader is referred to the following reference (Bagheri et al. 2015; Blake 1984; Sharma et al. 1990; and Bosschers 2018). Many marine propeller vendors have the means to assist in this process.

9.2.1.1 Calculation of Propeller Source Levels

Empirical predictions of broadband cavitation noise show a dependence on ship speed, propeller diameter, cavitation inception speed, number of blades, and submergence depth. Examples of predicted versus measured propeller source levels are provided in Fig. 9.17. The cavitation noise scales with the total number of blades (B), diameter (D, m) to the fourth power, rotation rate (N, rps) to the third power, and one over the square of the frequency of interest, as the following equation illustrates (Brown 2007):

$$L_s \text{ dB re } \mu\text{Pa–m}//\text{Hz} = 163 + 10\,\text{Log}\left(\text{B D}^4\,\text{N}^3/\text{f}^2\right) + 10\,\text{Log}\left(\text{A}_\text{c}/\text{A}_\text{d}\right) \quad (9.15)$$

where A_c is the swept area of cavitation on each blade (projected view) and A_d is the projected disc area.

This equation applies above the peak frequency, f_p, defined as:

$$f_p = \{28/(D(A_c/A_d))\}\sqrt{}(10.08/(\text{H} + 10.08)) \quad (9.16)$$

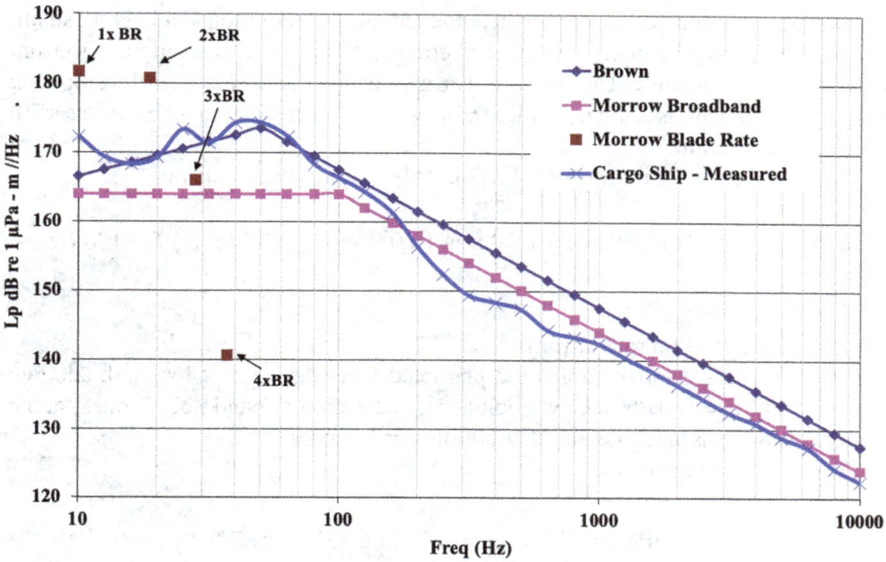

Fig. 9.17 Empirical versus measured propeller source level

where H is the submergence depth of the shaft in meters. Above and below the break frequency, the spectrum falls off at −1 dB per one-third octave band center frequency. The spectrum level can be converted to one-third octave bands by adding the following at each one-third octave band center frequency, f_{ctr}:

$$10\,\mathrm{Log}\,(0.23\,f_{ctr}) \tag{9.17}$$

If no tunnel[14] test data are available, the A_c/A_d term can be computed as:

$$A_c/A_d = \left(\left\{ N/N_i - 1 \right\} \big/ \left(N/N_i \right) \right) \tag{9.18}$$

where N is the operating rps and N_i is the inception rotation rate.

Below the break frequency, f_p, the continuous spectrum noise is more or less flat, again, in constant bandwidths (excluding blade rate components). The break frequency varies inversely with diameter and the normalized swept area and inversely with the square root of the absolute submergence pressure at the shaft or hub. In proportional bands, the break frequency becomes the peak frequency in the continuous frequency spectrum.

---

[14] Also bear in mind that cavitation in tunnel testing can usually be heard before there is visual evidence (Johannsen et al. 2014) Johannsen, C., O. Zarbock, and H. Bretschneider, "Experimental Determination of the Difference in Visual and Acoustic Cavitation Inception," A. Yucel Odabasi Colloquium Series, first International Meeting – Propeller Noise and Vibration, Istanbul, Nov. 2014.

Just above the inception speed, where the amount of cavitation is relatively small, its break frequency is high, and there is comparatively little continuous spectrum noise at low frequencies. The opposite is true when cavitation is well developed at higher speeds. This is shown in Fig. 9.26, where the break frequency drops with increasing cavitation.

The predicted level by Morrow 1987 is only slightly different:

$$L_s \text{ dB re 1 μPa–1m//Hz} = 34 + 10 \text{ Log} \left(BD^6R^6/4\right) - 20 \text{ Log } (f)\, f > 100\, Hz. \tag{9.19}$$

where R is the rotation rate in rpm.

Below 100 Hz, Morrow holds the predicted level flat, except for the blade rate components ($f_B = RxB/60$). Cavitation accentuates narrow band noise (tones) at the propeller blade passage rate and its multiples or harmonics.

These are computed as:

$$L_{s,n} \text{ dB re 1 μPa–m//Hz} = 106 + 20 \text{ Log} \left(D^3 f_B^2 n^2 e^{-n^2/2}\right) \tag{9.20}$$

where n is the nth harmonic of the blade rate. Per this approach, each higher harmonic is lower than the previous harmonic by over 12 dB.

The measured noise for the cargo ship at 16 knots is compared to both prediction methods in Fig. 9.17. The prediction per Brown is quite comparable to the measured radiated noise of the cargo ship.

Per Gray and Greeley 1980, the blade rate source level is:

$$L_{br} \text{ dB re 1μPa–m//Hz} = 140 + 10 \text{ Log} \left(f_B^6 H^2 V_{max}^2\right) \tag{9.21}$$

where H is the source depth, m, and V_{max} is the maximum cavity volume per blade, m^3.

In their sample study, V_{max} and H were determined to be:

$$V_{max} = 3 \times 10^{-5} D^3; H = 0.12 D^{1.7}; D = 0.16\, L^{0.75} \tag{9.22}$$

where L is the ship length in m. This method approximates the maximum cavitation volume expected based on H, D, and L. For instance, a 100 m long ship with a 5 m diameter propeller would be predicted to have a blade rate source level of 188 dB re 1 μPa–m.

Acoustic cavitation occurs prior to the visual appearance of bubbles, either in a cavitation tunnel or on the vessel itself. The impact of certain propeller designs on cavitation will be discussed in detail in the following sections. If the vessel operates below the cavitation inception speed of the propeller, the machinery-induced noise dominates the signature from low to high frequencies. The cavitation inception depends on the blade form, disc area, and propeller tip speed. The cavitation inception speed, v_{incep}, for large commercial vessels is (Jalkanen et al. 2018):

$$v_{incep} = min\{\max[(1.42 - 1.2\, c_B)\, V_d; 9]; 14\} \tag{9.23}$$

where c_B is the block coefficient and V_d is the design speed of the vessel in knots. Another rule of thumb for cavitation inception speed is:

$$v_{incep} = 7 + 0.001L, m/s \tag{9.24}$$

where L is the length of the vessel in meters (Heckl 1988).

A good design guide is a tip speed less than 18 m/s (60 fps). The inception speed for most vessels, unless special attention is given to the hydro-acoustic design, is between 9 and 14 knots.

Lifting line, lifting surface, and vortex lattice propeller noise prediction methods have been developed to provide more accurate noise prediction methods – both above and below cavitation inception speed. These methods require more details on the geometry – chord length, camber, thickness, and hub size – of the propeller. The most accurate methods, using Computational Fluid Dynamic (CFD) models, account for the expected wake field in the propeller plane and the propeller 3-D surface parameters (Kinnas et al. 2002).

Some propeller vendors have the capability to accurately predict the cavitation inception speed for the various types of cavitation (tip, hub, surface) and the broadband and narrow radiated noise characteristics. As will be discussed in the section on noise control, these programs can also account for propeller design changes such as skew and tip relief.

9.2.1.2 Acoustics of Various Types of Propulsion Systems

Propulsor systems considered herein include open or ducted propellers, water jets, podded systems, and bow/stern thrusters. The main concern with respect to operating in an environmentally friendly manner is to operate the vessel below its cavitation inception speed. Once the propellers cavitate, the radiated noise signature is high, with the potential to adversely affect sea life. Monitoring systems that measure the hull vibration or pressure pulses over the propeller can easily be installed to monitor the cavitation inception speed and provide feedback for operating below that speed.

9.2.1.2.1 Open/Ducted Propellers

Most of the equations and discussion above apply to open propellers. Ducted propellers come in two types of Kort-nozzles. Kort nozzles are used mainly to improve the efficiency of a heavily loaded propeller. Kort nozzles are used primarily on tug and tow boats, which typically are low-speed vessels. These nozzles have accelerating ducts and pump jets. Decelerating Kort nozzles are often used on higher speed

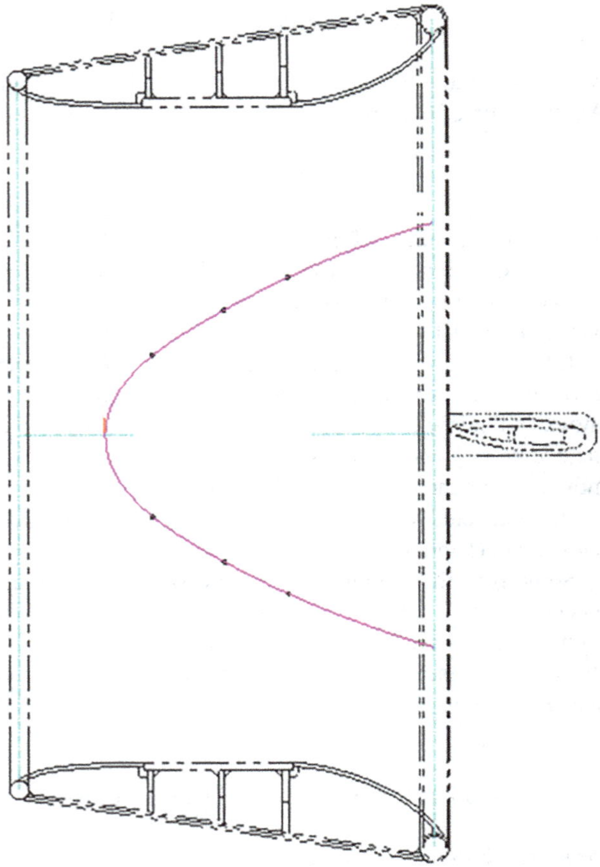

Fig. 9.18 Example of Kort Nozzle

vessels. Figure 9.18 shows an accelerating duct type. With this type of nozzle, the inflow velocity is increased, lowering the thrust and torque provided by the propeller. The duct itself provides some increased thrust. The duct in both types provides a partial baffle against higher frequency radiation athwartships of and below the vessel.

Ducted propellers have an additional factor – interaction of the tip with the duct. This close spacing may cause a vortex breakdown at the blade tip (Zhang et al. 2022). There may also be a strong interaction along this boundary. This interaction usually requires fairly decent hydroacoustic analyses (Haimov et al. 2010). Given the structural connection to the hull, the resonant modes of the duct itself and the cantilever connection to the hull should be investigated to avoid any resonance.

Per Morrow 1987, the radiated noise of a ducted propeller is 7 dB noisier than that of a standard propeller. Similarly, the radiated noise of a well-designed propeller is usually 20 dB lower than that of a standard propeller. The blade rate for a well-designed propeller is lower than that of a standard propeller by 3 dB for the first blade rate and 10 dB for the second blade rate.

9.2.1.2.2 Controllable Pitch Propeller

There is some evidence that fixed-pitch (FP) propellers are quieter than controllable pitch propellers (CPPs). However, a 'bucket' in the rpm/pitch operating condition may be found that minimizes the radiated noise. Furthermore, there are many CPP systems on naval vessels requiring reduced signatures, so a well-designed system may be as quiet as an FP system. Direct drive systems generally use CPP systems to reverse direction rather than stop the diesel and reverse its direction.

In a study by Zhu et al. 2022, a radiated noise minimum was found at 50% while varying the pitch from 20% to 82%. The changes in level and spectra are due in part to different cavitation characteristics. Another finding was that 'the cavitation noise at high RPM was significantly stronger than at low RPM."

Per Morrow 1987, the off-design for a controllable pitch propeller can be expected to be 6 dB noisier than that of a standard propeller of the same physical dimensions and operating speed.

9.2.1.2.3 Water Jet System

Water-jet propulsors contain an axial-flow pump with a highly loaded rotor/impeller – by propeller standards. Relatively tight tip-clearances can be maintained because these impellers tend to be smaller and have tight bearing clearances. The impeller is typically located above the waterline, and the jet discharges in air, although submerged water jets have been developed (Brown 2007; Fischer 2005).

Inlet ducting typically takes suction from a hull flush inlet and raises the water through two bends and a diffuser before reaching the impeller. While the pressure increase at the diffuser obviates the elevated location of the impeller, the resulting non-uniformity of inflow can be quite severe. Furthermore, there is usually a strong interaction between the stator and rotor, which also induces a time-varying inflow. Consequently, the radiated signature exhibits strong peaks associated with stator-rotor interactions (Parchen 2000). Cavitation has also been found to be severe to the extent of damaging the pump during off-design operation, such as accelerating the vessel (Fischer 2005).

Water jets have some of the highest reported overall radiated noise levels. The overall levels for the three vessels were as follows: 201 dB 1 μPa–m @ 27 kn (13.9 m/s), 210 dB 1 μPa–m @ 36 kn (18.4 m/s), and 197 dB 1 μPa–m @ 37 kn (19 m/s) (Rudd et al. 2014). The signatures were lowest at the bow aspect and equal for the abeam and stern aspects. Waterjets for naval operations are believed to be considerably quieter. On the other hand a parametric study showed that compared to a propeller with 'comparative absorbed power', the waterjet was 16 dB quieter (Yang et al. 2014).

As with any tunnel system, water jet propulsors have additional factors to consider. These include scattering and reflections by the duct and the center-body supporting the blades. The tunnel geometry may also introduce transverse modes and longitudinal resonances for a duct open at both ends (Anichkov et al. n.d.).

9.2.1.2.4 Podded Systems and Azimuthing Drives

Podded systems are being utilized on cruise and ice breaking ships. The major advantage with respect to acoustics is that there is no large wake faction. This should delay cavitation inception. A non-acoustic advantage is onboard space. Podded and azimuthing drives normally have a submerged electric motor directly coupled to the propeller without gearing. Z-drives, on the other hand, are driven by two right angle gears. These units are trainable, providing excellent maneuvering and backing power.

The podded propulsors are often operated with the propeller forward in the 'puller' configuration that can provide a minimum of disturbed flow into the propeller. This helps in forestalling propeller cavitation if the propeller tips are removed from the region of the hull boundary layer wake. Operating in the boundary layer has a major beneficial effect on fuel economy through the "wake fraction" factor. The use of such clear water-puller propulsors has been restricted to cruise ships where on-board quiet is essential for passengers. Conversely, podded units with the motor in direct contact with seawater are not necessarily friendly for underwater creatures. High radiated noise levels at low frequency due to the motor are evident in the podded systems tested to date.

Azimuthing drives can come in a few varieties, each of which has different effects on the underwater radiated noise. There are two main types of azimuthing drives, the type that uses electric motors and the kind that uses geared diesel propulsion. If the azimuthing thruster uses electric motors, this propulsion method can have many of the benefits of the diesel electric propulsion method. For the electric motor type, the motor can be either installed in the pod itself or installed above the pod in the ship. When the electric motor is installed in the pod, the underwater radiated noise due to the motor may be greater than if it were located above the pod. If the motor is located above the pod, drive linkage and gearing may be needed in the pod. This gearing will also radiate noise, especially mesh tones, the degree of which is determined by the particulars of the installation and machining tolerances. The gear mesh tones also typically increase in frequency and underwater radiated noise with increasing speed. It would be necessary to isolation mount the drive electric motor and any associated gearing to provide a relatively low radiated noise signature (Le Bert et al. 2002). A stator vibration isolation system inside the electric motor may provide beneficial radiated noise reductions.

In summary, a diesel geared azimuthing drive will usually have a slight increase in mechanically radiated underwater noise when compared to a conventional high-speed diesel geared propulsion system. An azimuthing drive using an electric motor has the potential to be nearly as quiet as a conventional diesel electric system if the

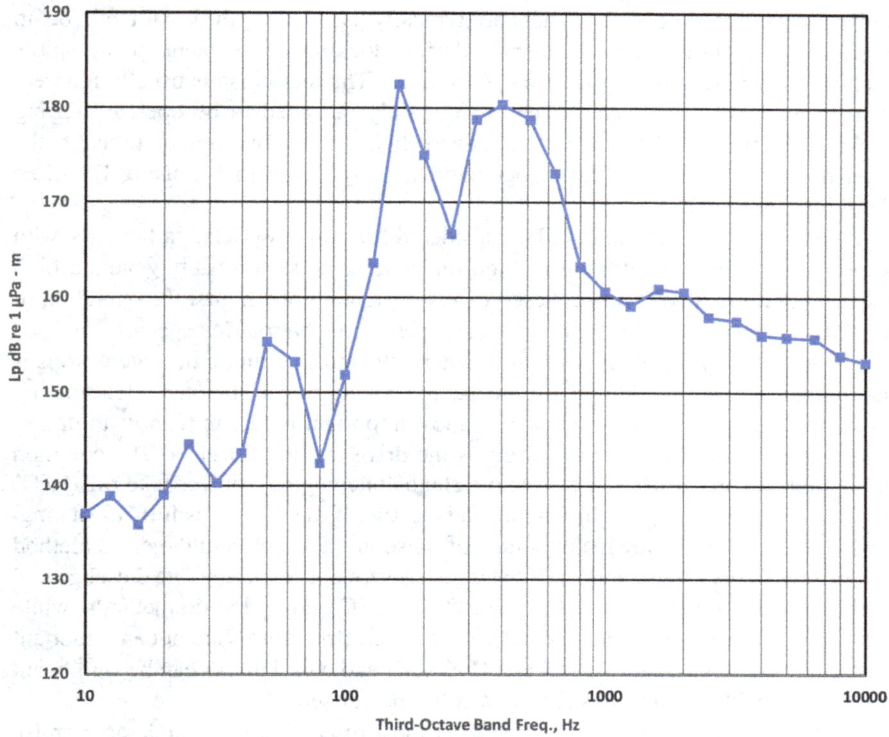

Fig. 9.19 Representative radiated noise spectrum for an Azipod vessel

motor is located within the ship. If the motor is located in the pod itself, the underwater radiated noise may be increased.

Resonant frequencies from podded systems couple and radiate well into the ocean. It has been found that some of these resonances are sensitive to bearing stiffness (Tian et al. 2022). Thus, if resonances are found, a change in the stiffness may provide some relief.

A representative radiated noise spectrum for an azipod-propelled vessel is provided in Fig. 9.19. Pusher units are found to be "some 10 to 20 dB higher than on a tractor unit" (Carlton 2019). This is related to the cleaner inflow into the tractor propeller. Noise and vibration control approaches for podded and azimuthing propulsion systems are provided in Sect. 9.5.1.

9.2.1.2.5 Bow/Stern Thrusters

Bow and stern thrusters can be extremely important noise sources. The majority of thrusters cavitate during operation, even at low thruster speeds. Once a thruster is cavitating, it will typically overwhelm any other noise source with respect to underwater radiated noise. Thrusters are typically used in two different methods: to provide lateral thrust as when in port during docking or to maintain the ship's position as in a dynamic positioning (DP) mode. Thruster noise is usually ignored, even in very quiet research vessels, if it is only designed to be operated during docking. In this case, the crew may be adversely impacted for a limited time. On the other hand, noise due to thruster operation for any vessel that is using DP often should be addressed.

Thruster noise should always be considered for cruise vessels, particularly with respect to docking. In addition to high underwater noise, thrusters generate high habitability noise levels. Since these vessels dock at all odd hours, the operation of the thrusters can significantly impact passengers and local marine species.

There are many different types of thrusters; the most common in large vessels is the tunnel thruster. Additionally, there are swing-down thrusters, telescoping thrusters, and azimuthing thrusters. The most important aspect in regards to underwater radiated noise from each variety is the drive method for each. The common drive methods are diesel geared (from a standalone engine), power-take off (PTO) from the propulsion engine, hydraulic, and electric. If the thruster is fully cavitating, the blade itself will be the major source of noise in the water, and the drive method will be a secondary noise source. If the thruster were not cavitating, the diesel geared method would be the noisiest, followed by the PTO and hydraulic methods, while the electric drive would be the quietest. Thruster electric motors are not an important source relative to cavitating impellers. Hydraulic and diesel drives can be significant sources, particularly for tones at low impeller power settings.

Bow and stern thrusters are standard on most research vessels and some cruise and work boats. Transverse tunnel thrusters are the most common type, although retractable thrusters are also utilized. Other hybrids include pump jets that take suction from below or from the side.

Unfortunately, most of these thrusters, particularly tunnel thrusters, cavitate at relatively low operational speeds. This cavitation is primarily due to poor inflow conditions associated with the thruster's locations in the hull, where the lines are rapidly changing. Another factor is blockage of the tunnel by the drive motor itself. Other dynamics are vortices and turbulence created by the inlet grating or motor support structure, too shallow an immersion depth, or flow interaction between adjacent tunnels or hull openings. The noise radiated by a tunnel thruster depends on the amount of cavitation (Brown 1999) and the directivity associated with the tunnel opening.

Per Morrow 1987, the radiated noise of a thruster is 7 dB noisier than that of a standard propeller. Example levels for thruster radiated noise are provided in Fig. 9.20.

As noted above, the flow uniformity into an impeller disk is a significant factor influencing noise generation. It has been found that noise is different depending on starboard or port trust, which is related to which side contains the impeller drive mechanism and the fact that nearly equal thrust must be developed regardless of which direction the impeller turns. The difference is significant (5–10 dB) in the low-frequency range, especially for frequencies coinciding with blade rate

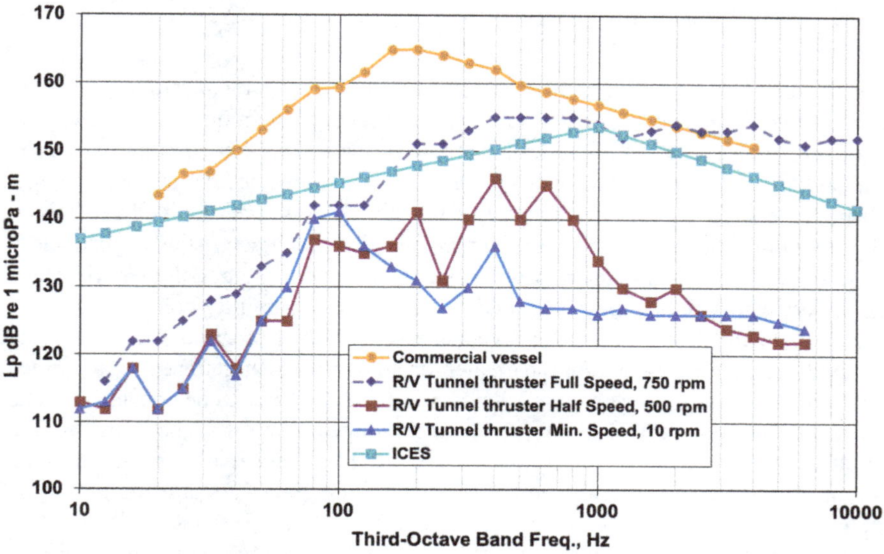

Fig. 9.20 Representative thruster-radiated noise

multiplies. The noise is higher if the struts and the drive mechanism are located upstream of the flow. This highlights the need to have thin and streamlined struts and hubs. It has been found to be helpful in reducing the nonuniformity of the inflow velocity, including the rounding of the tunnel inlet edges. A rule of thumb is that the radius of rounding should be not less than 0.05 *D*, where *D* is the tunnel diameter. If the cavitation inception speed can be significantly delayed, the radiated noise will be reduced.

Few regulatory agencies provide a noise goal to be met during thruster operations. The Indian Register of Shipping (IRS-G-DES-15) provides a thruster radiated noise requirement to be met for 40% load – "Normal Operation." Above 1000 Hz, the radiated noise in a third-octave band is to be below:

$$L_p \text{ dB re } 1\ \mu\text{Pa–m} = 198 - 10.4\text{Log}\,(f_{ctr}) \tag{9.25}$$

Noise and vibration control techniques for thrusters are provided in Sect. 9.3.1.

9.2.1.3 Singing Propeller

In addition to cavitation noise, another potentially high noise phenomenon is a 'singing propeller'. Singing occurs when a blade structural natural frequency corresponds to the vortex shedding frequency (Fischer 2008). This phenomenon is covered by Sect. 5.3.5.2.

9.2.2 Machinery Source Levels

The vessel's broadband signature is generally influenced by machinery-induced noise below the propeller cavitation inception speed. Low-frequency tones can be associated with either the propeller blade rate or the propulsion equipment rotation rate and their multiples. Methods to predict these source levels are presented in Chap. 5. Noise treatments with respect to habitability noise are provided in Chap. 8. The following subchapters discuss the parameters and characteristics of machinery noise relevant to underwater radiated noise. Conversely, methodologies to reduce the underwater radiated noise also tend to reduce the habitability noise.

As with habitability noise, the higher the source level of the machinery is, the higher the radiated noise and/or the greater the amount of treatment needed to reduce the radiated signature. The source level is not the only factor to be considered. The larger the 'source' area is, the greater the radiated noise for the same source vibration, for instance. Thus, a typical pump would radiate less noise than a large propulsion diesel even if both had the same vibration levels. As an example, if the footprint of a pump was 5 times less than the footprint of a diesel and the vibration levels were the same, the pump's contribution to the signature would be 7 dB less than that of the diesel. Conversely, the pump vibration could be 7 dB higher to have a contribution equal to that of diesel. Source level characteristics are provided in Chap. 5.

New or novel hybrid systems aside, propulsion machinery is generally diesel-geared, diesel direct drive, diesel-electric, gas turbine-geared, and gas turbine electric drive systems. Vibration levels of the hull surfaces in contact with seawater determine the absolute level of machinery-induced radiated noise. Thus, the lower the source vibration level and/or the greater the structural energy losses between the machinery feet and the wetted hull plate, the lower the resulting ship's signature. The characteristics of each type of equipment are discussed below and in Fischer et al. 1983, Fischer 2001, and Fischer 2005; Heckl 1988; Nilsson, 1981; and Bies, 1988. Fuel cells, solar and/or wind are examples of hybrid power generation that could significantly reduce machinery-induced radiated noise. Connecting to shore power while dockside will also reduce underwater (and airborne) noise while in-port.

Chapter 7 discussed machinery noise modeling with respect to habitability noise. As noted, detailed and accurate machinery induced noise models generally involve Statistical Energy Analysis (SEA) or Boundary Element Model (BEM) (Salio 2013; Liu et al. 2022). The following sections address the contribution of machinery-induced noise to the underwater signature of the vessel.

9.2.2.1 Diesel Drive

Of the three diesel systems, diesel electric systems are inherently quieter since the gensets can and are usually elastically mounted. For especially quiet research vessels, the gensets can be placed on a double isolation system. Except for military vessels, direct drive and diesel/geared systems are typically hard mounted directly to their foundations. In this case, structureborne noise from the diesel is readily transmitted to the nearby wetted hull plating and radiated into the ocean.

The vibration levels produced by medium- and high-speed diesels are typically higher than those produced by low-speed diesel. As shown in Chap. 5, diesel vibration source levels usually scale as (power/weight)^2 (Fischer et al. 1983); therefore, heavy low-speed diesels have lower source levels due to their lower power-to-weight ratio. Machinery source vibration levels are best obtained from the vendor or measurements on similar types of machinery. If vendor data are not available, there are source prediction techniques provided in Fischer et al. 1983.

9.2.2.1.1 Direct Drive Diesel

Direct drive low-speed diesel is the most common propulsion system found on large merchant vessels. These large diesel engines operate at such low RPMs that they do not need reduction gears when coupled to a shaft. Reduction gears can potentially be a large source of noise, so eliminating them is positive from a noise perspective. These engines are very massive. This can be both positive and negative as far as noise is concerned. Typically, a very massive engine will have lower vibration levels (and thus the structureborne noise is reduced) than a lighter engine of similar horsepower. On the other hand, even if an engine has lower vibration levels, these can potentially result in higher vibratory forces due to the large mass and contact area of the engine when compared to a smaller engine. The biggest downside to these engines is that their large size, mass, and rigid shaft coupling make them nearly impossible to isolate using conventional mounting techniques. The typical frequency spectrum associated with a low-speed diesel will have a large tonal component at the frequency of rotation, firing rate, and their harmonics. The authors have not seen whether Distributed Isolation Material (DIM) has been tried to reduce radiated noise. DIM would at least reduce the frequencies above 200 Hz by approximately 5 dB.

9.2.2.1.2 Medium & High Speed Geared Diesels

Another fairly common propulsion system is a high- or medium-speed diesel engine coupled to a reduction gearbox that in turn is coupled to the propeller shaft. These engines are typically much smaller than low-speed diesels. They have a much higher power-to-weight ratio. As discussed earlier, they generally have high vibration levels, especially for those with aluminum blocks. The use of a reduction gear is essential due to the high rotation speeds of the engines. These reduction gears are another source of noise. To assure a quiet reduction gear, the gear must be made to very tight tolerances and have a significant mass, both of which may increase the cost over a ‘normal’ reduction gear. Medium-speed diesels are generally quieter and have lower vibration levels than their high-speed counterparts, especially in the mid- to high-frequency range.

The higher vibration levels found on high- and medium-speed diesels (when compared to low speed) can be effectively mitigated by placing the engine on resilient mounts. This allows the engine to vibrate but keeps most of the energy from entering the ship's structure and then radiating into the water. Medium-speed diesels are more difficult to mount than high-speed diesels because their rotation rate is close to the resonant frequency of many isolation systems.

If an engine is going to be isolation mounted, all connections made to the engine must also be sufficiently resilient. This includes any piping, cabling, and especially the connection to the shaft. Resiliently mounting an engine increases the system cost and increases the vertical space occupied. The underwater radiated noise spectrum associated with higher-speed geared systems contains peaks at the rotation rate and at half the rotation rate (4-stroke diesels only). There are also smaller tonal components associated with firing rate and any gear-meshing frequencies.

The source vibration level of low- and medium-speed diesels scales as 5.5 log (kW) and directly with "rated rpm." The vibration also scales as 30 log (operating rpm/rated rpm) (Fischer-SNAME Supplement). Thus, the vibration level is quite speed dependent, with a 9 dB change for operating at half speed. High-speed diesels show little variation in vibration between half and full power. The overall vibration of high-speed diesel varies with power and inversely with weight.

Diesel generators are typically driven by medium-speed diesels. As stated above, these can be resiliently mounted as a unit. In many vessels, these units are located on upper decks of the engine room, farther removing them from connecting with the ocean.

9.2.2.1.3 Diesel Electric

Diesel electric is currently the selected propulsion system for some of the world's quietest ships – both with respect to underwater noise and habitability noise. Typically, these vessels are designed to perform research or some type of acoustic surveying. They employ either high-speed or medium-speed diesels that are coupled to generators. The electric power from the generator drives either AC or DC electric motors. There is no propeller shafting to be connected, so it can be put on very soft isolation mounts. In most cases, the diesel generators will be operating at a constant speed, so the mounting system can be designed to be especially effective at this speed, i.e., 1200 (20 Hz) or 1800 rpm (30 Hz).

Diesel-electric propulsion system consists of electric motors directly coupled to controllable or fixed pitch propellers. Some designs maintain propulsion machinery redundancy by having two electric motors on a single shaft. The electric motors themselves are generally very quiet. DC motors are usually quieter than AC motors, while DC motors are larger, more massive, and require more equipment aboard the ship to rectify the AC signal produced by the generators. Some ships employ both types of motors on the same shaft so that the quieter DC motor can be used during survey operations and a more powerful and noisy AC motor can be used during normal operations. A modern electric propulsion motor that has been manufactured to rigid tolerances will usually have such low vibration levels that it can be mounted directly to the ship's structure without the use of resilient mounts. Vibration-isolated stator coils have been developed to reduce motor vibrations.

The frequencies transmitted into the water from diesel electric propulsion systems have peaks at the diesel rotation frequency, half rotation if it is a 4-stroke engine, and firing rate. These will be radiated at a significantly lower noise level than if the engines were not resiliently mounted. There is also radiated noise associated with the electric motors themselves, although these are typically radiated at noise levels that are low in relation to the rest of the noise radiated from the ship.

Electric drive with two stage-raft mounted gensets is the method employed by fisheries research vessels built to meet stringent underwater noise requirements such as those imposed by the International Cooperative for the Exploration of the Sea (ICES) criterion (Fischer 2004).

9.2.2.2 Gas Turbine Drive

Gas turbine propulsion is becoming more common. This type of system typically has much lower vibration levels when compared to reciprocating engines. Thus, their structureborne-to-radiated noise contribution is lower. Gas turbines generally have an enclosure that surrounds the unit, which greatly reduces secondary structureborne noise and airborne noise. Gas turbine-geared propulsion systems require very high reduction ratio gearboxes. These shafting/gearing systems can be complicated. Any gas turbine machinery has extremely high velocity flows on the air intake side as well as the exhaust side. These high flows generate considerable noise, and care must be taken to ensure that they do not increase the radiated noise into the water due to casing vibration and/or hard ducting connections near the hull. The frequencies transmitted into the water from a gas turbine geared propulsion system are usually seen as a peak at the rotation rate of the turbine itself as well as moderate peaks at the final shaft rotation rate and the major gear mesh frequencies.

The benefits of a gas turbine generator driving electric motors are similar to those of a diesel electric propulsion system. The same type of drive motors can be used as with diesel electric.

9.2.2.3 Electric Drive

For vessels that need to operate in environmentally sensitive areas, electric drive offers many advantages – both in terms of feasibility and economics. Several gas turbines or diesel-driven generators can be grouped and isolation mounted on a single 'raft', which is then isolation mounted to the supporting hull foundation. This arrangement enlists the mass of all the engines and generators, as well as that of the raft, to reduce the 'above mount' vibration levels, particularly at less than full power, when several engines may be shut down.[15]

---

[15] It may be necessary to force lube oil through the non-operating machinery to prevent bearing wear.

Even large, direct drive electric motors are quieter when compared to reduction gears and piston engines, and as a general rule, they need not be isolation mounted. At part load, fuel efficiency can be maintained by shutting down some of the engine generator units while running the remaining units at full power. While electric drive components may cost more initially than mechanical or direct drive propulsion, the overall fuel economy may compensate.

Gas turbine-electric drive is becoming common in cruise ships, where podded propulsors are finding acceptance. However, bear in mind that the Azipod has the motor in direct contact with the ocean, and thus, the radiated signature could be higher than a vessel with the motor inside the hull.

9.2.2.4 Geared Drive Systems

Gears produce tones at the gear mesh frequency and its harmonics. This frequency is determined by the rotation rate times the number of teeth on the gear. Double reduction gearboxes have two separate mesh frequencies, one for the lower reduction system and one for the higher reduction system. Typical marine gearboxes have mesh frequencies in the range from 200 to 500 Hz. The quality of manufacturing of the gearbox has a large influence on the vibration at the mesh frequency. Gears with a lower tooth spacing error (higher quality) can have a vibration level 10 to 15 dB lower than that of a lower quality gearbox (Fischer 2001). The gearbox is usually hard mounted, especially when the gear contains the thrust bearing. This vibration is easily transmitted to the hull, where it radiates effectively. Close coupled diesel/gearboxes can be isolation mounted, which reduces the contribution of both the diesel and the gearbox. In this case, an external thrust bearing is needed or the isolation system must be capable of taking the propeller's thrust.

9.2.2.5 Auxiliary Machinery

As stated previously, propulsion machinery, generators, and propulsor dominate the radiated signature of most vessels. This is especially true when the machinery/equipment has particularly high noise or vibration levels. In this regard, cavitating pumps and reciprocating compressors are items that can easily contribute to radiated noise. Auxiliary equipment located below the water line and close to the hull should be considered when addressing their induced noise. Another example of auxiliary equipment that should be considered is any hydraulically driven unit. These typically produce strong tones at multiples of the pulsation rate of the pump.

Empirical methods to assess the acoustic source levels and controlling factors of any marine machinery are provided in SNAME T&R 3-37 (Fischer et al. 1983; Fischer 2001) and Heckl (1988). These references provide information that allows one to compare the acoustic source levels of typical marine equipment – both the overall level and frequency dependent levels. Scaling for changes in source level with power, rotation rate, and type of equipment are provided. General 'design' guidance, with the intent of minimizing building noise problems into the vessel, is also provided in the reference and in Chap. 8 of this text.

9.2.2.6 Underwater Exhaust

There are two components to the contribution from the underwater exhaust of a diesel. One is the direct noise from the exhaust combustion, and one is from bubbles produced by the gas/fluid mixture (Bie et al. 2015). The former has empirical prediction methods, and the latter, to the authors' knowledge, has limited modeling verification (Zhang et al. 1993). For gas bubble prediction, limited published data show that this component is potentially a serious problem over a limited frequency range. Transmission along the piping (Munjal 1987) and the effectiveness of dry or wet mufflers need consideration. The impedance difference between air in the exhaust pipe and the sea water will cause a high-end reflection, particularly at low frequencies. Note that sea-connected systems have an opening below a free surface that will radiate as a dipole with the strongest radiation downward. The radiation pattern will depend on the depth of the opening below the surface and the diameter of the opening.

The bubbles associated with an underwater exhaust can cause high radiated noise levels (Purshouse 1986; Norwood and Chen 2002; Norwood and Chen 2004). These references do not provide empirical models, yet they provide some ballpark levels. These levels for the expected flow rates are well below the ICES criterion. According to (Norwood and Chen 2004), "at higher flows, bubble-acoustic signals become irregular because multiple bubbles are involved, and no simple theory can predict the acoustic fields."

The low-frequency (plane waves in duct) radiated signature, $L_{p\ pipe}$ dB re 1 μPa–m, from an underwater exhaust, where $L_{p\ exh}$ dB re 20 μPa is the pressure level in the pipe, can be estimated as (Zhang et al. 1993):

$$L_{p\ pipe}\,\text{dB re 1 μPa–m} = 8\ \text{Log}\ (Q) + L_{p\ exh} + 16 \tag{9.26}$$

where Q is the mean volume flux, L/min, inside the pipe as it exits to the water.

9.2.3 *Flow Induced Noise*

There are two components to flow-induced noise. The first is re-radiation from the hull following excitation by turbulent flow. The other is direct radiation from the pulsation pressures associated with the turbulent flow. Flow-induced noise is generally considered significant above 16 kn; therefore, it is more important for naval vessels and high-speed craft. In one published study on a frigate (Lv et al. 2016), the overall level due to flow pulsations was 130 dB re 1 μPa and that due to re-radiation was 148 dB re 1 μPa (distance not provided). The flow-excited spectra are highest at low frequencies, sloping down at 3 to 6 dB per octave.

9.3 Primary Underwater Radiated Noise Paths

Airborne and structureborne acoustic transmission paths internal to the vessel are discussed in Chap. 6. Those for underwater radiated noise are illustrated in Fig. 9.21. Machinery-induced noise, whether transmitted over the internal airborne or structureborne path, essentially excites the wetted hull plate. As discussed in Chap. 6, the radiation efficiency of the hull plate determines the acoustic energy radiated into the water. This underwater radiated noise depends on the hull material properties -Young's Modulus and density, radiating area, and frame spacing and frame cross section geometry. The machinery-induced vibration and the above parameters are combined to determine the machinery contribution to the underwater signature. The simplest assumption is that this energy spreads out semi-spherically in water with limited or no directivity, especially abeam. However, as will be shown in the following sections, this assumption is often incorrect.

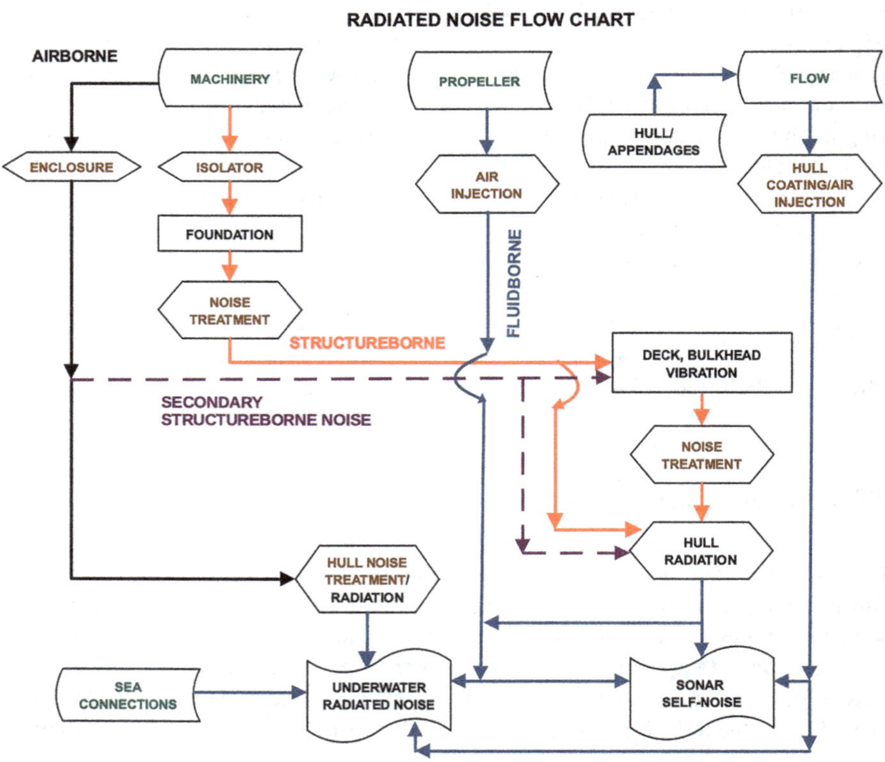

Fig. 9.21 Radiated noise sources/paths

The underwater radiated noise from fluid connections to the water is treated differently. The radiated noise depends on the in-line pressure pulsations and parameters of the opening. It also depends on how much of the in-line acoustic energy is reflected back into the pipe at the opening due to the impedance mismatch. The radiated noise also depends on the depth of the opening below the surface (Purshouse 1986), which is a pressure release boundary.

The propeller contribution is mainly from collapsing cavitation bubbles that directly produce underwater noise (see Sect. 9.2.1). This source can be considered directional, especially for bow and beam aspects. For bow aspect, the high area of cavitation (near the top-dead center) may be shielded in part by the hull. For aft and abeam aspects, the entire cavitating area is assumed to be seen directly. Vibration transmitted in either direction by shafting is not usually a concern for commercial vessels. Secondary re-radiation from the hull plate due to propeller excitation is not often a concern.

The main acoustic path for underwater radiated noise is obviously the fluid connected path between the vessel's sources discussed above and the 'receiver'. The simplest assumption is hemi-spherical spreading of any of the vessel's sources to this receiver, whether the receiver is a hydrophone, a fish or a mammal. Spherical spreading means that a very noisy ship, with a signature on the order of 190 dB re 1 μPa–m, operating near an environmentally quiet area with an ambient noise level of 90 dB re 1 re 1 μPa, would need to remain over 100 km (54 n.mi.) away to have no impact at all. However, if operating in shallow water, where there is typically cylindrical spreading, the vessel would almost always have an impact. In the ocean, these simple assumptions of spherical or cylindrical spreading only apply to distances of less than 1 km from the vessel. Past that distance, the transmission path in the water is affected by the geometry of the bottom, the water temperature vertical profile, salinity, and many other factors (see Sect. 9.7) (Potter and Warn-Varnas 1990).

9.3.1 Airborne to Radiated Underwater Noise Transmission

Using the worst-case assumption that normally incident airborne noise is transmitted into the water without the hull impedance accounted for, the following equations can be utilized. First, the power transmission coefficient, $\prod_{trans}$, is

$$\Pi_{trans} = 1 - \left[(z_{wtr} - z_{air}) \Big/ (z_{wtr} + z_{air}) \right]^2 \tag{9.27}$$

where z_{wtr} is the impedance of water $= 1.5 \times 10^6$ rayls and z_{air} is the impedance of air $= 414$ rayls.

Then, the relation between the transmitted underwater noise and incident airborne noise is:

$$L_p \text{ dB re } 1\mu\text{Pa–m} - L_p \text{ dB re } 20\mu\text{Pa} = 62 + 10 \text{ Log } (\Pi_{trans}) \tag{9.28}$$

A more theoretical approach is provided in Kim et al. 2011. There are two components – the radiation from the plate as excited by the airborne noise, Δ_{rad}, and the transmission loss through the hull plate, Δ_{trans}. These are computed as follows:

$$\frac{\langle p_2^2 \rangle}{\langle p_1^2 \rangle} = \left(\frac{r^2}{1 + {}^{\eta_2}/_{\eta_{21}}} \right) \left(\frac{n_2}{n_1} \right) \frac{VA\rho_2^2 c_2^2 \sigma_2}{2\pi M \rho_1 c_1^2} \tag{9.29}$$

where σ_1 is radiation efficiency of dry plate, σ_2 is radiation efficiency of wet plate, V = volume of room, A area of interface, M mass of plate, t is the thickness, ρ_1 is the density of the air, ρ_2 is the density in water, c_1 and c_2 are the sound speed in air and water, respectively. Equation 9.29 can be rewritten as:

$$\begin{aligned}\Delta_{rad} = 73 + 10 \text{ Log} \left(\frac{1}{1 + {}^{\eta_2}/_{\eta_{21}}} \right) + 10\text{Log} \left(\frac{n_2}{n_1} \right) + 10\text{Log } (VA) + 10 \text{ Log } (\sigma_2) \\ - 10 \text{ Log } (M) + 20 \text{ Log } (\rho_2 c_2) - 10 \text{ Log} \left(2 \pi \rho_1 c_1^2 \right)\end{aligned} \tag{9.30}$$

where n_1, n_2, η_2, and η_{21} are computed as:

$$n_1 = \omega^2 V / 2\pi^2 c_1^3$$

$$n_2 = \sqrt{3}\, A / 2\pi c_L t$$

$$\eta_2 = \eta_{int} + \left({}^{\rho_2 c_2 \sigma_2 A}/_{\omega M} \right)$$

$$\eta_{21} = {}^{(\rho_1 c_1 \sigma_1 A)}/_{(\omega M)}$$

Another approach is to compute Δ_{trans} as:

$$\tau = \frac{4}{\left(1 + \left({}^{m 2\pi f}/_{\rho_2 c_2} \right)^2 \right)} \tag{9.31}$$

$$L_{water} - L_{air} = \Delta_{trans} = 10\text{Log}(\tau) - 30 \tag{9.32}$$

where m is the surface mass of the plate and c_L is the longitudinal wave speed in the plate.

This path is more critical for isolation mounted propulsion equipment without enclosures.

9.3.2 Hull Radiation

The machinery-induced radiated noise is composed of the vibration level reaching the wetted hull and the 'radiation efficiency' of the hull. This is another frequency-dependent parameter, depending on the radiating area, thickness of the hull plate, material properties of the hull plate (Young's modulus and density), frame spacing, and frame section modulus. The hull plate is a good radiator of underwater noise at high frequencies (above a parameter called the critical frequency) and does not radiate well at low frequencies. The critical frequency for a steel-water loaded plate of 9 mm thickness is 25,000 Hz. Thus, the hull radiates very well in the region of sonars and much less so in the low frequencies. The other significant factor controlling the amount of noise is the volume velocity due to motion of the entire hull at very low frequencies, typically below 50 Hz. This factor has an opposite frequency dependence from the radiation efficiency; it is higher at low frequencies and drops off quickly with increasing frequency.

As discussed in Sect. 2.7, hull radiation efficiency describes the coupling between the vibration level and radiated power. Above the critical frequency of the plate, the radiation efficiency is 1. The radiation efficiency below coincidence or critical frequency of the ribbed hull is predicted according to Fraser and Swindell (1988).

$$\sigma = \left(8\lambda_C\lambda/\pi^4 S\right)g_1(\alpha) + \left(p\lambda_c/4\pi^2 S\right)g_2(\alpha) \tag{9.33}$$

where λ_c is the coincidence wavelength of the panel, m; λ is the acoustic wavelength in the water, m; S is the panel area, m^2; h and l are the panel dimensions, m; p is the panel perimeter, m; $\alpha = (\mathrm{f}/\mathrm{f_c})^{1/2}$; $\mathrm{f_c}$ is the coincidence or critical frequency of the water-loaded panel; and k_p is the plate wavenumber, m^{-1}. The remaining terms are defined as:

$$g_1(\alpha) = \alpha^2[\cot 2\psi_1 + \cot 2\psi_2]\left[1 + \frac{\sin kh}{kh} + \frac{\sin kl}{kl} + \frac{\sin k\left(h^2 + l^2\right)^{1/2}}{k\left(h^2 + l^2\right)^{1/2}}\right] \; for\; k_p > k \; and\; f < \frac{f_c}{2} \tag{9.34}$$

$$g_2(\alpha) = \frac{1}{\left(1 - \alpha^2\right)^{1.5}}\left[\left(1 - \alpha^2\right)\mathrm{Log}\left[\frac{1 + \alpha}{1 - \alpha}\right] + 2\alpha\right] \; for\; k_{px} > k \; and\; k_{py} < k \tag{9.35}$$

$$If\; k < \frac{\pi}{l}:\ \sin\psi_1 = \frac{\pi}{lk_p} : k > \frac{\pi}{l}:\ \sin\psi_1 = \frac{k}{k_p}$$

$$If\; k < \frac{\pi}{h}:\ \sin\psi_2 = \frac{\pi}{hk_p} : k > \frac{\pi}{h}:\ \sin\psi_2 = \frac{k}{k_p}$$

The assumption for most of these models includes a distributed source acting on the hull. When the hull is directly impacted by a point force, the radiation efficiency and radiated noise are computed differently (Morrow 1987). For vessels with a low radiated noise limit, the best approach is not to have point attachment to the wetted hull plate. This will avoid a potential noise problem and reduce the analytical modeling needed.

The following estimates of radiation efficiency are based on experiments (Uchida et al. 1986):

$$10\ \mathrm{Log}\ \sigma = 10\ \mathrm{Log}\left(\rho_s\sqrt{B}/S\right) - 78\ f \le f_1 \tag{9.36}$$

where $f_1 = 0.25\, f_0$ and $f_0 = 700\ \left(\rho_s\sqrt{B}/S\right)^{0.2}$

$$10\ \mathrm{Log}\ \sigma = 16.7\ \mathrm{Log}\ \left(4f/f_0\right) + 10\ \mathrm{Log}\left(\rho_s\sqrt{B}/S\right) - 78\ f_1 < f \le f_2$$

where $f_2 = 2\, f_0$

$$10\ \mathrm{Log}\ \sigma = 50\ \mathrm{Log}\left(f/16{,}000\right) - 10\ f_2 < f \le f_3$$

where $f_3 = 16{,}000$ Hz

$$10\ \mathrm{Log}\ \sigma = -10\ \mathrm{f3} < \mathrm{f}$$

where B is the plate bending stiffness:

$$B = \left(Et^3/\left(12\left(1 - \upsilon^2\right)\right)\right)$$

Additional information on radiation efficiency is provided by Pan et al. 2017.

9.4 Control Approaches

Renilson Marine Consulting in a report to the International Fund for Animal Welfare (IFAW) has "identified that a reduction in hydro-acoustic noise of 3 dB for vessels which exceed mean noise levels of 175 dB by one standard deviation (16% of vessels) would result in a reduction of 40% in the area ensonified to 120 dB (assuming a standard deviation of 5.3 dB for assemblages of vessels as found by Scrimger and Heitmeyer 1991). It also identified that a 6 dB reduction would reduce the corresponding area by 60%. Therefore, great gains can be made by reducing the noise output from the noisiest vessels." This report also addresses some of the cost implications and provides examples of practical changes for a containership, large tanker, and fast passenger ferry. The following sections address methodologies and steps that can be taken to achieve these goals.

The basic approach to designing a very-quiet vessel will usually include one or more of the following features: quiet propellers with a high cavitation inception speed; diesel-electric propulsion; vibration isolated diesel generators and auxiliaries; and insulated machinery rooms. Designing quieter commercial vessels does not require all these extensive approaches. Reductions of 3 to 6 dB can be achieved with the purchase of 'quiet machinery', the use of single-stage resilient mounts where possible, and attention given to the propeller design to increase its cavitation inception speed. However, if the vessel is already built, there are other approaches that can be considered. In this case, the only way to affect the signature is to slow down to operate below the cavitation inception speed and shut off any unnecessary equipment. Safely securing compressors and hydraulic systems may reduce both tonal and broadband noise. Flow improvements can be added to reduce the velocity profile into the propeller.

When a vessel is being modified or its life extended, the treatments discussed below should be considered in the redesign process. This could include a more efficient, quieter propeller and change of propulsion equipment.

Some of the 'non-engineering' methods to reduce anthropogenic noise are to consider trade-offs between slowing down or speeding up[16] (McKenna et al. 2013), reducing the density of vessels (Dunn et al. 2021) (particularly in environmentally sensitive areas), and timed entry into ports.

The remaining sections only address in detail those treatments specific to underwater radiated noise. Chapter 8 details such treatments as isolation mounts, damping, absorption, etc. Noise control options applicable to airborne and structureborne noise components may also reduce underwater radiated noise.

9.5 Noise Control Approaches

This section covers controls applicable directly to reduction underwater radiated acoustic levels. Habitability noise and vibration treatments are discussed in Chap. 8.

9.5.1 Propulsor Noise Control[17]

Ideally, the propeller should turn as slowly as possible. To generate the same thrust as a faster-spinning propeller, slow propellers generally have to be much larger in area. Fortunately, larger propellers reduce thrust loading and improve efficiency. A

[16] Particularly for fixed pitch propelled vessels as lower speeds for changeable pitch propellers (CPP) may be 'off-design' and cavitate.

[17] Also see Sect. 8.12, propeller noise control with respect to habitability noise.

balance must be made because the larger the propeller diameter is, the faster the tips of the blades move, which can, again, cause cavitation. In addition to these parameters, the number of blades and the blade design generally affect performance and acoustics. For good efficiency as well as acoustics, it is important to have a smooth inflow into the propeller. There are many different propeller designs that aim to accomplish this.

As mentioned earlier, if a propeller cavitates, it will be the dominant underwater noise source. All propellers eventually cavitate, so a quiet propeller basically delays this process. A well-designed propeller will have a higher cavitation inception speed than a standard propeller. Different propeller designs are inherently quieter than others, e.g., skewed propellers.

9.5.1.1 Effect of Changing Propeller Speed

Significant noise reduction can be achieved by slowing the propeller and/or ship speed (see Figs. 9.22 and 9.23). In Fig. 9.22, A_c is the area of cavitation on the propeller disc associated with the assigned speed. A_d is the disc area. As shown, as this ratio increases from 0.09 to 0.66, the overall radiated noise increases by 35 dB as the speed shifts from 6 to 13 kn. More importantly, in the low-frequency range, there is an additional gain from the peak frequency shift. At 100 Hz, the difference between these two A_c/A_d extremes is 47 dB. Therefore, slowing the vessel down

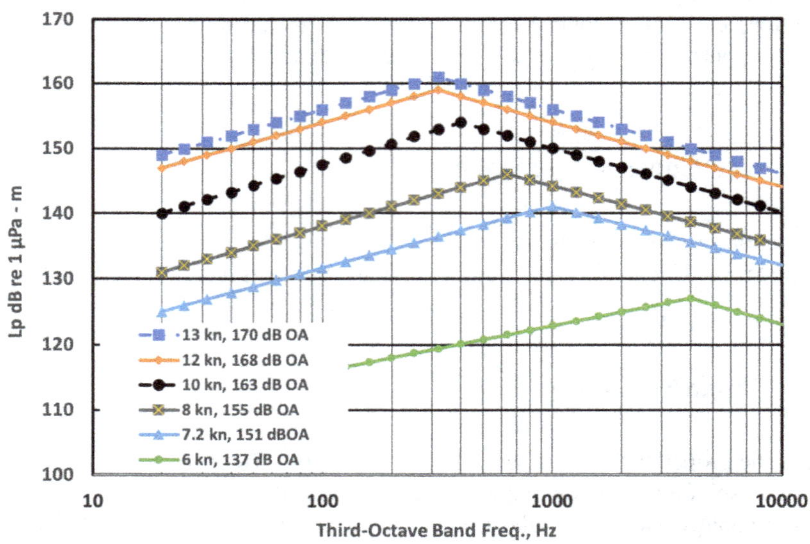

Fig. 9.22 Predicted scaling of propeller radiated noise with speed

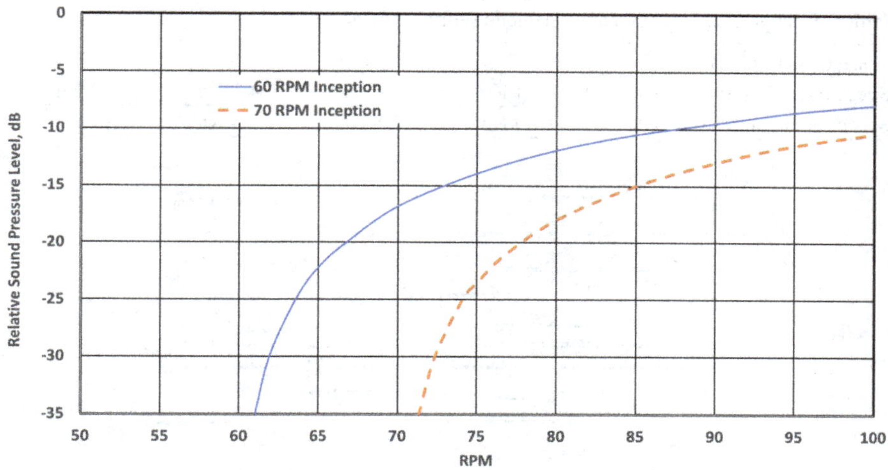

Fig. 9.23 Noise reduction with changing cavitation inception speed

Table 9.7 Low-frequency signature reduction for change in speed

Noise Reduction, dB	3	6	10	12
Speed reduction from 14 Kn	0.4 Kn	0.8 Kn	1.1 Kn	1.2 Kn
Speed reduction from 25 Kn	2.7 Kn	4.8 Kn	7.0 Kn	7.8 Kn

can decrease the cavitating area, reduce the peak noise level, and decrease the low-frequency noise. This would be a significant change in the environmental impact.

As shown in Fig. 9.23 for cavitation versus speed, a propeller with a 70 rpm inception speed operating at 80 rpm will be at least 6 dB quieter than that same propeller with a 60 rpm inception speed operating at 80 rpm.

Reducing speed yields an immediate, large reduction in propeller cavitation noise. A small disadvantage in sensitive areas is that subject animals will be exposed for a longer time. A benefit is that fuel consumption will be reduced as speed is reduced. However, crew costs will be increased with longer transit times. Most important is that accomplishing fewer trips per year when speed reduction is mandated in certain areas will reduce shipper revenue. This could be a major impact that must be quantified. High-speed ferries will be most adversely affected by mandated speed reductions. They don't work well at any but cruising speed. A 10–15 dB reduction in the propeller radiated noise can be achieved by dropping the speed from 12 knots to 8 knots. Tables 9.7 and 9.8 (Brown 2007) show the low-frequency and high-frequency noise reductions that can be achieved in the cavitating noise signature by slowing down a vessel.

Real-time routing of ships to limit noise impacts to acceptable levels while also minimizing operational impacts is achievable. High-speed routes must avoid current

Table 9.8 High-frequency signature reduction for change in speed

Noise reduction, dB	3	6	8	10
Speed reduction from 14 Kn	2.4 Kn	3.4 Kn	4.3 Kn	5.2 Kn
Speed reduction from 25 Kn	4.0 Kn	5.8 Kn	7.4 Kn	8.9 Kn

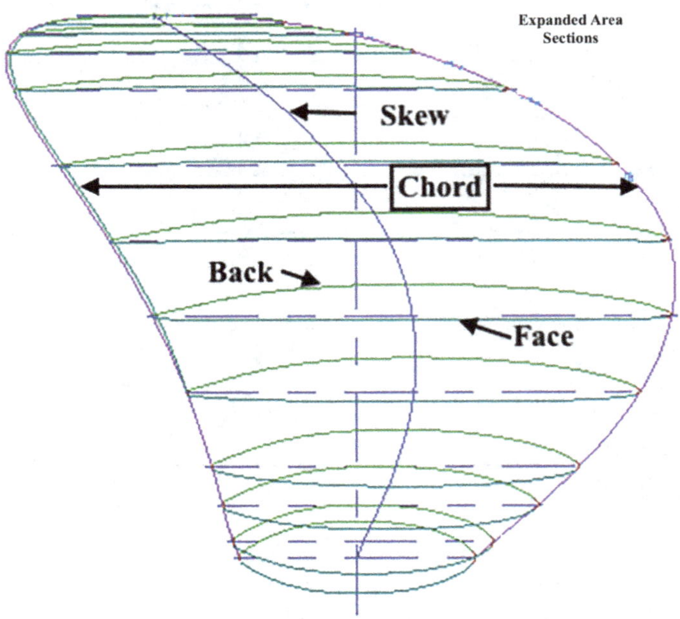

Fig. 9.24 Illustration of propeller skew

marine mammal locations, and reduced speed "legs" where mammals are present must be kept to minimum length. This may require tracking marine mammal concentrations, monitoring ambient noise levels, and concurrent modeling to determine radiated noise levels for acceptable exposure. Routing is now commonly performed for the minimization of weather and sea-state impacts on commerce for a fee. This is also an achievable means for noise mitigation. Only the route tracking function needs new development (Brown 2007).

MacGillivray et al. 2019, showed that slowing vessel speed is "an effective method for reducing underwater radiated noise from commercial vessels." Reductions of "11.5 dB for container ships,10.5 dB for cruise ships..., and 5.9 dB for bulkers" were measured.

9.5.1.2 Blade Form and Skew

Modern propeller design methods allow the steps necessary to maximize the speed at which cavitation occurs (Greeley and Kerwin 1982; Kinnas et al. 2002). By properly shaping the blades, particularly their outlines in terms of their 'sweep' (as in a swept

Fig. 9.25 Forward Skewed Propeller for Survey Vessel, Clockwise rotation

wing) or in marine terms, their 'skew,' and by leading-edge contour details, the cavitation sensitivity to water inflow variations may be reduced. Figure 9.24 shows the geometry of a skewed propeller. To maximize the efficiency, propellers must operate in the wake of the hull, which is unfortunately highly nonuniform. It is the sensitivity to this nonuniformity that compromises the cavitation performance of propellers. Thus, wake-adapted propellers for a particular hull form are more desirable.

For cavitation noise and vibration avoidance or mitigation, an advanced design impeller may be advantageous. Such impellers, with an unconventional forward skew of the leading edges toward the blade tips, whose skew slope is related to local leading-edge radii, have been shown to greatly increase inception speed and show a tolerance to inflow nonuniformity. Based on simple low-aspect ratio approximations, forward skew can be shown to reduce the lift/incidence slope at the outer radii, whereas common back skew increases it. Forward-skewed blades have been used with good results on axial propeller-type and mixed-flow pump impellers, advanced water jet impellers and several ducted thrusters (Brown 2007). Forward skew has been used on a ducted CRP main propulsion screw with a notable reduction in radiated noise, all without loss of capacity or efficiency (see Fig. 9.25). Ducted CRP propellers installed in a Kort-nozzle have an additional benefit from forward skew. The cavitation is reduced as well as an increase in efficiency due to the lower sensitivity-to-incidence at the tip (Brown 2007).

The noise reduction, NR, dB, achieved by adding (aft) skew to a propeller design is shown in Table 9.9 (adapted from Sasajima et al. 1986). The speed for the full-scale measurement was 4 kn. At this speed, there should not be a significant amount of cavitation. In spite of this, the reduction at low frequencies was significant.

Research is being carried out on the effect of varying the tip rake angle (Ghassemi et al. 2018). Higher positive and negative rakes were found to significantly change

Table 9.9 Skew noise reduction (NR), dB

Freq. (Hz)	100	125	160	200	250	320	400	500	630
NR, dB	**12**	**19**	**19**	**18**	**17**	**13**	**9**	**8**	**6**
Freq. (Hz)	800	1000	1260	1600	2000	2500	3150	4000	5000
NR, dB	**7**	**4**	**5**	**4**	**5**	**5**	**4**	**7**	**6**
Freq. (Hz)	6300	8000	10,000	12,600	16,000				
NR, dB	**4**	**6**	**8**	**8**	**11**				

the radiated noise with a slightly lower impact on efficiency – "When the rake angle is −6 the SPL reduces 40%[18] and efficiency 6%".

Advanced design propellers may be applied as original equipment on newly built ships at little cost increase. They may also be substituted for otherwise acceptable propellers on existing ships during overhaul – at a more substantial cost. Because of their robustness against flow disturbances, advanced design propellers are capable of reducing noise at a given ship speed beyond inception. The reductions will be found to be small, although worthwhile at full power/speed. A modest reduction in speed can yield a major reduction in noise.

There need be no loss of efficiency. As propellers are generally priced by their weight, there is no major cost either. Additional benefits of reduced cavitation are reduced hull vibration, lower structural fatigue, and a reduced amount of propeller erosion.

9.5.1.3 Overall Performance Based on Propulsion System Type

Table 9.10 lists numerous types of propulsion systems with differing levels of overall performance with respect to efficiency, maneuvering and acoustic performance (adapted from Jebsen 2004).

Items that affect fixed pitch propellers include rings around the tips and 'wings' on the tips. (Gao et al. 2019). Another way of unloading the tips is with Contracted and Loaded Tip (CLT) (de Jong 1991). Note that consistent data on the acoustic effectiveness of many of these treatments are not yet available.

Usually, a "quiet" mode for operating a controllable pitch propeller can be established during builder trials. Figure 9.26 shows how the pitch/speed combinations affect the radiated signature of a Fisheries Research Vessel (FRV) (adapted from Traverso et al. 2015). The signature peaks in the mid-frequency range for the 57%/6 kn condition. For 100%/10 kn, cavitation noise starts to dominate above 2000 Hz but is lower than the 6 kn signature in the mid-frequencies. Above 125 Hz, the 77%/8 kn condition is quieter by up to 20 dB than either of the other operating conditions; below 125 Hz, the performance varies.

The following is general design guidance for quiet propellers or thrusters:

[18]Percent reduction is not a good metric, 40% reduction in pressure is −4 dB; only −2 dB if pressure squared.

Table 9.10 Examples of performance characteristics of various propulsors

Type	Efficiency	Acoustics	Maneuvering	Comments
Shaft & Strut Props	Good	Good	Fair	Shaft inclination could cause inflow variation; good cavitation inception speed
Asymmetric pre-swirl stators	Improved over shaft/strut	Good		Improved cavitation performance over shaft/strut; potentially degraded turbulence ingestion noise
Contra-rotating	Very good	Very good cavitation performance		Advanced shafting development required
Azimuthing pusher pod	Good (depends on pod size)	Poor – Low cavitation inception speed	Excellent	Favorable hull interaction; strut wake impinges on rotor
Skeg	Good	Poor	Fair	Skeg creates severe wake deficiency, high blade rate noise, low cavitation inception speed
Azimuthing tractor pods	Fair to good	Very good	Excellent	Efficiency depends on pod size; no turbulence ingestion, best cavitation inception speed (machinery in pod can radiate tonal noise)
ECO-Cap[a]	2.9% energy saving	−6 dB		
Ultimate rudder Bulb[a]	3.9% energy saving	−7 dB		
Neighbor Duct & Ultimate Rudder Bulb[a]	8.7% energy saving			

[a]Tachikawa et al. 2019

- impeller tip speed should be less than 18 m/s
- increased impeller pitch may produce the same thrust but at a lower rotation rate, which is equivalent to a lower tip speed
- avoid a straight radial edge, a skewed impeller shape has significantly better low-frequency performance than one with a radial trailing edge (see Fig. 9.24)
- for ducted impellers, consider trade-off between increased tip clearance, which reduces unsteady load at tips that operate in the fin wake, and effect on tip vortex cavitation and thrust performance
- adjust blade pitch distribution to unload tip
- number of blades should be as large as practically possible, considering all the other hydrodynamic properties – speed, thrust, direction of rotation, pitch distribution, position of propeller relative to the rudder, and hull wake

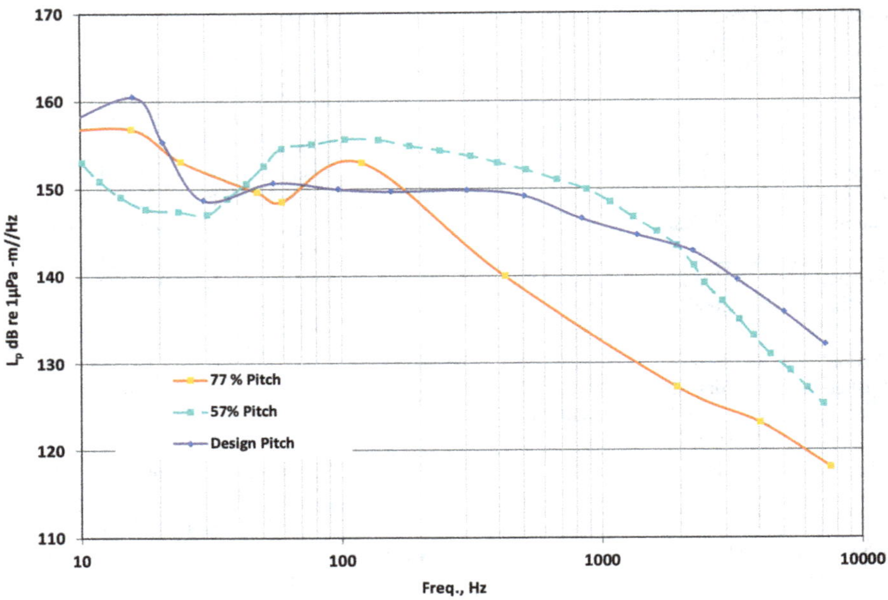

Fig. 9.26 Effect of pitch/speed operating conditions on a vessel's radiated signature

- consider an odd number of blades, which are susceptible to lower amplitude wake harmonics, therefore generating lower forces
- avoid tip vortices by using a low tip loading and wide blade
- thruster impeller should have symmetrical blade sections and well-rounded edges for equal performance in either direction

For a propeller to have good acoustic performance, it should have sufficient clearance between the blade tips and the hull of the ship, operate below cavitation inception speed and have smooth inflow of water into the propeller. It should also be free of any acoustic anomalies, such as when flow over the propeller induces a resonant response in the propeller blade (this is known as a singing propeller and is not uncommon).

Low-frequency cavitation noise from water-jet propulsors might be attenuated by out-of-round cross-sections in an air-backed inlet duct, resulting in reduced wall impedance. This mixed flow may affect the efficiency of the system. There is a splash-noise component where the jet stream impacts the sea surface behind the vessel. The noise levels generated depend on droplet size and relative velocity as well as the air content in the vessel wake.

There are several developing technologies that may be beneficial, and most of them are propulsion systems that provide fuel savings and noise reduction. Chief among these are the "RIM" systems, where the permanent magnet motor is an integral part of the nozzle. This 'shaftless' drive system should benefit from a uniform inflow, thereby reducing the hydro-acoustic propeller-induced noise and improving the overall system efficiency. This system also has the benefit of no tip clearance and the opportunity to optimize the blade geometry. Furthermore, this system takes less space on board the vessel and provides some flexibility in the ship's arrangements. Several of these systems are being built for thrusters and azimuth drives (Yakovlev et al. 2011; Lee 2005; and Yan et al. 2017).

Compared to metallic propellers, propellers made of composite materials may offer acoustic and efficiency advantages. Carlton (2019) noted that composite propellers can likely reduce radiated noise by 5 dB. Given the inherent properties of composites, these types of propellers may be significantly lighter, and it may be possible to elastically tailor the blade to achieve improved performance (Motley and Young 2011; Liu and Young 2009). This approach has already been undertaken in aerospace and for wind turbines. It has been shown that passive shape adaptation of blades for wind turbines can improve their fatigue life and efficiency (Lee and Flay 2000; Lobitz and Veers 2003).

As already noted, the cleaning and maintenance of propeller and propeller surfaces is critical. Propellers with dings, missing sections, barnacles, etc., can be significantly louder than a clean propeller in its original form. These flow disruptions can degrade vortex shedding, turbulence, and cavitation phenomena.

Recently, it has been demonstrated on one vessel (1) that a clean hull and (2) a hull and propeller coating paint will effectively reduce the radiated signature (Gaier et al. 2022). This difference between 'clean hull' and 'coated hull' is shown in Fig. 9.27. The coating reduces the signature from 4 to 17 dB depending on the frequency. This coating is not like a decoupling coating,[19] which is a relatively thick, soft material providing an impedance mismatch. The paint in this case is a hard polymeric layer. This study recommends not placing paint on the trailing edge of a propeller; in other areas, it potentially decreases cavitation. The peak performance between 60 and 80 Hz was not explained.

Renilson 2009 also addresses hull coatings and decoupling materials. This report also suggests that hull coatings, at least anti-fouling coatings, can provide measurable reductions for some loading conditions and frequencies. It was noticed (and expected) that when cavitation was significant, the reduction was smaller. These reports show that treatments for underwater noise are an ever-changing landscape.

Rudder/propeller systems that improve flow and reduce turbulence should also have a favorable effect on underwater noise. Such systems usually consist of an integrated propeller and the rudder that are considered as one propulsion unit, designed for optimum propulsive efficiency.

[19] In the past, hull coatings tended to have adhesion problem, are subject to tearing when hitting debris, and may adversely impact hull resistance. Some current coatings claim to decrease hull resistance.

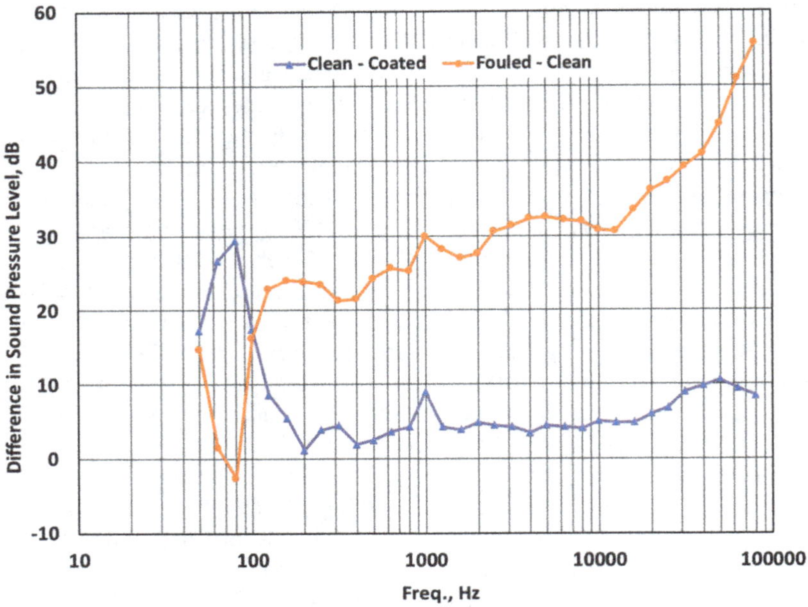

Fig. 9.27 Effect of clean versus fouled hull and coating versus clean hull. (Adopted from Gaier et al. 2022)

The manufacturing of quiet propellers is highly dependent on the surface finish (ISO 484-1:2015) and defects on the surface, particularly on the leading edge (Gospodnetic 2022). "The 0.5-mm defect tested is one of the tightest ISO 484-1 propeller manufacturing tolerances, yet it has been demonstrated that it affects cavitation inception significantly and detrimentally. The rules need tightening up," said Gospodnetic in Professional Mariner, Journal of the Maritime Industry, Feb 21, 2023.

For counter-rotating (CR) propellers, "the power is distributed over two propellers, which results in a better performance." Within the CR unit, both propellers rotate in opposite directions so that the second propeller compensates for the whirling of the water caused by the first propeller. As a consequence, rotation loss is minimized with maximum efficiency gained from the available power and propeller diameters." Another study on both cavitation and fuel efficiency is presented in Hallander 2013 for a chemical tanker with controllable pitch propellers. The results show "that lowering the RPM instead of reducing the pitch requires less power, produces less cavitation and generates lower URN[20] when operating at reduced speed…".

[20]URN – underwater radiated noise

In general, energy saving devices affecting propulsion may result in a lower radiated noise signature at the same time as providing cost and operational benefits. These devices generally improve flow into the propulsor and/or improve propulsor design. Improvements in efficiency usually result in lower noise due to reductions in turbulence ingestion, vortex shedding, and delaying the onset of cavitation. These 'systems' generally account for the interaction of the hull, wake into and out of the propulsor, and propulsor/rudder interaction. Detailed design studies are needed to achieve the reductions in noise and cost associated with these energy-saving devices.

A summary of propeller treatments; their effectiveness – level and frequency; cost estimate; and availability – now/future/far future are summarized in Table 9.11 (Spence et al. 2007).

A summary of the 'maturity and estimated technology readiness level' of various noise control treatments to control underwater radiated noise is provided in Table 9.12 (Smith and Rigby 2022).

An extensive, and mainly qualitative, summary of potential 'quieting treatments' versus effectiveness, cost, feasibility, measurability/verifiability, co-benefit, and relevant information is provided in a report by Hemmera 2016 to the Vancouver Fraser Port Authority. Renilson Marine Report 2009, and Spence et al. 2007, provide additional details and information on propeller noise and possible treatments.

9.5.2 *Machinery Noise Control*

All of the machinery, foundation, and damping treatments presented in Chap. 8 with respect to habitability noise would apply to the reduction of underwater radiated noise. Selecting when and how to operate a specific piece of machinery can also be a form of noise control. If a commercial vessel is entering a zone that is environmentally sensitive, they may choose to transit at a slower speed, only use bow/stern thrusters up to a certain percent load, or secure any non-pertinent equipment. This is a common practice on fisheries research vessels, for example, that only needs to be extremely quiet during surveying missions.

In terms of new technology, 'superconducting' electric motors should not only have low noise and vibration but also potentially have significant non-acoustic advantages. These include lower weight, smaller size, higher torque, and responsiveness over a large load range.[21] New sinuous AC motor drives are being developed that have lower vibration than the typical commercial motor. Some AC motors are being provided with isolation mounted stators.

New hybrid electric drives that are being utilized for their fuel efficiency could also reduce the vessel's noise signature. The electric drive would be used for slow speed propulsion with the main diesel secured. Power for the AC propulsion motor

[21] http://www.oceanpowermagazine.net/2009/08/02/electric-ships-using-superconducting-motors-run-quiet-and-clean/

Table 9.11 Propeller and thruster treatments v. effectiveness, cost, and availability

Treatment Name	Brief description	Effectiveness		Cost Estimate	Availability (N) Now, (F) Future, (FF) Far Future
		Reduction	Frequency Range		
Low noise propeller design	Propellers can be designed to have reduced noise relative to conventional designs without the need for additional treatments	3–10+ dB	All	No additional material cost. Machining costs may be increased	N
Low noise thruster design	Thrusters can be designed to have reduced noise relative to conventional designs without the need for additional treatments	3–10+ dB	All	No additional material cost. Machining costs may be increased	N
Ducted propellers	A duct is located around the propeller to potentially reduce noise and increase efficiency	Varies depending on implementation	All	Increased costs associated with manufacture of duct	N
Forward-skew blade	Propeller and thruster blades incorporate a forward skew instead of a conventional Backward skew	3–20+ dB	All	No additional material cost. Machining costs may be increased	N
Ring propellers	A full or partial ring is used around the edge of the propeller. Leads to reduced noise	N/A	N/A	Assumed increased cost relative to open propeller	N
Drop thruster	A retractable thruster is used in place of a conventional tunnel thruster	5–15+ dB	All	2-6x tunnel thruster designs	N
Z-drives and podded systems	Systems that locate the propeller in an area with improved flow characteristics	5–10+ dB	All	Approx. \$200,000 – \$300,000 per unit	N

(continued)

Table 9.11 (continued)

Treatment Name	Brief description	Effectiveness Reduction	Frequency Range	Cost Estimate	Availability (N) Now, (F) Future, (FF) Far Future
Waterjets	Alternative to conventional propeller systems for high speed vessels	5–20+ dB estimated	All	Higher than conventional propulsion systems	N
Rim drive propulsion	New thruster design that can reduce noise	5–10 dB	O/A	Approx. $150,000 per unit	N/F
Voith Schneider propulsion	Alternative to trainable propeller systems for maneuvering – Used on minesweepers.	N/A	N/A	Higher than conventional propulsion systems	N
Propac rudder	Rudder design with claimed increase in efficiency	N/A	N/A	Increased cost is likely	N
Foul release coatings	By applying this coating to a propeller some beneficial effects on noise are seen.	0–10 dB 0–10 dB increase in sound for some conditions.	50–10,000 Hz >1000 Hz	N/A	F/FF
Composite material propeller	Propellers made from composite materials	N/A	N/A	N/A	F
Propeller bubble emission	Air bubbles are emitted from the propeller blades into the water	>10 dB 0–10 dB increase	>500 Hz 20–80 Hz	High	N
Thruster bubble emission	Air bubbles are emitted into the stream of an operating tunnel thruster	0–20 dB Possible increase	>100 Hz <100 Hz	Est $20,000 – $50,000	N
Air bubble masker	An air bubble layer is emitted at or near the vessel hull and issued to block sound radiation from propellers	>10 dB 0–10 dB increase	>500 20–80	Est $30,000 – $75,000	N

(continued)

Table 9.11 (continued)

Treatment Name	Brief description	Effectiveness		Cost Estimate	Availability (N) Now, (F) Future, (FF) Far Future
		Reduction	Frequency Range		
Anti-singing edge	A special design is used for the trailing edge of the propeller blade to reduce or prevent singing propellers	Removal of singing tone when problem exists	Frequency of tone	No additional material cost	N
Regular maintenance	Growth and damage to propellers can significantly increase underwater radiated noise. Regular maintenance can reduce this impact	Varies. Large reductions are possible for highly fouled or damaged propellers/ thrusters	All	Small when performed with other scheduled maintenance	N
Toroidal shaped propeller[a]	Low cavitation, reduce or eliminate tip vortices	N/A	N/A	N/A	F to FF

[a]Item added by authors

would be provided by the ship service diesel generator(s), which should be isolation mounted, or by batteries.

At this point, there are no off-the-shelf active noise or vibration systems that are proven to be effective for general commercial use for large ocean-going vessels. However, active mount isolation systems have been demonstrated on research vessels and yachts (Yang et al. 2012; Winberg et al. 2005; and Moon et al. 2013). These systems can reduce the diesel tones but require extensive space for the actuators and need control systems that accurately track the frequencies of interest. Where there are multiple diesel sources and variable speed engines, these systems do not have an impeccable track record.

While many of the systems discussed above should provide reduced noise on commercial vessels, there is little-to-no acoustic data to date to directly support these assertions. However, some of these techniques are being used successfully on military craft with low signatures. These include partial tunnels above open propellers, air layers, and composite blades. The European AQUO project has identified 'practical guidelines and solutions in order to predict and to mitigate underwater noise footprint' (Audoly et al. 2015).

For example, an "active control" cavitation-reducing propulsion control system was tested at full scale onboard an M-frigate class ship of the RNLN. This system was installed "to demonstrate the effect of the developed new propulsion control system on the propulsion system in general and on propeller cavitation in particular"

Table 9.12 Propeller treatment maturity and estimated technology readiness level (Smith and Rigby 2022)

Technology	Noise Source	Description of maturity	TRL[a]
Vortex generator	Sheet and bubble cavitation, blade-rate noise	Widely used in many industries for controlling flow. Experimental and numerical studies demonstrate effectiveness for improving wake uniformity and noise reduction. Performance also demonstrated at full-scale. Effectiveness and optimal design will be vessel specific and future studies should also consider the impact of the self-noise of the vortex generator.	7–9
Water/polymer injection	Tip vortex cavitation	Numerous experimental and numerical studies indicate viability of such an approach but no known full-scale trials at the time of writing.	3–4
Flexible thread	Tip vortex cavitation	Experimental studies at model scale indicate viability of technology. Further studies assessing integration issues and effectiveness at full-scale needed.	3–4
Winglets	Tip vortex cavitation	Widely used in the aerospace sector and the theory is well understood. Some experimental and numerical studies show that winglets can be effective in reducing TVC. Further research needed to understand the overall impact on noise and efficiency.	4–5
Roughness applied to tip/edges	Tip vortex cavitation	Demonstrated numerically and experimentally at model scale. Further research needed to understand the trade-offs between noise reduction and efficiency loss.	3–4
Propeller boss-cap fin/ eco-cap	Hub vortex cavitation	Widely used and multiple numerical, experimental and full-scale studies demonstrate their effectiveness at reducing or removing hub vortex cavitation. Noise reduction characteristics associated with this have also been confirmed experimentally.	6–7
Energy-saving devices to reduce propeller rotation rate	Propeller noise (cavitating and non-cavitating)	Multiple experimental studies and full-scale trials demonstrating the effectiveness of such technology for generating thrust and improving propulsive efficiency. However, their impact on noise reduction has not been widely assessed.	1–3

(continued)

Table 9.12 (continued)

Technology	Noise Source	Description of maturity	TRL[a]
Serrated leading/ trailing edges	Leading-edge and trailing edge noise	Widely researched and many experimental and numerical studies demonstrate their effectiveness. Technology now utilized in renewable energy sector. Further research needed to understand effectiveness for reducing trailing edge and interaction noise on marine vessels.	4–6
Porous surfaces/ edges/tip	Tip vortex cavitation, leading-edge and trailing-edge noise	Numerous experimental and numerical studies show effectiveness in reducing phenomena such as trailing edge noise and TVC. Further research is needed to understand how they could be effectively applied to propeller or appendage design and understand impact of fouling on performance.	3–4

[a]TRL 1 – Basic principles observed; TRL 2 – Technology concept formulated; TRL 3 – Experimental proof of concept; TRL 4 – Technology validated in lab; TRL 5 – Technology validated in relevant environment; TRL 6 – Technology demonstrated in relevant environment; TRL 7 – System prototype demonstrated in operational environment; TRL 8 – System complete and qualified; TRL 9 – Actual system proven in operational environment: per https://en.wikipedia.org/wiki/Technology_readiness_level

(Vrijdag 2009). In a comparison of the existing propulsion control system versus the active system, noise reductions up to 12 dB are reported at speeds below 12 kn. Between 12 and 18 kn, no noise reduction was found; however, in the range from 18 to 24 kn, a reduction of up to 3 dB was found. The exact conditions under which these gains were found are not reported.

9.5.3 Path Noise Control

Once a certain source is selected, controlling the different paths of noise is the next step. Again, there are three main paths of noise from a piece of machinery: the airborne component, the structureborne component, and the secondary structureborne component. Effective and efficient noise control relies on the ability to treat the most offending path first and then treat the less important paths, if needed. Generally, the greatest offending path for most machinery sources is the structureborne path. This is because modern ship construction techniques readily transmit vibration throughout the ship, to the hull, and into the water. The secondary structureborne and airborne paths both directly benefit from any treatment that lowers the airborne noise. Any treatments that are directly applied to the hull or ship structure that are designed to affect structureborne noise will also affect the secondary noise. Chapter 8 – Noise treatments contains information on the usual treatments for habitability noise. Those utilized mainly for underwater noise are discussed below.

While not a treatment per se, double hulls have the advantage of reducing noise transmitted via the airborne path into the water. They also afford an optimum location for damping of the floors. Fluid-filled tanks below the propulsion machinery may have a detrimental impact on radiated noise (Levi 1989). Smith (1990) showed that "full internal fluid containers have the effect of increasing the external sound radiation from surface ships in the frequency range of 64 Hz to 1.5 kHz."

9.5.3.1 Damping

Constrained layer damping on hull plating and machinery foundations in the engine room has been specified for quiet research vessels (Fischer 2005). Detailed studies showed that aluminum-constrained plates could be used in lieu of the specified steel plates, resulting in significant weight savings (25 tons) with little impact on the radiated noise. A different damping layer was also recommended based on optimization studies carried out with *DesignerNOISE®*.

9.5.3.2 Hull/Propeller Air Emitter

Air emissions have been used on naval vessels to reduce their underwater radiated noise or signature. These have been placed on the hull – emitter belts at various locations along the hull forward of the engine room, on the leading edge of propellers, on the stabilizer, and on the rudders. In the past, these systems have had significant maintenance issues with marine growth in the orifices. Bleed air from gas turbines has been used to supply pressurized air. Otherwise, additional compressors, themselves noisy, must be used.

Air-cavity ships (ACS) are advanced marine vehicles that use air injection or an air layer on the hull to improve a vessel's hydrodynamic characteristics. This system is intended to reduce drag, improve propulsion efficiency, and reduce greenhouse gases. The impact on the underwater radiation has not been proven; however, the air cushion theoretically could reduce the underwater radiated noise (Matveev 2004).

Research is always being conducted on quieting methods, and one should always be tuned to their application and measured effectiveness.

9.5.3.3 Summary of Radiated Noise Treatments for Quiet Research Vessels

The recommendations for a 'quiet' research vessel assuming a diesel-electric vessel are listed in Table 9.13.

Table 9.13 Recommended radiated noise treatments for quiet vessels

Double-stage isolation of gensets,
Enclosures for the gensets,
Isolation mounting of all rotating or reciprocating machinery,
Distributed isolation material (DIM) mounting for all transformers and all static propulsion power equipment
Design of high-impedance foundations for all isolation mounted equipment
Selection of low source level equipment, e.g., low speed, heavy, rotating in lieu of reciprocating, etc.
Use of filters on the electric drive system,
Design of a cavitation-free propeller up to 11 knots
Hydroacoustic design of hull form – bow to stern
Structural damping of machinery foundations and hull plating
Treatment of all sea-connected systems with flexible hoses and decoupling material near hull openings
Design and installation of low-speed hydraulic system with hydroacoustic muffler resiliently attached to ship's structure.

9.5.4 Fluidborne Noise Control

Noise transmitted along piping systems into the water can be reduced by following many of the recommendations outlined above for machinery – namely, having low source level equipment and using resilient elements along the transmission path. In terms of low source level equipment, rotary (screw or gear) pumps should be selected over piston pumps. The pumps, especially those over 18 kW (25 hp), should be vibration isolation mounted with flexible hoses to reduce structureborne noise. The pump and control valves should be cavitation free to minimize their contribution to the radiated signature. A flexible hose also provides an 'impedance" mismatch, reducing the fluidborne noise transmitted along the piping system and into the water (Purshouse 1986). If possible, a flexible hose connection should be added close to the hull opening. In-line hydraulic silencers should be used to attenuate pump pulsations – for example see Fig. 9.28 (Wilkes & McLean).

Furthermore, hull openings below the water line should be flush and as deep as feasible. Inlets of the NACA type are recommended. Discharges should make an angle of no more than 10 degrees with the local hull surface. Inlets and discharges should be aligned with the flow. Another treatment is to line the sea-chest cavity with a rubber lining such as Rubetex® and use flexible hose connections, as shown in Fig. 9.29.

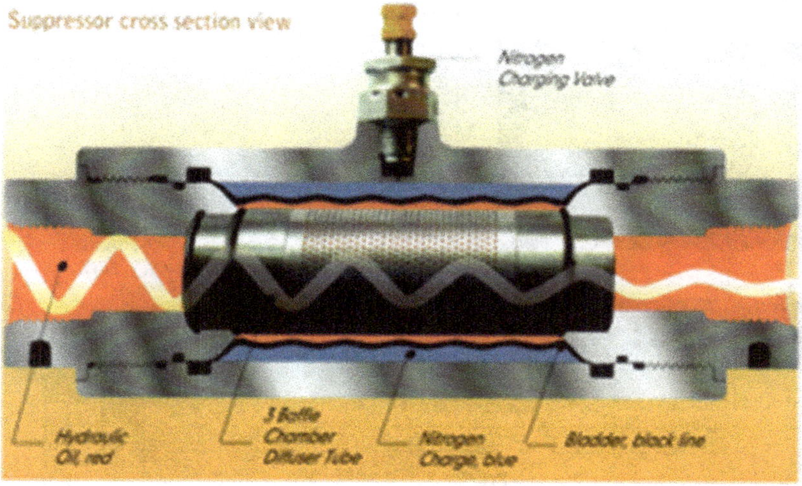

Fig. 9.28 Example of Wilkes Mclean Suppressor. (Provided by Wilkes & McLean)

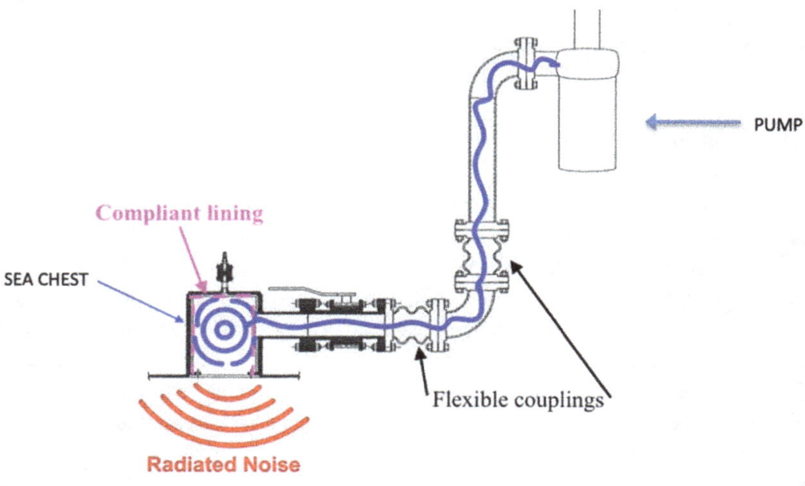

Fig. 9.29 Sea-chest treatments

9.5.5 *Monitoring System*

There are a multitude of possible monitoring systems for ship structure, machinery, logging performance, etc. Two systems related to the radiated noise signature are monitoring of underwater radiated noise or monitoring of machinery vibration levels.

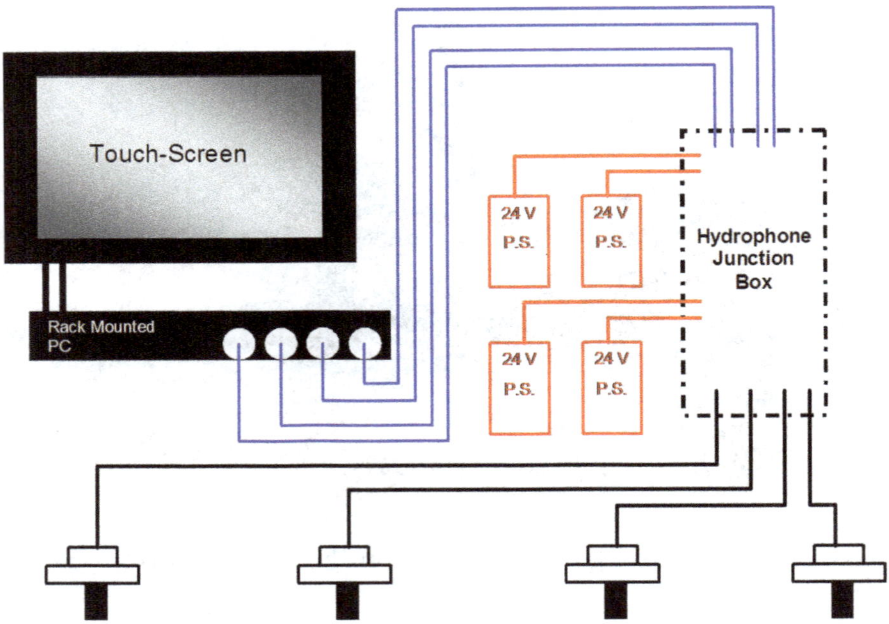

Fig. 9.30 Sound guard system diagram

9.5.5.1 Monitoring Underwater Radiated Noise

There are systems available to monitor the vessel's radiated (or platform) noise. This can be accomplished either by on-board hydrophones or hull mounted accelerometers or a combination of both. One such system is *SoundGuard,* developed by Noise Control Engineering LLC. It is a PC-based data acquisition system for the evaluation of underwater noise from any vessel but has been installed mainly on research vessels (R/V). It acquires underwater noise levels from "self-noise hydrophones" (i.e., hydrophones located in the ship's hull) to evaluate the acoustic condition of the vessel. The system can log underwater noise levels for documentation purposes. With this system, underwater sound levels are compared to the noise limits specified in the International Council for the Exploration of the Seas (ICES) Cooperative Research Report 209 (CRR 209). A *SoundGuard* system diagram is shown in Fig. 9.30. A screenshot of the collecting and logging function is shown in Fig. 9.31.

Another similar system, based on accelerometer measurements over the propellers, is covered in Palomo et al. 2023. This type of system is typically used to monitor the cavitation inception speed.

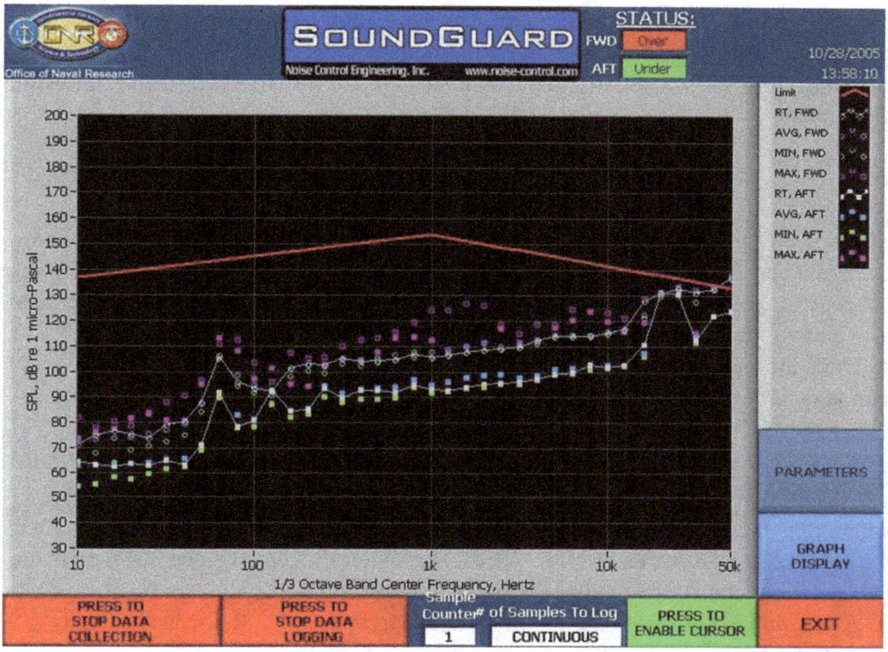

Fig. 9.31 Collecting and logging data screen shot

9.5.5.2 Vibration Monitoring

Another way to monitor the radiated signature is to track the vibration of critical machinery and/or the hull (Garnier et al. 1997). Machinery monitoring would concentrate on propulsion equipment. Hull vibration monitoring was performed on a wetted hull plate in the engine room. Again, a baseline data set would be needed to determine changes in machinery or hull vibration, set alarms, or provide alerts on significant changes to the vessel's signature (Gerhard 2013). If these data were correlated with underwater radiated signature measurements, the system could provide a real-time evaluation of the vessel signature. New wire-less systems would facilitate installation of such a system. Furthermore, such a monitoring system could be used for machinery health diagnostic purposes (Smiley 1984).

9.6 State of the Art Technologies for Underwater Noise Predictions and Assessment

To predict the noise radiated from a ship, one must first decide on a modeling technique. The method can be as simple as determining how much a specific plate on a ship will vibrate and then using the plate's area to predict how much noise will be emitted from the plate. The simplest approximation is that the radiated acoustic power is proportional to the area that is radiating and the square of the velocity of the radiating surface. The radiated acoustic power, Π_{rad}, is controlled by the following parameters:

$$\Pi_{rad}(f), W = \rho c\, A\, \sigma(f) \langle v^2(f) \rangle \tag{9.37}$$

where σ = radiation efficiency with water loading, which is different for spheres, cylinders and flat panels, depending on frequency (Mace 1981); v = vibration level, m/sec; A = radiating area, m^2; c = sound speed in seawater, = 1500 m/sec; and ρ = density of seawater, 1026 kg/m^3.

Once the radiated power is computed, the radiated sound pressure level, p_{rad}, at 1 m is determined as:

$$p^2_{rad}(r = 1m) = \prod\nolimits_{rad} \rho c Q / 4\pi \tag{9.38}$$

where Q is the directivity, assumed to be 2, for radiation into a hemisphere and p is the radiated pressure – Pa or N/m^2. Thus, the higher the plate velocity or the greater the radiating area is, the greater the radiated noise.

An empirical equation relating the radiated pressure, p^2_{rad}, to the hull velocity, v^2, is given by Fraser et al. 1987:

$$p^2_{rad}/v^2 = \left((\rho c)^2/2 \right) \times \mathrm{Log}\left[1 + \left(2R^2/d^2\right)\right] * \sigma \tag{9.39}$$

where d is the distance from the hull, R is the average dimension of the radiating area, and σ is the water-loaded radiation efficiency. In terms of radiated noise and hull velocity, the following equation may be used:

$$\left(L_p \text{ dB } \mu\text{Pa–m}\right) - \left(L_v \text{ dB re } 5 \times 10^{-8}\ \frac{\text{m}}{\text{s}}\right) =$$

$$20\,\mathrm{Log}(\rho c) + 10\,\mathrm{Log}(\sigma) + 10\,\mathrm{Log}\left[1 + 2R^2/d^2\right] - 29 \tag{9.40}$$

In terms of radiated noise and hull acceleration, the following equation may be used:

$$\left(L_p \text{ dB } \mu\text{Pa–m}\right) - \left(L_a \text{ dB re } \mu\text{G}\right) =$$

$$20\,\mathrm{Log}(\rho c) + 10\,\mathrm{Log}(\sigma) + 10\,\mathrm{Log}\left[1 + 2R^2/d^2\right] + 1 - 20\,\mathrm{Log}(f) \tag{9.41}$$

Some example values are presented in Table 9.14. In this example, the radiating area is 37 m^2, the radiation efficiency is representative, and d is equal to 10 m.

Javier et al. 2023 presents some measured and predicted radiation efficiencies and radiation transfer functions. The measured data are consistent with the radiation efficiency presented in Sect. 9.3.2. The average and standard deviation of three

Table 9.14 Example radiation transfer function

Frequency, Hz	125	250	500	1000	2000	4000	8000	16,000
10 log σ	−44	−36	−28	−21	−14	−9	−7	−6
$L_p - L_v$	51	59	66	74	80	86	87	89
$L_p - L_a$	39	41	42	44	44	44	39	35

L_p dB re μPa–m; L_v dB re 5 × 10^{-8}; L_a *dB* re μG.

Table 9.15 Measured radiation transfer function, $L_p - L_a$ dB re 1 $\mu Pa^2/\mu G^2$

Freq., Hz	10	12.3	16	20	25	31.5	40	50	63	80
$L_p - L_a$	63	59	56	46	45	51	43	42	41	39
Std. Dev, dB	7	7	3	10	7	14	17	8	10	11
Freq., Hz	*100*	*125*	*160*	*200*	*250*	*320*	*400*	*500*	*640*	*800*
$L_p - L_a$	40	38	36	36	33	29	30	32	35	34
Std. Dev, dB	10	13	12	12	11	12	8	5	4	3
Freq., Hz	*1000*	*1250*	*1600*	*2000*	*2500*	*3200*	*4000*	*5000*	*6400*	*8000*
$L_p - L_a$	30	26	31	32	29	33	35	37	42	43
Std. Dev, dB	6	5	6	7	4	3	5	1	1	1

one-third octave band measurements of the radiation transfer function, $L_p - L_a$ dB re 1 $\mu Pa^2/\mu G^2$, are provided in Table 9.15 (adopted from Javier et al. 2023).

There are several techniques that are applied commercially that more accurately predict the noise due to multiple pieces of machinery and the three major paths of noise discussed earlier. Nearly all these methods employ the use of specialized computer programs and algorithms to address the complex interactions and mathematics that are present when looking at real-world cases. Additionally, most methods have other benefits, such as being able to predict the noise and/or vibration levels in the rest of the ship interior. With these advanced tools and an understanding of the acoustic environment, optimal noise control approaches can be selected, reducing the usual impact on cost, space and weight.

9.6.1 3-D Computer Aided Design (CAD) Modeling

One such method is Statistical Energy Analysis (SEA) (Plunt 1980a; Fahy 1994; Jensen 1976; Burroughs et al. 1997; and Tso and Hansen 1997). SEA predicts the vibration by starting with the known noise and vibration levels or power (F × v) from a piece of machinery. From there, the user builds a simple model of the ship, including all pertinent structural elements. The program computes the propagation of the energy flow from the equipment through all the structural connections of the ship. The effectiveness of the various installed treatments, e.g., isolation mounts, will be included in the computations. At completion, the program has predicted the vibration of the hull plates and/or the noise in ship compartments below the waterline. The program can then calculate the noise in the water due to the vibration of the hull, accounting for the radiation parameters of the hull plating.

Ship structures are quite complicated, and the propagation of acoustic energy through the entire ship system requires an accurate accounting of acoustic loss – including radiation coupling to the air and water, coupling through intersections of deck and bulkheads, and inherent structural and 'added[22]' damping effects. The coupling of the machinery source to the structure also requires careful consideration. The source is not the typical 'point' source considered in most acoustic problems; rather, it is distributed. Thus, extended sources require different coupling functions for the structure.

There are several commercial software packages that implement or partially implement acoustic CAD modeling. As shown in NOAA's Vessel Quieting Technology conference, Session 2, *DesignerNOISE*® can be used to determine the hull vibration from all shipboard sources (Fischer 2004). Then, estimated or measured transfer functions can be applied to relate the hull vibration-to-far-field radiated noise (Morrow 1987; Morrow et al. 1987; and Fraser et al. 1987).

More sophisticated software such as Energy Finite Element Analysis (EFEA) methods are being developed to provide even more detailed modeling methods (Gilroy and Smith 1999; Layton et al. 2005; Vlahopoulos et al. 2005; Zhang et al. 2001; and ICSV 2008). These methods combine the best attributes of SEA and FEA. Commercial design and regulatory agents are also developing in-house tools that account for the radiated noise from vessels (MARIN (The Netherlands), Germanischer Lloyd (GL), Det Norske Veritas (DNV), ViPac (Melbourne, Australia)).

SEA models are relatively quick to build when compared to some other modeling methods. Because of their straightforwardness and the fact that they only use the structural elements of the ship, SEA models can be performed at a very early stage in the ship's design. SEA on typical ship structures has reliable results from 125 Hz – 8000 Hz. SEA modeling has been shown to have an accuracy of 3–4.5 dB between 63 Hz and 2000 Hz (Hynna et al. 1995). The authors have seen very accurate results out to 10,000 Hz. The accuracy of these modeling methods is highly dependent on receiving accurate noise and vibration levels for every relevant piece of machinery. If the data cannot be measured, they must be predicted. This will generally lead to inaccuracies.

9.6.2 Finite Element Modeling

Finite Element Analysis (FEA) is another widespread modeling method. An FEA model of a ship is much more detailed than an SEA model. FEA models encompass

---

[22] Damping associated with fittings such as joiner bulkheads, furniture, bunks, piping systems, etc.

all the structural elements on a ship, including small stiffener characteristics, stanchions, and doorways that SEA usually does not include. This exponentially increases the time to make a model. The benefit and limitation of FEA is that it is only usable at fairly low frequencies. The typical frequency range for FEA is 1 Hz–100 Hz. FEA is also capable of determining if the ship has any strong reactions to specific frequencies (local resonances). The results of the FEA model are similar to those of the SEA model in that the vibration of the hull will be known.

Novel analysis methods are being developed but have not been put to general use in the commercial field to date. This includes Energy Finite Element Analysis (EFEA) (ICSV 2008; Layton et al. 2005; Zhang et al. 2001; and Vlahopoulos et al. 2005) and Boundary Element Modeling (BEM) (Beskos 1987). Hybrid methods are being developed. These consist of a mixture of FEM, CFD, and a process called PML – Perfectly Matched Layer (Chaturvedi et al. 2020). PML considers 'the absorption of radiated/scattered waves in computational acoustics and its effectiveness as a nonreflecting boundary. These tools are being used primarily in military applications, and their applicability to commercial vessels may come shortly.

9.7 Underwater Radiated Noise Propagation Modeling

Noise propagation in the ocean depends on many variables (Erbe et al. 2022). Parameters such as water temperature, salinity, wave height, ocean depth, and seafloor characteristics all play a role in determining how noise will travel once it is in the water. There are additional factors, and their effect on the noise varies over frequency. For instance, the absorption value increases by several orders of magnitude over the frequency range of 1–1000 Hz (Lurton 2002). This makes trying to predict how the noise will propagate in a given situation very complex. An example for the propagation losses (TL in this graph) versus distance and for multiple frequencies is shown in Figs. 9.32. Note that the difference in transmission loss at 100 m between 31.5 Hz and 500 Hz is 20 dB. Furthermore, the spreading from 10 to 100 m at 31.5 Hz is greater than spherical, 28 dB and vice a predicted 20 dB. Similarly, at 500 Hz in this example, the spreading between 10 and 100 m is less than that predicted by spherical spreading. Other factors, such as ice cover, sea floor properties, bubble content, and shallow water, significantly impact transmission loss (Meyer and Audoly 2020). Attenuation within rivers and for different sea states can be quite problematic.

Software to predict sound propagation affects has been developed (Siderius and Porter 2008; Küsel 2019). This includes ray tracing and normal mode propagation. There are numerous propagation prediction programs available from both commercial and government sources (Heitmeyer et al. 2006; Etter 2003). It is always best to measure the propagation losses at the test site whenever possible (Arsene and Borda 2018). In general, a calibrated underwater noise source and a distant receiver hydrophone are used to calibrate the test site over the frequency range of interest.

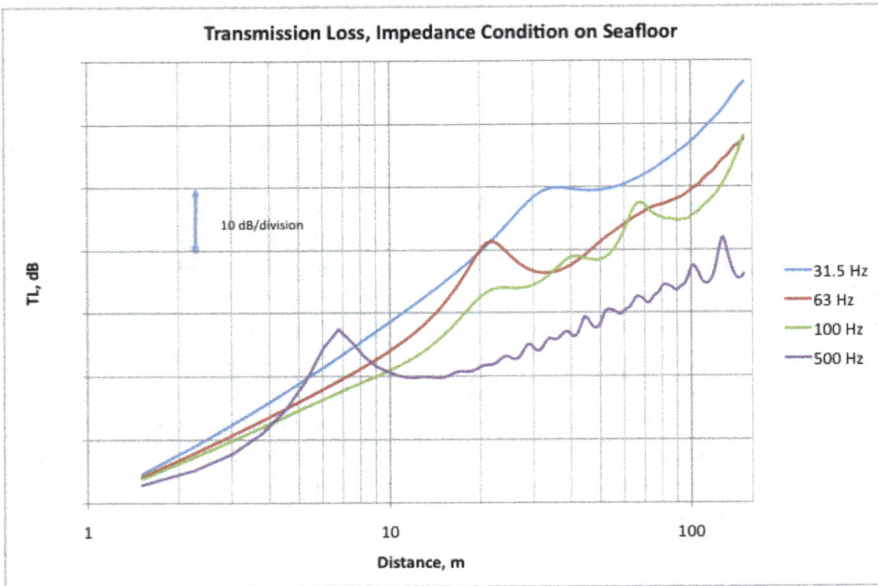

Fig. 9.32 Propagation Loss (TL) v. Impedance of the Seafloor

In most cases, factoring in any of the above characteristics will serve to lower the noise more than the simple spreading (either spherical or cylindrical) method alone. This means that if one simply uses the spreading metric discussed previously, it will be a nonconservative estimate of the actual noise in most cases. Modeling techniques and methods other than simple spreading are beyond the scope of this document.

Another factor to be considered is the impact of Lloyd's mirror. If the ocean surface is a perfect reflector, it will cause an interference pattern and decrease the measured level at a hydrophone to a set depth and distance from the ship source (Audoly 2017). This Lloyd's Mirror Interference Pattern (LMIP) can be approximated by (Ainslie 2010).

$$LMIP = 10\text{Log}\left(4\ \sin^2\left(kd_sd_r/_r\right)\right) \tag{9.42}$$

where k is the wavenumber $k = (2\ \pi\ f/c)$, f is the frequency of interest, c is the speed of sound, d_s and d_r are the source and receiver depths, respectively, and r is the source-receiver range.

Lloyd's mirror is especially important at low frequencies (see Fig. 9.33). In this figure, the direct path, spherical spreading attenuation at all frequencies is −43 dB. At 10 Hz, for the given geometry of depression angle, depth of the source, and horizontal range, the measured level is −56, which is 14 dB below that for the direct path. If the surface is rough from wind and waves, this impact is less significant.

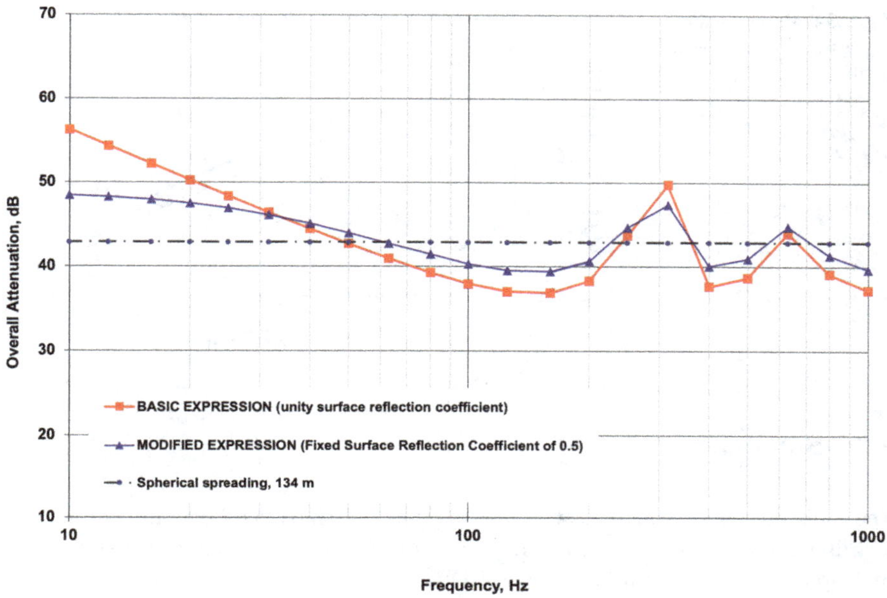

Fig. 9.33 Effect of Lloyd's mirror v. infinite ocean medium with spherical spreading

The Lloyd mirror also affects how marine life senses the ship's radiated noise. For instance, the apparent signature of sources near the surface, such as jet skis and small watercraft, is greatly reduced in the far field. This effect extends even to high frequencies (private communication from C. Audoly).

In shallow water, there is another consideration. In this case, the ocean acts as a waveguide. Consideration of the normal mode cutoff frequency, f_{co}, is important. Below this frequency, sound will not propagate. The cutoff frequency is computed as:

$$f_{co} = \left(\frac{c}{4h}\right) \Big/ \sqrt{1 - c_w^2/c_{sed}^2} \tag{9.43}$$

where c_w is the speed of sound in water, c_{sed} is the speed of sound in the sediment, and h is the water depth. Representative cutoff frequencies are shown in Fig. 9.34.

An empirical equation for long-distance propagation in shallow water is:

$$L_{p\infty} \text{ dB re } 1\mu\text{Pa@r} = L_{p\ source} - 10\ \text{Log}(r) - 10\ \text{Log}\left({}^{H}\!/_{2r}\right) \tag{9.44}$$

where H is the water depth.

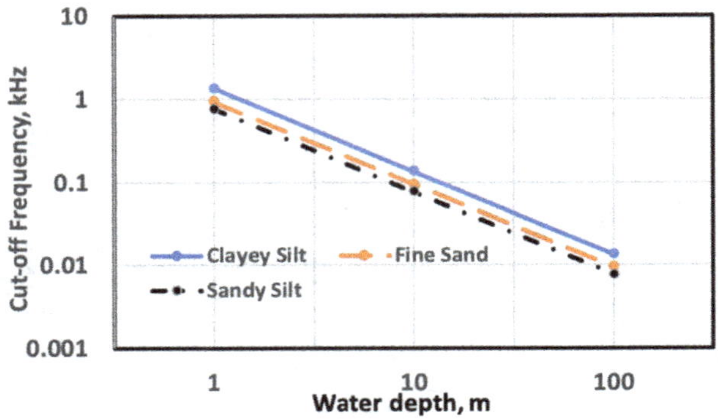

Fig. 9.34 Representative cutoff frequency vs. water depth

9.8 Summary

The radiated noise associated with various types of ships can be significant and is distributed throughout the oceans. Estimates of the upper expected broadband noise by ship type are as follows (Annon 2001):

- super tankers (approximately 127 vessels at sea at any time): 187 to 232 dB
- freighters, bulk carriers, large tankers (approximately 23,000 at sea at any time): 185 to 200 dB
- tankers, merchant ships (approximately 100,000 at sea at any time): 155 to 190 dB
- medium-small motor-powered vessels (approximately hundreds of thousands at sea at any time): 150 to 160 dB

These underwater noise levels have the potential to cause hearing loss (permanent threshold shift), communication masking, stress/annoyance, and habitat abandonment among ocean life. These factors are a concern that requires further determination of long-term impact.

Technology exists to evaluate and control the radiated noise from these commercial vessels. This technology can be applied to vessels that plan to operate in environmentally sensitive areas. The primary noise sources are the propulsion drives and the propulsors. The propulsion machinery should be selected based on having low vibration source levels and/or utilizing vibration isolation mounts where feasible. The propulsor should be selected based on its expected cavitation inception speed for the operating scenario while in the environmentally sensitive area and for its absolute level when cavitating. Bear in mind that a ship with a lower radiated noise level generally also has lower habitability noise and vibration levels, affecting its crew and passengers.

Efforts are underway to design ecological and sustainable[23] ships – addressing not only environmental impacts but also engineering and cost to provide an optimal

[23] ABS Guidance to Assist the Marine and Offshore Industries in Achieving Sustainability.

design.[24] These types of efforts are needed, especially for high-speed ferries, which may operate in many undeveloped areas close to sensitive breeding grounds. High-speed, lightweight ferries can have significant underwater noise. Similarly, there is a concern for LNG and container ships, which have targets of higher speed and power and are becoming larger and more numerous. These may eventually be operating in more environmentally sensitive areas of the oceans.

It is not necessary to apply naval combatant silencing techniques to commercial vessels. What is needed is a concerted effort to apply good acoustic design techniques and treatments to those vessels expected to have significant underwater noise that need or plan to operate in truly sensitive marine mammal areas. Technologies exist that can reduce the radiated noise from commercial ships.

The amount or level of noise control is dependent on bio-acoustic objectives that are yet to be identified completely. For instance, lowering a ship's speed may reduce its radiated signature, while at the same time, it may affect the risk of whale strikes (Killeen et al. 2022).

While not directly related to ship noise, a review of existing and future potential treatment for reducing underwater noise from offshore oil and gas activities is provided by Spence et al. 2007 and Wyatt 2008.

References

Abrahamsen, K.: The ship as an underwater noise source, ECUA 2012, 11th European Conference on Ship Acoustics, Edinburgh, Scotland, July 2012

Ainslie, M.A.: Principles of Sonar Performance Modeling. Springer-Praxis, Chichester (2010)

Ainslie, M., de Jong, C.A.F., Dol, H., et al.: Assessment of Natural and Anthropogenic Sound Sources and Acoustic Propagation in the North Sea., TNO Report TNO-DV 2009 C085,, The Netherlands (Feb. 2009)

ACCOBAMS, Agreement on the Conservation of Cetaceans of the Black Sea, Mediterranean Sea and Contiguous Atlantic Area (ACCOBAMS), Resolution 3.10 "Guidelines to Address the Impact of Anthropogenic Noise on Marine Mammals in the ACCOBAMS Area", 2007 and Resolution 2.16 "Impact Assessment of Man-Made Noise", (2004)

Anichkov, S.L., Vilensky, G. G., Kuz'min, A. G.: Broadband Acoustic Radiation by Ducted Marine Propeller (n.d.). http://www.math.chalmers.se/Conf/ECMI98/standard/Abstracts/Vehicles-aero dynamics/kuzmin.pdf

Au, W.W.L., Green, M.: Acoustic interaction of humpback whales and whale-watching boats. Mar. Environ. Res. **49**(5), 469–481 (2000) ISSN 0141-1136.

Audoly, C.: Parametric Model for Underwater Radiated Noise of Fast Ships and Comparison to Other Ship Categories, Fast 2017, Nantes-France, Sept. 2017

Audoly, C., Meyer, V.: Measurement of Radiated Noise from Surface Ships – Influence of the Sea Surface Reflection Coefficient on the Lloyd's Mirror Effect. Proceedings of ACOUSTICS 2017 Perth, Perth, Nov. 2017

Audoly, C., Rizzuto, E., Mullor, R., Hallander, J.: Mitigation measures for controlling the ship underwater radiated noise, in the Scope of AQUO Project. *OCEANS* 2015 – Genova, (2015), pp. 1–6, https://doi.org/10.1109/OCEANS-Genova.2015.7271381

[24] www.ecoship.com

Audoly, C., Rousset, C.: AQUO Project – Modeling of ships as noise source for use in an underwater noise footprint assessment tool. InterNoise 2014
Annom: *Final environmental impact statement for the North Pacific acoustic laboratory*. Office of Naval Research (May 2001)
Arsene, D., Borda, C., et.al.: Procedures for measuring and modeling underwater acoustic field parameters. Acoustics and Vibration of Mechanical Structures – AVME-2017, Springer Proceeding in Physics, 198, (2018)
Arveson, P., Vendittis, D.: Radiated noise characteristics of a modern cargo ship. J. Acou. Soc. Of Amer. **107**(1), 118–129 (Jan. 2000)
Bagheri, R., Seif, M.S., Mehdigholi, H., Yaakob, O.: Analysis of noise behavior for marine propellers under cavitating and non-cavitating conditions. Ships Offshore Struct. **12**, 1–8 (2015)
Beskos, D. (ed.): *Boundary Element Methods in Mechanics* First Series: Computational Methods in Mechanics, North-Holland, Amsterdam (1987)
Bie, H.-Y., Ye, J.-J., Hao, Z.-R.: Effect of nozzle geometry on characteristics of submerged gas jet and bubble noise. J. Lab. Autom. **1-8** (2015)
Bies, D., Hansen, C.H.: Engineering noise control. Unwin Hyman Ltd., London (1988)
Blake, W.K.: Aero-Hydroacoustics for Ships, Vol. I & II, David Taylor Naval Ship Research and Development Center, DTNSRDC-84/010, June 1984
Bonetti, E., Calcagno, P.: Low Noise as the First Task of a Vessel — A Description and Some Remarks About Acoustic Quieting Design Criteria and Features of SACLANT ASW Centre Oceanographic Vessel. Shipboard Acoustics, Nartinus Nijhoff Publishers, Dordrecht (1986)
Borisov, S., Gritsenko, A., etal.,: Results of Studies Between Molikpaq and Piltun and Near Chayvo Bay. V.I. Il'icev Pacific Oceanological Institute, Sakhalin Russian Federation, July 2003
Bosschers, J.: A semi-empirical prediction method for broadband Hull-pressure fluctuations and underwater radiated noise by propeller tip vortex cavitation. J. Mar Sci Eng. **6**(49) (2018)
Bradley, D. L., Stern, R.: *Underwater Sound and the Marine Mammal Acoustic Environment: A Guide to Fundamental Principles,* Prepared for the U.S. Marine Mammal Commission, July 2008
Breeding, J., Pflug, L., Bradley, M., Hebert, M. & McBride, W.: RANDI 3.1 Physics Description. Naval Research Laboratory Technical Report NRL/FR/7176--95-9628 (1996)
Brown, N.: Cavitation noise problems and solutions. In: Janssen, J.H. (ed.) International Symposium on Shipboard Acoustics. Elsevier Scientific Publications, Amsterdam (1976)
Brown, N. A., Norton, J. A.: Propeller Blade Structures and Methods Particularly Adapted for Marine Ducted Reversible Thrusters and the Like for Minimizing Cavitation and Related Noise. United States Patent No. 3,972,646, Aug. 3, 1976
Brown, N.: Thruster Noise. In: Dynamic Positioning Conference, Houston (Oct. 1999)
Brown, N.A., Dai, T.K., Althouse, T., Fischer, R., Petersen, E.: DP Bow Thruster Noise Remediation in R/V ROGER REVELLE, AGOR-24. Dynamic Positioning Conference, Sept. 2001
Brown, N.: Existing/Future Technology to Address Radiated Noise by Modifying Vessel Propulsion and Operating Parameters. In: Presentation at NOAA Ship Quieting Symposium, Washington DC (2007)
Brown, N. A., Fischer, R., Eckart, J.:. Ship Silencing to Enable Deep Water Bathymetric Survey. *Subsea Survey 08*, Galveston (2008)
Burroughs, C., Fischer, R., Kern, F.: An introduction to statistical energy analysis. J. Acoust. Soc. Am. (JASA). **101**, 1779–1789 (1997)
Carlton, J.S.: Marine Propellers and Propulsion, 4th edn. Butterworth Heinemann, Oxford (2019)
Chaturvedi, A., Joga, R. K., Dhavalikar, S., Girish, N.: Advanced simulations of Underwater Radiated Noise (URN) predictions for naval ships. 15th International Naval Engineering Conference & Exhibition, Proceedings of INEC (2020)
Coates, R.: Underwater Acoustic Systems. Wiley, New York (1989)
Cruz, E., Lloyd, T., Bosschers, J., Lafeber, F.H., Vinagre, P. Vaz, G.: Study on inventory of existing policy, research and impacts of continuous underwater noise in Europe. EMSA report EMSA/

NEG/21/2020. WavEC Offshore Renewables and Maritime Research Institute Netherlands. (2021)
Dames & Moore: https://openlibrary.org/books/OL23941679M/PTFI_environmental_audit_report, microfiche, (1996)
de Jong, K.: On the optimization and the design of ship screw propellers with and without end plates. University of Groningen, Department of Mathematics, The Netherlands (1991)
de Jong, K.: Characterization of Ships as Sources of Underwater Noise. NAG-DAGA (2009)
De Robertis, A., Wilson, C., Willamson, N.: Do Silent Ships See More Fish?." AFSC Quarterly Report. J. Acoustical Soc. Am, 121 (Oct-Dec 2008)
Dunn, C., Theriault, J., Hickmott, L., Claridge, D.: Slower Ship Speed in Bahamas Due to COVID-19 Produces Dramatic Reduction in Ocean Sound Levels. Front. Mar. Sci. **8** (2021)
E.U. 2008: The European Union Marine Strategy Framework Directive (Adopted in June 2008). Available on http://ec.europa.eu/environment/water/marine/index_en.htm
Erbe, C., Duncan, A., Vigness-Raposa, K.J.: Introduction to sound propagation under water. In: Erbe, C., Thomas, J.A. (eds.) *Exploring Animal Behavior Through Sound: Volume 1*. Springer, Champions (2022). https://doi.org/10.1007/978-3-030-97540-1_6
Erbe, C., McMauley, R., McPherson, C., Gavrilov, A.: Underwater noise from offshore oil production vessels. JASA-EL J. Acoust. Soc. Am. **133**(6), EL465–EL470 (June 2013)
Etter, P.C.: Underwater Acoustic Modeling and Simulation, 3rd edn. Spon Press (2003)
Fahy, F.J.: Statistical energy analysis: a critical overview. Philos. Trans.: Phys. Sci. Eng. **346**(1681), 431–447 (1994)
Fischer, R., Courtney, B., Nelson, D.: *Design Guide for Shipboard Airborne Noise Control*. Society of Naval Architects. and Marine Engineers (SNAME). T&R Bulliten 3–37, Jersey City (1983)
Fischer, R., Boroditsky, L.: *Supplement to the Design Guide for Shipboard Airborne Noise Control*. Society of Naval Architects and Marine Engineers (SNAME). T&R Bulletin 3-37, Jersey City (2000)
Fischer, R.: Factors Affecting the Radiated Signature of Underwater Vehicles. Underwater Intervention 2001 Proceedings, Tampa, (2001)
Fischer, R.: Controlling Machinery Induced Underwater Noise. NOAA Vessel Quieting Technology, Washington (Sept. 2004)
Fischer, R.W.: Acoustic design, construction, and testing of NOAA's fisheries research vessel. In: Lloyds Ship Noise and Vibration Conference, London (June, 2005)
Fischer, R., Brown, N.: Factors affecting the underwater noise of commercial vessels operating in environmentally sensitive areas., MTS/IEEE Oceans '05 Symposium, Washington DC (Sept. 2005)
Fischer, R.: Singing propellers – solutions and case histories. Soc. Nav. Archit. Mar. Eng. (SNAME), Mar. Technol. **45**(4), 221–227 (2008)
Fischer, R., Yankaskas, K.: A program approach to reducing shipboard noise on naval vessels. INCE Noise-Con 2011, Portland OR (July 2011)
Frank, V.: Underwater Noise from Commercial Shipping: A Look at the Arctic. International Fund for Animal Welfare. www.ifaw.org
Fraser, A. J., Swindell, R. J.: A Mathematical Model for Calculating the Acoustic Power Radiated from a Ship's Hull. Proceedings, Internoise 88, Avignon (1988)
Fraser, A.J., Morrow, R.T., Swindell, R.J., Dean, C.: Methods for Predicting Underwater Noise Levels Radiated from a Ship's Hull, British Maritime Technology LTD., Project 14403, (April 1987)
Gaier, M., AlGermozi, M., Ingham, R., Rodionov, I.: Using Paint to Reduce Underwater Noise for the Shipping Industry. Paint & Coatings Industry (February 7, 2022)
Gao, H., Zhu, W., et al.: Effects of various winglets on the performance of marine propellers. Appl. Ocean Res. **86**, 246–256 (May 2019)
Garnier, B., de la Croix, D., Webster, D.: From vibration monitoring to acoustic signature monitoring of a ship. 5th International Congress on Sound and Vibration, Dec. 1997

Gerhard, K.: Non-Intrusive Vibration Monitoring in US Naval and US Coast Guard Ships. United States Naval Academy, Thesis (June 2013)

Ghassemi, H., Gorji, M., Mohammadi, J.: Effect of tip rake angle on the hydrodynamic characteristics and sound pressure level around the marine propeller. Ships Offshore Struct. **13**(7), 759–768., Taylor & Francis (2018). https://doi.org/10.1080/17445302.2018.1457207

Gilroy, L., Smith, M.: Comparison of SEA and PFFEA predictions of radiated noise from a stiffened box structure. Inter-Noise 99, Ft. Lauderdale Dec. 1999

Gorji, M., Ghassemi, H., Mohamadi, J.: Calculation of sound pressure level of marine propeller in low frequency. J. Low Freq. Noise Vibr. Act. Control. **37**(1), 60–73 (2018)

Gospodnetic, S.: *Impact of Manufacturing Tolerances on Propeller Performance Investigation 1: 2D Foil Section in the Rectilinear Flow.*, Report RD36-TC-01. Dominis Engineering Ltd., Ontario (2022)

Gray, L.M., Greeley, D.S.: Source level model for propeller blade rate radiation for the world's merchant fleet. J. Acoust. Soc. Am. **67**, 516–522 (1980)

Greeley, D.: Calculation of Low Frequency Cavitation Source Strength of Marine Propellers. Bolt, Beranek, and Newman TM No. 459, July 1978

Greeley, D., Kerwin, J.: Numerical methods for propeller design and analysis in steady flow. SNAME Trans. **90** (1982)

Haimov, H., Bobo, M., Vicario, J., del Corral, J.: Ducted Propellers. A Solution for Better Propulsion of Ships. Calculatoins and Practice. First International Symposium on Fishing Vessel Energy Efficiency, E-Fishing, Vigo (May 2010)

Hallander, J.: Prediction of underwater noise from ships propellers. SSPA. (2013)

Hannay, D.: Source Level Measurements from 2004 Acoustics Programme, vol. 66. Sakhalin Energy (2004)

He, X., Liu, A., et al.: Numerical calculation of the radiated noise of contra-rotating propellers. J. Phys. Conf. Ser. **1600**, 012004 (2020)

Heckl, M., *Theoretical and Experimental Investigations on the Air- and Body-Sound Conduction from a Ship's Plating into the Surrounding Water.*, Bureau of Ships Translation No. 935, (1974)

Heckl, M.: Entwicklung Von Methoden aur Schallpegelprognose fur den Unterkunfts – *und Maschinenraumbereich von Schiffen*, in German, Bericht Nr. 193/1988, Technisch Universitat Berlin, Hamburg (1988)

Heitmeyer, R.M., Wales, S.C., Pflug, L.A.: Shipping noise predictions: capabilities and limitations. Mar. Technol. Soc. J. **37**, 54–65 (2004)

Heitmeyer, R., Waler, S., Pflug, L., Hayward, T.: Ship-induced Noise Predictions in the Atlantic and the Pacific: A Comparison of Two Noise Models. Naval Research Laboratory, NRL/FR/7121-06-10, 123, May 2006

Hemmera: *Vessel Quieting Design, Technology, and Maintenance Report, Options for Potential Inclusion in EcoAction Program,* Hemmera Envirochem Inc.. Report 302-045.03, B.C., Canada, April 2016

Hynna, P., Klinge, P., Vuoksinen, J.: Prediction of structureborne sound transmission in large welded ship structures using statistical energy analysis. J. Sound Vib. **180**(4), 583–607 (1995)

ICES Cooperative Research Report No. 209, Underwater Noise of Research Vessels Review and Recommendations, (1995)

International Congress on Sound and Vibration (ICSV) 15. Exterior noise predictions of commercial ships using energy flow analysis methods. Korea, July 2008

IMO Particularly Sensitive Sea Areas.; https://www.imo.org/en/OurWork/Environment/Pages/PSSAs.aspx

IMO 2009. Noise from Commercial Shipping and its Adverse Impacts on Marine Life, Report from the Correspondence Group (MEPC59/19), April 09

IMO 2008a. Minimizing the introduction of incidental noise from commercial shipping operations into the marine environment to reduce potential adverse impacts on marine life. Paper MEPC 58/19 submitted by the United States to 58th session of IMO Marine Environment Protection Committee. October 2008

IMO 2008b. Report of the 58th Session of IMO Marine Environment Protection Committee. October 2008

Ishii, Y., Miyawaki, K., et al.: Underwater noise prediction and development of special purpose ships. Mitsubishi Heavy Ind. Tech. Rev. **57**(1) (March 2020)

Jalkanen, J.-P., Johansson, L., Liefvendahl, M., et al.: Modeling of ships as a source of underwater noise. Ocean Sci. **14**, 1373–1383 (2018)

Javier, R.R., Jaime, P.P., et al.: Analysis of the underwater radiated noise generated by Hull vibration of the ships. Sensors. **23** (2023)

Jebsen, G.: Surface ship noise control. Presented at National Oceanic and Atmospheric Administration (NOAA) International Symposium, Shipping Noise and Marine Mammals: A Forum for Science, Management, and Technology, Arlington, May 2004

Jensen, J.: Calculation of Structureborne Noise Transmission in Ships Using the 'Statistical Energy Analysis' Approach. Computer Science (1976)

Johannsen, C., Zarbock, O., Bretschneider, H.: Experimental determination of the difference in visual and acoustic cavitation inception. A. Yucel Odabasi Colloquium Series, 1st International Meeting – Propeller Noise and Vibration, Istanbul, Nov. 2014

Junger, M.C., Feit, D.: Sound, Structures and Their Interactions. The MIT Press, Cambridge, MA (1986)

Killeen, H., Fields, J., Fangue, N.: Manages and unmanaged whale mortality in the California current ecosystem. Mar. Policy. **140**, 105039 (June 2022)

Kim, H.-S., Kim, J.-S., Kim, B.-K., et al.: Effect of airborne noise from ship machinery on underwater noise. J. Soc. Nav. Archit. Korea. **48**(6), 569–574 (Dec. 2011)

Kinnas, S., et al.: Prediction of cavitation performance of single/multicomponent propulsors and their interaction with the hull. Trans. SNAME. **110**, 215–244 (2002)

Kipple, B.: Southeast Alaska Cruise Ship Underwater Acoustic Noise, NSWC Report 71-TR-2002/574, Oct. 2002

Kipple, B. M., Gabriele, C. M.: Glacier Bay watercraft noise – noise characterization for tour, charter, private, and government vessels: report to Glacier Bay National Park by the naval surface warfare cent-detachment Bremerton. Technical Report NSWCCD-71-TR-2004/545 (2004)

Kipple, B., Gabriele, C.: Underwater Noise from Skiffs to Ships. Proceedings of the 4th Glacier Bay Science Symposium, U.S. Geological Survey Scientific Investigations Report 2007–5047, (2007)

Küsel, E.T., Siderius, M.: Comparison of propagation models for the characterization of sound pressure fields. IEEE J. Ocean. Eng. (2019)

Lanni, F., Brown, N.: Impeller for Marine Waterjet Propulsion, U.S. Patent No. 6,135,831, Oct. 24, 2000

Layton, G., Brennan, D., Boroditsky, L., Fischer, R., Koko, T.S., Akpan, U.O., Dunbar, T.E., Mae, S.: Energy Finite Element Analysis (EFEA) Algorithms and Software, Contract No. N00014-04-C-0448, NCE Report, 2005

Le Bert, J.F., R. Barbarin, et al.: Pod Motor Vibroacoustic Simulation. Transport Noise and Vibration, St. Petersburg, June 2002

Lee, A., Flay, R.: Compliant blades for passive power control of wind turbines. Wind Eng. **24**, 3–11 (2000)

Lee, C. W.: The Design of Rim-Driven Propulsor and Flow Characteristics. Thesis, Systems and Naval Mechatronic Engineering, 2005. in Chinese

Lee, K., Lee, J., Kim, D., Kim, K., Seong, W.: Propeller sheet cavitation noise source modeling and inversion. J. Sound Vib. **333**, 1356–1368 (2014)

Levi, A.: Effect of Internal Fluid Containers on Sound Radiated Externally from a Surface Ship. MIT Department of Ocean Engineering, MS Thesis, Jan. 1989

Li, D.-Q., Hallander, J., Johansson, T., Karlsson, R.: Cavitation Dynamics and Underwater Radiated Noise Signature of a Ship with Cavitating Propeller. In: Salvatore, F., Broglia, R., Muscari, R. (eds.) VI International Conference on Computational Methods in Marine Engineering. MARINE (2015)

Liu, X., Yang, D., Li, Q., Liu, J.: Energy-averaged method for ship mechanical noise prediction under kinematic excitation in the full frequency domain. Ocean Eng. **246**, 110615 (Feb. 2022)
Liu, Z., Young, Y.L.: Utilization of bend-twist coupling for performance enhancement of composite Marine propellers. J. Fluids Struct. **25**, 1102–1116 (2009)
Lobitz, D., Veers, P.: Load mitigation with bending/twist-coupled blades on rotors using modern control strategies. Wind Energy. **6**, 105–117 (2003)
Lurton, X.: An Introduction to Underwater Acoustics, Principles and Applications. Springer, Praxis Publishing, Chichester (2002)
Lv, X., Liang, Q., Shi, D., Wang, Q.: Prediction of the hydrodynamic noise of one certain frigate. Inter Noise '16, Hamburg, (2016)
Mace, B.R.: Sound radiation from fluid loaded orthogonally stiffened plates. J. Sound Vib. **79**(3), 439–452 (1981)
MacGillivray, A., Li, Z., Hannay, D.: Slowing deep-sea commercial vessels reduces underwater radiated noise. J. Acoust. Soc. Am. **146**(1), 340–351 (July 2019)
Matveev, K.: On the Influence of Artificial Cavitation on Underwater Noise Radiation from a Ship Hull. Proceedings of IMECE04 2004 ASME international mechanical engineering congress and exposition, Anaheim, November 13–20, 2004
McCarthy, E.: International Regulation of Underwater Sound: Establishing Rules and Standards to Address Ocean Noise Pollution. Kluwer Academic Publishers, Boston (2004)
McCauley, R.D., Cato, D.H.: Patterns of fish calling in a nearshore environment in the Great Barrier Reef. Philos. Trans. R. Soc. Lond. Ser. B-Biol. Sci. **355**, 1289–1293 (2000)
McDaniel, S.T.: Sea surface reverberation: a review. J. Acoust Soc. Am. (JSA)., 0001-4966. **94**, 1905–1922 (1993). https://doi.org/10.1121/1.407514
McDonald, M.A., Hildebrand, J.A., Wiggins, S.M.: Increase in deep ocean ambient noise in the Northeast Pacific west of san Nicolas Island, CA. J. Acoust. Soc. Am. **120**(2), 711–718 (2006)
McKenna, M.F., Wiggins, S.M., Hildebrand, J.A.: Relationship between container ship underwater noise levels and ship design, operational and oceanographic conditions. Sci. Rep. **3**(1) (2013)
Meyer, V., Audoly, C.: Accounting for sea floor properties in the assessment of underwater noise radiated from ships in shallow water. International Conference on Underwater Acoustics, 40, ICUA 2020
Mitson, R.B.: Underwater noise of research vessels: review and recommendations. ICES Coop. Res. Rep. **209**, 1–61 (1995)
MMO (Marine Management Organization), *Technical Annex: Modeling Mapping of Continuous Underwater Noise Generated by Activities,* MMO Project No. 1097, August 2015
Moon, S.-J., Choi, S.-M., et al.: An inertia-type hybrid mount combining a rubber mount and a piezostack actuator for naval shipboard equipment. Int. J. Nav. Archit. Ocean Eng. **5**(62), 62–80 (2013)
Morrow, R.T., Ship Noise Design Methods, Final Report, SEA Modeling for Airborne Transmitted Environmental/Underwater Noise and Associated Noise Control, British Maritime Technology Ltd., Project No. 14403, April 1987
Morrow, R.T., Fraser, A.J., Swindell, R.J., Middleton, E.H., Dean, C.: Noise Design Methods for Ships, Environmental and Underwater Noise, British Maritime Technology Ltd., Project No. 14403, April 1987
Motley, M.R., Young, Y.L.: Performance-based design and analysis of flexible composite propulsors. J. Fluids Struct. **27**, 1310–1325 (2011)
Mousavi, B., Rahrovi, A., Kheradmand, S.: Numerical simulation of tonal and broadband hydrodynamic noises of non-non-cavitating underwater propeller. Pol. Marit. Res. **3**(83), 46–53 (2014)
Munjal, M.L.: Acoustics of Ducts and Mufflers. Wiley Interscience Publication, New York (1987)
Musel, E., Siderius, M.: Comparison of propagation models for the characterization of sound pressure fields. IEEE J. Ocean. Eng. **44**, 598–610 (Jan. 2019)
Nilsson, A.C., Tyvand, N.P. (eds.): Noise Sources in Ships: I: Propellers. Nordforsk, Miljovardsserien (1981)

NRC – 2003, National Research Council of the U.S. National Academies (NRC): Ocean Noise and Marine Mammals, p. 192. National Academy Press, Washington, DC (2003)

NRC – 2005, National Research Council of the U.S. National Academies (NRC): Marine Mammal Populations and Ocean Noise: Determining when Ocean Noise Causes Biologically Significant Effects, p. 126. National Academy Press, Washington, DC (2005)

Norwood, C.: Diesel Exhaust Noise Mapping and Noise Reduction. Maritime Platforms Div., DSTO Melbourne, Victoria

Norwood, C., Chen, L.: Underwater Exhaust Noise. Acoustics 2002, Innovation in Acoustics and Vibration, Annual Conference of the Australian Acoustical Society, Adelaide, 2002

Norwood, C., Chen, L.: Water Injection for Bubble Noise Reduction. Proceedings of Acoustics 2004, Gold Coast, Australia, Nov. 2004

Oossanen, P. van: Trade-Offs in Sub-Cavitating Propeller Design. Symposium High Powered Propulsion of Large Ships, Publication 490, Netherlands Ship Model Basin, December 1974

Palomo, P., Mendez, R., Piqueras, L.: The great challenge of propeller cavitation in shipbuilding. Continuous control with innovative Non-intrusive Cavitation Detection System (Ni-CDS). Ship Sci. Technol. **17**(33), 21–32 (July 2023)

Pan, X., MacGillvray, I., Trinh, V., Forrest, J.: Investigations on the total radiation efficiency of ribbed plates floating on water, Acoustics 2017 Perth, Proceeding of Acoustics (2017)

Parchen, R.: Noise production of ship's propellers and waterjet installations at non-cavitating conditions. Acoustics Div./Flow Acoustics, Report No. P2000-7 Delft, 2000

Peng, Z., Fan, J., Wang, B.: Analysis and modeling on radiated noise of a typical fishing boat measured in shallow water inspired by AQUO Project's model. Arch. Acoust. **4**(2) (2018)

Plunt, J.: Methods for Predicting Noise Levels in Ships; Experiences from Empirical and SEA Calculation Methods. Department of Building Acoustics, Gothenburg (1980a)

Plunt, J.: Methods for Predicting Noise Levels in Ships. Chalmers University of Technology, Gothenburg (April 1980b)

Possenti, L., Reichart, G.-J., de Nooijer, L., et al.: Predicting the contribution of climate change on North Atlantic underwater sound propagation. PeerJ. **11**, e16208 (2023). https://doi.org/10.7717/peerj.16208

Potter, J., Warn-Varnas, A. (eds.), *Ocean Variability & Acoustic Propagation.* Proceedings of the Workshop on Ocean Variability and Acoustic Propagation, La Spezia, Springer Science+Business Media, B.V., (1990)

Purshouse, M.: Underwater noise radiation due to transmission through the cooling water system of a Marine diesel engine. Buiten, J. (ed.), Shipboard Acoustics, Proceedings ISSA '86, Martinus Nijhoff Publishers, Dordrecht, 1986

Pylkkanen, J.: Review of Propeller Hydrodynamics for "LIKKUTEHO", VTT Industrial Systems, Research Report VAL37-023352, April 2002

Renilson Marine Consulting Pty Ltd., *Reducing Underwater Noise Pollution From Large Commercial Vessels*, Commissioned by the International Fund for Animal Welfare, March 2009

Richardson, W.J., Greene, C.R., Malme, C.I., Thomson, D.H. (eds.): *Marine Mammals and Noise*, vol. 576. Academic Press, New York (1995)

Ross, D.G.: Mechanics of Underwater Noise. Pergamon Press, New York (1976)

Rudd, A., Richlen, M., Stimpert, A., Au, W.: Underwater sound measurements of a high speed jet-propelled marine craft; implications for large whales. Pac. Sci. **69**(2), 155–164 (Oct. 2014)

SAE Technical Paper #952154: Fluid Power, Applications, Standards, Noise, Lubricants, and Testing (SP-110). Society of Automotive Engineers, Inc., Warrendale (1995)

Salio, M.P.: Numerical Assessment of Underwater Noise Radiated by a Cruise Ship, vol. 10, pp. 308–327. Ships and Offshore Structures, Taylor & Francis (2013)

Sand, O., Karlsen, H., Kundsen, F.: Comment on "Silent research vessels are not quiet". J. Acoust. Soc. Am. **123**(4), 1831–1833 (April 2008)

Santos-Dominguez, D., Torres-Guijarro, S., et al.: ShipsEar: an underwater vessel noise database. Appl. Acoust. **113**, 64–69 (2016)

Sasajima, T., Nakamura, N., Oshima, A.: Model and full scale measurements of propeller cavitation noise on an oceanographic research vessel with two different types of screw propellers. In: Buiten (ed.) Shipboard Acoustics., Proceedings ISSA '86. Narinus Nijhoff Publishing (1986)
Scrimger, P., Heitmeyer, R.: Acoustic source-level measurements for a variety of merchant ships. J. Acoust. Soc. Am. (JASA). **89**(2), 691–699 (1991)
Sharma, S.D., Mani, K., Arakeri, V.H.: Cavitation noise studies on marine propellers. J. Sound Vib. **138**(2), 255–283 (1990)
Siderius, M, Porter, M.: *Effects of sound on the marine environment,* Portland State University Report, 2008
Simmonds, M., Dolman, S., Weilgart, L. (eds.): Oceans of Noise. Whale and Dolphin Conservation Society, Wiltshire (May 2003)
Smiley, R.: Expert system for machinery fault diagnosis. Sound Vib. **18**(9) (Sep. 1984)
Smith, A.: Effect of surface ship internal fluid containers on external acoustic intensity measurements. MIT MS Thesis, May 1990
Smith, D. R., Slater, J. E.: The Geometry of Marine Propellers. *Defense Research Establishment Atlantic, Canada*, Technical Memorandum 88/214, November, 1988
Smith, T., Rigby, J.: Underwater radiated noise from marine vessels: a review of noise reduction methods and technology. Ocean Eng. **266**, 112863 (2022)
Southall, B., Hatch, L.: OSPAR Convention for the Protection of the Marine Environment of the North–East Atlantic. Draft of Preliminary Comprehensive Overview of the Impacts of Anthropogenic Underwater Sound in the Marine Environment, Module 5: shipping, Administrator of the Oslo and Paris Conventions, Aug. 2008
Southall, B., Finneran, J., et al.: Marine mammal noise exposure criteria: updated scientific recommendations for residual hearing effects. Aquat. Mamm. **45**(2), 125–232 (2019)
Spence, J., Fischer, R., Bahtarian, M., et al.: Review of Existing and Future Potential Treatments for Reducing Underwater Sound from Oil and Gas Industry Activities, Noise Control Engineering LLC Report 07-001, Billerica, Dec. 2007
Tachikawa, T., Okada, Y., Katayama, K., et al.: Efficiency improvement effect and water noise reduction by energy saving devices. In: Bensow, R., Ringsberg, J. (eds.) VIII International Conference on Computational Methods in Marine Engineering. MARINE (2019)
Tian, Y., Zhang, C., Yang, L., et al.: Analysis of vibration characteristics of podded propulsion shafting based on analytical method. J. Mar. Sci. Eng. **10** (2022)
Traverso, F., Gaggero, T., Rizzuto, E., Trucco, A.: Spectral analysis of the underwater acoustic noise radiated by ships with controllable pitch propellers. Oceans 2015 MTS/IEEE, Genoa Conference, June 2015
Traverso, F., Gaggero, T., Tani, G., et al.: Parametric analysis of ship noise spectra. IEEE J. Ocean. Eng. **42**, 424–438 (Aug. 2016)
Trevorrow, M.V., Vasilyeva, B.: Directionality and maneuvering effects on a surface ship underwater acoustic signature. J. Acoust. Soc. Am. **124**(2), 767–778 (Aug. 2008)
Tso, Y., Hansen, C.: The prediction of structureborne noise transmission in ships using statistical energy analysis. Acoust Australia. **25**(3) (1997)
Uchida, S., Yamanaka, Y., Ikeuchi, K.: Prediction of underwater noise radiated from ship's hull. Bull. Soc. Nav. Architect. Japan. **686** (1986)
UN Convention on Migratory Species (CMS), Resolution 9.19: Adverse Anthropogenic Ocean Noise Impacts on Cetaceans and other Biota. (2008)
Urick, R.: Principles of Underwater Sound for Engineers, 3rd edn. Peninsula Publishing, Los Altos (1983)
Vercuil, J.B., Terwisga, P.F.: Testing a New Full Scale Cavitation Observation System on board Hr. Ms. TYDEMAN.," NTC '50, Newcastle upon Tyne (April 2000)
Vlahopoulos, N., Wang, A., Wu, K.: An EFEA formulation for computing the structural response of complex structures. Proceedings of IMECE 05, ASME IMECE, November 5–11, 2005, Orlando FL, Paper Number IMECE2005-81113, (2005)
Yakovlev, A.Y., Sokolov, M.A., Marinich, N.V.: Numerical Design and Experimental Verification of a RIM-Driven Thruster. Second International Symposium on Marine Propulsors, Hamburg (June 2011)

Yan, X., Liang, X., et al.: A review of progress and applications of ship shaft-less rim-driven thrusters. Ocean Eng. **144**, 142–156 (2017)

Yan, X., Song, H., et al.: Review of research results concerning the modeling of shipping noise. Pol Marit Res. **28**, 102–115 (2021)

Yang, T.J., Suai, Z.J., et al.: Active vibration isolation system for a diesel engine. Noise Control Eng. J. **60**(3), 267–282 (2012)

Yang, Q., Wang, Y., Zhang, Z.: Numerical prediction of the fluctuating noise source of waterjet in full scale. J. Mar. Sci. Technol. **19**, 510–527 (2014)

Yatsenko, S.: *Potential Treatments for Reducing Underwater Sound to Protect Marine Life,* Appendix D: Reducing Underwater Sound to Protect Marine Life, CSL Australia, July 2012

Verkuyl, J.B., van Terwisga, P.F.: Testing a New Full Scale Cavitation Observation System on Board Hr. Ms. TYDEMAN. Newcastle upon Tyne (April 2000)

Vrijdag, A.: Control of Propeller Cavitation in Operational Conditions, Thesis – Technische Universiteit Delft, VSSD, Delft The Netherlands, June 2009

Winberg, M., Johansson, S., et al.: Active vibration isolation in ships: a pre-analysis of sound and vibration problems. Int. J. Acoust. Vibr. **10**(4) (2005)

Wittekind, D.K.: A simple model for the underwater noise source levels of ships. J. Ship Prod. Design. **30**(1), 7–14 (2014)

Wyatt, R.: *Joint Industry Programme on Sound and Marine Life*, Review of Existing Data on Underwater Sounds Produced by the Oil and Gas Industry, Issue 1, Seiche Measurements, Ltd., Great Torrington, Aug., 2008

Zhang, W., Ma, Q., Zhang, Z.: Estimation of low frequency noise radiated from submerged exhaust exit. In: Chapelle, P., Vermeir, G. (eds.) Proceedings Inter-Noise 93. Belgium, Noise Control Foundation, Poughkeepsie (1993)

Zhang, W., Wang, A., Vlahopoulos, N., Validation of the EFEA method through correlation with conventional FEA and SEA results SAE Paper no. 2001-01-1618, SAE J. Passeng. Cars – Mech. Syst., 110, pp. 2104–2119, 2001

Zhang, X., Liu, Z., Cao, L., Wan, D.: The clearance effect on the tip leakage vortex evolution and wake instability of a ducted propeller. J. Mar. Sci. Eng. **10**(8) (2022)

Zhu, C., Gaggero, T., Makris, N., Ratilal, P.: Underwater sound characteristics of a ship with controllable pitch propeller. J. Mar. Sci. Eng. **10** (2022). https://doi.org/10.3390/jmse10030328

Resources

NRC: Ocean Noise and Marine Mammals. National Academy Press, Washington, DC (2003)

NRC – 2005: Marine Mammal Populations & Ocean Noise, National Research Council, National Academy Press, 2005

Ocean Noise and Marine Mammals, National Research Council, (2003)

Reynolds, J. & Jasny, M.: Implementing Solutions to Shipping Noise. Natural Resources Defense Council, Symposium on Shipping Noise and Marine Mammals, Arlington VA May 2004

Scrimger, P., Heitmeyer, R.: Acoustic source-level measurements for a variety of merchant ships. SACLANT SR-143

Southall, B., 2005. Shipping noise and Marine mammals: a forum for science, management, and technology. Final report of the National Oceanic and Atmospheric Administration (NOAA) International Symposium. U.S. NOAA Fisheries, Arlington, Virginia, May 18–19, 2004, 40 pp.

Cummings, J.: Regulating Ocean noise: entering uncharted waters. J. Int. Wildl. Law Policy. **10**, 101–107 (2007)

Haren, A.M.: Reducing noise pollution from commercial shipping in the Channel Islands National Marine Sanctuary: a case study in marine protected area management of underwater noise. J. Int. Wildl. Law Policy. **10**, 153–173 (2007)
Anon 2008. International Workshop on Shipping Noise and Marine Mammals. Okeanos – foundation for the sea, Hamburg, Germany. (21–24 April 2008)
Medwin, H.: Sound in the Sea. Cambridge University Press, Cambridge, UK (2005)

Chapter 10
Design Guidance, Do's and Dont's, and Optimized Solutions

Universal control of noise and vibration is still somewhat of a novelty. Between existing and pending noise and vibration regulations, it is one of the most difficult responsibilities for the shipbuilding industry – including regulators, designers, builders, and owner/operators. The great majority of noise and vibration treatments for ships carry significant weight, space, and cost impacts. This is the major reason given for noise control being one of the last issues considered and implemented during a ship design or refit.

Another reason is the perceived inaccuracy of existing prediction tools. Comprehensive computer-aided tools have only been developed during the last two decades. Accurate noise prediction algorithms are essential to the successful implementation and *optimization* of acoustic treatments (Fischer et al. 2013). ANY optimization/trade-off/what-if? of noise and vibration treatments critically depends on what component, airborne or structureborne, predominates and over which frequency bands it prevails. Furthermore, everyone would benefit if the design was not overtreated or undertreated.

Guidance is provided in this chapter with respect to the 'tricks' of designing a quiet vessel. The intent here is to reduce the level of treatments needed and/or ensure they are correctly installed. The latter are covered by a listing of Dos and Don'ts. Following these steps will minimize the need for add-on treatments and ensure that noise and vibration control treatments are selected and utilized properly. Many prediction tools assume this guidance, as a minimum, is implemented in the design and carried out in the building of the vessel. The costs of noise and vibration controls will only be minimized if these steps are applied early in the design process. In addition, concept design is not considered *too* early. Illustration of the 'ability to impact' and 'cost of design changes' is shown in Fig. 4.4.

These noise and vibration approaches compete with other design objectives. These include trade-offs and what-if analyses, which are needed to adjudicate the best overall approach to a ship design. As stated in the Noise/Vibration Control Plan Chap. 4, applying the Design Guidance and following the Do's and Don'ts will facilitate this process. Costly treatments will result after the keel is laid out if due

R. Fischer, L. Boroditsky, *Noise and Vibration Control on Ships*,
https://doi.org/10.1007/978-3-031-55170-3_10

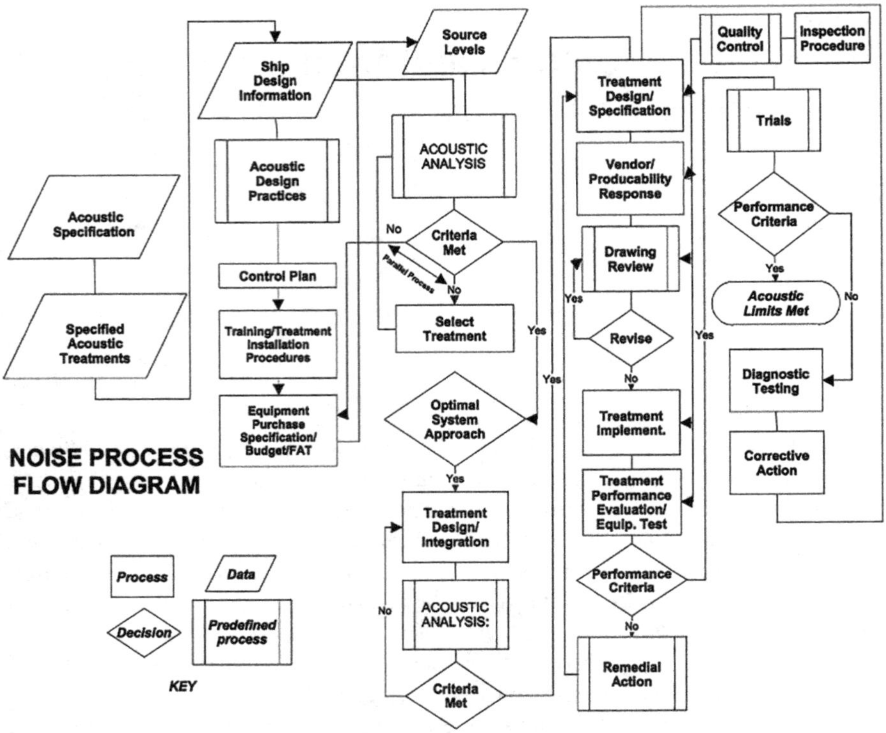

Fig. 10.1 Noise process flow diagram

attention is not paid to all the steps in the Noise Control Plan. A slight variation of the Plan shown in Chap. 4 is repeated in Fig. 10.1 as a reminder of the number of integrated steps needed.

To successfully design and build a quiet ship takes a concerted effort. It starts with a good set of noise goals and ends with certification of meeting those goals. In between, an unambiguous specification needs to be written or applied,[1] acoustical analyses must be iteratively performed, optimal abatement methods must be selected and detailed, and treatment installation inspections must be performed. As with any systems integration project, the owner, naval architect, equipment and material vendors, and shipyard all need to understand these steps and their implications to effectively deliver a quiet ship. The critical parameters for a successful noise control project are detailed in this chapter.

---

[1] One of the first steps in risk reduction is a careful evaluation of acoustic criteria, their applicability to the vessel being built, any potential error in the written criteria, and the implications on the extent of noise and vibration controls needed to meet the requirement, and critical paths needed to achieve the desired results.

Various acoustic criteria can be applied to ships. The most prevalent are hearing conservation and habitability noise criteria for crew and passengers (see Chap. 3). Others include habitability vibration, machinery and structural vibration, underwater noise, and platform or sonar self-noise. The latter are not only relevant for military vessels but may also apply to research vessels or commercial vessels operating in environmentally sensitive areas. Methods to meet these criteria – whether regulatory, owner or operator generated – can have a significant impact on the design and construction of a vessel. To obtain an optimal 'system' design, one needs to understand the implications couched in meeting not only the acoustic (noise and vibration) requirements but also the potential impact on space, weight, cost, and operation of the vessel. While all of these steps are appropriate for noise-sensitive designs, such as a research vessel, the process may be scaled back for less sensitive designs.

The implementation of treatments can also have a large bearing on the overall ship system design. As an example, incorporating isolation mounts may require a flexible shaft coupling for propulsion systems and additional space, especially vertical. To reduce the need for any treatment, 'quiet' sources or inherently quiet sources should be utilized. Due to many global acoustic regulations, vendors are beginning to produce lower acoustic source levels for such items as diesels. Inherently quieter equipment would the choice of rotary equipment over reciprocating equipment. Often, such items may have longer periods between overhaul or replacement. Additionally, it is beneficial to optimize the noise mitigation treatments during the initial ship design stage rather than after the construction phase is completed. After the initial construction is complete, the retrofit of noise control treatments may cost as much as ten times the costs had treatments been selected by implementing an evaluation and optimization process during the design phase (Yankaskas et al. 2017). One ***important*** difficulty in optimizing acoustic treatments is when the ship's specification has conflicting hardware and performance specifications. These conflicts should be resolved as early as possible, or multiple design revisions may be necessary, often at great expense.

The optimization considered in this chapter is intended to reflect the impact of selecting treatments based on their overall effectiveness on noise or vibration, their cost – both in dollars and impacts on weight; space and maintenance; and possible Return On Investment (ROI). Furthermore, in some cases, acoustic treatment has beneficial non-acoustic impacts. For instance, some spray-on damping treatments also offer thermal or anti-sweat characteristics. Some acoustic insulation can also double as fire and/or thermal insulation.

At each stage of design and construction, the expected acoustical parameters of the vessel should be monitored based on the Noise and Vibration Control Program Plan (see Chap. 4) and noise prediction procedures outlined in Chaps. 5, 6, 7, 8, and 9. The Noise Control Team should identify additional engineering, construction, Quality Assurance (QA), and testing needed. QA is the insurance policy needed to safeguard that the design and construction are going according to Plan. Challenging areas include the design and implementation of effective and advanced treatments such as compound mounting systems for gensets, the optimal placement of damping, and whether some 'active control' systems may be needed. Risk issues should be flagged. Possible work-around solutions need to be considered with respect to when and how they would be implemented, if necessary.

Reducing noise, in an optimal manner, has generally been relegated to acoustical experts. Currently, there has been an emphasis in the naval community to (1) reduce hearing loss and make ships quieter, (2) improve predictive tools, (3) implement advanced control treatments and (4) understand, to a greater degree, the impact of noise on both hearing loss and the crew's overall performance. This effort has led to a new US Navy noise standard, MIL-STD 1474E 2015.[2] These technologies, if utilized on new ship designs and during retrofits of existing vessels, can enable the crews to perform better in an improved acoustic environment. Furthermore, hearing loss impacts and their associated costs will be reduced (see Fig. 10.2 (Fischer and Yankaskas 2011)). Optimizing based on underwater radiated noise concerns will generally lead to a quieter habitability environment.

A study by Bowes et al. 2006, was undertaken to 'address the escalating cost of noise and hearing loss by developing a life cycle cost model of noise on a Navy Platform.' This study showed a 15:1 to 17:1 Return On Investment (ROI) by investing in noise control methods with respect to hearing loss compensation paid by the U.S. government.

10.1 Design Guidance and Do's and Don'ts

The following information provides generalized Do's and Don'ts that are applicable to any ship design regardless of the specified acoustic environment. Other guidance is provided by Morrow 1988, and Noonan 1990.

10.1.1 Layout

The overall airborne and structureborne noise mitigation approach is controlled by the vessel layout. Simply stated, it is cost-effective to separate noise-sensitive compartments from machinery spaces and thus reduce the extent of treatments needed to achieve acoustic compliance. Thus, place lockers, tanks, and storerooms as areas that surround machinery compartments, thus maintaining "buffers" between these and noise/vibration-sensitive living and working areas. The inverse is not to place noise-sensitive spaces – command, classrooms, medical, berths, etc. – directly adjacent to or over a main engine room, auxiliary engine room, or bow thruster room.

[2]The latest US Navy noise standard as of 2016.

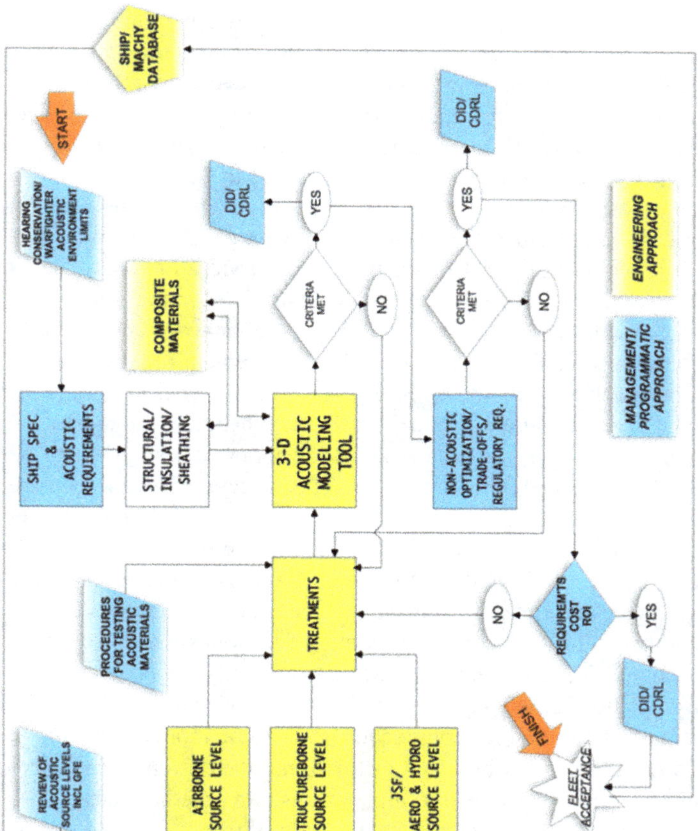

Fig. 10.2 Optimization process. DID/CDRL Documents for Contractor Data Requirements List, ROI Return on Investment

Finally, noise 'piles' up in corners. Bunk head positions are often located in corners. To reduce the noise at head positions, consider reversing the head and feet positions. Other possibilities are low-frequency absorbers in the corners or active low-frequency noise control.

10.1.2 Equipment Source Levels

One basic concept for noise control is to specify or select equipment with inherently low noise and/or vibration source levels. Equipment with a low source level may eliminate the need for extensive treatments. Of course, there will always be a trade-off between noise and the weight, cost, and operation of the vessel. However, one beneficial side effect of using quiet equipment is generally reduced maintenance requirements/costs. More details on Buy Quiet are presented in Sect. 8.1.

Accurate acoustic prediction software exists. As with any modeling process, and thus the decision process, the results are predicated on having good acoustic source-level information. Suitable, frequency dependent, noise and vibration source-level data that accurately describe the actual machinery items included in the ship design may not be readily available. Good 'source' data can be obtained from measurements acquired at the supplier's factory as part of the procurement requirements or prior to installing the machinery in the ship. These source-level data should be measured and reported in accordance with accepted ISO or other accepted regulatory body acoustic test standards. It is not uncommon for vendor acoustic data to be optimistic, providing low noise or vibration levels. This is the case when data are based on a small sampling and the variability in tolerances and manufacturing quality are not captured appropriately. "Historical" data on similar units or vessels may also be a reasonable source. If neither of these datasets are available, there are analytical approaches that can be used to approximate the acoustic source levels; see Chap. 5. This is the largest limiting factor in predicating decisions on how to control noise. The 'risk' of the project attaining the criteria is critically dependent on the quality of the acoustic source-level data. If source levels are inaccurate or questionable, then the selection of noise control approaches needs to be more aggressive or 'fallback' treatments need to be identified and factored into the design/build schedule. The fallback approach may often lead to higher costs and schedule delays.

Owner-provided equipment, e.g., electronic equipment or sonars, are and can be noisy. Thus, they should not be neglected in the modeling process. Unexpected or unaccounted for operational conditions, such as icebreaking, station keeping with high thruster power, bow wave impact at high speed, singing propellers, inflow interaction with the hull, etc., may be another significant noise source that can cause excessive noise if not included in the modeling.

10.1.3 Acoustic Insulation

The purpose of this section is to provide general design guidance for the use of acoustic insulation and acoustic absorption materials. Guidance on the composition of the materials themselves is also provided. Acoustic insulation is a generic term when used within this text. It can refer to any of the vendors and materials providing 'absorptive' characteristics. Material, such as mineral wool, polyurethane foam,[3] and ceramic fibers, can be considered. In general, fiberglass is fireproof and offers both thermal and acoustic insulation. Foams are lightweight and economical[6]. They tend to absorb sound in the higher frequency range. Rockwool is moisture and fire resistant. It is easy to fit. Rockwool typically out performs fiberglass in terms of sound absorption. Ceramic fibers can provide absorption in high-temperature locations, such as gas turbine exhausts or exhaust pipe wrapping.

10.1.3.1 Insulation Type & Usage

Table 10.1 summarizes the material descriptions and suggested usage. The final determination of which treatments are used should be developed via a detailed noise prediction.

Material Requirements

Acoustic insulation can consist of layers of two or three materials used for different purposes (Simon and Pfretzschner 2004). Absorption material for marine constructions generally consists of fiberglass or mineral wool. To provide absorption, the material should be exposed (not covered by solid sheathing) within the compartment of interest. For instance, fiberglass insulation placed between a structural bulkhead and a joiner bulkhead is not considered acoustic insulation (though it will help increase the airborne transmission loss, not the absorption). All acoustic materials should have a known Noise Reduction Coefficient, NRC, greater than 0.8. Thermal or fire insulation, if left as the finish (interior) surface,[4] may also qualify as acoustic insulation treatments.

Acoustic insulation, such as high transmission loss (HTL) material -Type III, comprises either higher density mineral wool or a composite of a limp mass layer[5] at least 4.9 kg/m^2 (1 lb./ft^2) sandwiched midway between two absorptive materials, minimum 50 mm (2″) thick.

If the absorption material is to be covered by a sheathing, it should be perforated. The perforated lining usually has 3.2 mm (0.125 inch) holes in the staggered center with a minimum open area of 30%. The sheathing should be painted or anodized prior to installation. The builder should cover and protect perforated linings during

[3]If smoke, fire, and toxicity requirements are met.

[4]Covered by either a vapor barrier and/or perforated sheathing.

[5]Meeting regulatory marine fire, smoke and toxicity requirements.

construction to assure that paint and debris do not plug the perforations. If the absorption needs to be covered with an unperforated sheathing as a vapor barrier, the material should be at most 0.0254 mm (0.001″) thick. Acoustic ceiling panels, if used, should have sufficient suspension attachments to preclude any sagging, vibration or buckling.

In heavy traffic and work areas, the treatments should be protected by perforated aluminum sheathing. Type II treatment should be installed external to any other treatment on plane surfaces and should be the treatment exposed to the airflow in ducts. Type II treatment should not be painted and should be marked as follows:

NOISE CONTROL SURFACE DO NOT PAINT

Seams of both Type I and Type II acoustic absorptive treatments should be covered with 50 mm wide tape as recommended by the manufacturer. Exposed edges of insulation should be sealed with adhesive when installed on the overhead.

As noted in Table 10.1, Type I or Type II acoustic insulation should be installed on the overhead and bulkheads of all spaces, as needed, to meet the noise requirements.

Table 10.1 Summary acoustic insulations and possible usage

Type	Material description	Recommended use	Possible compartment usage
I	Perforated face acoustic board. **Minimum NRC[a] = 0.80**	Should not be used in Food Prep, medical, machinery or other "wet" spaces.	Overheads of Storerooms, Electronics Rooms, Fan Rooms, Bow Thruster Room.
II	Incombustible board w/fire-, oil-, & water-resistant facing. **Minimum NRC = 0.80**	Used in Food Preparation & Consumption, Medical, Machinery and spaces exposed to oil, waters or vapors.	Overheads of the Main & Auxiliary Machinery Rooms, Labs, & Galley[b]
III	Double layer incombustible board separated by a limp septum[c] > 7.3 kg/m^2. The inner fiberglass layer is to be NRC 0.66 or greater. The outer fiberglass layer can be acoustic board or fiberglass blanket. **Minimum NRC = 0.80**	Used on bulkheads and overheads where high transmission loss (HTL) noise reduction is needed.	Internal & External Bulkheads in the Main and Auxiliary Machinery Rooms. This treatment can also be used to minimize hull transmitted radiated noise.
IV	Overhead lay-in panel in conjunction with overhead sheathing grid system. Panels to be 3 mm thick corrugated and perforated aluminum panels. **Minimum NRC = 0.65**	Acoustic overhead	Overheads of Messroom, State Rooms or Cabins, Dry Lab, Computer Lab, Acoustic Lab, Offices, Conference Room, Lounge, Pilothouse.

[a]NRC Noise Reduction Coefficient, a measure of the absorption coefficient
[b]Galley may require metal sheathing finish for health reasons
[c]Material must meet regulatory requirement – fire, smoke, and toxicity

Type II acoustic absorptive treatment should be installed on the noise-producing side of boundaries separating machinery spaces and workshops from living and messing spaces. Only Type II acoustic absorptive treatment should be installed in machinery spaces and in spaces such as the plenum chambers into which machinery space exhaust systems discharge and where exposed absorptive treatments could act as collectors of combustible materials.

When Type III treatments are to be installed in wet spaces such as showers, washrooms and water-closets, they should be sheathed with polished CRES, with a minimum of 1.6 mm (14 gage). Fasteners for securing the sheathing should be isolated from the treated bulkhead.

Type I or II treatment should be placed on the warm side of vertical surfaces bounding uptake enclosures and should extend from deck to the overhead. Treatment should be installed behind deck coaming. In all other areas, Type I or II treatment on vertical surfaces should extend from 150 mm (6″) above the deck to the overhead. The lower exposed edges should be sealed with 100 mm (4″) wide tape.

All seams in the septum and inner and outer layers of Type III transmission loss treatment should be sealed with 50 mm wide tape as recommended by the manufacturer (see Fig. 8.22). All penetrations, to include studs, should be sealed with sealing compound. Seams in the wire reinforced lead layer of Type III transmission loss treatments, if used, should overlap a minimum of 75 mm.

Type III treatment should extend from the deck to the overhead. A coaming should be installed on the deck, and the transmission loss treatment should be installed behind the coaming. When no other insulation is needed or specified at the boundary of the Type III treatment, the Type III treatment should be extended 300 mm beyond the indicated boundary on bulkheads and overheads. When other insulation is specified at the Type III treatment boundary, the high surface density septum should be extended 300 mm beyond the boundary on the bulkheads and overheads.

Where Type III insulation is being used, it should be installed behind status boards, control or switching panels, or equipment mounted on bulkheads, unlike Type I or II treatment. Exposed edges of acoustical treatment around air ports, doors and other locations where such edges are subject to damage should be protected by light "Z" or flat bars. Exposed edges not subject to damage should be protected with tape as recommended by the manufacturer.

Type IV treatment, where needed, should be installed as lay-in panels (in lieu of sheathing panels) in a sheathing grid system.

10.1.4 Vibration Isolation

One of the most effective treatments is the utilization of low-frequency isolation mounts. Obviously, not every item can be mounted. However, liberal use of a relatively low-cost Distributed Isolation Material can significantly reduce noise induced by motors, pumps, compressors, fans, and other auxiliary equipment, including diesels in some cases. Low-frequency isolation is essential on diesel generators, compressors, and propulsion machinery where feasible.

As discussed in Chap. 8, there are generally six types of mounting systems:

1. Hard mounted – machinery feet connected directly to foundation with or without Chockfast type material[6]
2. Distributed Isolation Material (DIM) pads – 8 mm to 25 mm thick pads that provide mid- to high-frequency isolation
3. Low-frequency isolation – spring or rubber mounts that attenuate 20 dB or so above 500 Hz.
4. Two-stage isolation provides significant (<40 dB) attenuation above 50 Hz.
5. Active isolation system – provides significant attenuation at the tuned frequency for the system.
6. Hybrid isolation system – combinations of low frequency/raft/DIM mounting.

Guidance is provided in the following sections on the design, Do's and Don'ts, and installation of various mount types. Table 10.2[7] lists the suggested types of isolation systems based on machinery type. This list should be considered for 'quiet' vessels, such as Research Vessels.

10.1.4.1 Single-Stage Isolation Systems

For any of the mount systems described above, the degree of isolation versus system natural frequency is shown in Table 10.3.

The effectiveness of isolation mounts depends on both the systems natural frequencies and the amount of damping,[8] ζ, inherent in the mount itself (see Fig. 10.3):

A damping loss factor, ζ, of 0.05, is considered the minimum value for significant damping. Most elastomeric materials used for isolation mounts have this level of damping. The transmissibility of a mounting system is determined by:

$$T = \sqrt{\frac{\left(1 + (2\zeta f_d/f_0)^2\right)}{\left(\left[1 - \left(f_d/f_0\right)^2\right]^2 + \left[2\zeta f_d/f_0\right]^2\right)}} \tag{10.1}$$

where ζ is the damping ratio, f_d is the drive frequency, and f_0 is the system natural frequency:

$$f_0 = (1/(2\pi))\sqrt{k/M} \;\; Hz.$$

where k is the total spring stiffness, N/m, and M is the equipment weight, kg.

---

[6] Chockfast material, a pourable epoxy grout, may slightly attenuation vibration above 4000 Hz.

[7] Information in Table 10.2 similar to that in Table 8.12; however, some parameters are different.

[8] The loss factor, η, is 2ς

Table 10.2 Suggested isolation systems for machinery

System and associated noise control		Iso. Mount	High Imp.[a] Fnd.	Fnd.[b] Damping	Flex connection	Resilient pipe hangers	FAT[c]
Propulsion	Propulsion motor	DIM		Yes			Yes
	Motor cooling pump	1-stage	Yes		Yes		
	Motor cooling fan	1-stage					
	Motor lube oil pump	DIM	Yes			Yes	
	DC Filter	1-stage					
	Propulsion trans.	1-stage					
	Propulsion thyristor	1-stage					
Electrical	Diesel generator	2-stage	Yes	Yes	Yes	Yes	Yes
	Diesel Exh. piping	1-stage				Yes	
	Diesel Exh. mufflers	1-stage			Yes	Yes	
	Transformers, solid state	DIM					
	Frequency converters	DIM					
	Power supply cooling	DIM					
	Cable					Yes	
Auxiliary	Fuel oil purifier	DIM			Yes		
	Sea water cooling pumps	1-stage			Yes	Yes	
HVAC	A/C supply fans	1-stage			Yes	Yes	
	A/C refrigeration plants	2-stage	Yes	Yes	Yes	Yes	Yes
	A/C chill water pumps	1-stage			Yes	Yes	
	Ventilation fans	1-stage			Yes	Yes	
	Fan coil assemblies	DIM			Yes		Note [e]
	Fan coil units	DIM			Yes		Note [e]
Domestic fresh water	Fresh water pumps	1-stage			Yes		
	Hot water Circ pumps	1-stage			Yes		

(continued)

Table 10.2 (continued)

System and associated noise control		Iso. Mount	High Imp.[a] Fnd.	Fnd.[b] Damping	Flex connection	Resilient pipe hangers	FAT[c]
Sanitary	Sewage treatment plant	1-stage			Yes	Yes	Note [d]
	System pumps	1-stage			Yes	Yes	
Refrigeration	Ship service refrigeration plants	2-stage	Yes	Yes	Yes	Yes	Note [d]
Compressed air	Air compressors	2-stage	Yes	Yes	Yes	Yes	Note [d]
Steering gear	Hydraulic plant	1-stage	Yes	Yes	Yes	Yes	Note [d]
Bow thruster		Maybe	Yes	Yes			Note [e]

[a]Imp Impedance
[b]Fnd. Foundation
[c]FAT Factory Acceptance Testing
[d]Airborne noise testing only
[e]FAT or good vendor data

Table 10.3 Degree of isolation effectiveness vs. ratio of drive frequency to mount natural frequency

Ratio of drive frequency to mount natural frequency	Degree of vibration isolation
Below 1.4	None (Amplification possible)
1.4–3	Negligible
3–6	Low, < 5 dB
6–10	Fair, <10 dB
Above 10	High, >15 dB

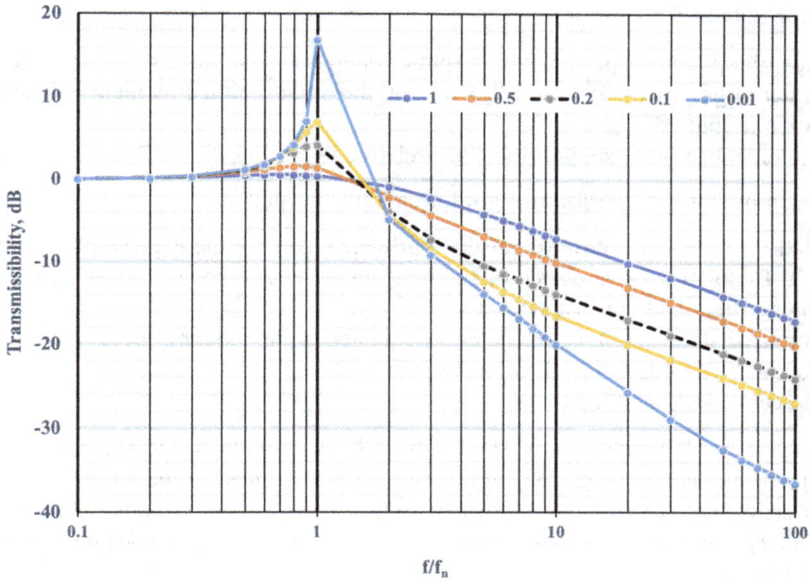

Fig. 10.3 Isolation mount transmissibility with various levels of damping

Based on Fig. 10.3, resilient mounts should be sized so that the isolation system's vertical natural frequency is less than or equal to half of the equipment's operating frequency to avoid a resonant or amplified response. The machine's operating frequency is usually its rotational rate, which, when expressed in rpm, is divided by 60 to determine its frequency in Hertz (Hz) or cycles per second.

The above equation can also be computed for the transverse and longitudinal uncoupled natural frequencies, where the total stiffness is that of the mounts in the appropriate direction. For rotational degrees of freedom, most FEA codes can be utilized, assuming that rotational moments of inertia are available for the equipment. Guesstimates of the rotational moment of inertia can always be determined empirically. Knowledgeable mount vendors can provide six-degree-of-freedom calculations.

Table 10.4 Variation in stiffness and damping v. load for a 6E2000 Navy standard mount

Load	Dynamic stiffness, k_d	Natural frequency	Loss factor η	Static stiffness, k_s
N	N/m	Hz		N/m
3114	262	5.7	0.08	344
5115	402	5.5	0.07	340
7117	559	5.5	0.06	338
8896	650	5.3	0.07	337

For the design of isolation systems, the following information from the vendor must be obtained:

- type of equipment, power, design rpm or operating rpm range if variable speed
- center of gravity (CG), total ***WET*** weight, three translational moments of inertia, as a minimum
- equipment (mount) support locations relative to the CG

For each mount type the following information is needed.

- Static and *dynamic* stiffness for three principal axes for the expected load range (see Table 10.4 for an example) and temperature environment (especially for the engine room and exhaust systems)
- Expected static and dynamic natural frequency of the mount system
- Maximum and minimum load range
- Expected creep and replacement height

An experienced mount vendor should be able to provide the loss factor versus frequency and temperature. This will vary with the hardness of the mount.

In sizing resilient mounts, the total weight on the mounts should include the machinery, any subbases (above the mounts), associated fluids, and weight of piping carried by the mounts.[9]

Ideally, the resilient mounts should be located equidistant from the isolated equipment's center-of-**gravity** (CG) in both horizontal directions (longitudinal and transverse). In many cases, mounts are located equidistant from the machinery's center-of-**geometry**. This is only acceptable if the center of gravity coincides with the center of geometry. Otherwise, mounts with different load and spring stiffnesses may have to be used on the same unit, which generally causes installation mix-ups and spare equipment problems.

It is preferable to locate the plane of the resilient mounts at the same vertical height as the center of gravity. Shipboard this is generally not possible. In this case, the vertical distance from mounts to the center of gravity should be minimized, as shown in Fig. 10.4. For design guidance, keep the isolation mounts to a reasonable number. It is preferable to have fewer rather than more mounts. Four is usually the minimum number of mounts. Avoid using unsymmetric layouts for mounts, e.g.,

[9] Usually, the piping weight up to the first flex coupling or first isolator supporting the piping.

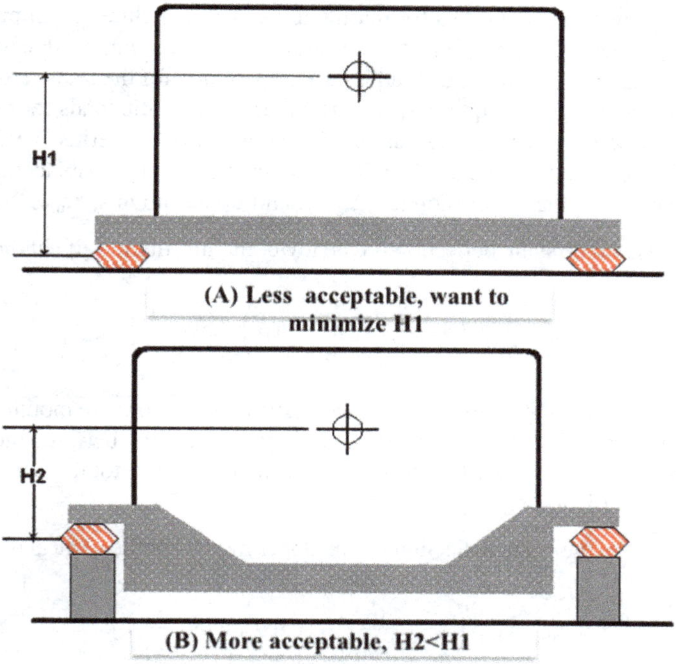

Fig. 10.4 Recommended vertical spacing for mounts

two on the port side and one on the starboard. For any odd number design, different mount stiffnesses would be needed on each side.

A minimum clearance of at least a 50 mm envelope should be maintained around the mounted equipment to prevent the unit from striking adjacent ship structures, adjacent machinery, or other objects during maximum deflection due to ship motions. A formula detailing this clearance is provided in the NAVSEA Resilient Mount Handbook n.d.

As a minimum, the vertical natural frequency should always be determined. If the translational stiffnesses[10] are available and to ensure a correct overall design, the six natural frequencies – three translational and three rotational – should be determined. None should coincide with hull, machinery rotational frequencies, or propeller blade rates. For two-stage isolation mounted systems, the twelve degrees of freedom should be computed – and these usually cover a greater frequency range than the single-degree-of-freedom system but offer significantly improved isolation above the second resonant frequency.

[10]Bear in mind that spring isolators have a very low lateral stiffness compared to the vertical stiffness. In a seaway, spring isolators tend to short out unless a resilient pad is used to limit this motion.

There are other considerations for marine units. These include the impact on the load from the torque, as appropriate for a propulsion unit, and impact due to motions in a seaway. The maximum torque load is easily computed for the transverse spacing of the mounts and engine torque v. speed. The increase in static loads for motions in a seaway may be significant. These are computed as follows (Veritec 1985):

Using the maximum allowed ship roll, θ, compute the added displacement in the vertical direction on the lower mounts, Δ_{lower}, and upper mounts, Δ_{upper}:

- The increase in the static deflection on the lower mounts in the vertical direction is:

$$\Delta_{lower} = M\left(1 + \frac{d_{v\,cg}}{d_{t,spacing}}\tan\theta\right)\cos\left(\frac{\theta}{2n}\right)k_v \tag{10.2}$$

where M is the mass of the diesel; $d_{v\,cg}$ is the vertical offset from the mount top to the diesel vertical centerline; $d_{t,\,spacing}$ is half the transverse spacing between the mounts; n is the number of mounts fitted to each side; and k_v is the total stiffness of the mounts in the vertical direction.

- The increase in the static deflection on the upper mounts in the vertical direction is:

$$\Delta_{upper} = M\left(1 - \frac{d_{v\,cg}}{d_{t,spacing}}\tan\theta\right)\cos\left(\frac{\theta}{2n}\right)k_v \tag{10.3}$$

- the rotation of the engine relative to the seating is

$$\left(\Delta_{lower} - \Delta_{upper}\right)/2d_{t,spacing} \tag{10.4}$$

One cannot go wrong placing Distributed Isolation Material (DIM) below the mounting feet of any vibrating equipment. These pads provide approximately 5 dB of isolation above 250 Hz. DIM pads are easily sized and installed, following guidance in the NAVSEA DIM Handbook, 1982. DIM pads typically carry loads of 17,500 kg/m^2 (25 psi) to 35,000 kg/m^2 (50 psi). DIM installations should also have washers made of DIM between the bolt head and equipment foot. The body of the bolt should be isolated from the ship's structure. Typically, rubber tubing is used for this isolation. Components mounted on DIM do not require flexible piping connections if adequate piping flexibility is utilized.

Foundations supporting resiliently mounted machinery and equipment should not utilize open truss type structures unless adequate, direct support for each resilient mount is provided. Foundation systems designed and qualified as open truss type structures should not be cut or altered to accommodate piping or electrical cable interferences. Using a doubler or billet plate under each isolator significantly improves the local impedance and therefore the mount performance (see Fig. 8.10).

As a final check on the suitability of an isolator design a Campbell diagram should be made like the one shown in Fig. 10.5. This plot can be used to determine if there is a resonance that needs to be avoided. In this example, the mount vertical

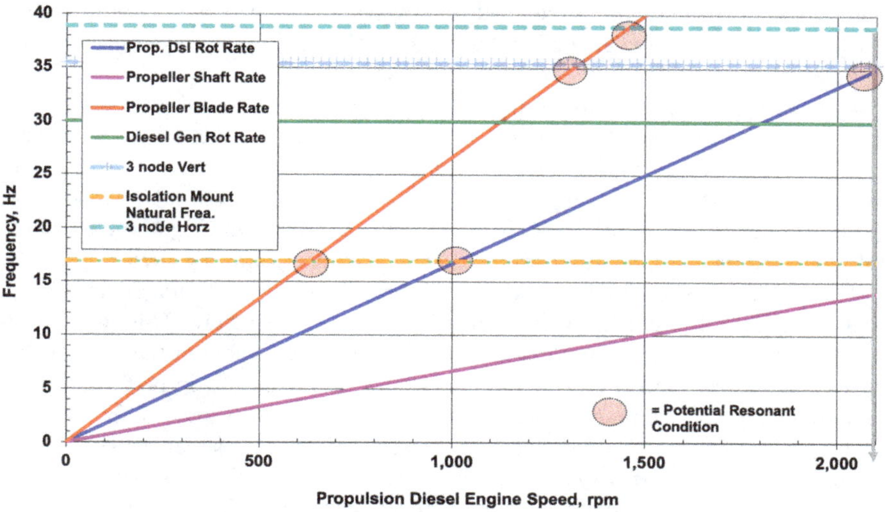

Fig. 10.5 Campbell diagram – potential resonant conditions

frequency was calculated to be 17 Hz (relatively stiff mount). This is a plot of the engine rpm from low to full speed versus the various ship resonances and excitation rates, such as shaft rate. In this type of plot, every intersection between a horizontal line (resonance) and a diagonal line due to an operating condition represents a potential resonance. In this case, the mount resonance coincides with the propeller blade rate excitation at approximately 580 engine rpm and again with the propulsion diesel shaft rate at 1000 rpm. Five other coincidences occur between (1) propeller blade rate and the diesel generator rotation rate, (2) propeller blade rate and the hull 3rd mode vertical, (3) the propeller blade rate and hull 3rd mode horizontal, (4) the propulsion diesel shaft rate and the diesel generator rotation rate, and (5) the propulsion diesel shaft rate and the hull 3rd mode (not all coincidences highlighted in figure). Given the coincidences for the diesel isolation system below half power of the diesel, the possibility of a strong resonant response is not expected. If resonances cannot be avoided by design changes and the response is expected to be strong, a barred speed range might be the only solution.

Shafting torsional resonances (Zoul and Kovac 2018) should also be included in the Campbell diagram. In this way, one can identify or exclude any strong[11] excitation of the shafting system.

There are several vendors of thrust capable mounts. Some all-metallic spring mounts are used on smaller vessels where there is no thrust bearing. One set of

[11] A "strong" excitation is one with a high 'modal participation factor' and a high modal mass.

springs takes vertical loads, while another set takes longitudinal loads due to thrust. Elastomeric mounts have also been designed to take thrust while isolating vertical vibrations.

Another approach typically used on naval vessels with a low signature requirement is to place the propulsion and/or auxiliary machinery on a cradle or U-shaped support, which is then attached to the deckhead above rather than the foundation below. The cradle is either hard mounted or isolation mounted to the overhead for additional attenuation (Fusil 2004). This provides a longer path length to the wetted hull plate and an additional intersection between the deck and side shell.

Theoretical transmission losses addressing only the stiffness, *k*, of a mount would predict significantly increasing transmission loss with frequency. This theoretical effectiveness, even once damping is included, is not achievable in the real world. Two other factors are that spring isolators have a significant dip in performance at their 'standing wave' frequency and rubber isolators have 'wave guide' frequencies or 'internal resonances.' These standing waves or internal resonances (IRs) may significantly limit the total system performance (Dylejko et al. 2014; Heckl 1976). These resonances can be expected above 100 Hz. As shown in Fig. 10.6 (Dylejko

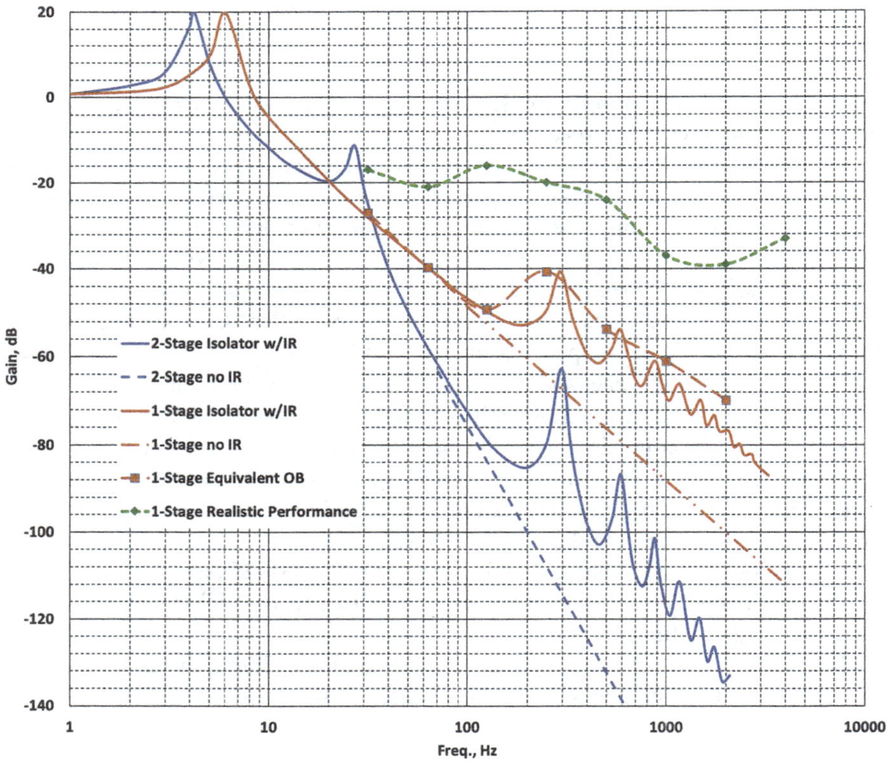

Fig. 10.6 Impact of internal resonances on single- and two-stage isolation mounting systems. IR internal resonance

et al. 2014), these internal resonances for a single-stage system can reduce the transmission loss by as much as 25 dB at the first resonance, ~290 Hz. They also reduce the performance across the frequency bands above this first internal resonance. For a two-stage system, the first internal resonance theoretically reduces the transmission loss by over 50 dB. Above this internal resonance of approximately 300 Hz, the broadband performance is reduced by over 30 dB. Fortunately, this degradation in loss at the mid- to high-frequencies is not serious enough to cause a significant structureborne noise problem unless this mount internal resonance matches a forcing frequency. If the input is broadband, the expected gain in terms of octave band performance is shown in the curve labeled '1-Stage Equivalent OB.' The curves for 1- and 2-Stage, with and without IR, are 'theoretical' curves that are not achievable in realistic installations. Typical gains for a 1-stage system installed in a ship are shown by the curve '1-Stage Realistic Performance.'

10.1.4.2 Compound (Double-Stage) Isolation Systems

Often, diesel generator sets, compressors, and refer plants need to be compound (two-stage) mounted with an intermediate raft. This raft should be designed to behave as a rigid mass up to at least 100 Hz; otherwise, structural resonances will degrade the overall isolation effectiveness. For optimum performance, the raft should have a weight equal to that of the equipment or machinery. As a minimum, the weight of the intermediate raft under each piece of machinery should have a total weight equal to or greater than 0.4 times the weight of the machinery, including the subbase to which the equipment is attached. An example of two-stage isolation is provided in Fig. 10–7.

The two vertical natural frequencies of a two-stage system are computed as follows (Beranek and Ver 1992):

$$\left(f_n/f'_1\right)^2 = C \pm \sqrt{C^2 - R^2} \tag{10.5}$$

where $2C = R^2 + 1 + k_1/k_2$; $R = f'_2/f'_1$; $f'_1 = 1/\left(2\pi\sqrt{M_1(1/k_1 + 1/k_2)}\right)$; and

$$f'_2 = (1/2\pi)\sqrt{(k_1 + k_2)/M_2}$$

where M_1 is the upper mass; k_1 is the total stiffness of the upper mounts; M_2 is the lower mass (intermediate raft); k_2 is the total stiffness of the lower mounts; f'_1 is the natural frequency of M_1 supported on k_1 and k_2 in series without consideration of M_2; and f'_2 is the natural frequency of M_2 between the two springs when M_1 is fixed in space.

The higher natural frequency, f_n, obtained by using the plus sign in Eq. 10.5 is always greater than the larger of f'_1 and f'_2. An example of these two natural frequencies is shown in Fig. 10.6 for the T_f (2-Stage) curve. The lowest frequency is approximately 6 Hz, and the second frequency is around 28 Hz.

The first- and second-stage resilient mounts should be selected such that all mounts are at less than their maximum rated load limits. The 12-degree-of-freedom (vertical and translational) resonant frequencies of the system should be equal to or less than 70% of the machinery's rotation rate. This is the diesel rotation rate for a 2-stroke and half rotation rate for a 4-stroke. Low-speed diesels, 900 rpm or less, are difficult to correctly isolate due to the likely coincidence between the rotation rate and the multiple degrees of freedom for any isolation system. For the same reason, high-speed diesels are less complicated to isolation mount. The locations of the mounts on the intermediate raft shall be such that the resonant pitch and roll frequencies of the raft do not coincide with any of the exciting machinery tones.

The intermediate raft should block the transmission of airborne noise from the machinery from being incident on the tanktop below. Otherwise, the isolation system will be flanked or shorted by the airborne excitation of the tanktop. As a proof of concept, using a plywood barrier below the raft showed a 5 to 10 dB reduction in vibration on the tank top above 500 Hz (Fischer 2006). As expected, the barrier was ineffective below this frequency. Additional guidance on the design of two-stage isolation systems is provided by Zhiqiang et al. 2016, and Muravyov and Hutton 1997 (Fig. 10.7).

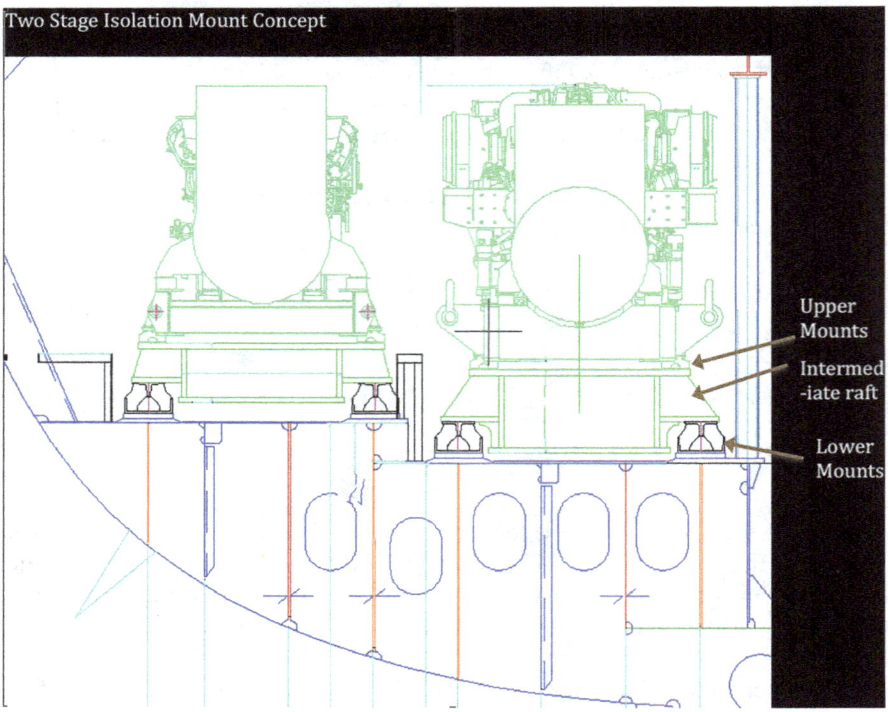

Fig. 10.7 Two-stage isolation system cross-section

10.1.4.3 Mount Installation Guidance

Knowledgeable mount vendors provide mount installation and maintenance guidance. Lacking this, the following guidance should be followed. The actual height of each isolator should be measured before and after[12] the mounts are completely loaded. This determines the mount deflection and whether it is within the vendor's requirements. Equal deflection indicates equal loading. Deviations of greater than 15% from equal loading should be investigated. The mount should not be cocked by more than 5 degrees. The mount height should be measured again after 10 days to determine the amount of creep. Mounts that are very soft may have high creep amounts, which could adversely affect shaft alignment on propulsion systems with close coupled gearboxes. Shims may be used under mounts to correct for small deviations in mount height or foundation levelness. Some mount types have internal height adjustment mechanisms.

The resilient rubber elements should never be painted. A note to this effect should be put on appropriate drawings. Welding or flame cutting near the mounts should be finished prior to installation of any mount. Again, a note to this effect should be put on appropriate drawings.

For maintenance purposes, where possible, locate resilient mounts or pipe hangers such that they can be inspected regularly without having to remove structural, pipe, or duct work. Provision should be made for periodic removal and replacement of resilient mounts or pipe hangers. Vendors should provide replacement criteria and expected service life. This requirement should be reflected in the headroom space and lateral clearances.

10.1.4.4 Material Requirements, Isolators

There are many suppliers of vibration isolation mounts. Mounts should be designed to have captive elements (i.e., so equipment does not come loose if mount fails for any reason). Choose from vendors who make mounts that are designed for marine applications. US Navy-type rubber mounts should be considered along with commercial manufacturers of spring or neoprene/rubber marine mounts. Christie Grey, Barry Controls, Lord, Rubber Design, Shock-Tech, and Metalastik are a few manufacturers of marine mounts. The isolators on propulsion systems should have inherent limit stops. If not, external, adjustable-limit stops will need to be designed into the system. Limit stops should not have metal-to-metal contact. Rather, a rubber pad, such as DIM, should gradually engage given the motion of the supported system. In the area of active research, magnetic suspension active vibration isolation systems are being developed for use on shipboard two-stage isolation systems (Yang et al. 2017).

[12] The after measurement should occur after at least 24 hours after the mount is fully loaded or within manufacturer recommendations.

For applications in oily environments, such as the Engine Room, the resilient element should be made of neoprene or nitrile rubber. Natural rubber may be used if protective housing is provided. Mounts used on hot equipment, such as exhaust systems, should be protected from the direct radiation of heat unless the mount is rated to take the expected heat. Thermal break material (see Fig. 10.10) should be installed between the hot side and the mount to limit the mount temperature to the manufacturer's tolerance. Mounts used in wet areas, where there is a possibility for more than occasional immersion of the mount in sea or bilge water, should be installed using noncorrosive hardware. Fasteners and ratings for fasteners should follow the directions of the mount manufacturer.

All drawings for isolation systems should note the load direction and orientation for the mounts; operational load and expected deflection; static and dynamic stiffness; rated load; and replacement deflection. During inspections, the orientation of the mount should be checked as mounts, especially small mounts such as pipe hangers, have too often been installed upside down.

Resilient pipe hangers are necessary when flexible hose connections are utilized. As shown in Figs. 10.8, 10.9, and 10.10, resilient pipe hangers should be attached to heavy, rigid parts of the ship structure for the isolation treatment to be most effective. Pipe hangers should never attach to plating sections between stiffeners. For the best results, resilient pipe hangers should be used for the first rigid connection. Do not attach pipe hangers to a machinery subbase or piping directly to rigid ship structures without a rubber grommet isolator or pipe hanger. Downstream of the pipe hangers, the piping should be held in rubber lined grommet supports attached to structural frames, especially for hydraulic systems. The pipe system/resilient support system should be checked to determine if there is any coincidence with the system excitation

Fig. 10.8 Double arch flex hose with control rods (left side of photo)

Fig. 10.9 Dog-leg configuration with two sections of double arch flex hose and control rods

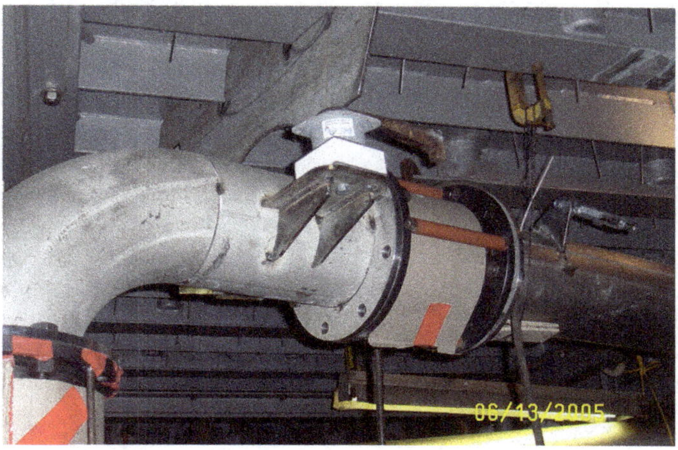

Fig. 10.10 Exhaust pipe with thermal break below the isolator

frequencies. Resilient pipe hangers should have a natural frequency less than or equal to that of the mounts supporting the machinery to which the associated piping is connected.

Regulatory agencies, such as DNV GL, have recommended practices for structural analysis of piping systems.

10.1.4.5 Flex Connections to Mounted Equipment

Resiliently mounted equipment should have flexible hoses, flexible exhaust, and/or flexible cable connections. Synthetic reinforced flexible hoses should be used wherever system requirements and regulatory bodies permit. The intent is to use as compliant a hose as possible based on system constraints. To obtain freedom of motion in two planes, the preferred approach is to use a flexible hose in a dog-leg or right-angle configuration. However, the use of a double arch type flexible hose can be used for hoses over 65 mm (2.5″) diameter and system pressures less than 861 kPa (125 psi).

For dogleg configurations, the free length (that length of hose unconstrained by clamps, fittings, nipples, spigots, etc.) should be at least equal to 18.0 cm (7″) plus 4 hose diameters for each leg. While dog leg configurations are preferred, single right angle hose configurations are permissible provided they are at least equal in length to the dog leg configuration and the hose manufacturer's minimum bend ratio is not exceeded. Renewable end fittings should be used whenever possible. These fittings should be compatible with and similar to the material requirements of the associated piping system.

The preferred dog leg installation is with the hoses in the vertical plane with the elbow downward. Depending on the specific arrangement and hose design, configurations with the elbow up may require resilient support at the elbow (see Fig. 10.8). Horizontal configurations should have support at the elbow.

All hoses should be at least one size larger than the piping to which they are connected to match the inside diameters as closely as possible. Where needed by regulatory bodies, flex hoses should have fire protection sleaving. Loose fitting and flexible add-on non-metallic sleaving is the preferred methods. When and where solid sleaving is needed, it should be arranged in such a manner as to not impede the free motion of the equipment under ship slamming conditions and should not provide the opportunity for a shorting path between the mounted equipment and ship structure.

Tie rods on flexible arch hoses should only be used as limit stops and should be supplied with rubber grommets to prevent metal-to-metal contact. Electrical connections should be made with generous service loops.

10.1.4.6 Foundation and Subbase Design Guidance

The foundations used for vibration-isolated equipment have a different criterion from hard-mounted equipment. As mentioned in Sect. 10.1.4, the foundation shall be able to accommodate mounts located equidistant from the center-of-gravity, not equidistant from the center-of-geometry.

Machinery foundations should be designed based on two parameters. First, they should not have a structural resonance that matches the rotation rate or primary forcing frequency of the machinery, e.g., pump passage rate. Second, they should be

Fig. 10.11 Notched raft to provide additional height clearance

'dynamically' stiff – which in general terms can be equated to 1.5 to 2× the general practice for plating thickness for a foundation – including both the top plate and vertical floors. Using gussets local to the attachment point will also increase the local stiffness or impedance. For critical designs, a finite element analysis approach should be used to check both of these design parameters.

Local stiffeners should be added to points where mounts attach to the foundation. This should generally consist of gussets on either side of the isolation mount. The goal is to have a local stiffness at least one hundred (100) times the vertical stiffness of the resilient mount. The minimum acceptable ratio of local to mount stiffness is ten (10).

One way to overcome the height restriction for isolation systems is to notch the raft above the isolator. This is illustrated in Fig. 10.11. This could shave 50 to 75 mm (2″ to 3″) off the overall height and provide sufficient clearance between the raft and foundation.

Built-up foundations should be anchored to hull framing. Noisy machinery should be attached to stiffened bulkheads or hull frame-stringer intersections (i.e., deep frames). Foundations should be designed to stress structural members in tension or compression; bending and torsional stress are undesirable. For the same force input, F, torques (F × moment arm) input more power into the structure. The foundation design should take into account the machinery excursions of various appendages of the mounted equipment (such as the oil pan, cantilevered motors, and control panel).

If a resiliently mounted machinery item consists of two or more components connected by shafting, a common subbase under both units should be considered as a means to provide proper alignment. A subbase may also be used when it is necessary to secure resilient mounts at some position other than that of the mounting feet provided with a piece of machinery. Subbases should be rigid. Again, for critical designs, an FEA will confirm whether the entire isolation system, subbase and foundation are appropriately designed.

Avoid cantilevered foundation off the ship's vertical framing, as these induce high bending moments into the structure. As noted above, moment coupling will contribute significantly to the structureborne noise. Cantilever foundations, if used, should be attached to flanges of frames and not plating. Foundations should not be attached to surfaces that are common with noise-sensitive spaces, e.g., accommodations, offices, and passageways in those areas.

10.1.4.7 Resiliently Supported Diesel Exhaust Systems

Resiliently mounted exhaust systems should be connected to their prime movers and between various parts of the system with flexible metallic bellows connections. Metallic bellows should be of small corrugations, extremely flexible type such as Lo-Corr by Flexonics. They should be of sufficient length to allow free motion of all resiliently mounted elements under all ship conditions. To restrain motion and allow thermal loads, a V configuration is preferred for the resilient hangers past the first flexible connection to the diesel. Here, the pipe is at the base of V, and each leg contains an isolation mount.

Particular care should be exercised in the location of the bellows and the resilient supports of the system. The intent is to prevent imposing undue loads and torques on the bellows pieces due to ship motion and thermal loads. To minimize side loads, the first flexible joint should be located at the exhaust flange of the machinery.

Exhaust system ducting or piping should be mounted resiliently to rigid ship structures. This includes deep frames of stiffened bulkheads, not to plates between stiffeners. Care should be taken to ensure that the exhaust system does not resonate and that the proper hot and cold loads and motions due to sea state are placed on the mounts. Insulation in the way of bellows pieces should be prevented from filing or jamming corrugations.

Existing CAE software[13] that models the gravitational, ship motions, and thermal loads; the loads on support locations; and accounts for the resilient support characteristics should always be used to prevent system resonances or twisting of the flexible connections, resulting in failures.

[13] Caepipe, Ceasar II, and Autopipe.

10.1.4.8 Electrical System Attachments

As with piping and ductwork, wiring to resiliently mounted equipment should be arranged in such a manner as to not restrain the equipment under any ship's motion. Wiring should run in a 90-degree or Z configuration with a minimum free length of 30 cm (12″) for outside diameters of up to 20 mm (0.75″), 60 cm (24″) for diameters from 20 mm to 40 mm (1.5″) and 1 m (3.3′) for diameters greater than 40 mm (1.5″). At least 75 mm (3 in) of slack should be provided to allow for motion of the equipment under ship slamming.

10.1.4.9 Shafting Systems

For isolation mounted machinery, selection of the proper flexible shaft coupling should be determined by the load carried by the shaft, by the angular and parallel misalignment that is anticipated, and by the variation in the axial separation between the two coupled shaft ends. Many standards exist for reasonable shaft balance, alignment, and runout. These should be invoked as appropriate. Room should be allocated for any potential flexible coupling early in the design, as large diameter flexible coupling may be needed for softly mounted systems.

The American Bureau of Shipping (ABS) 2004 has guidance notes on propulsion shaft alignment. Most regulatory agencies, e.g., Class NK, have similar documents. Kozousek and Davies 2000, provides additional input on the design and alignment of propulsion systems.

Cardan shafts historically have been a source of high vibration at low frequency, typically with a 2x rotation rate (Cowper and Kolomojcev 2003). These shafts are often used for direct diesel-driven Z-drive systems. Alignment criteria such as offsets, maximum and minimum load on bearings, and maximum shaft bending stress should minimize potential problems. One advantage of Z-drives is that the propulsion equipment can be resiliently mounted to the hull without thrust capable mounts and seals.

If there is a requirement to take thrust on the diesel mounting feet, the selection of a proper flex coupling becomes critical. Numerous types of coupling exist, and vendors should have inputs to the selection of a proper coupling. The stiffness, both static and dynamic, of the coupling – axial and athwartships – should be included in the detailed analysis of the isolation system. As with rubber isolation mounts, couplings with rubber elements can have standing waves. Measurements by Van Bakel 1986, show that the axial standing wave for several different coupling vendor couplings varies from 400 Hz to 1250 Hz.

10.1.4.10 Isolation Mounted Superstructures

It is especially important that the isolators used below superstructures or accommodation modules be:

- Capable of carrying the rated static loads and maximum expected dynamic loads (typically approximately 5 g)
- Mounts should be uniformly loaded, generally to within 20% of rated load.
- Mounts should have a captive, fail safe design
- It is recommended that the vertical natural frequency be below 4 Hz
- Snubbers should be built-in and provide progressive and positive snubbing in all three translational directions.
- All metallic mounts are preferred to rubber to minimize creep and meet ambient temperature and fire requirements.
- Mounts should be pre-compressed from the vendor to facilitate installation and removal.

All connections across the deck and superstructure bottom should be flexible. These should be able to accommodate the expected movements of the superstructure.

10.1.4.11 Stanchion Location

Do not land stanchions on foundation top plate near high vibration source machinery. If needed in this location, isolation mount the machinery. If the stanchion path is problematic, it may be possible to isolation mount the top of the stanchion with a Distributed Isolation Material (DIM) (Fischer 1996). This control needs careful review by regulatory agencies.

10.1.5 Damping Treatment Applications

Some general guidelines for damping tiles are as follows[14]:

- Areas to be treated are approximately 70% of the center of the plating area between frames or stiffeners. Stiffeners or angle flanged plates do not require damping.
- Do not install damping tiles on curved surfaces having a radius of 46 cm (18″) or less.
- Do not install damping closer than 2.5 cm (1″) from the stiffeners.
- Do not install damping to locations where tile would be less than 20 cm (8″) wide or 413 square cm (64 sq. in.) of area.

[14] Spray-on damping treatment—be sure to follow vendor prep and application directions.

- All areas of coverage should have a single damping tile unless the area of coverage is greater than 30 × 30 cm (12"x12″) (e.g., more than one piece of scrap damping tile cannot be used to treat any area).
- Do not install damping tiles over any weld beads
- Damping tiles are subject to damage by excessive heat. Installed tiles should be protected from minor welding/burning splatter in the area. No major welding/burning should be allowed in these areas. Any damaged tiles should be replaced.
- Tiles on vertical, sloping and overhead tiles should be braced to hold in place while adhesive sets.
- Damping tiles are not typically applied to the inside of fuel and drinking water tanks, unless specifically approved by the material vendor and approved by regulatory authorities.

10.1.6 Piping System Design

NASA has a document covering gas flow design guidance (Nelson 2005).[15] This includes designs to reduce gas flow-induced noise in pipes.

Fluid systems should be designed with the lowest flow velocities consistent with other system requirements. A maximum velocity of 1.8 m/s (6 ft./s) should be used as the design parameter for all water systems, with 4 m/s (13 ft./s) being the maximum flow velocity allowable. The radius of pipe bends should be a minimum of five (5) diameters for low noise generation. Abrupt changes in flow direction and pressure should be avoided. Ill-fitting gaskets and other flow discontinuities should be avoided.

Orifice plates in a piping system should be avoided wherever possible. Do not use single, high pressure drop orifice plates for system balancing. Use multiple plates in series at mismatched distances. Follow orifice plates with straight pipe for at least seven (7) diameters. An alternate to single hole orifices is multiple hole orifice plates. Orifices in noise critical systems should not produce pressure drops greater than 2.8 kg/cm^2 (40 psi).

Globe and angle valves are known to cause noise problems. These should only be used where it is impractical to use any other system and/or on systems that do not affect the acoustic environment of the vessel. Throttling valves are also known to cause severe noise problems under all conditions where local cavitation can occur. If throttling is necessary in a system, the quietest method is the use of multiple pressure drops.

In terms of the pipe, the following design guidance will minimize pipe-induced vibration into the ship's structure (Wachel and Bates 1976).

---

[15]NASA also has an Excel worksheet to compute "Gas-flow-noise-estimation-2005."

Fig. 10.12 Examples of mounts used to support piping

- The natural frequency of pipe spans should not match excitation frequencies, e.g., pump rotation or pulsation.
- With reciprocating compressor and pump piping, the natural frequency of pipe spans should be twice the highest pulsation component
- Do not use unreinforced branch connections
- Support the pipe on each side of the bend and connect to a stiff ship structure. The support should be resilient – either clamp with rubber liners (50 mm (2") thick with 40–45 Shore A Durometer) or isolation mounts; hanger should be as short as possible (see Figs. 10.12 and 10.13)
- Eliminate as many bends as possible
- Where piping isolation is used, the hangers should be in a V-configuration on horizontal runs.

A water-proof flexible piping penetration of a bulkhead is shown in Fig. 10.14. A fireproof stuffing would be needed for a fire rated bulkhead. Penetrations through Class A divisions by cables, pipes, ducts and trunks need USCG Approval 164.128 or Council Directive 96/98/EC A.1/3.26.

10.1.7 Hull Form and Appendages

The following section describes generalized hydrodynamic design guidance with respect to a quiet vessel. Computational Fluid Dynamics (CFD) should be used to optimize the recommended values given herein.

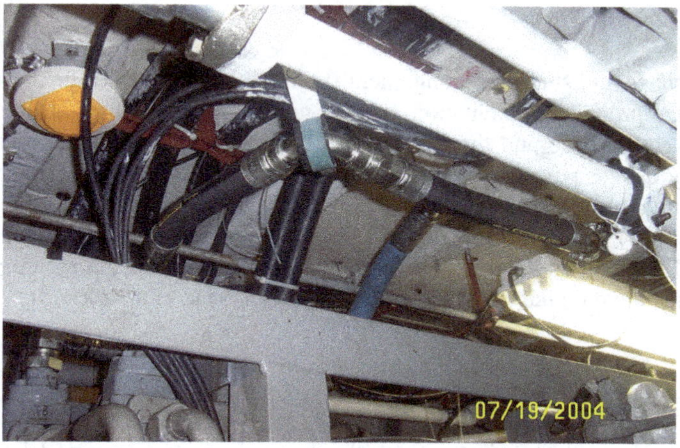

Fig. 10.13 Flex hose isolation at elbow

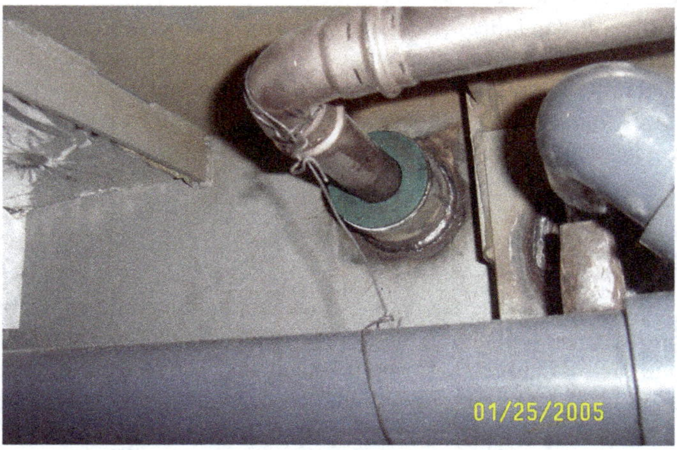

Fig. 10.14 Water-proof bulkhead penetration

10.1.7.1 Bow Form

The half entrance angle should not exceed 7.5° to prevent extensive bow wave breaking and attendant bubble sweep-down. The sonar dome shape and location have a large effect on the bow wave as well. If the sonar dome is to be bow-mounted, then its location and shape and the shape of the bow should be determined through experimental and/or analytical procedures.

10.1.7.2 Stern Form

The stern shape and shafting arrangement should not compromise the quality of the inflow to the propeller if cavitation is to be minimized. The angle at the propeller between the shaft line and the overhead buttock lines should not exceed 10° for preliminary design purposes to ensure that the tangential and radial wake components are suitably small. A detailed examination of the cavitation performance of the propeller should be performed using the wake field of a similar ship, tunnel tests, or detailed CFD analyses. Vortex generators may reduce propeller pressure pulses by reducing the wake deficit at the top of the propeller (Han et al. 2004).

10.1.7.3 Bilge Keel and Skeg

The bilge keel should be aligned with the mean flow, as determined from model experiments or CFD studies. The leading and trailing ends of the bilge keel should be gradually and smoothly tapered. The tip of the bilge keel should have a radius not less than 6% of its outward projection.

The skeg has a strong influence on the hydrodynamic coefficients of the ship. The skeg should be designed to provide only as much directional stability as is needed. Excessive stability requires extra loading on the rudder and a greater likelihood of cavitation of the rudder. The skeg should be streamlined in form and incorporate a trailing edge detail similar to that of the rudder (see below).

10.1.7.4 Rudder, Shafts, and Struts

The actual angle of attack in a steady turn should not produce a tip vortex or leading-edge cavitation. As preliminary guidance, the rudder should have a thickness, t, to chord, c, ratio (t/c) of between 0.15 and 0.20 and a section shape from the NACA[16] 66 or 65 family. The trailing edge should incorporate a 25° chisel to suppress coherent vortex shedding (singing).

The shafts and struts should be designed to provide as little disturbance to the propeller inflow as possible. Shaft bossings are not recommended. The shaft angle should be as flat as feasible. Struts should have section shape, thickness, and trailing edge detail as recommended above for the rudder. In addition, the struts should be aligned to the local flow angle as determined from model experiments or CFD analysis. The struts should be raked to increase propeller tip-strut clearance and suppress flow-induced vibrations.

[16]NACA – National Advisory Committee for Aeronautics.

10.1.8 Propeller Design Guidance

A large unknown in many designs is the hydro-acoustic contribution from the propeller or propulsors. This factor is not as important if the noise-sensitive spaces are located remote from the plane of the propeller and/or underwater radiated noise is not a concern. If the cavitation inception speed and cavitating area can be accurately modeled or measured during tunnel tests that are performed early in the design stage, then this contribution can be more accurately included in the acoustic model.

Propeller cavitation can be 'delayed' and minimized by appropriate CFD methods, matching the propeller geometry to the hull's wake. Tunnel tests can accurately predict the expected wake from a hull. A harmonic analysis can be performed on this wake and used to determine an appropriate blade geometry—mainly skew (Blake 1984). The operation of a CPP propeller can be optimized during design or tunnel tests (Bertetta et al. 2012) and during sea trials. The objective is to minimize noise, whether radiated underwater or resulting onboard noise, at different pitches, rpm, and loads (Van Beek 2004). For example, at a constant rpm and for a variation in pitch from 20% to 82%, the measured noise level varied by over 30 dB (Zhu et al. 2022). The overall signature was minimized at 51% pitch; however, there were differences in which the frequency range dominated due to 'the different kinds of cavitation taking place' (Traverso et al. 2017).

To maintain higher values of the cavitation number – in order to avoid or minimize cavitation – use large diameter propellers and turn them slowly. Twin screws can do this much more readily than single screws (Brown 2007). Aziopods have hydrodynamic benefits – excellent wake fields that lead to improved efficiency, less cavitation, reduced propeller-induced vibration, and reduced noise levels (except at drive system tones). A pulling propeller is always better than a pushing propeller regarding cavitation and pressure pulses. Support struts disturb the flow into the propeller. They increase the danger of cavitation inception, especially while steering. Cycloidal propulsion can be quiet, as shown by their use on minesweepers (Jurgens et al. 2004).

"The simple way to increase the efficiency is reduction of blade area, increase load near the tip and use a larger diameter with low RPM. However, the vibration and noise will be the first obstacle when a designer tries to increase the efficiency of the propeller in these ways (Bodger et al. 2014)." Thus, a balance between efficiency and increased cavitation inception speed and smaller cavitating area will necessitate the use of explicit hydro-acoustic design methods. A proficient and experienced propeller design expert may be needed if a very quiet propeller is needed.

One way to affect the inflow on a twin screw vessel is to check whether the inflow is better for inboard or outboard turning propellers. Table 10.5 shows the reduction in vibration on the rudder post for changing to inboard turning vice outboard turning screws.

For waterjets, it has been found that impeller inflow can be markedly improved by the simple expedient lengthening of straight ducting before the impeller to form a settling section. This adds significantly to the effective weight and length of the

Table 10.5 Vibration reduction changing from outboard to inboard turning propellers

Octave Band Freq., Hz	31.5	63	125	250	500	1000	2000	4000	8000	OA
Reduction, dB	1	3	2	3	<1	3	6	2	9	4

propulsor system. However, this is usually prohibited in high-performance craft. As with propellers, waterjet impellers can benefit from skew – either forward (Lanni et al. 2000) or backward.

Some vendors will supply air injection systems with their thruster. These 'Prairie' type masker systems can reduce the radiated noise level of the system. If designed correctly, this air would be pulled into the thruster and mixed with the cavitation, resulting in a gaseous cavitation-ventilation mixture. This mixture would cushion the bubble collapse and thereby reduce the radiated nose and induced vibrations but may affect the thrust of the system (Fischer 2000).

Another new technology is PressurePores™ (Aktas et al. 2020; Knight n.d.; Altar et al. 2021). Small, strategically bored holes in the propeller blades were shown to reduce cavitation noise by up to 5 to 21 dB without affecting efficiency.

In terms of the optimization of thrusters, the following ranking is provided:

Tunnel – loudest (quieter with RIM drive)
Drop down – medium
Omni thruster – quietest

Omnithruster has achieved a quiet design for use on mine-sweepers. This system incorporated a patented Omniphase noise masking system (Maritime Reporter and Engineering News, Sept 2000).

Other design guidance is as follows:

- A high manufacturing accuracy of propeller – Class S or 1 versus 11 or 111 is preferred (Kuiken 2008).
- Shaft inclination – smaller the better (Aktas et al. 2014)

A rule-of-thumb for the tip clearance is nominally 20% of the propeller diameter. Other factors influencing the hull pressure pulses are the propeller diameter, rotation rate, depth to shaft line, and ship speed. In terms of the hull response to these pressure pulses, the hull area and hull impedance control the vibration response. The smaller the area excited and the thicker the hull plate/moment of inertia of the frames, the lower the response. Thus, if the clearance is less than 20%, trade-offs can be made on the hull side. The parameters controlling the pressure impulse at blade frequency, Δp_c (Pa), for a cavitating propeller are:

$$\Delta p_c \propto \left((ND)^2 V\right)/\sqrt{h_a}\right)\left(1/(d/R)\right)^{K_c} \quad (10.6)$$

where d is the distance from $r/R = 0.9$ to the hull plating, R is the propeller radius, and $K_c = 1.7 - 0.7\,(d/R)$. To convert to a percentage clearance, let $d = C \times 2R + 0.1 \times R$, where C is the percent tip clearance. The parameters controlling the pressure impulses for a non-cavitation propeller are, Δp_O (Pa):

Table 10.6 Change in hull pressure vs. tip clearance

Tip Clearance, % Diameter	Increase in pressure, dB
5	14
10	8
15	4
20	0
25	−3

$$\Delta p_0 \propto (((ND)^2)/B^{1.5})(1/(d/R))^{K_0} \tag{10.7}$$

where $K_0 = 1.8 - 0.4 \times (d/R)$.

As shown in Table 10.6, going from a 20% tip clearance to a 15% tip clearance increases the hull pressure by 4 dB. As noted above, this can be compensated for by increases to the hull plate thickness, improvements to the wake, and/or good hydro-acoustic design of the propeller.

10.1.9 HVAC System Design

This section provides guidance for the quiet design of heating, ventilation and air conditioning (HVAC) systems. HVAC noise is generally of concern for compartments with noise limits of 60 dB(A) and lower. Particular attention should be directed toward the design and installation of an HVAC system with minimal airborne and structureborne sound transmission. With respect to noise criteria, noise from HVAC systems should be designed to be at least 5 dB below the compartment's specified noise limits. Noise from HVAC systems located in machinery spaces is generally not a concern because the ambient machinery noise levels are higher than those typically created by HVAC equipment. HVAC-induced noise in machinery spaces may be more of a concern during maintenance periods with operating machinery secured.

Noise generation in HVAC systems occurs from four sources: fans/blowers, flow noise generated at branches, flow noise generated at turns, and flow noise generated at diffusers. Flow noise generated by perforated metal duct linings is typically ignored except for very high flow speeds. The following information will assist in the design of a quiet system (also see latest ASHRAE[17] Applications Handbook, Sound and Vibration Control):

1. Air flow velocities should be kept as low as possible. A target for quiet systems should be velocities no higher than 5.1 m/s (1,000 ft/min.), and never exceed 12.7 m/s (2,500 ft/min.). The desired terminal velocity limit is 2.0 m/sec (394 ft/min.) in noise critical spaces. Higher velocities require extensive

[17] ASHRAE – American Society of Heating, Refrigerating and Air-Conditioning Engineers

Table 10.7 Maximum airflow to meet various noise limits

Duct type	Compartment noise limit, dB(a)	Max. airflow in duct, m/s (fpm)
SQUARE TURN	75	12.2 (2400)
	65	8.1 (1600)
	55	4.6 (900)
RADIUS SQUARE TURN w/SHORT VANES	75	14.7 (2900)
	65	10.2 (2000)
	55	5.1 (1000)
SQUARE TURN, LONG VANES	75	16.3 (3200)
	65	12.7 (2500)
	55	6.1 (1200)
RADIUS TURN w/LONG VANES	75	16.8 (3300)
	65	13.7 (2700)
	55	6.1 (1200)

treatment, particularly near each diffuser terminal. Table 10.7 shows the recommended maximum airflow velocity for various installations.

2. The fan inlet air flow should be straight for at least three duct diameters. Turns closer than three diameters cause air flow turbulence that increases noise.
3. The fan discharge should be straight for a distance of at least three major duct widths to avoid turbulence. Noise will significantly increase due to turbulence from elbows too close to the discharge (or inlet).
4. Five to ten diameters of straight duct are needed for turbulence to die out and flow to equalize. If the air flow does not become smooth before the next fitting or terminal device, the flow noise generated in the next element will increase. For high-velocity systems, it is critical that offsets or transitions near the fan be avoided.
5. Design the HVAC system so that fans operate near their peak efficiency. Fans that do not operate near peak efficiency will have higher noise levels than fans operating at peak efficiency. Often, excessive noise found onboard is due to fans operating off peak efficiency. The noise difference between operating at peak efficiency and off design has been found to be as high as 15 dB.
6. All ventilation fans should be resiliently mounted, and all connections to the fan, including ducting and electrical cables, should be flexible and at least as effective as the vibration isolators. Similarly, all fan coil assemblies and units should be isolation mounted.
7. If possible, fans should not be attached to bulkheads, decks and overheads that form the boundaries of noise-sensitive spaces.
8. Square turns, mitered elbows and zero radius elbows should generally be avoided. Transforming sections should be symmetrical. Avoid using abrupt changes in cross-section or duct turns. Noise increases when expansion in a transition section exceeds a 7° angle from the straight ahead.
9. Where possible, use circular metallic ducts instead of rectangular ducts. Circular ducts are more effective in preventing "breakout noise" from the duct walls than

Table 10.8 Required straight length of duct

Turn Size (Degrees)	Diameters of straight duct
30	2
45	3
60	4
90	5

rectangular ducts. In addition, circular ducts attenuate noise moving down the duct better than rectangular ducts. Finally, low-frequency rumbles caused by turbulence are less likely to occur in circular ducts than in square ducts because circular ducts are usually more rigid.

10. To avoid cross-talk between rooms served by the same duct system, provide duct layout offsets and 90° lined turns.
11. The use of vane turns is advised when the turn is within five duct diameters of a fan. If the bend is more than five diameters from the fan, regular elbows with ample radii should be used, as they cause a lower pressure drop.
12. The area of the duct that engages a fan should be at least equal to, if not greater than, the active area of the fan. Additionally, fan intakes should be kept symmetrical with respect to the fan.
13. Acoustically treated plena should be used to reduce noise created by the fan or air flow, especially when outlet ports are near fans, branches or turns. The plena should be designed so that the inlet and outlet ports do not line up. Additionally, the plenum should be made as large as possible and acoustically lined.
14. The inside of the ducts should be as smooth as possible to avoid turbulent flow. Seams should be faired, and protruding objects should be avoided. Improperly fitted gaskets and flex couplings should be avoided as well. The leading edges of dampers, splitters and deflectors should be rounded or folded back. When using vaned or un-vaned elbows, Table 10.8 shows the values for the straight length of the duct preceding and following the fitting.
15. Splitters should be located as far from the terminal as possible. The flow rate over the splitter should be as low as possible. The noise generated by the splitter varies as the sixth power of the flow rate. Thus, a doubling of speed corresponds to an 18 dB increase in noise. Orifices are preferred for balancing systems.
16. External cladding used as an acoustic treatment is not generally satisfactory. Internally lined ducts, with 25 mm or 50 mm (1″ or 2″) thick fiberglass, covered by vapor barriers and perforated metal—25 to 35% open—should be used extensively for noise control. In lieu of this, many vendors have specially designed internal duct linings.
17. Rat guards on flexible couplings should not bridge the flexible coupling. The guard should be rigidly attached to only one end of the ductwork and folded back at the other end, providing a 6 mm (¼") gap.

18. Diffuser terminals should have a manufacturer's noise rating that is at least 10 dB lower than the compartment's noise criterion. Acoustical treatment of all air mixing boxes should be considered.
19. Noise from HVAC systems should be designed to be at least 5 dB below the compartment noise specification so that when combined with other noise sources, the noise specification is not exceeded.
20. The use of dampers to balance the system is discouraged. If needed, the dampers should be placed as far from sensitive compartments as possible. Use two dampers, each at 20% closed, rather than one at 40% closed.
21. A system designed around two-speed fans will have approximately 10–12 dB lower noise at lower fan speeds.

As already noted, using a 'Buy Quiet' program, where the fan coil units are purchased with their casing radiated noise level at least 5 dB below the compartment's requirements, will facilitate meeting the compartment's criteria while underway.

High-velocity air should be adequately expanded in a controlled manner – velocity should be tightly controlled. Two-stage expansion chambers with air distributions of 20 to 30 m/s (4,000 to 6000 ft./min) along with flow noise control were quiet (Neale 2006).

10.1.10 Bow Thruster Design

The following rules of thumb should be followed with respect to the location and design of tunnel thrusters (Fischer 2006):

1. A centerline of the tunnel should be a minimum of one duct diameter above the baseline.
2. A centerline of the tunnel should be a minimum of one duct diameter below the lightest water line.
3. The minimum axial distance from the lower opening edge to the impeller is one duct diameter.
4. The minimum length of the tunnel at the centerline is two duct diameters, or there is potential for turbulent flow, which also reduces thrust.
5. The hull should be dished downstream of the tunnel opening. The major axis of the dish should be approximately 15 degrees below the horizontal (or follow flow lines), and the angle from the tunnel opening to the edge of the un-dished hull should not exceed 15 degrees (see Fig. 10.15).
6. The radius of the inlet along the hull interface should be 0.05 times the duct diameter.
7. The grills themselves should not be flat bars but aerodynamically shaped with rounded leading edges. Grills should take up as little free area as possible. The maximum opening blockage should be 10% to 15%.

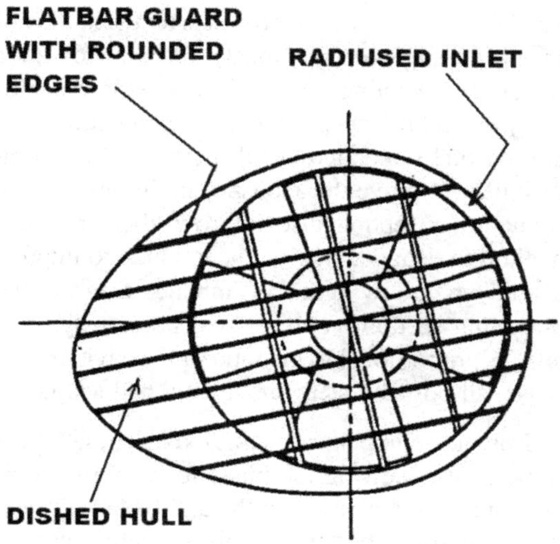

Fig. 10.15 Grill Alignment (Fischer 2006a)

8. The axial spacing between the grill and impeller edge should be a minimum of 0.3 times the duct diameter from the impeller.
9. Struts holding the impeller should be faired as much as possible to reduce turbulence, especially when thrusting in a direction where the flow hits the struts before the impeller.
10. Reduce strut cross-sectional area as much as possible.
11. Place generous fillets between the strut and tunnel wall to reduce vortices.
12. For twin tunnel thrusters, the distance between centerlines should be 1.5 duct diameters to 2 diameters for fixed pitch impellers (FPs); the distance should be as small as possible for controllable pitch impellers (CPs).

The following guidance applies to the thruster impeller (Fischer 2006):

1. Increased impeller pitch may produce the same thrust but at a lower rotation rate, which is equivalent to a lower tip speed.
2. The impeller tip speed should be less than 30 m/s (100 fps).
3. A skewed impeller shape has significantly better low-frequency performance than one with a radial trailing edge. Skew for "reverse" thrust since this should not harm "ahead" thrust; requires careful hydroacoustic design. For side-by-side tunnels, the skew could be in opposite directions.
4. Consider the trade-off between increased tip clearance, which reduces the unsteady load at tips that operate in the strut wake, and the effect on tip vortex cavitation and thrust performance.
5. Adjust the blade pitch distribution to unload the tip.
6. Consider an odd number of blades, which are susceptible to lower amplitude wake harmonics, therefore generating lower forces.
7. The impeller should have symmetrical blade sections and well-rounded edges for equal performance in either direction.

The best noise control approach is to attack the noise at the source. The source in this case is the impeller. Obviously, the best condition is a non-cavitating impeller over the entire operating range. This is seldom achieved. The next best approach is to delay cavitation inception to as high an operating power as possible. Since thrusters are not often operated at full power but rather at partial power, this approach can definitely improve the acoustic conditions onboard. Often, improvements that delay cavitation inception also improve the thrust developed by the system. The best method to delay cavitation inception is to improve the inflow conditions to provide a uniform inflow into the impeller disc. These methods were described in the preceding section. As discussed previously, once the impeller cavitates, its noise level is proportional to the ratio of the cavitating area to the impeller disc area. In this sense, all bow thrusters are equally bad actors.

Low noise impeller designs exist. These typically make use of skewed blades. In some cases, skewing the blade forward may relieve tip overloading and provide a more uniform loading over the entire blade. The next best non-engineering approach is not to place berths or noise-sensitive compartments within one deck or two compartments longitudinally of the thruster room. Place storage or refrigeration lockers in this area.

10.1.11 Sonar Self-Noise

The following features help to minimize sonar self-noise.

Sonar transducers should be arranged to minimize interference associated with local flow noise. All hull appendages, including the centerboard, sonar transducers and their housings, and the hull within 3000 mm (10′) of sonar transducers should be cavitation free, aligned with water flow and fair within 3 mm (1/8″) over any 600 mm (24″) span and should have a 0.0635 mm (2.5 mil) finish when painted.

Within 6000 mm (20′) of sonar receive transducers, all piping and ducting, except in fuel tanks, should be supported by resilient hangers. Air flow in ducting should not exceed 10.0 m/s (2,000 ft./min). Piping penetrations for fluid systems should be located at least 6000 mm (20′) from sonar receive transducers or at least 3000 mm (10′) if the structural path to the sonar from bulkhead penetration and pipe is damped.

The bow thruster inlet and discharge openings should be designed to be cavitation free and minimize turbulence at the sonar transducer locations. The design should prevent the accumulation of air in the thruster.

Hull openings for main sea suctions and discharges should be located at least 9000 mm (30′) from all sonar transducers. Except for bow thruster openings and uncontaminated seawater system intakes, all sea connectors and underwater shell openings should be aft of sonar transducers. Orifices and other flow control devices for fluid systems should be located a minimum of 6000 mm (20′) from sonar receive transducers.

In terms of limiting bubble sweep-down, flow fence and diverting fins and a drop-down gondola have very good effectiveness; cow catcher fairing and U-shaped forebody have good effectiveness; and a flat mount keep fairing has fair effectiveness (Rolland et al. 2010)

10.2 Optimized Solutions

Trade-off and What-If studies now can be carried out in a matter of minutes rather than days as with the 'manual' implementation of some 'cookbook' type design guides, e.g., SNAME Design Guide. This is especially true with existing 3-D CAD software. All acoustic prediction software, such as *DesignerNOISE®*, relies on having reasonable estimates for machinery acoustic source levels. Then, these programs allow complex noise issues to be addressed early and throughout the design. These software tools can be used to strike a balance between the selection of machinery with different source levels and noise control treatment types. The result is a lower overall cost and quicker turn-around design time, lower crew noise exposure, fewer hearing loss cases or impacts on the environment, and a lower total cost of ownership.

Optimization and trade-off studies need to consider the effectiveness of the suite of noise and/or vibration treatments versus impacts on weight, cost – material and installation, and maintenance/replacement. The treatments that effectively address both airborne and structureborne noise are usually the best choice over treatments that are effective against just one of these components. Controlling low-frequency noise is usually more difficult than controlling mid- to high-frequency noise; thus, treatments that control this component usually have a higher priority than those controlling only mid- or high-frequencies. All control materials must also meet marine regulatory requirements, usually fire, smoke, and toxicity; isolation mounts must be 'captive' to avoid damaging equipment if a fire burns the resilient elements; and producibility concerns must be addressed. This information is provided in Chap. 8.

The results of a trade-off study for the impact of several different designs are provided in Fig. 10.16. As shown, the span from 'As built' to full treatment – resilient mounts, engine room insulation, damping and floating floor – is 11 dB for the A-weighted noise. At 1000 Hz, this difference is 26 dB. The As-Built has an A-weighted noise level 27 dB above what could be achieved with extensive treatments.

Appropriate design tools are needed to leverage multiple approaches and predict all critical acoustic environments – airborne and structureborne noise, machinery and structural vibrations, underwater noise, and sonar self-noise. In this manner, a unified model can be used for many different purposes. Likewise, acoustic treatments impact performance across multiple analyses will enable optimization of the overall system performance while minimizing impacts on cost, weight, space and operation and performance of the vessel. With the development of this type of toolset, the crew and passengers will be subject to an acceptable acoustic

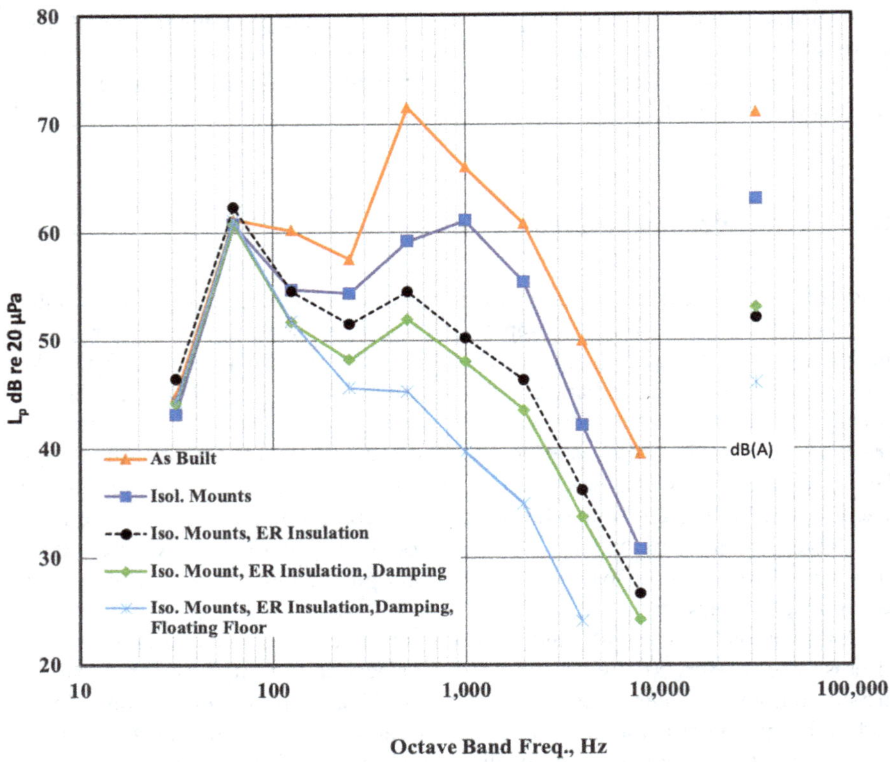

Fig. 10.16 Noise comparisons for various approaches used to optimize treatments

environment, the vessel will operate more efficiently, and the vessel impact on its environment will be minimized (Fischer et al. 2008). The availability of computer aided design (CAD)[18] programs facilitates this process of 'engineering out the noise.' Additionally, the speed of solution of most of these programs facilitates 'trade-off studies' to select and optimize various suites of treatments versus their impact on space, weight and Total Cost of Ownership (TCO) (Bowes et al. 2006).

Often, special testing – either dockside, in the lab, or at vendor sites – may be needed to ensure the accuracy of critical or "risk appropriate" assumptions made in the design/prediction process. Post sea trials can also be used to further identify any changes to treatments to optimize both the acoustic environment and non-acoustic impacts. Finally, this optimization is critical for the design of quiet research vessels, naval combatants, and lightweight, high-speed craft.

[18] DesignerNOISE®, developed by Noise Control Engineering, LLC under a SBIR grant from the USN Office of Naval Research.

10.2.1 Source Optimization

The use of 'low' noise and vibration sources was discussed in Chaps. 5 and 8. The optimal design of low sources such as diesels is being carried out by various vendors. This includes engine block, oil pan, and valve cover changes to reduce vibration response and radiation or the application of close-fitting cladding or partial barriers. Many vendors of gensets supply enclosures.

Not every muffler/silencer works for every diesel/gas turbine system. Many on-deck and internal noise environments have been problematic due to poor matching of the muffler/silencer to the exhaust system back pressure and expected insertion loss. Many vendors provide invalid (overly optimistic) insertion loss data due to improper and/or outdated testing. Gas turbine exhaust stacks need special attention to the flow paths, which typically cause separated flow and high regenerated noise. Many systems can be optimized by the application of a computer aided flow (CFD) study – see Fig. 10.17 for the optimization of a right-angle turn on a gas turbine system (courtesy M&I Power Technology, Canada).

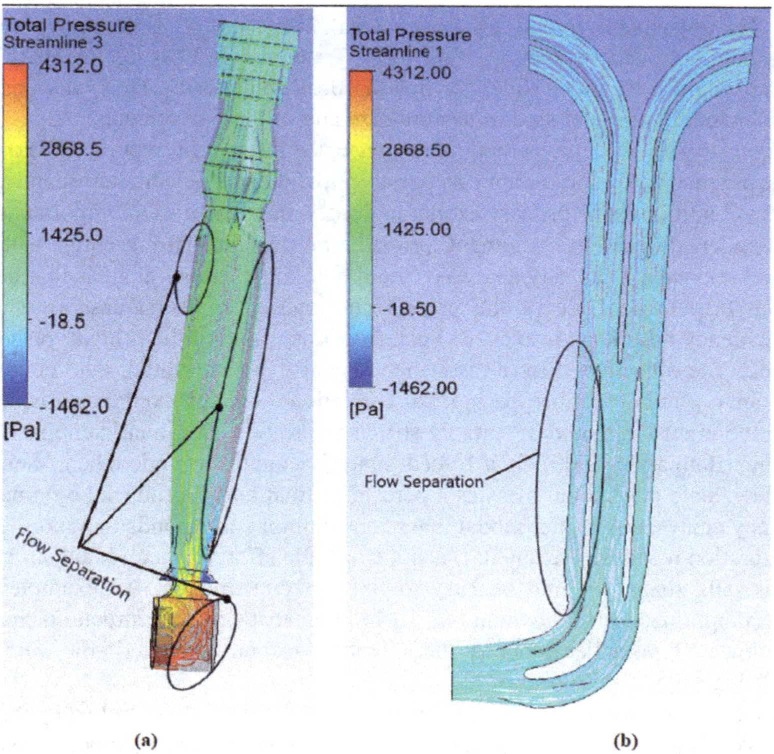

Fig. 10.17 Modifications based on CFD to improve flow and reduce noise; (**a**) original, (**b**) modified

Not only is the noise reduced with the application of CFD, but other factors, such as improvement in efficiency and reduced back pressure and vibration, can also result. Furthermore, for gas turbine systems, reduced carbon emissions and elimination of splitter silences are achievable in many cases (private communications with M&I Power). In one study to optimize a gas turbine exhaust, the total pressure was reduced from 3341 Pa to 180 Pa, and the power consumption was reduced from 369 kW to 20 kW.

The selection of appropriate duct insulation needs to be tuned to the major frequencies produced by the gas turbine (Liu et al. 2014). One note is that the exhaust temperature will have a significant influence on sound propagation due to a change in sound speed with temperature (Sun et al. 2010). At 21 °C (70 °F), the speed of sound is 344 m/s (1,130 fps), while at 200 °C (400 °F), it is 438 m/s (1437 fps).

10.2.2 Path Optimization

Noise: The objective in optimizing airborne path noise controls is not to overtreat or undertreat the acoustic insulation or use High Transmission Loss treatments (HTL), including the use of joiner panels or floating decks/bulkheads. Thus, any analysis should include 'trade-off' studies to minimize any of these treatments.

Structure: To reduce any structural response, the three most important factors are to have as small an excitation force as possible, avoid 'flexible' structures, and avoid resonant conditions. The primary excitation forces, in order of usual importance, are the propeller, propulsion equipment, propeller/propulsion/shafting equipment system, and sea state. Thus, any analysis should determine structural areas that have a resonant response to one of the propulsion 'tones. The worst case is when a low-frequency hull mode is excited. Therefore, this is one of the primary responses to check. The objective, once a resonance response is determined, is to conduct a forced analysis to determine the optimal correction – change excitation frequency, reduce the input force, and/or suitably stiffen or provide additional damping to the structure (Bathiarian 2000). If a forced analysis cannot be undertaken, then one should evaluate the 'modal mass as a percent of total mass" produced by a natural frequency analysis. A higher modal mass contribution corresponds to a more likely high vibration response if excited. It is not worth the effort to check small structures such as rails since it would be easy to fix post construction. An example of a structural optimization is shown in Fig. 10.18. Note that while the structural changes and a lowered propeller reduced the vibration response, the criteria were still exceeded at several ship locations.

As noted in the previous chapters, the propeller excitation is controlled by the hull form/wake field, number of blades and rpm, propeller blade geometry, and tip clearance. Based on most studies, 5 bladed propellers produce lower hull amplitudes than 4 or 6-bladed propellers. This is true for a 1× blade rate and a 2× blade rate. However, an even number of blades produce lower bending moments.

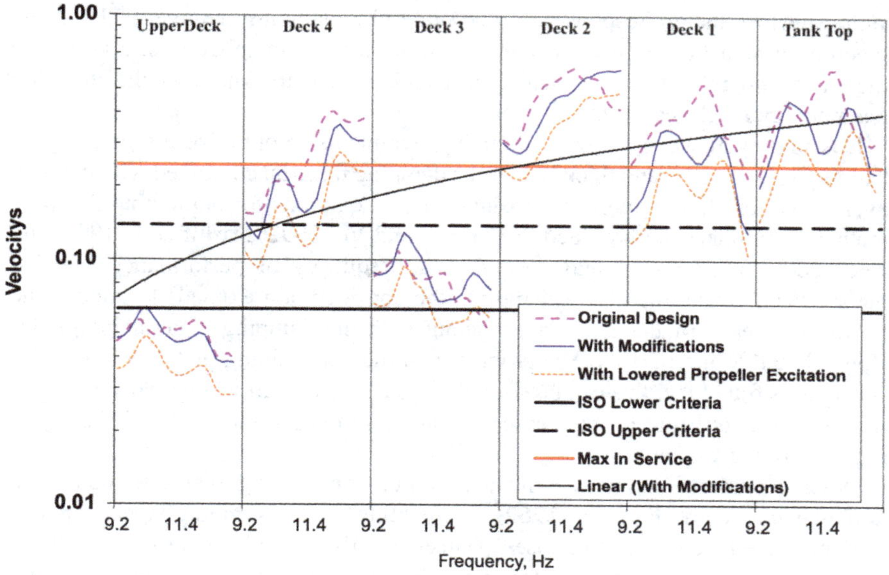

Fig. 10.18 Optimization of the vibration response

As a minimum, resonances should be checked for the hull girder, deckhouse, and masts. This analysis should include large cranes, and other structures having large cantilevers.

Machinery-induced vibrations are controlled by the free unbalanced forces and moments, type of machinery (reciprocating v. rotary), foundation impedance (high stiffness), and bearing/shafting inputs. These factors should be considered in selecting machinery type, bearing types, and forcing frequencies.

Resonance avoidance is the objective in optimizing a ship's design. If not, the fixes to the structure post-construction can cost profoundly. Again, the worst case is structural damage due to a resonant response (Ungar 1962).

10.2.3 Damping Optimization

Many options exist relative to damping materials. This includes damping tile – constrained and unconstrained; trowel-on; and spray-on. Due to regulatory requirements, some of these materials can only be used in machinery spaces. Others cannot go behind acoustic or thermal insulations. All of these treatments have a diversity of attachment methods and restrictions. For spray-on treatment, this includes dry times and avoidance of sagging due to overspray, if not properly applied. It is important to note in any optimization modeling that damping treatments need to account for their impact on airborne transmission loss, radiation efficiency, acceptance, and composite loss factor (response of base plate plus damping treatment). Even the selection of areas to be treated needs optimization (Subramanian et al. 2004). One rule of thumb is that it is typically best to have the thickness of the damping treatment approach that of the base plate.

It is important when selecting damping treatments to optimize the modulus E, damping loss factor η, and thickness of the damping material combined with the base layer – steel or aluminum or composite – plate. Equations for optimizing damping treatments on plates are provided in Beranek and Ver 1992; Nashif et al. 1991; and Jones 2001. Temperature and desired peak frequency of performance are also factors. With constrained layer damping, the consideration expands to include the thickness of each of the three layers along with the damping material properties (Zoghaib and Matteri 2014). The properties of the constraining plate are important. The authors found that this can profoundly impact the overall weight when this type of treatment is utilized. In most cases, an aluminum constraining plate is nearly as effective as a steel plate.

Other considerations are segmented or cut constraining plates to tune to a particular frequency (Plunkett, 1969). Segmenting the constraining layer increases the damping efficiency in some cases (Tian et al. 2016). Another consideration is the use of slurry-type ballast treatments to provide modest damping. One such treatment, Ballast Crete, has been tested and shown to provide up to 13 dB of vibration reduction in the mid- to high-frequencies. Another variable is including ceramic balls in the elastomeric material to improve its insulation/coating ability. Bear in mind, other treatments, such as acoustic insulation and floating decks, can have a marginal impact on system damping. One further way to increase or optimize the damping treatment is to use a spacer between the base layer and damping treatment, whether it is constrained or unconstrained (Zhao and Wang 2015). While Zhao developed this approach for train rails, it may prove useful for ship ribs and frames.

In a composite, damping materials have also been successfully used as a visco-elastic layer in the core, which can provide high inherent damping (Li and Crocker 2003). Damping treatments can also be applied to equipment closures, lockers, diesel oil pans and valve covers, subbases, and foundations.

Some vendors of damping treatments have suggestions on the application and optimization of their particular material. The results of one investigation into light-weight spray-on damping are shown in Fig. 10.19. Above 1000 Hz, the damping treatments all start to increase the airborne transmission loss of a steel bulkhead. As discovered by the authors, considering only one parameter of the impact of damping may lead to erroneous conclusions.

For an example of an optimization case, Table 10.9 shows that approximately 1100 m^2 of damping will produce significant noise reductions (8 to 13 dB); however, an additional 300 m^2 of treatment will only provide another 2 to 3 dB of further reduction – hardly noticeable to most ears.

Table 10.10 shows the calculated and measured noise levels in these compartments for different operating speeds. Without the Option 1 treatment, these compartments would exceed their criteria. Therefore, the Option 2 treatment with an added weight of over 4000 kg was unnecessary.

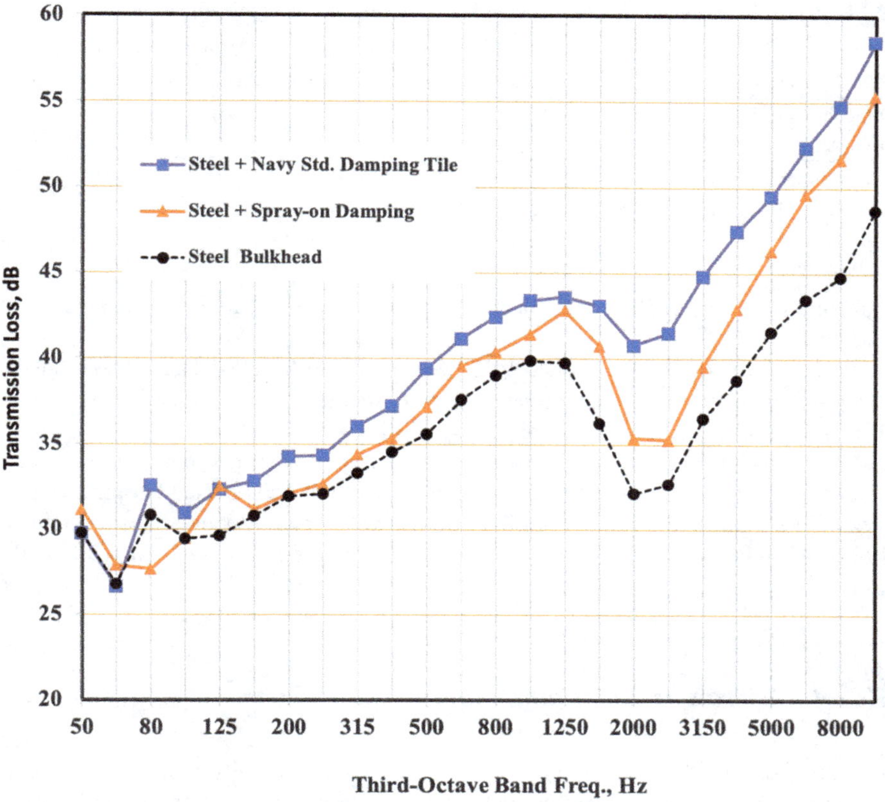

Fig. 10.19 Transmission loss with various damping treatments on steel bulkhead

Table 10.9 Investigation of damping options

	Option 1	Option 2
Location:	Noise reduction, dB	
Machinery Control Room	8	11
Mission Bay	13	15
Damping parameters:		
Area, m^2	1120	1400
Weight, kg	17,470	21,800

As noted above for the speed of sound variation with temperature, the performance of many damping treatments is also temperature dependent. Thus, any optimization would need to reflect the best and worst-case temperature ranges for the application of damping layers. The difference between temperatures on machinery foundations, wetted hull plates, and interior accommodations is significant. More

Table 10.10 Calculated and measured noise levels for option 1 damping, dB(A)

	Diesel driven 20 knots		Turbine driven 46 knots	
Compartment:	Measured	Calculated	Measured	Calculated
Machinery ctrl. Space	81	78	82	82
Bay Area	74	73	78	79
Mess, lounge	59	60	74[a]	67

[a]Controlled by an amplitude modulating source of unknown origin

than one damping treatment application has failed due to failure to consider this variation.

Weight reductions of up to 40 to 50% have been achieved using 'perforated' damping treatments. Further studies are recommended to optimize the size and distribution of holes. (Akanda et al. 1996).

FEA tools have been shown to be useful in the optimization of damping layers to reduce structural vibration (Shan et al. 2019). SEA tools can be used to optimize the areas to be treated to reduce (optimize) the weight impact (Akanda and Goetchius 1999). Akanda provided a discussion of representing the multilayer composite with an equivalent layer.

10.2.4 Vibration Isolation Optimization

To optimize any ship isolation system, it is necessary to fully understand mount properties. A real isolator does not behave like a simple spring. Furthermore, while the theoretical isolation performance of a single-stage isolator increases with frequency, the real-world limit is on the order of 40 dB for one stage and 55 dB for two stages. It should also be noted that the performance achieved in a lab or factory test is seldom achieved on vessels. This is due in part to various flanking paths through piping and cable connections, differences in the mount static and dynamic stiffness and unloaded height, differences in foundation impedance, and slight differences in the uniformity of the foundation height between the factory and the ship. Furthermore, the mount properties for the same batch may vary by as much as 5 to 10%.

As noted in Fig. 10.6, isolators, such as spring or metal isolators, have internal resonances or 'waves', which limit their performance (Du et al. 2003). Rubber isolators also have some non-linear characteristics. Complex frequency-dependent stiffness properties are examined in Muravyov and Hutton 1997, Hutton 2001, and Alkhatib 2013. Often, these procedures require the use of finite elements to completely describe the elastic system.

There are isolators constructed to have 'dual or tunable behaviors' where over some range they are 'soft' to reduce force transmission and over others where they are hard to limit relative displacement (Marzbani et al. 2014). In some cases, the tuning is based on a magnetorheological fluid's variable viscosity rate (Southern

2008). These have been dubbed hydraulic mounts or tunable magneto-rheological fluid-elastic mounts. This type of dual behavior would be ideal in the marine environment; however, to date, it has not been utilized to any great extent.

Any isolation system is optimized by providing a high impedance (dynamic stiffness) above the mount, a low impedance of the mount, and a high impedance foundation. The rule of thumb for this impedance difference is at least a factor of ten. This optimization process implies that a good forced FE program is needed to ensure the lowest vibration levels on the foundation for a given machinery excitation. The variables are the number, type, and performance characteristics of the mounts.

An isolation method not to use is a 3-point isolation system. This system would require different mounts inboard and outboard, with different stiffnesses and load capabilities.

As with damping, the performance of elastomeric mounts may be temperature dependent. As such, these parameters should be reflected in any optimization study, especially for mounts used on exhaust systems and on diesels.

Two-Stage Mount Systems

Two-stage mounting systems obviously have multiple parameters to consider to optimize the overall performance, avoid system resonances, and ensure that the mounts are not flanked by airborne noise. A very theoretical approach to some of these parameters is covered by Zhiqiang et al. 2016. In terms of optimization, the mass of the intermediate raft should equal the mass of the equipment. In reality, this is difficult to achieve for most marine machinery. Most systems can provide a reasonable isolation performance with an intermediate raft equal to 40 to 50% that of the equipment.

Based on experience, it is better to have one subbase under the machinery item rather than separate, individual mass in each of the four corners. This approach creates too many degrees of freedom, and it is difficult to decouple individual modes of the independent mass from the excitation from above.

Figure 10.20 provides the average transmission loss across a genset mounting system – that is, the above mount vibration minus the foundation vibration. For most quiet research vessels, the critical frequency range is below 250 Hz. The best performer in this range is a single stage mount system attached to a 100 mm (4″) thick steel billet attached to the foundation top and running the length of the mounting system on the inboard and outboard mount string. The second best is the classical two-stage mounting system with a single, intermediate raft. This raft is approximately 40% the weight of the genset. The hybrid two-stage system has diesel on two stages of isolation, while the generator is on one stage of isolation. The lowest performance is achieved with an intermediate raft made of independent masses – without a structural connection between the masses. The dip in performance in the 250 Hz band for this system has been shown to be due to airborne noise exciting the tank top, basically flanking the effectiveness of the mounts. There are significant weight and space (height) impacts for each of these choices of mount

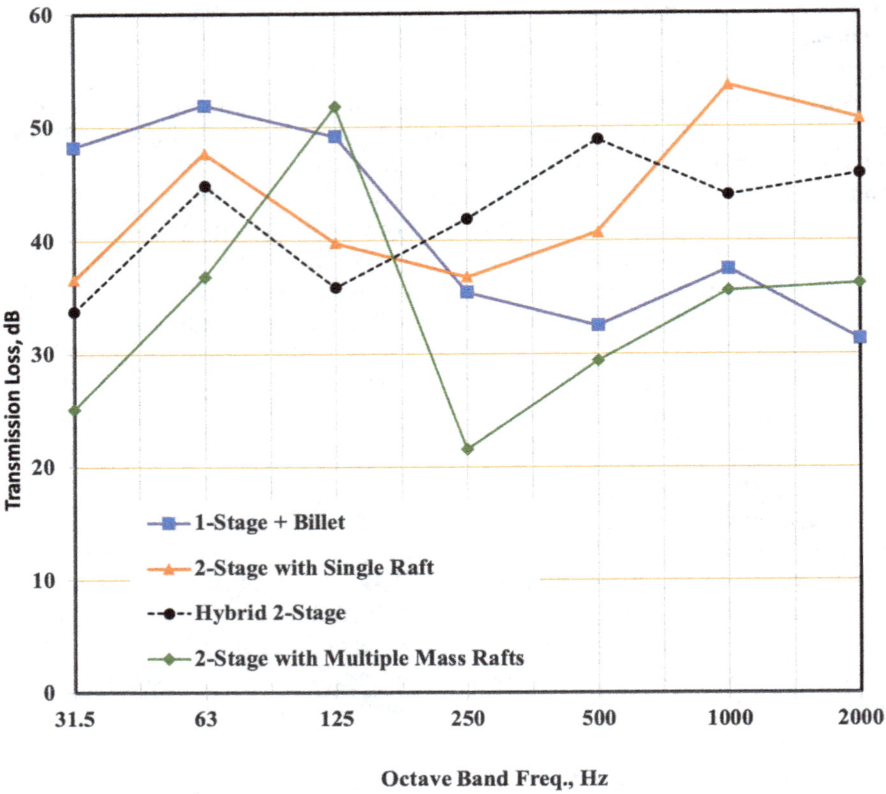

Fig. 10.20 Isolation mount system performance comparisons

systems, not to mention the complexity of installation and maintenance. This comparison indicates that the simplest approach is actually the best performer at low frequencies.

10.2.5 Absorption/High Transmission Loss Optimization

The use of acoustic insulation generally has two impacts on the noise environment: reducing reverberant noise and affecting airborne transmission loss. Low frequencies are usually the dominant noise component shipboard. Absorption is most effective in the mid- to high-frequencies and thus has minimal effect on low-frequency noise. Its primary use is to reduce the A-weighted noise and improve the ability to communicate in the SIL bands of 500, 1000, 2000 Hz, and 4000 Hz.

Anything that can be done to improve the low-frequency absorption will assist in optimizing the noise reduction. In the range of thicknesses used on ships and for low frequencies, the flow resistivity does not significantly impact the absorption coefficient. The use of perforated facings (and a vapor barrier) on top of the acoustic material and/or air spaces between the structure and insulation will improve its low-frequency performance.

It does not pay to select a material with a high absorption coefficient just because it is lightweight. It is very likely that the 'lightweight' material will have a poor transmission loss, and the net effect in an adjacent space may be a significant increase in airborne noise. In terms of weight optimization, it is best if fire or thermal insulation is utilized as the acoustic absorptive material.

To achieve the most useful results, the acoustically absorptive material should be placed as the internal 'finished' surface(s) of the compartment. Note that the insulation is not effective for absorption if it is placed behind a solid sheathing or joiner panel. However, this will increase airborne transmission loss if placed in this void. Insulation should always be placed in the void space of any joiner work around noisy machinery spaces. The expected noise reduction associated with increasing the absorption in a compartment may be on the order of 2 to 4 dB. NB.: increasing the absorption in a machinery compartment will not affect the noise in the vicinity of a machinery item since the absorption only impacts the 'reverberant noise' and not the direct noise component.

To significantly increase the airborne transmission loss at a noisy machinery space boundary, a High Transmission Loss (HTL) treatment should be considered on the bulkhead boundaries and the overhead (if a noise-sensitive space is over the machinery space). In some cases, to reduce secondary airborne noise, it may be advantageous to treat the overhead regardless of the noise sensitive space above. The entire HTL treatment must overlap and fit snugly between the ship's frames.

In some cases, acoustic insulation is used over fire or thermal insulation. The combined acoustic performance with respect to absorption and transmission loss can be evaluated and optimized using methods presented by Guy 1989 or Hong et al. 2002.

If fire protection insulation is used in lieu of any absorptive treatment, the following recommendations are made by Lee and Liu 2006 with respect to materials meeting a Class A-30 and A-60:

- The sound absorption capability of rock wool 60 k[19] with a thickness of 50 mm (2″) is better than that of a ceramic fiber 80 k with a thickness of 25 mm (1″).
- The sound transmission loss of a steel panel with fire protection material results primarily from the steel panel

[19] k value represents the material's thermal conductivity

If the use of acoustic material, e.g., fiberglass or mineral wool, is not allowed,[20] then it is possible to design a perforated panel over a void or an air space to provide some absorption (Schultz 1986). Another alternative is to use a honeycomb panel (Pan et al. 2005). Furthermore, perforated facings over a void can be tuned to absorptive over a particular frequency range (Sakagami et al. 2008). This is useful is there is a room resonant mode or tone in the room.

Note that most vendors only test absorption coefficients from 125 to 4000 Hz. One reason is the difficulty in testing low frequencies. Extrapolating below 125 Hz is difficult at best. The SNAME Design Guide does cover lower and higher frequencies as a guide. Some marine vendors, such as Rockwool and CertainTeed, have also extended the frequency range for absorption coefficients.

Sometimes it is optimal to select the transmission loss needed for a particular situation, e.g., reverse engineer the selection of treatment. In this case, it is useful to have information from vendors that provide a 'menu' of treatments with their transmission loss values. Dietz 2004 provides such a list for Rockwool material, showing various composites and thicknesses.

Bear in mind that the total transmission loss of a ship's structure is limited by any flanking paths – including louvers and doors – especially undercuts for ventilation returns and penetrations.

Cored bulkheads – with a core material between two faces – are often used for free standing, non-structural bulkheads. The transmission loss is controlled by the Young's modulus of the core and inherent damping. While this is something the naval architect cannot control, it is something the vendor ought to be able to supply. Metamaterials may provide some lightweight, high transmission loss performance. For instance, non-structural bulkheads may be composed of a 'honeycomb acoustic metamaterial'. In one case, it was shown that such a panel, at 1.3 kg/m^2 (0.3 lbs./ft^2), achieved a low-frequency sound transmission loss greater than 45 dB (Sui et al. 2015). Metamaterials can be 'tuned' to perform over a certain frequency range, providing a useful optimization approach (Hao et al. 2021).

10.2.6 Receiver Treatment Optimization

In general, noise control in a receiver space may be needed if:

- Noise controls applied to the source and noise path are not sufficient
- The source/path treatments are too expansive to reach the acoustic goal, or
- Given the distribution of sources, it is more effective to implement receiver treatments.

Once it has been determined that receiver treatments are needed, two kinds of receiver spaces should be distinguished: adjacent to the source room and remote from the source room. These two groups differ in their contributions of airborne and

[20] Due to high temperatures or stringent hygienic requirements

structureborne noise. First is an adjacent room that has airborne noise transmitted through the partition between the source and receiver rooms. Second are remote compartments receiving purely structureborne noise. A further breakdown within the room itself is also necessary. This is a separation between the vibration reduction of the room's boundary and reduction of the ability of the surfaces to radiate noise. Within these distinctions, treatments such as floating room/deck, HTL, isolation mounted joiner bulkheads, and increased absorption can be selected to optimize the receiver treatments.

If privacy is a concern, each stateroom and cabin should be sound insulated for privacy along all boundary bulkheads. An acceptable design standard for privacy is an airborne sound insulation index (Ia)[21] of 20 dB. Joiner vendors should be made aware of this requirement as soon as possible. The divisional bulkhead should be continuous from deck to deckhead. If not, curtain plates or limp mass layers should be used between the top of the joiner and the deckhead.

To provide adequate privacy, the construction needs to present a uniform transmission loss from deck to deckhead and side to side. To ensure privacy in critical areas, penetrations of the bulkhead should be minimized. If penetrations are needed, they should be grouped and sealed light tight.

10.2.7 Miscellaneous Do's and Don'ts

Large, low-speed propulsion diesels often have lateral restraints near the top of the diesel. As a minimum, hydraulic shock absorbers should be used to prevent point forces from being transmitted into the ship structure. Each side of the shock absorber should be anchored in high impedance structures.

References

Akanda, A., Goetchius, G.: Representation of Constrained/Unconstrained Layer Damping Treatments in FEA/SEA Vehicle System Models: A Simplified Approach. SAE Technical Paper, 1999-01-1680 (1999)

Akanda, A., Onsay, T., Goetchius, G.: Perforated Damping Treatment; A Novel Approach to Reduction of Weight. Society of Automotive Engineers, 99NV-152 (1996)

Aktas, B., Turkmen, S., et al.: Systematic cavitation tunnel tests of a propeller in uniform and inclined flow conditions as part of a round Robin test campaign. In: 1st International Meeting – Propeller Noise & Vibration, Istanbul, Nov. 2014.

Aktas, B., Yilmaz, N., Atlar, M., Sasaki, N., Fitzsimmons, P., Tayor, D.: Suppression of tip vortex cavitation noise of propellers using PressurePores™ technology. J. Mar. Sci. Eng. **8**, 158 (2020)

Alkhatib, F.: Techniques for Engine Mount Modeling and Optimization, Thesis, A Dissertation, The University of Wisconsin – Milwaukee, Dec. 2013

[21] Also known as the Sound Reduction Index, R, per IEC 801–31-39.

Altar, M., Fitzsimmons, P., et al.: Underwater Noise Measurements with a Ship Retrofitted with PressurePores™ Mitigation Technology and Using HyDrone™ System. In: 11th International Symposium on Cavitation – CAV 2021, Daejon, Korea, May 2021

American Bureau of Shipping: Guidance Notes on Propulsion Shafting Alignment, Houston TX, 2004

Bathiarian, M.: Diagnosing and Curing Global Ship Resonance. Institute of Environmental Science & Technology (May 2000)

Beranek, L.L., Ver, I. (eds.): Noise and Vibration Control Engineering, Principles and Applications. John Wiley & Sons, Inc, New York (1992)

Bertetta, D., et al.: CPP propeller cavitation and noise optimization at different pitches with panel code and validation by cavitation measurements. Ocean Eng. **53** (2012)

Blake, W.K., Aero-Hydroacoustics for Ships, Vol. I, David Taylor Naval Ship Research and Development Center (DTNSRDC) Report – 84/010, Bethesda, MD, June 1984

Bodger, L., Helma, S., Sasaki, N.. Vibration control by propeller design. In:1st International Meeting – Propeller Noise & Vibration, Istanbul, Nov. 2014

Bowes, M., et al, Computing the Return on Noise Reduction Investments in Navy Ships: A Life-Cycle Cost Approach, CAN CRM D0014732.A2/Final, Sept. 2006

Brown, N.A.: Existing/Future Technology to Address Radiated Noise by Modifying Vessel Propulsion and Operating Parameters. NOAA Vessel Quieting Symposium (2007)

CertainTeed, https://www.certainteed.com/technical-insulation/products/ultimate-marine-insulation/

Cowper, B., Kolomojcev, A.: Solutions to failures of shaft bearings on vessels fitted with Z-drive thrusters. In: ICMES 2003, Helsinki University, May 2003

Dietz, J.: Measurements of Sound Reduction Index, Steel Bulkhead with Various Types of Insulation, Odegaard & Danneskiold-Sansoe A/S, Report 04.1618, Dec, 2004

Du, Y., Burdisso, R.A., Nikolaidis, E., Tiwari, D.: Effects of isolators internal resonances on force transmissibility and radiated noise. J. Sound Vib. **268** (2003)

Dylejko, P., Macgullivray, I., Skvortsov, A.: Isolator Internal Resonance and Radiated Noise from Ships. In: InterNoise 2014, Australia, 2014

Fischer, R.: Noise and vibration control program for the M/V Sulfur Enterprise. In: Noise Con 96, Bellevue, WA, Nov. 1996

Fischer, R.: Bow thruster induced noise and vibration. In: Dynamic Positioning Conference, Houston Tx, Oct. 2000

Fischer, R.W.: Addressing thruster noise and vibration. In: World Maritime Technology Conference, London, Mar. 2006

Fischer, R., Yankaskas, K.: Noise Control on Ships – Enabling Technologies. In: ASNE Intelligent Ship IX, May 2011

Fischer, R.W., Boroditsky, L., et al.: Airborne noise flanking of Shipboard vibration isolation systems. Sound and Vib. **40**(12), 19–22 (2006)

Fischer, R., Boroditsky, L., Spence, J.: Acoustic design tools – present and future integrated software. In: 9th International Naval Engineering Conference & Exhibition, IMarEST, London, April 2008

Fischer, R.W., Yankaskas, K., Page, C.: A Roadmap for Future Noise Control in Acquisition: Acoustical Engineering Controls and Estimated Return on Investment for DOD Selected High Noise Sources. In: 31st International System Safety Conference, Boston, MA, 2013

Fusil, E.:Submarine signature and vulnerability. In: ASW Asia 2004

Guy, R.W.: A preliminary study model for the absorption or transmission of sound in multilayer systems. Noise Control Eng. J. **33**(3) (1989)

Han, J.M., Kim, J.J., Seo, J., Hwangbo, M.: Application of flow control devices to improve propeller cavitation performances. In: Proceeding PRADS, Luebeck-Travemuende (2004)

Hao, J., Guo, S., et al.: Research on the sound insulation properties of membrane-type acoustic metamaterials. In: E3S Web of Conferences, vol. 252. EDP Sciences (2021)

Heckl, M.: The different ways of sound transmission in and around resilient mounts. In: Janssen, J.H. (ed.) International Symposium on Shipboard Acoustics 1976: Proceedings of the

International Symposium Organized by the Technisch Pysische Dienst TNO-TH (Institute of Applied Physics), The Netherlands (1976)
Hong, K.H., VanRuiten, M., Raveendra, S.T.: Computation of transmission loss and absorption coefficient of multilayer sound absorbers with perforated facing. In: Inter-Noise 2002, International Congress and Exposition on Noise Control Engineering, Dearborn, MI, Aug. 2002
Hutton, S.G.: Optimization of vibration mount properties for application to shipboard diesel engines – Phase III, Defense Research Establishment Atlantic, Report DREA CR 2000-077, Jan. 2001
Jones, D.I.G.: Handbook of Viscoelastic Vibration Damping. Wiley (2001)
Jurgens, D., Moltrecht, T., Flipse, M.: Cycloidal propulsion – the quiet maneuvering propulsion for large yachts. In: 18th International HISWA Symposium on Yacht Design and Yacht Construction, Amsterdam, Nov. 2004
Knight, H.: PressurePores propeller tech cuts underwater noise pollution. The Engineer https://www.theengineer.co.uk/pressurepores-propeller-sea
Kozousek, V., Davies, P.: Analysis and Survey Procedures of Propulsion Systems: Shaft Alignment; Lloyd's Register's Technical Association: London, UK, 2000
Kuiken, K.: Ship propulsion and propellers. In: diesel engines II for ship propulsion and power plants, target global energy training. Onnen. **38–113**, 368–395 (2008)
Lanni F. et al.: United States Patent 6,135,831 – Impeller for marine waterjet propulsion apparatus, Oct. 2000
Lee, J.-C., Liu, T.-Y.: Experimental study of the performance of ship fire protection materials. J. Mar. Sci. Technol. **14**(1) (2006)
Li, Z., Crocker, M.: A study of the damping in sandwich structures. In: 10th International Congress on Sound and Vibration, Stockholm, Sweden, July 2003
Liu, S., Chen, W., Zhang, Y.: Design optimization of porous fibrous material for maximizing absorption of sounds under set frequency bands. Appl. Acoust. **76** (2014)
Marzbani, H., Jazar, R.N., Fard, M.: Hydraulic engine mounts: a survey. J. Vib. Control. **20**(10) (2014)
MIL-STD-1474E, Department of Defense Design Criteria Standard Noise Limits, 15-April 2015
Morrow, R.T.: Noise design methods for ships. In: Meeting of The Royal Institution of Naval Architects, April 1988
Muravyov, A., Hutton, S.: Optimization of Vibration Isolation Mounts for Application to Shipboard Diesel Engines, Defense Research Establishment Atlantic (DREA), Contractor Report 97/437, April 1997
Nashif, D., Jones, D.I.G., Henderson, J.: Vibration Damping. John Wiley & Sons, NY (1991)
NAVSEA, *DIM Mount Design Handbook*, (NAVSE S9078-AA-HBK-010-DIM), 1982
NAVSEA, *Resilient Mount Handbook* (NAVSEA 0900-LP-089-5010)
Neale, J.R., Experimental Investigation of Noise Generation from the Expansion of High Velocity HVAC Flows On Board Ocean Going Fast Ferries, PhD Thesis, The University of New South Wales, Sydney, Australia, June 2006
Nelson, D.: Reduced-Noise Gas Flow Design Guide, Revised. NASA Glen Research Center, July 2005
Noonan, E.: Ship Vibration Design Guide, Ship Structure Committee Report SSC-350, US Coast Guard Headquarters, Washington DC, 1990
Noonan, E.F., Feldman, S.: State of the art for shipboard vibration and noise control. In: Ship Vibration Symposium. SNAME, NJ, 1978.
Pan, J., Guo, J., Ayres, C.: Improvement of Sound Absorption of Honeycomb Panels. In: Proceedings of Acoustic 2005, Busselton, W. Australia, Nov. 2005
Plunkett, R., Lee, C.: Length Optiminization for Constrained Viscoelastic Layer Damping, Dept. Aeronautics and Engineering Mechanics, Minnesota U., AD0859392, 1969
Rockwool, https://rti.rockwool.com/applications/marine-and-offshore/

Rolland, D., Clark, P.: Reducing bubble sweepdown effects on research vessels. In: Oceans'10 IEEE SYDNEY, Sydney, NSW, Australia, 2010, pp. 1–7. https://doi.org/10.1109/OCEANSSYD.2010.5603903

Sakagami, K., Morimoto, M., Yairi, M.: Application of microperforated panel absorbers to room interior surfaces. Int. J. Acoust. Vib. **13**(3), 120–124 (2008)

Schultz, T.J.: Acoustic Uses for Perforated Metals: Principles and Applications. Industrial Perforators Association, Inc (1986)

Shan, Y., et al.: Vibration characteristic analysis and control of hospital ship. IOP Conf. Ser.: Mater. Sci. Eng. **544**, 012046 (2019)

Simon, F., Pfretzschner, J.: Guidelines for the acoustic Design of Absorptive Devices. Noise Vib. Worldwide. **35**(1), 12–21 (2004)

Southern, B.M.: Design and Characterization of Tunable Magneto-Rheological Fluid-Elastic Mounts, Master of Science Thesis, Virginia Polytechnic Institute and State University, Blacksburg, VA, April 2008

Subramanian, S., Surampundi, R., et al.: Optimization of damping treatments for structure borne noise reduction. Sound Vib. **38**(9), 14–19 (2004)

Sui, N., Yan, X., et al.: A lightweight yet sound-proof honeycomb acoustic metamaterial. In: Applied Physics Letters, vol. 106. AIP Publishing LLC (2015)

Sun, F., Cheb, H., Wu, J., Feng, K.: Sound absorbing characteristics of fibrous metal materials at high temperature. Appl. Acoust. **71**, 221–235 (2010)

Tian, S., Xu, Z., et al.: Dimensionless analysis of segmented constrained layer damping treatments with modal strain energy method. Shock. Vib. **2016**, 8969062 (2016)

Traverso, F., Gaggero, T., Tani, G., Rizzuto, E., Trucco, A., Viviani, M.: Parametric analysis of ship noise spectra. IEEE J. Ocean. Eng. **42**(2), 424–438 (2017). https://doi.org/10.1109/JOE.2016.2583798

Ungar, E.: Maximum stresses in beams and plates vibrating at resonance. J. Eng. Ind. **84**(1), 149–155 (1962)

Van Bakel, J.: Acoustic Transfer Functions of Flexible Shaft Couplings: Measurement method and results. In: Buiten, J. (ed.) Shipboard Acoustics. Martinus Nijhoff Publishers, Bordrecht (1986)

Van Beek, T.: Technology guidelines for efficient design and operation of ship propulsors. The Ship (2004)

Veritec: Vibration Control in Ships, Hovik, Norway, 1985.

Wachel, J.C., Bates, C.L.: Escape piping vibrations while designing. Hydrocarb. Process. **55**, 152–156 (1976)

Yang, B., Hu, Y., Vicario, F.: Improvements of magnetic suspension active vibration isolation for floating raft system. Int. J. Appl. Electromagn. Mech. **53**, 193–209 (2017)

Yankaskas, K., Fischer, R., et al.: Engineering out the noise. Hear. Res. **349** (2017)

Zhao, C., Wang, P.: Theoretical modeling and effectiveness study of slotted stand-off layer damping treatment for rail vibration and noise control. Shock. Vib. **2015**, 1–12 (2015)

Zhiqiang, L., Lin, H., Changgeng, S.: Optimization of natural frequencies of large-scale two-stage raft system. J. Phys. Conf. Ser. **744**, 012180 (2016)

Zhu, C., Gaggero, T., et al.: Underwater sound characteristics of a ship with controllable pitch propeller. J. Mar.Sci. Eng. **10**, 328 (2022)

Zoghaib, L., Matteri, P.O.: Modeling and optimization of local constraint elastomer treatments for vibration and noise reduction. J. Sound Vib. **333**(26) (2014)

Zoul, V., Kovac, P.: A brief overview about the development of torsional vibration calculation and education of methods for their calculation. Sci. J. Silesian Univ. Technol. Ser. Transport. **99**, 205–211 (2018)

Chapter 11
Unique Acoustic Characteristics of Ships and Their Control

Ships come in various lengths, widths, and construction materials depending on their intended utilization. Sizes range from small power boats to large aircraft carriers and super tankers. There are many different types of ships and hull features (mono-hull, catamaran, trimaran, etc.), structural construction materials – metal v. composite, joiner/interior materials, main propulsion plants, propulsors (open or ducted, water jet, pump jet, etc.). Most of these factors will have varying impacts on the ship's acoustical characteristics. Currently, oceanographic, research, cruise, and most military vessels typically require the greatest attention with respect to a quiet design. The selection of the prime propulsion system – diesel direct, diesel geared, gas turbine, diesel-electric, hybrid electric or hydrogen, battery – is usually performed early in the ship's design. It also has the greatest impact on the level of treatment needed to meet given acoustic goals.

This chapter will highlight some of the salient acoustic differences between commercial vessels and these specialized vessels or acoustic sources. These unique features and their possible impact control approaches will be discussed. Some differences between the airborne, vibration, and underwater radiated noise were already assessed in previous chapters.

11.1 Station Keeping/Thrusters

Many workboats and research vessels have various types of thrusters. In one instance, noise and vibration criteria for these types of vessels are contained in ABS[1] Crew Habitability on Workboats (see Chap. 3). Furthermore, there are

[1] American Bureau of Shipping 2008, 2012, and 2016

R. Fischer, L. Boroditsky, *Noise and Vibration Control on Ships*,
https://doi.org/10.1007/978-3-031-55170-3_11

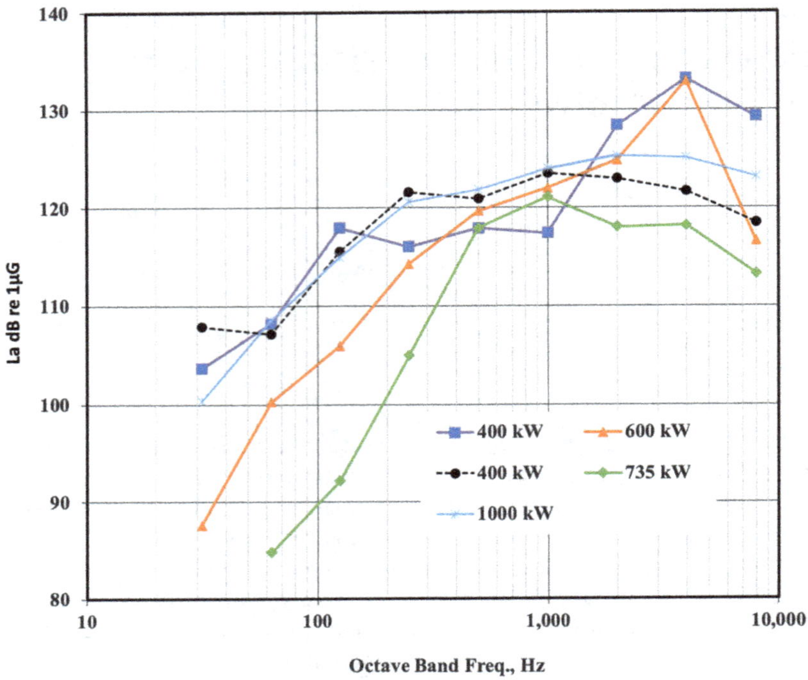

Fig. 11.1 Vibration levels for various thrusters

requirements to be met during dynamic positioning.[2] During dynamic positioning, the noise criterion for workboats is only 5 dB above that for Transit Conditions for HAB++ in Cabins, Dining Spaces, Indoor Recreation Spaces, and Medical/First Aid Center. The noise limit to qualify for HAB or HAB+ in these spaces is the same for transit as for dynamic positioning. Thus, greater attention should be given to any thruster installation if ABS Hab++ criteria are to be met.

The increase in noise with thruster operations can easily exceed the 5 dB allowed by ABS. Noise data for offshore service vessels report an average level of 80 dB (A) for cabins on the lowest deck near the bow thruster. The HAB++ noise limit is 60 dB(A), and that for HAB is 65 dB(A). The vibration levels on the thruster equipment and on the tunnel walls are usually very high, often equivalent to that of propulsion diesel (see Fig. 11.1). Peak levels for diesels are on the order of 120 dB re µG. Additional information on thruster acoustic source levels are provided in Sect. 5.2.2 and treatments covered in Sect. 8.10. Thruster impact on underwater radiated noise is provided in Sect. 9.1.1. Dos and Don'ts with respect to thrusters are provided in Sect. 10.1.10.

[2]Det Norske Veritas and other regulatory agencies also have habitability noise and vibration criteria based on vessel type.

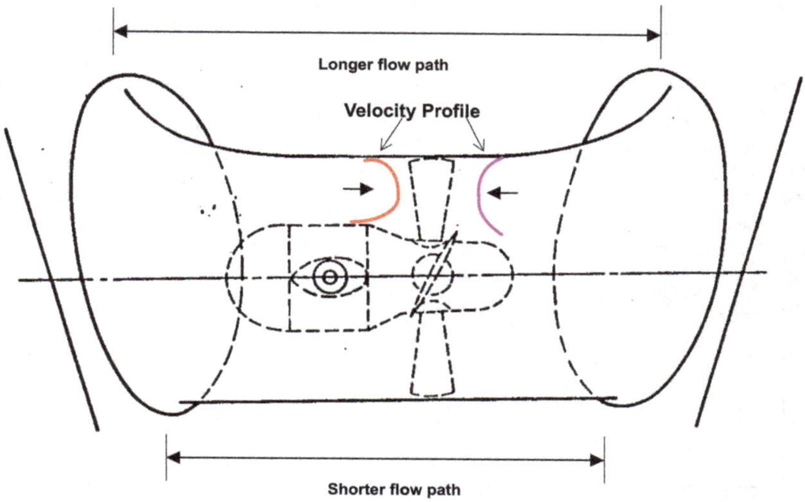

Fig. 11.2 Bow thruster inflow

The ABS Guide also presents allowed habitability vibration levels in accommodations on workboats above and below 60 m (200 ft) in length and for HAB, HAB+, and HAB++. As with the noise criteria, these will not typically be met without attention given to the type of thruster and location of the accommodation relative to the thruster itself.

As covered in Chaps. 5 and 9, once a propeller cavitates, it creates very high underwater noise levels. Bow and stern tunnel thrusters, in particular, almost always cavitated when loaded over 40% to 50% power. This is partly due to operating in a Bollard condition – high thrust while maintaining a zero or low speed. Even a well-designed[3] and installed bow thruster is bound to cavitate well below full power. Part of the reason is the high percentage blockage caused by the drive motor and the struts holding this motor (see Fig. 11.2). Another reason is the difference in the shorter path on the lower section versus the longer flow path on the upper section (Fischer 2000). Thus, it will be difficult to meet a stringent habitability noise/vibration or radiated signature while station keeping or while conducting straight line operations using the thruster. The above concerns apply to almost all types of thrusters – drop-down, azimuth pump-jet, and mixed-flow impellers.

It has been found that below 50% of the motor load, a Fixed Pitch (FP) impeller offers a slight (> 3 dB) advantage over a Controllable Pitch (CP) impeller operating at constant speed (Andresen n.d.). However, a variable speed combined with a CP impeller has between a 2 to 6 dB advantage up to 75% motor load.

[3] See Chap. 9

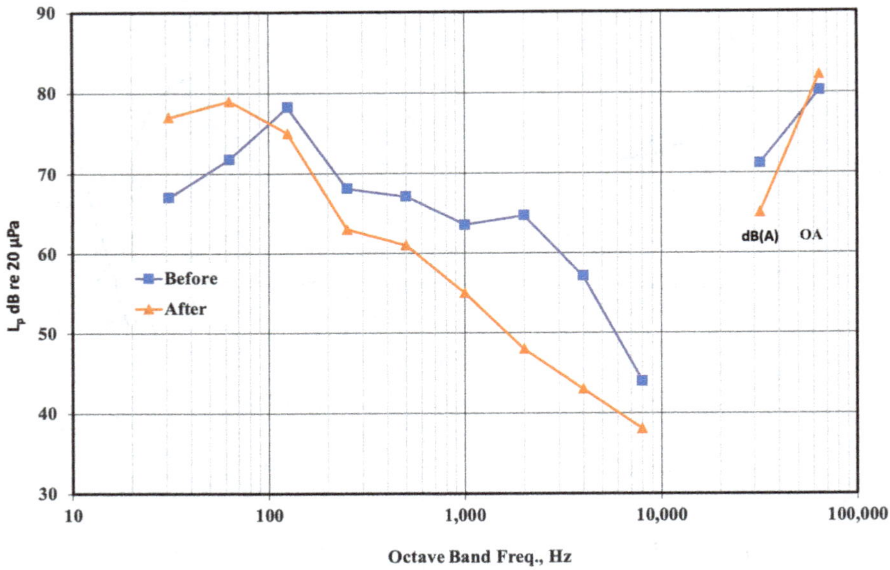

Fig. 11.3 Berth noise before/after bow thruster treatments

Bow thrusters are typically operated intermittently or in a station keeping mode, which increases their noise impact on the crew, especially those located in the forward compartments. The A-weighted noise level in these compartments can increase by as much as 20 dB with operation of the thruster. This impact sometimes worse for cruise boats and ships, especially when the bow thruster is operated at night or dockside. The noise from thrusters peaks in the 125 to 2000 Hz range.

Treatments that are typically applied include floating decks and/or rooms in accommodation, air injection into the tunnel or through the blade edges (which may reduce thrust), compliant layers inside the tunnel, tunnel resiliently mounted to the ship via a larger diameter tunnel, and resilient mounting of the tunnel or its components (see Figs. 8.32 and 8.33).

Pump jets that take suction from a sea chest can have treatments installed inside, such as decoupling material (compliant closed cell foam) and damping material. An example of before and after such treatments is shown in Fig. 11.3. The quietest type of thruster is that developed for use on mine-sweepers. These typically have air injection sites and quiet impellers.

When a tunnel inside a tunnel has air injection, the insertion loss can range from 9 to 20 dB over the frequency range (Nilsson 1986). Doors fitted on each end of a tunnel thruster will reduce turbulence noise generated with operation of the vessel at service speeds.

Another approach is an RIM drive system that may reduce the noise impact. However, there is still a strong dependence on the tunnel design – location, in-flow, etc. (Yakovlev et al. 2011). Noise reductions of 4–8 dB in airborne noise, 6–12 dB in structureborne noise, and 5–6 dB in underwater noise have been measured by Johnson 2015, for permanent magnet thrusters.

As noted in Chap. 9, the operation of a thruster is likely to interfere with on-board sonars and will likely cause high underwater radiated noise, especially once it cavitates.

Drop-down thrusters, while an improvement over tunnel thrusters, in the sense they have a likely higher cavitation inception speed, may still cause interference with sonars and increase the vessel radiated signature once they do cavitate. In our experience, standard tunnel thrusters start to cavitate above 50% power. Some vendors, such as Schottel, maker of a 'rudderpropeller' for minesweepers, can provide radiated noise signatures for their equipment.

Bow thrusters are known to produce high structureborne noise in forward compartments. Therefore, as a minimum, one should perform analyses to show that there is no resonance of the foundation with the motor and impeller rotation. A comprehensive noise analysis should help and identify and separate airborne, structureborne, and fluidborne noise transmitted into forward compartments (see Fig. 8.32).

11.2 Water Jet Propulsion

Waterjet acoustic source levels are discussed in Sect. 5.2. For cavitation noise and vibration avoidance or mitigation, an advanced design impeller is usually needed. Such impellers, with an unorthodox forward skew of the leading edges toward the blade tips, have been shown to greatly increase inception speed and tolerance to inflow non-uniformity (Lanni and Brown 2000). Low-frequency cavitation noise from water-jet propulsors might be attenuated by out-of-round cross-sections in an air-backed inlet duct, resulting in reduced wall impedance. Optimistically, it may be possible to approach 30 kts. in water-jet propelled military craft with reduced cavitation and radiated noise.

There is a splash-noise component where the jet stream impacts the sea surface behind the vessel. The noise levels generated depend on droplet size and relative velocity as well as the air content in the vessel wake.

At speeds where conventional propellers would be fully cavitating, a water jet may not be doing so (Yang et al. 2014). Therefore, it would induce lower noise and vibration over a greater operating speed. Furthermore, these units are located aft, away from most critical noise spaces. In addition, jets transmit no thrust to the drive equipment, facilitating mounting the diesel or gas turbine. Source/path treatment should consider spray-on damping to conform to the compound surface of the waterjet.

11.3 Various Ship/Vessel Types

This section addresses the special acoustic characteristics of various types of ships/vessels, construction, or usage.

11.3.1 Fast Craft

Fast crafts are usually lightweight. The typically consist of fast ferries, crew transfer boats, special naval operations boats, hydrofoils, etc. In addition to mono-hulls, their hull forms include catamarans, trimarans, and surface effect ships. Given their high power-to-weight ratios, they are aluminum and sometimes glass reenforced plastic (GRP) constructions. Most of these factors adversely impact habitability and underwater noise.

The lightweight construction does not bode well for high impedance foundations. The propulsion engine rooms in catamarans or trimarans act as small enclosures. As discussed in Sect. 8.4, this limited volume will increase the sound pressure inside the engine room relative to the free source level of the machinery. The short path from the foundation to the wetted hull plate will increase the hull vibration.

The airborne transmission loss for the lighter materials is usually less than for the equivalent steel construction, on the order of 5–10 dB lower. However, the radiation efficiency for the dry plate is similar given that the longitudinal wave speeds are similar. The radiation efficiency of the wetted hull plate increases due to the increased impact of water loading. Another adverse impact is the lower loss factor associated with aluminum constructions. The one acoustic advantage is the longer path from the foundation to the Main Deck. This path is amenable to damping.

Table 11.1 shows the calculated and measured noise levels on the X-Craft/*Sea Fighter* (Yankaskas and Fischer 2005). This experimental craft has the following

Table 11.1 Sea fighter general parameters

Parameter	Value
Overall Length	79.9 m
Length at Waterline	73.0 m
Overall Beam	22.0 m
Beam at Waterline	5.06 m
Depth at Waterline	2.99 m
Design Displacement	1100 tonnes
Max Speed	50 Knots
Propulsion Machinery:	(2) GE LM2500 GT; hard mounted.
	(2) MTU 16 V 595 TE 90 Diesels; isolation mounted
Compound gearbox	Hard mounted
Propulsor	(4) Kamewa 125SII Waterjets, 72,000 hp

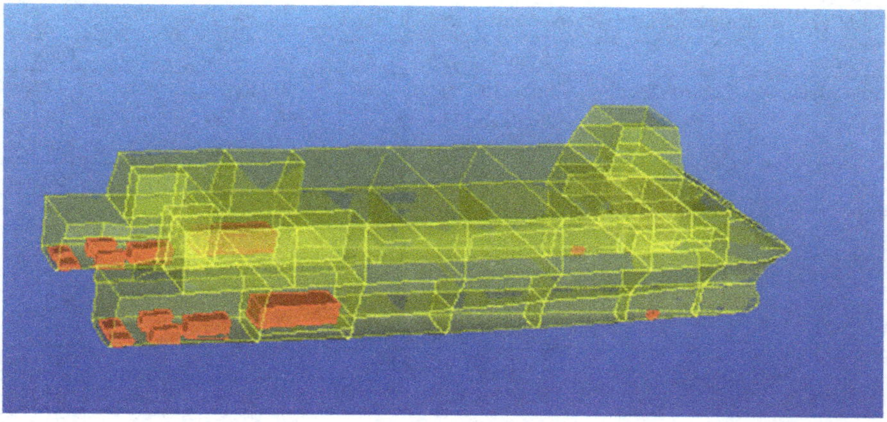

Fig. 11.4 3-D DesignerNOISE® model of Sea Fighter

Table 11.2 Calculated and measured noise levels, dB(A)

Compartment	Diesel driven, 20 knots		Turbine driven, 46 knots	
	Measured	Calculated	Measured	Calculated
Machinery Ctrl Space	81	78	82	82
Mission Bay	74	73	78	79
Mess, Lounge	59	60	74[a]	67

[a] Controlled by an amplitude modulating source of unknown origin

general parameters and the *DesignerNOISE®* model shown in Fig. 11.4. Sources are shown in red – gas turbine, propulsion diesels, gearboxes, and waterjets – port and starboard aft, and control fins forward.

Noise predictions were only carried out after construction had begun, so the opportunity to add significant noise controls was limited. The predicted versus measured noise levels are provided in Table 11.2 (Fischer et al. 2007).

Table 11.3 shows a breakdown of the airborne (AB), structureborne (SB), secondary structureborne (SSB), and total noise expected for the major sources shown in Fig. 11.4 (Fischer et al. 2007).

One significant source for fast craft is the vibration and noise at the bow associated with passing through waves. For example, in forward berthing on frigates or cutters, speeds above the 20 kn noise level are close to those in the compartment adjacent to the engine room. Figure 11.5 shows the vibration level of the outer shell of aluminum ferries measured on three vessels. All sensors were located at the water line area and near the forepeak bulkhead. The speed of the vessels was approximately 30 knots.

One can see that the vibration level on the outer shell is over 110 dB re μG, which is close to the vibration on a hull structure near propulsion machinery. Even in moderate sea waves, such as Sea State 1 or 2, the impact of bow waves leads to a

Table 11.3 Calculated noise level in compartments at 20 kn

Compartment	Source	AB	SB	SSB	*Total*
Stbd Machinery Control Space	Inboard Genset	60	55	74	74
	Outboard Genset	60	59	74	74
	Inboard Water Jet		58		58
	Outboard Water Jet		60		60
	Prop Diesels	58	51	72	72
	All Sources	66	66	78	78
Mission Bay	Inboard Genset	–	44	69	69
	Outboard Genset	–	47	69	69
	Inboard Water Jet		47		47
	Outboard Water Jet		47		47
	Prop Diesels	–	40	67	67
	All Sources	–	54	73	73

This table shows that secondary structureborne noise dominates over secondary structureborne or airborne noise.

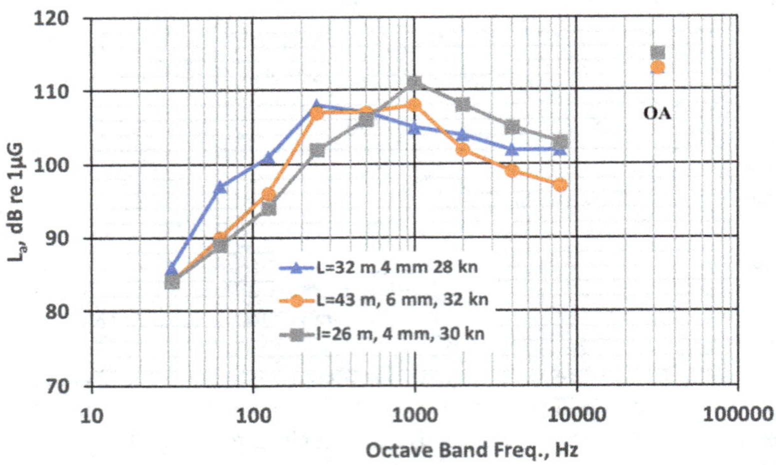

Fig. 11.5 Vibration level on the bow shell plate, operating at high speed

remarkable noise increase. For example, if the vessel speed is 12 m/sec (24 knots), the noise level will be higher by 6 dB at a wave height of 0.5 m relative to operating in calm water conditions. As noted in Sect. 5.3.1, the speed, wave height, and vessel direction relative to wind all influence the induced hull vibration.

Not only is the bow affected by high-speed flows, foils and control surfaces can produce high hull vibration and internal habitability noise levels. The spectrum shape and vibration levels produced by the foils on X-Craft are shown in Fig. 11.6.

Fast craft often have composite hulls. However, noise measurements have shown that this construction does "not demonstrate any acoustical advantage of a composite

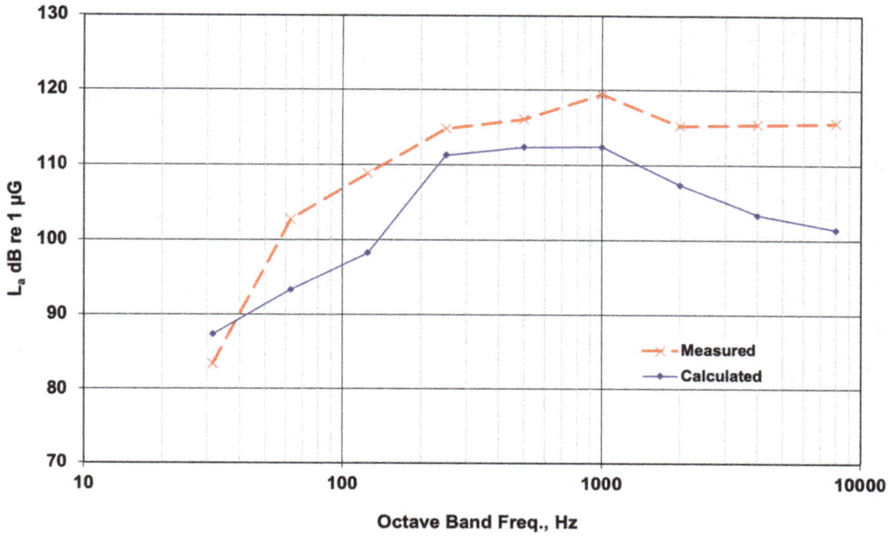

Fig. 11.6 Vibration levels on the pontoon shell near the foil attachment area

Fig. 11.7 Time trace of vibration while operating at 10 kn in an 80% ice field

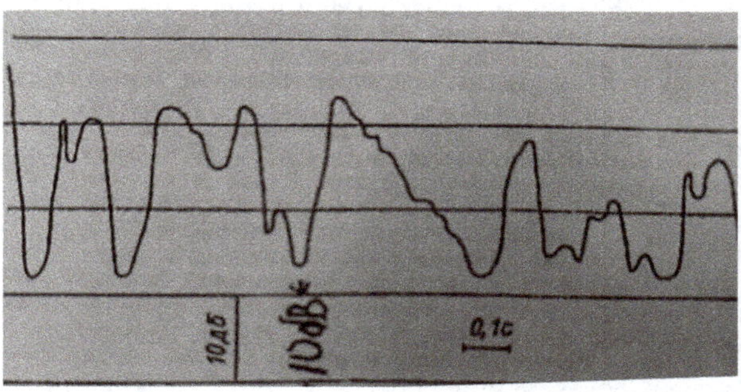

hull in spite of significantly higher material damping loss factors for composites" (Boroditsky and Fischer 2010). One reason for this phenomenon is that there is a higher coupling loss factor at connection points and a higher radiation efficiency.

11.3.2 Icebreaking

The major concern is an additional noise source – ice breaking – close to accommodations. Elevated vibrations are expected when cracking ice. The induced vibration depends on many factors, including ice thickness, type of hull, ship speed, contact area, and whether the stern or bow is used for the breaking process. Ice breaking noise is not as steady as other machinery sources. It may change in value by 10 to 15 dB during fractions of a second. Figure 11.7 shows a time trace of the

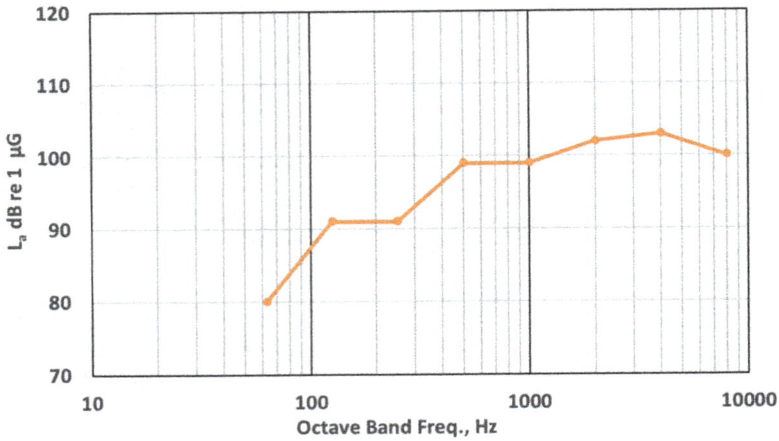

Fig. 11.8 Vibration measured on ice belt, operating at 12 kn

vibration on the outer shell recorded at 10 kn in an 80% ice coverage field (Boroditsky 1985). This variability plus the high noise and vibration induced while breaking ice make it very annoying.

This record shows that the outer shell is excited by random impulses as ice contacts the hull. The force of ice impact depends on ice quality and vessel speed (Boroditsky 1985). During the 1970s and 1980s, the Russian Research Institute of Shipbuilding Technology under Boroditsky's supervision measured noise and vibration on icebreaker with propulsion power from 8.9 MW to 55 MW. Multiple measurements show that despite operating in a large variety of ice fields, the broad band noise and vibration spectrum shape in the frequency range between 63 and 8000 Hz were practically stable. A stable spectrum shape was confirmed for icebreaker speeds ranging between 4 and 12 knots. Figure 11.8 illustrates the measured octave band vibration level on a 50 mm thick ice belt at 12 knots. One can see that vibration levels are of the same order as machinery vibration.

Habitability vibration levels while breaking ice typically exceed the ISO 6954 standard for habitability vibration criteria, resulting in 'harmful vibration.' Vibration levels predicted by Pang et al. 2016 are shown in Table 11.4:

Pang et al. 2016 and Pang 2022, in a numerical simulation during icebreaking, showed that 'ice-breaking force will yield not only significant global vibration but also local vibration in the area where ice and ship hulls collapse. . .' As shown above, their simulation predicted "Harmful Vibration" from the Bow Deck to the Aft Deck to the crew living room on the B Deck. Furthermore, the characteristics of ice class propellers are affected by the wake considering the interaction with ice (Pustoshny et al. 2017).

Table 11.4 Habitability vibration response while breaking ice

Area	Response, mm/s	Standard, mm/s	Result
Bow Deck	24.6	≤ 8	**Harmful**
Aft Deck	8.3	≤ 8	**Harmful**
Aft Deck 2	6.475	≤ 8	Harmless
Compass Deck	1.59	≤ 8	Harmless
Crew Living	1.69	≤ 6	Harmless
Capt. Living	1.59	≤ 4	Harmless
Hanger Deck	7.68	≤ 8	Harmless
Hoist Machine	7.39	≤ 8	Harmless

The hull's natural frequencies tend to be excited by the ice breaking process, so it is important to address the response of the hull to impulsive forces (Heyn and Skjetne 2017). See Suarez et al. 2022, for examples of design criteria with respect to reducing icebreaking induced vibrations.

With respect to underwater radiated noise, the highest estimated noise for the USCG *Cutter Healy* was 195 dB 1 μPa–m//Hz while breaking ice going forward and 200 dB backing and ramming heavy ice (Roth et al. 2013). The *Canmar Supplier* had measured levels of 181 and 195 while in open water and while breaking ice, respectively, for broadband noise from 20 to 1000 Hz (OCS EIS/EA 2008). For the *M/S Voima,* a similar radiated level was measured, 177 dB in open water at 50–60% power and 190 dB when icebreaking full astern (OCS EIS/EA 2008).

The ice cover may also change the underwater propagation characteristics. The strongest decay is with a 'free-surface' (open water), and the weakest decay is with a 'rigid cover' (ice sheet) (Streckwall and Bretschneider 2014). Furthermore, sea ice creates a strong time- and space-varying surface, which will affect propagation. The sea ice and varying sound speed profiles may reduce the range of a direct path transmission.

The propeller is the greatest concern with respect to underwater radiated noise. These propellers will have a thickness and shape depending primarily on strength while operating in ice. When operating at full thrust and low forward (or backward) speed, cavitation is likely.

Another concern is the effective isolation mounting of the machinery. Any isolated system will need to take the shock load and motions imposed while breaking ice.

11.3.3 Dredging

Large pumps and extensive piping often increase the airborne noise level in the main machinery space and in compartments adjacent to dredging pipes. To minimize this, these pipes should be resiliently supported to stiff dredge structures.

A UK study of the underwater radiated noise on six dredging vessels concluded the following (MALSF 2011):

- Source levels below 500 Hz are similar to those for a cargo ship traveling at modest speed
- Source levels above 1000 Hz show higher broadband noise depending on the aggregate extraction and aggregate type

Since many dredges operate in shallow water, the 'cutoff low frequency' limit would apply. Below this limit, propagation is not expected in the far field. For example, for depths of 20 to 45 m, this frequency is approximately 30 Hz. Above this frequency, propagation is usually calculated using a 'source-image' model. As noted in Sect. 9.7, the type of seabed has a large influence on the attenuation losses.

ERDC 2001, provides radiated noise levels and dominant frequencies for various distinct bucket dredge events such as bottom contact, bucket digging, bucket closing, winch in/up, and barge loading. Of these, the bottom contact is generally the loudest and has a frequency range from 60 to 370 Hz. The estimated source level, taking into account either 20 log r spreading or 10 log r spreading, is between 167 and 143 dB re 1 μPa–m//Hz, respectively. The real signature is likely somewhere in between. Cutter Head dredges have been measured to have source levels between 151 and 157 dB re 1 μPa–m//Hz (ERDC 2014). The peak frequencies are between 20 and 1000 Hz. The sound is primarily produced by the pumps and generators during hydraulic cutterhead dredging. It was noted that there were 'additional sounds from the rotation of the cutterhead in the substrate and movement of the material through the pipeline.' Thus, there can be a higher noise level from the pipeline depending on the coarseness of the material on the ocean bottom. Wang 2011 also noted that 'the presence of aggregate pumped through the pipe is a major source' of broadband noise above 1000 Hz. As one would expect, the signature is lower in transit. At the typical speeds of dredges, the propeller is unlikely to be cavitating unless it is heavily loaded (bollard condition). However, some of the broadband noise exceeds that of merchant vessels traveling at slow speeds (Wang 2011).

11.3.4 Seismic Surveying and Towing Arrays

The noise from a seismic vessel while towing an array can be noisier than that during transit due to the extra thrust on the propeller. That is, the propeller design has been optimized to operate at that load and inflow (lower speed). Usually, the vessel radiated noise does not interfere with the return signal from the air-gun excitation. The DNV's underwater radiated overall noise limit for its Silent-S class for seismic vessels is 175 dB re 1 μPa–m. In each one-third octave band between 3.15 and 315 Hz, it is 168 dB re 1 μPa–m.

The goals for the allowable radiated noise from the towing vessels are usually established by the allowed signal-to-noise at the towed array. The array is on the order of 200 m behind the vessel. In most cases, this received level should be below 112 dB re 1 μPa. This corresponds to a ship source level in the aft direction of 158 dB re 1 μPa–m.

Low frequency is of concern on these vessels (below 315 Hz). These vessels with air guns have powerful air compressors. Careful isolation of these units and the propulsion machinery will be needed.

Seismic exploration itself, while a significant source of underwater noise, is outside the scope of this book. However, EIS/OEID[4] in a study funded by the National Science Foundation (NSF), defines a 'low-energy' as a system with a level below 180 dB re 1 μPa//Hz −100 m. BOEM 2014 has been investigating methods to reduce the impact of air guns, including marine vibriosis, low-frequency acoustic source (patented), deep-tow acoustics/geophysical system, low-impact seismic array, underwater tunable organ-pipe, bubble curtains, parabolic reflectors, and air gun silencer.

11.3.5 Hovercraft, Surface Effect Ship (SES) or Air Cushion Vessel

Pass-by noise from hovercraft is expected to be higher than that from other marine craft due to their propulsion and lift fans. Measured levels are on the order of 79 dB at 305 m (1000′) for a Hovercraft and 94 dB for a surface effect ship at idle. Noise measured near an air cushion vessel was 120 dB (A).

Underwater noise from hovercraft was found to be caused primarily from the propulsion fan, although contributions from the lift fans are also possible. Underwater noise is created through direct airborne transmission to the water. Therefore, treatments that reduce airborne noise will also reduce underwater noise. Modifications to the fan blades and inflow/outflow characteristics have been identified as most practical. It is noted that hovercraft produce underwater sound levels that are much lower than those for conventional vessels. Thus, the use of hovercraft instead of conventional vessels may be an appropriate form of noise control for some applications (Spence et al. 2007).

Some of the acoustic advantages of an SES type vessel on-cushion are (1) the equipment is elevated, (2) composite hull, (3) reduced wetted surface area, and (4) reduced signature at high speed if water jets are used. Estimates for the radiated noise signature of a large surface effect ship (SES) with water jets are shown in Fig. 11.9.

Based on limited data, a small hovercraft produces an underwater signature of 140 to 148 dB re 1 μPa–m//Hz, depending on whether cylindrical or spherical spreading is used (Blackwell and Green 2005). The noise of lift fans and drives – possibly trapped in a reverberant lift chamber – couples well to underwater noise as

[4]Environmental Impact Statement (EIS)/Overseas Environmental Impact Statement

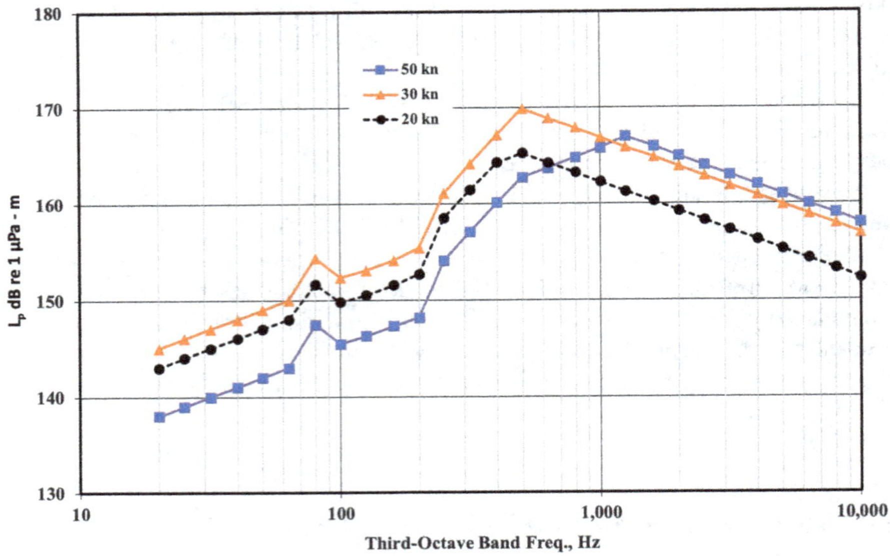

Fig. 11.9 Estimated underwater radiated noise signature for Large SESs at various speeds

airborne noise, which is incident within a vertical cone of 7 degrees. Within this cone, airborne noise couples well to underwater noise with just the impedance difference reducing the incident noise.

11.3.6 Hydrofoil

In addition to possible radiated noise from the propulsor, the foil used for lift on a hydrofoil will likely cavitate. When not cavitating, the foil will produce noise via the trailing edge and tip vortices. Even when cavitating, radiation from the tip and trailing edges plays an important role in generating the radiated signature. Once cavitating, the noise phenomena include sheet, large-scale cloud, and wake cavitation.

The inception point for the foil is influenced by the development of the boundary layer. Motion through the water (pitching with sea state) will also affect cavitation inception. The size and shape of the foil also impact its radiation efficiency. The foil will radiate in a dipole manner.

Due to its method of operation, a hydrofoil has three different stages of concern acoustically. These are with the hull in water, becoming foilborne, and fully foilborne, with the hull out of the water. During the first two stages, the machinery-induced noise discussed in Chaps. 8 and 9 still apply. With the craft on foils, the propulsor and foils were the dominant sources.

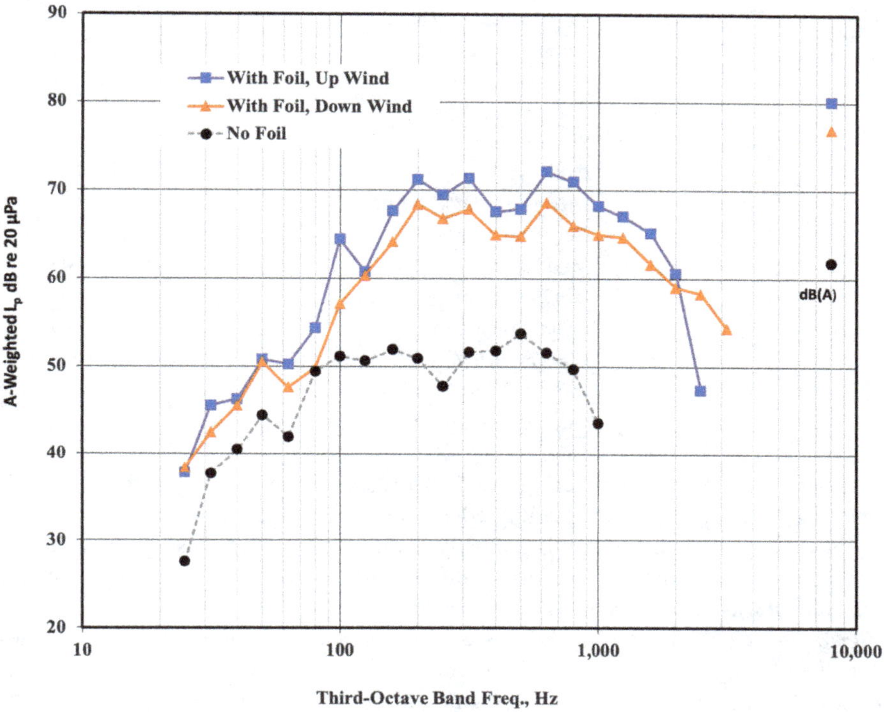

Fig. 11.10 Airborne noise in cabin on-foil and off-foil at full speed

In addition to the higher underwater radiated noise expected from hydrofoils, they may also exhibit a higher foil vibration, leading to a significant hull vibration. This results in increased on-foil noise, as shown in Fig. 11.10. Measured data on another ferry with foils amounted to 79 dB(A) in the Salon – Forward and 84 dB(A) in the Salon – Aft. For another ferry, noise levels of 75 and 72 dB(A) were achieved at 40 kt in the Lower and Upper Saloons, respectively (Bonafoux and Higgins 2015).

The measured broadband (20–1000 Hz) underwater radiated noise of three hydrofoils operating at 30 kn ranged from 179 to 197 dB re 1 µPa–m (Li et al. 2019). For comparison, in the same bandwidth, container ships at 20.6 kn had noise levels of 185 dB; vehicle carriers at 15.2 kn were at 180 dB; bulk carriers at 13.8 kts at 186 dB; and product tankers at 15.6 kn at 182 dB. In a wider bandwidth, 10 to 48,000 Hz, the source level ranged from 185 to 205 dB. The above data were provided while foil-borne. The noise signature is lower for operating below being foil-borne and in transition.

As with propeller blades, an increase in the surface roughness of the foil will produce a significant increase in the radiated noise over a wide frequency range. As shown for propellers, passive controls would consist of changes to the flow field pressure and velocity distribution. These would be accomplished by putting

Table 11.5 Drillship Compartment noise levels, Lp dB re 20μPa

Compartment/Location	Max dB(A)
Pumps	118
Generator Room	117
Compressor Area	113
Shale Shaker	109
Drill Area[a]	104
Diesel Generator Room	107
Generator Room Supply Fans	106
Mud Pit	104
Gas Production by valves	101
Injection Platform	100
Gas Production	99
Gas Scrubber	97

[a]Often controlled by brake drum squeal

obstacles in the suction surface of the foil, reshaping the surface of the foil, and changing the geometry of the leading edge (Simanto et al. 2022).

Since the greatest excitation is typically at the leading edge, this component has been resiliently attached to the rest of the foil (Brungart, 2005). Based on tunnel tests, reductions of 5 to 10 dB were achieved.

Delaying cavitation on the foils is as important as delaying cavitation on the propellers for this type of vessel. Furthermore, it has been shown during numerical studies that the radiated noise from the foil increases significantly with cloud cavitation (Hwang et al. 2021). The effects of serrated trailing edges, wavy leading edges (Pendar et al. 2021; Simanto et al. 2022), and perforated edges have been studied. However, these are still in the research stage with limited results. Similarly, the use of spanwise micro-grooved surfaces has been studied numerically (Dang et al. 2019). Theoretical reductions of less than 10 dB are expected.

11.3.7 Drillship/Offshore Structures

Airborne noise criteria for offshore installations are covered by the ABS Guide to Crew Habitability on Offshore Installations[5] and the Guide to Crew Habitability on Mobile Offshore Drilling Units.[6] As shown in Table 11.5, the following are the noisiest areas on a drillship (HSE 1999):

[5]Applies to tension leg platforms, floating production (TLPs), storage, and offloading (FPSOs), floating, storage and offloading (FSOs), spars, fixed platforms and any other buoyant or non-buoyant structure supported by or attached to the seafloor.

[6]Applies to drill ships, drill barges, self-elevating drilling units (SEDUs), column stabilized drilling units (CSDUs), or any other vessel used for the purposes of drilling.

Recommended controls start at the source, as noted in previous chapters. In order of effectiveness for most offshore rigs, the following should be considered:

- Enclosures – around equipment, with isolation to rig a secondary concern.
- Pipe lagging – combination of flexible isolating layer covered by outer flexible, mass layer
- Cladding – in lieu of enclosures for equipment, combination of flexible isolating layer covered by outer flexible, mass layer
- Barrier – partial enclosures around noisy sources on the drilling platform or on-deck
- Isolation mounting of quarter's structure

In terms of controlling noise from flares on offshore structures, injecting water into a high-pressure gas jet can reduce shock-associated noise. In one test, the noise was reduced by 13 dB (Bussman and Knott 2000). One additional note. The accidental hydrocarbon release and resulting fire on an offshore platform can generate high intensity noise, sufficient to cause hearing damage and equipment failures (Goyer and Worraker 1996). The maximum level calculated for an unignited release is 157 dB re 20 μPa.

The reported underwater radiated noise for drill ships is as follows (OCS 2016):

- 154 to 176 dB re 1 μPa–m, drilling
- 188 dB re 1 μPa–m, with thrusters

11.3.8 Direct Drive Commercial Vessels

Many large commercial vessels are directly driven. Given the size and weight of the diesel and the requirements for the foundation/inner bottom, there are few controls that are effective. However, auxiliary equipment can contribute to the radiated signature, especially at their rotation and/or firing rate tones. These can and should be treated by isolation mounting this equipment.

However, the propeller is usually the most significant source on these vessels. As noted in Chap. 9, this source can be attenuated by proper design and by maintenance of the hull and propeller surfaces.

11.3.9 Unmanned and Autonomous Underwater Vehicles – UUV and AUV

One major advantage of autonomous or unmanned underwater vehicles is their low radiated noise signature. This signature can be well below the ICES radiated noise limit for performing fish stock assessment.

Predicting and controlling the radiated noise of small, unmanned underwater vehicles (UVVs) or autonomous underwater vehicles (AUVs) is a complicated task (Fischer, 2001). The greatest advantage with respect to their signature is that the propulsors typically do not cavitate. This is due in part to the low operating speed, increased operating depth, and the opportunity for uniform inflow into the propulsor blades. These vehicles have many "sources" that produce acoustic energy and thereby contribute to the underwater signature of the vehicle. In many cases, these sources are closely coupled to the water without any significant 'path length' available to attenuate the acoustic energy. In addition to the sources already discussed for ships, propulsors, motors, and pumps, these vehicles also have contributions from speed reducers, controllers, high-pressure air systems, ballast pumps, control fins, and electronics.

The "shorted" transmission path from the source to the ocean also needs to be defined and understood. Both fluidborne and structureborne paths must be considered; airborne paths are usually less of a consideration. For many underwater vehicles, sonar self-noise may be an important design factor. Sensors are located closer to acoustic sources given the limited length of most AUVs. In one case, the operator had to utilize a towed array since on-board sensors could not provide the signal-to-noise SNR needed to operate successfully.

Acoustic models would need to identify the important paths and frequency ranges. These usually extend beyond those normally considered for ships because of high-frequency acoustic systems on most underwater vehicles. To optimize the signature and reduce the adverse impacts on space, payload, and cost of typical treatments, one must have a good understanding of the process of modeling the vehicle's radiated noise.

In order of importance, the primary sources are (1) propulsion motor vibration, (2) propulsor discrete frequency forces, (3) motor controller vibration, (4) ballast pump vibration, especially pulsations, and (5) high pressure air systems, if utilized. Energy that is not dissipated internally in the vehicle will eventually become radiated noise. On a large vessel, there is more opportunity for attenuation by these internal losses.

The main assumption in terms of modeling is that the vehicle is 'wet'; that is, sources are not enclosed in a large encompassing pressure hull such as a submarine. Simple models assume that all vibrating sources (except the propeller or thruster) are spherical or cylindrical objects in direct contact with the water. These smaller vehicles are more amenable to finite element (FE) and boundary element (BE) modeling (Frandsen 2001; DeBiesme 2000). Hydro-dynamic noise models can be based on large eddy simulation for flow-induced noise and re-radiation from the shell (Wang et al. n.d.).

For "wet" motors, the source would typically be a cylindrical casing. Radiation from a cylinder must address both axial radiation from end caps and cylindrical radiation, including pulsations, from the body. In cases where a pressure housing is used, the assumptions are that the casing will have vibration levels comparable to those of the machinery they encase. However, given enough space inside the

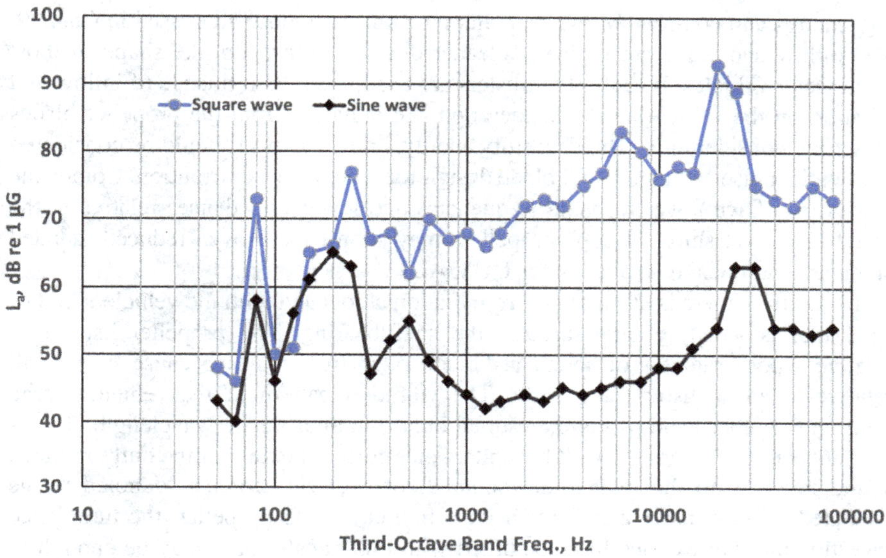

Fig. 11.11 Source vibration – motor driven by Square or Sine Waves

enclosure, isolation mounts and internal absorption would positively impact this assumption. Vibration transmitted from the sources to the hull must account for transmission through frames to the hull shell.

Most of the tones in the signature will be dominated by the electric motors and propulsion gearing. Tones belong to pole and motor slot passage (Timar 1989). Motor drive frequencies can be in the mid-thousands of Hertz. Motors controlled by electronically switched current sources typically show high vibrations. As an example, Fig. 11.11 shows that the difference can be as high as 30 dB in the radial vibration of a motor driven by a square wave and one driven by a sine wave. In terms of control, the best approach is to ensure a clean waveform of the drive current. In regards to speed reducers, as already noted in Chap. 5, spur gears produce more noise than helical gears. Next would be tones from the pumps at vane or piston rates.

Again, the propeller/thruster is generally the noisiest source and is closely coupled to the frame. It requires careful hydro-acoustic design to be quieter than the other machinery sources. Propellers are generally nonmetallic and have different re-radiation parameters than ship propellers. The vibration levels of the surfaces in contact with water control their relative contribution to radiated noise. The lower the source vibration levels are, the lower the resultant radiated signature.

The measured overall level for Autosub at 3 kn is approximately 130 dB re 1 µPa–m. The spectrum is dominated by low frequencies (Griffiths et al. 2001).

Features and controls for the propeller are similar to those addressed in Chap. 9. The inflow and wake are typically cleaner due to the torpedo-like shape of most underwater vehicles. If multiple thrusters are used, then the influence of inflow and location on the vehicle need consideration. Re-radiation from the propeller blades may need consideration. The flexibility and dynamic response should be considered in both the noise estimation and blade design for their noise contribution. Composite blades may have lower resonant frequencies and lower impedances relative to the metal blades on ships. "Loop" propellers may soon have proven reduced radiated signatures and be advantageous for UUV/AUV.[7]

In addition, there is an 'indirect' re-radiation associated with the vehicle's hull or shell that is structureborne through the shaft/bearing. The propeller itself is a relatively poor radiator of sound at "blade rate" frequencies because it is small relative to an acoustic wavelength. The vibration pattern of the vehicle's large hull, on the other hand, is commensurate in size to an acoustic wavelength. Therefore, it "couples" relatively well for underwater noise. The ratio of the hull's radiated sound intensity to the magnitude (squared) of a driving force is denoted by its "radiated noise transfer gain." Even for a non-cavitating propeller, the turbulence ingestion and vortex shedding combined with this transfer gain may be enough to cause a significant signature. Similarly, the excitation of vortex shedding or turbulence ingestion by the fins can couple to the hull structure by this radiated noise transfer gain.

The radiation efficiency for an underwater vehicle is not like a ship made of ribbed plating. The radiation efficiency is controlled by the shape of the source or canister containing the source. Thus, the radiation efficiency can either be that of a point source or sphere or a cylinder driven radially or axially (Junger, 1986). Each has a different, frequency-dependent value. Furthermore, the hull and propeller can be made of various composites, so their vibration and radiation would be dependent on their material properties – Young's Modulus and density.

In terms of noise control for underwater vehicles, they are more amenable to decoupling, cladding, or flow control than ships. For instance, twisting and cambering of tail control fins can provide swirl in the propeller flow opposite of the direction of propeller rotation. This straightens the flow into the propeller, reducing noise and improving propeller efficiency. Figure 11.12 shows this concept.

Other treatments for consideration are cladding, decoupling and damping, as shown in Fig. 11.13. Note how channels have been maintained for water cooling. Zimmerman, 2005 showed that the tail cone could have a significant impact on the vehicle's radiated noise. On one vehicle, the original tail cone came with a ducted fan, vectored thrust system. Other treatments included:

- Lower propeller speed by using a propeller with greater thrust; reduced radiated noise and hull vibration
- Replacing the motor/gearbox with a direct drive, brushless DC motor

[7] https://www.bdoutdoors.com/boating/boat-accessories/loop-propellers-prop-innovation-sharrow-marine/

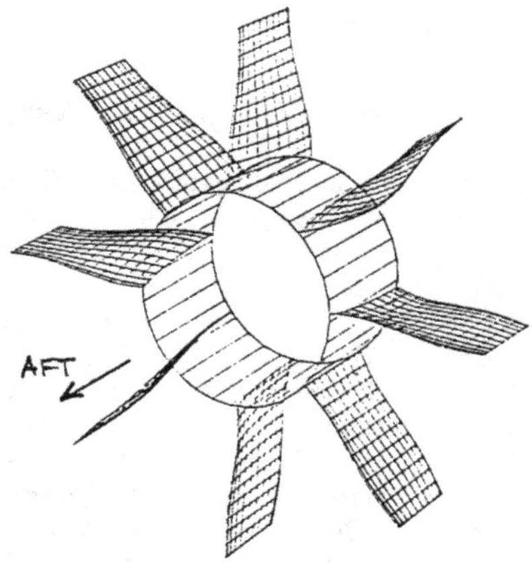

Fig. 11.12 Tail fins for pre-swirl

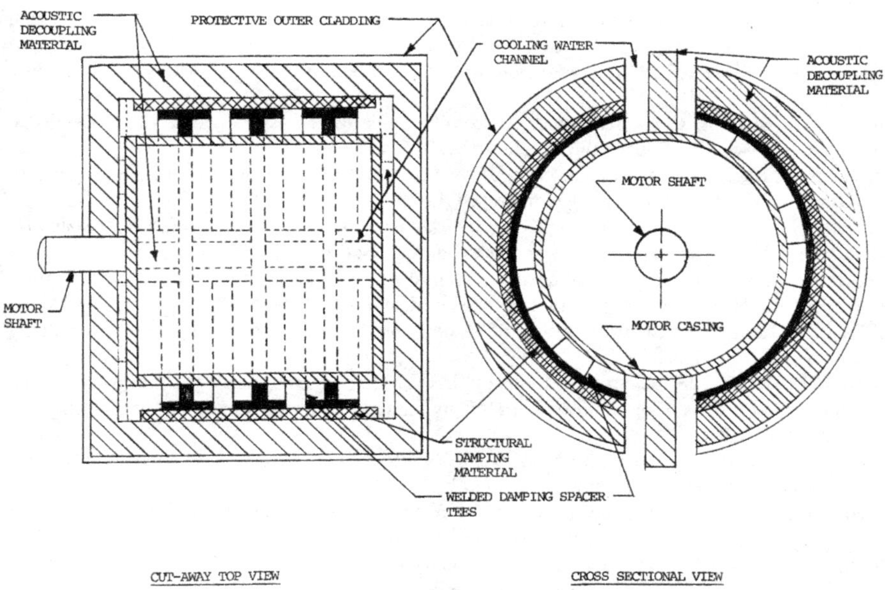

Fig. 11.13 External treatments for motor canister

- A sine wave drive was used instead of a trapezoidal voltage drive.
- Isolation mount hydrophones

These treatments provided a 20 dB decrease in the radiated noise across the frequency range up to 15 kHz.

Counterintuitively, the water loading can short out the effectiveness of an isolation mounting system. This is similar to the flanking by airborne noise of a two-stage isolation system (Chap. 8). In this case, the water provides the vibration short circuit.

References

American Bureau of Shipping: ABS, Guide to Crew Habitability on Workboats. ABS, Houston TX (2008)

American Bureau of Shipping: ABS, Guide to Crew Habitability on Mobile Offshore Drilling Units. ABS, Houston TX (2012)

American Bureau of Shipping: ABS, Guide to Crew Habitability on Offshore Installations. ABS, Houston TX (2016)

Andresen, K.: Thrusters for Auxiliary Propulsion and Manoeuvrability. Ship Propulsion Systems, Brunvoll Thrusters, n.d..

Blackwell, S.B., Green, C.R.: Underwater and in-air sounds from a small hovercraft. J. Acoust. Soc. Am. (JASA). **118**(6), 3646–3652 (2005)

BOEM (Bureau of Ocean Energy Management): Quieting Technologies for Reducing Noise During Seismic Surveying and Pile Driving Workshop – Summary Report, OCS Report BOEM-2014-061, CSA Ocean Sciences, Inc., Stuart, FL, March 2014

Bonafoux, J., Higgins, G.: Development of the first 50+ knot ferry in North America. Pentamaran (2015)

Boroditsky, L.: Acoustic Vibration of Ice Breaker Shell Plating. Journal "Sudostroenie", 1985, N.5, p. 9-11 (in Russian)

Boroditsky, L., Fischer, R.: Noise Prediction for Composite Hull Vessels. Inter Noise 2010, Lisbon, Portugal, June, 2010

Brungart, T., Myer, E., Capone, D., et al., Hydrofoil vibration and noise reduction with leading edge isolation. Noise-Con 2005, Minneapolis, MI (2005)

Bussman, W.R., Knott, D.: Unique concept for noise and radiation reduction in high-pressure flaring. In: Offshore Technology Conference (OTC), vol. 12160. OTC (2000)

Dang, Z., Mao, Z., Tian, W.: Reduction of hydrodynamic noise of 3D hydrofoil with spanwise microgrooved surfaces inspired by sharkskin. J. Mar. Sci. Eng. **7**, 136 (2019)

DeBiesme, F.-X.: Modeling the Acoustic Signature of an Ocean Explorer Class Autonomous Underwater Vehicle, MS thesis, Florida Atlantic University, 2000

ERDC TN-DOER-E14: Characterization of Underwater Sounds Produced by Bucket Dredging Operations, Aug. 2001

ERDC TN-DOER-E38: Characterization of Underwater Sounds Produced by a Hydraulic Cutterhead Dredge during Maintenance Dredging in the Stockton Deepwater Shipping Channel, Calif., March 2014

Fischer, R.W.: Bow Thruster Induced Noise and Vibration. Marine Technology Society, Dynamic Positioning Committee, Houston, Oct. 2000

Fischer, R.: Factors affecting the radiated noise of underwate vehicles. Underwater Intervention 2001 Proceedings, Marine Technology Society (2001)

Fischer, R., Boroditsky, L., Yankaskas, K.: Controlling Noise on High Speed Naval Craft. In: ASNE High Speed Performance Ships Crafts Workshop, 2007

Frandsen, S.: Experimental Analysis of the AUV Acoustic Signature, Thesis –Florida Atlantic University, May 2001

Goyer, H.G.D., Worraker, W.J.: Jet Noise Effects During Accidents Report OTH 94 455, AEA Technology. HSE Book, Oxfordshire, UK (1996)

Griffiths, G., Millard, N., Enoch, P.: On the radiated noise of the Autosub autonomous underwater vehicle. ICIES J. Mar. Sci. **58**, 1195–1200 (2001)

Heyn, H.-M., Skjetne, R.: Time-frequency analysis of acceleration data from ship-ice interaction events. Cold Reg. Sci. Technol. **156**, 61–74 (2017)

HSE: Offshore Technical Report – OTO 1997 012, Collection and Analysis of Offshore Noise Data, Health & Safety Executive, May 1999

Hwang, H.-S., Paik, K.-J., Lee, S.-H., Song, G.: Numerical study on the vibration and noise characteristics of a Delft Twist11 Kydrofoil. J. Mar. Sci. Eng. **9**(2), 144 (2021)

Johnson, G.: Permanent Magnet Thrusters, Why PM Thrusters, Rolls-Royce, 2015

Junger, M., D. Feit.: Sound, Structures, and Their Interactions, 2nd Ed., The MIT Press, Cambridge, Ma. (1986)

Lanni, F., Brown, N.: Impeller for Marine Waterjet Propulsion. U.S. Patent No. 6,135,831, Oct. 24, 2000

Li, J., White, P., et al.: Underwater radiated noise from hydrofoils in coastal water. J. Acoust. Soc. Am. (JASA). **146**(5), 3552–3561 (2019)

Marine Aggregated Levy Sustainability Fund (MALSF): Marine Environmental Protection Fund (MEPF): Measurement of underwater noise arising from marine aggregate dredging operations, MEPF Ref. No: MEPF 09/PI08, Feb. 2011

Nilsson, A.C.: Noise reduction of bow thrusters. In: Buiten, J. (ed.) Shipboard Acoustics Proceedings ISSA '86. Martinus Nijhoff Publishers, Dordrecht (1986)

OCS – Outer Continental Shelf, Alaska Outer Continental Shelf, MMS 2008-0055, Beaufort Sea and Chukchi Sea Planning Areas, Draft Environmental Impact Statement, Vol. II Ch. 4.4, Nov. 2008

OCS – Outer Continental Shelf, Alaska Outer Continental Shelf, OCS EIS/EA BOEM 2016-069, Cook Inlet Planning Area, Final Environmental Impact Statement, Vol. I Chs. 1–5, Dec. 2016

Pang, F.Z.: Ice-induced vibration and noise of ships. In: Cui, W., Fu, S., Hu, Z. (eds.) Encyclopedia of Ocean Engineering. Springer, Singapore (2022). https://doi.org/10.1007/978-981-10-6946-8_109

Pang, F., Shen, Z., et al.: Research on Ice-Breaking Induced Vibration Characteristic of a Ship. Inter-Noise, Hamburg, 2016

Pendar, M-R, Pascoa, J., Roohi, E.:Cavitating flow structure and noise suppression analysis of a hydrofoil with wavy leading edges. In: CAV 2021, 11th International Symposium on Cavitation, Daejon, Korea, May, 2021

Pustoshny, A.V., Darchiev, G.K., Frolova, I.G.. The problem of propeller design for high ice class transportation ships. In: 5th International Symposium on Marine Propulsors, Espoo, Finland, June 2017

Roth, E.H., Schmidt, V., Hildebrand, J., Wiggins, S.M.: Underwater radiated noise levels of a research icebreaker in the Central Arctic Ocean. J. Acoust. Soc. Am. (JASA). **133**(4), 1971–1980 (2013)

Simanto, R., Hong, J.-W., Kim, K.-S., Ahn, B.-K., Shin, S.: Experimental investigation on cavitation and induced noise of two-dimensional hydrofoils with leading edge protuberances. JASA. **34**(12) (2022)

Spence, J., Fischer, R., et al.: Review of Existing and Future Potential Treatments for Reducing Underwater Sound from Oil and Gas Industry Activities, Noise Control Engineering Report 07-001, Prepared for Joint Industry Programme on E&P Sound and Marine Life, Billerica, MA, Dec. 2007

Streckwall, H., Bretschneider, H.: Prediction of tonal underwater noise pattern from Cavitating propellers with special attention to ice cover effects. Pol. Marit. Res. **21**, 35–40 (2014)

Suarez, J., Krewsky, W., Kris, K., et al.: Design Criteria for Vibration Mitigation for Icebreaking Vessels. In: SNAME Maritime Convention, Houston, Tx., Sept. 2022

Timar, P.L.: Noise and Vibration of Electrical Machines. Elsevier Science Publishers, Amsterdam (1989)

Wang, L.S.: Measurement of Underwater Noise Arising from Marine Aggregate Dredging Operations. Marine Aggregate Levy Sustainability Fund (MALSF) Report, Crown Copyright, 2011

Wang, K., Zhang, T., et al.: Numerical Simulations of Hydrodynamic Noise of an Underwater Vehicle. IEEE Xplore, n.d.

Yakovlev, A., Sokolov, M., Marinich, N.: Numerical Design and Experimental Verification of a RIM-Driven Thruster. In: Second International Symposium on Marine Propulsors, SMP'11, Hamburg, Germany, June 2011

Yang, Q., Wang, Y., Zhang, Z.: Numerical prediction of the fluctuating noise source of waterjet in full scale. J. Mar. Sci. Technol. **19**, 510–527 (2014)

Yankaskas, K., R. Fischer: X-craft: noise prediction modeling and hearing improvement: An HIS Application. In: American Society of Naval Engineers, Human Systems Integration Symposium 2005, Arlington VA, 2005.
Zimmermann, R., D'Spain, G., Chadwell, C.: Decreasing the Radiated Acoustic and Vibration Noise of a Mid-size AUV. IEEE Journal of Oceanic Engineering 30 (1), Jan 2005

Chapter 12
Cost and Return on Investment (ROI)

This chapter covers the possible payback or Return On Investment (ROI) of treatment versus paying hearing loss compensation or potential harm to the environment. Cost in the sense used here addresses treatment impacts on space and weight in addition to the material costs. Benefits with respect to hearing loss compensation are based solely on available US Navy and Veterans Administration data. This book does not consider cost with respect to ocean species; however, the reader is referred to IMO MEPC.1/Circ. 833 (2014) and Merchant 2019.

For example. Some materials normally installed onboard ships also have good, or at least minimal, acoustic properties. Thermal and fire insulations are the easiest to identify. Deck coverings are another as they provide some additional damping. Thus, taking their acoustic properties into account has minimal additional impact on space, cost, and weight. Occasionally, a higher performing acoustical material may be needed to go over the fire or thermal insulation to achieve the desired acoustic performance. In this case, there is an 'additional' cost in terms of weight, space and material. Another often utilized noise treatment is isolation mounting of equipment. Here, the material cost of the isolation mounts is sometimes minimal with respect to the other design impacts. The mounts typically raise the feet of the equipment, causing a space impact, at times creating exhaust pipe routing problems. In the marine environment, a clearance envelope around the mounted equipment is needed to avoid damage while operating in a high sea state, thus taking up additional space. The use of mounts often necessitates additional engineering and material costs for a high stiffness foundation below the mounts and the addition of flexible connections. Furthermore, elastomeric mounts have a life expectancy of 7–10 years.

These 'costs' vary with treatment type and ship acoustic goals. Studies have shown that the Return On Investment with respect to hearing conservation can be relatively short – much less than the life expectancy of the vessel. These variables are covered in this chapter.

R. Fischer, L. Boroditsky, *Noise and Vibration Control on Ships*,
https://doi.org/10.1007/978-3-031-55170-3_12

For the Armed Forces, hazardous noise is the only known occupational hazard with exposure levels exceeding protection capability (Fischer and Page 2013). The US Navy and OSHA believe that "noise controls provide significant long-term advantages over personal hearing protection" (Technology for a Quieter America 2010). Based on studies carried out for the U.S. Navy, the authors have shown that this engineering and its associated cost provide a positive Return on Investment (ROI) (Fischer and Page 2013) with respect to compensation for hearing loss paid by the US Veterans Administration. Furthermore, given the latest efforts to reduce the environmental impact of ship noise on the oceans, one regulatory group – DNV – claims that "moderate noise emission is possible without increasing costs" (Collins 2010). Not only is there a 'cost' associated with human response to noise, but a similar impact on ocean species has been quantified for Southern Resident Killer Whales (Griffin 2006). The 'cost' is related to the energy cost on the echolocation ability of the whale. Whale watching vessels cost 18–23% of this energy.

Finally, in one study carried out for a gas turbine silencer effort (M&I 2018), it was expected that between better flow conditions and an improved exhaust silencer, millions of dollars per year could be saved on fuel cost for certain naval vessels. These changes would reduce noise and CO levels (M&I Power Technology 2018). This cost saving takes into account the mission profile, e.g., operating speed vs. percentage of time at that speed.

Sometimes, the cost savings involve more than just reduced material cost, but also in weight benefits and lower installation and maintenance costs. An accurate and easily applied integrated approach to noise/vibration control reduces the system weight, impact on space, impact on vessel operations, and cost. The cost may be in terms of a penalty (or incentive) for delivering a vessel meeting acoustic requirements. Retrofitting a vessel to reduce noise excesses can be as high as ten times that needed from conducting a noise/vibration analysis early in the design and installing recommended controls. Further non-marine treatment/cost benefits are provided in Berry and Flindell 2009.

There are other 'cost' impacts not specifically included in this chapter (Berglund and Lindvall 1995; World Health Organization, Europe 2005). These are hard to quantify in terms of actual dollars.

- retention of crew
- missed or mis-understood commands
- annoyance
- health
- sleep disturbance and after-effects of noise disturbance by noise
- stress-related effects
- Cardiovascular effects
- bodily fatigue
- concentration, task interference, productivity, motor skills.

12.1 Hearing Loss

As of 2020, more than 1.3 million US veterans received hearing loss compensation; over 2.3 million took compensation for tinnitus. These two losses were the most prevalent among all Veterans Administration types of losses. It has been estimated that these two losses cost the government over $1.2 billion annually (Alamgir et al. 2016). This is much higher than the equivalent civil losses estimated by the CDC to be $242 million annually for twenty-two million workers.

Noise remains the most prevalent occupational hazard associated within the defense systems and their operations. To reduce this noise in the future, US Navy acquisition programs must apply proven control measures and understand the impact of potentially even noisier systems. This would require a greater emphasis on risk management, improved technical knowledge on the feasibility and effectiveness of acoustic controls, and the real potential on ROI from acoustic treatments (Fischer and Page 2013).

Recently, the Defense Safety Oversight Council's Acquisition and Technology Task Force sponsored a project to identify common noise sources affecting multiple defense systems. A system engineering risk management process was applied to review key noise sources within the Department of Defense (DoD).

Estimates were made of exposed populations, the range of their occupational exposures and potential risk, and the fiscal cost of hearing loss (Fischer and Page 2013).

The parameters considered in computing the Return-on-Investment were (1) time weighted noise level, (2) number of systems and crew, (3) service life of the systems and crew, (4) effectiveness of the derated hearing protection device, (5) cost of audiograms, hearing aids, and veterans' disability, and (6) estimated effectiveness and 'cost' of treatments, materials and installation. The algorithms developed considered hearing loss per ANSI S3.44, ISO 1990:1999, the expected noise-induced permanent threshold shift (NIPTS), compliance with the DoD Hearing Conservation Program, and VA policies and payout.

Costs related to impact on crew performance or health and social impacts were not considered. The potential benefits of longer equipment life, lower maintenance, 'buying-quiet', increased efficiency and beneficial side effects of better designs, e.g., using Computation Fluid Dynamic (CFD) modeling, are other factors that cannot be directly computed. Thus, both the costs and benefits are underestimated, making the ROIs an underestimation. The return on investment is most easily dealt with respect to hearing loss since these costs can be quantified. The costs of material, engineering, installation, and maintenance are more difficult to track and vary from yard-to-yard and vendor-to-vendor.

For example, Table 12.1 provides a notional study with respect to the cost benefit and hearing loss by Fischer and Page 2013:

Table 12.1 Cost benefit study

DoD source	Return on investment	Potential NIHL cost reduction	Untreated TWA	dB(A) reduction	Service years
Shipboard Diesel Driven Systems	0.2:1–4:1	$774,708,120	110 dB(A)	33	40
Shipboard Gas Turbines	0.2:1–2:1	$38,509,074	90 dB(A)	8	35
Ships and High Speed Craft	1:1–3:1	$49,218,444	97 dB(A)	17	22
Shipboard Equipment	11:1–40:1	$3,889,987,680	95 dB(A)	7	40

NIHL Noise Induced Hearing Loss, *TWA* Time Weighted Average

12.2 Habitability Treatments

These 'costs' vary with treatment type and the ship's acoustic goals. Studies have shown that the return on investment with respect to hearing conservation can be relatively short. These variables are covered in Tables 12.2 for treatments in general and in Table 12.3 for treatments used to reduce diesel noise (Spence et al. 2007). Figure 12.1 shows the range of noise induced by diesel with the range of treatments presented in Table 12.3.

12.3 Underwater-Radiated Noise Treatments

Renilson 2009 provides 'approximate costs' changes to the propulsion system with the intent of reducing the underwater signature. The cost to change the propulsion systems on a 20,000 dwt containership, 250,000 dwt large tanker, and a fast passenger ferry are contained in Tables 12.4, 12.5 and 12.6.

These tables consider the cost of design studies, construction of a propeller, and add-on treatments, including Propeller Boss Cap Fins (PBCF), Propeller Cap Turbine (PCT), and vortex generators. Gains for the PBCF are on the order of 3–5% (Mewis and Hollenbach 2006).

Per Renilson 2009, these changes will result in an increased efficiency of 5–10%. The annual savings are estimated to be on the order of US $500 k to US $1000 k. Thus, the ROI is on the order of one year.

Renilson, 2009 states "the increase in efficiency is likely to be approximately 5–10%, resulting in a daily savings of 2–4 tonnes. . . .with an annual savings of US $400k–US $800k." The return on investment in this case is on the order of a year to two and a half years.

The cost for the installation of just a PBCF on various vessels is provided in Table 12.5 (Renilson 2009). Mitsui OSK Techno-Trade Ltd. experiments in a cavitation tunnel showed an unverified reduction of 3–6 dB for frequencies above

Table 12.2 Treatment cost and noise reduction

Type of treatment:	Treatment location:	Impact on location:	Approx-imate cost, material only:	Path affected[a]	Noise reduction, dB between 63 to 500 Hz	Noise reduction, dB between 500 to 8000 Hz
Engine enclosure	Source room	All rooms	\$100–\$200 sq. m.	AB, SSB	5–10	10–20
Absorptive layers	Source room or compartments	Source room, compartment	\$10–\$50 sq. m.	AB, FSB, SSB	2–4	4–8
High Transmission Loss (HTL)	Source room	Adjacent compartment	\$1–5 0.1 sq.m	AB, SSB	3	5–25
One stage resilient mounts	Source room	All compartments	\$80–\$4000 per unit; 4 mounts	FSB	3–10	10–20
Two stage resilient mounts	Source room	All compartments	\$5000–\$30,000 per unit	FSB	6–15	15–35
Silencers	Source room	Source room	\$500–\$2000	AB	10–20	15–30
Damping Tile	Source room	All compartments	\$50–\$180/sq. m.	AB, FSB, SSB	3–5	4–8
Damping Tile	Compartments	Compartments	See above	AB, FSB, SSB	3–6	4–8
Floating floor	Compartments	Compartments	\$150–\$400 sq. m.	FSB, SSB	6–8	7–9
Joiner panel	Source room	Adj. Compartment	Insulation costs above	AB, SSB	1–4	3–6
Joiner panel	Compartments	Compartment		AB, FSB, SSB	2–5	4–7

[a] *AB* Airborne Path, *FSB* First Structureborne Path (structural), and *SSB* Secondary Structureborne Path

Table 12.3 Example of potential treatment impacts for diesels (Fischer 2013)

Treatment	Cost	Space	Weight
Enclosure	$15,000	Volume of equip +0.6 m	Approximately 0.63 kg/sq. m.
Buy-quiet	$100,000–500,000 (10% of original)	Minimal impact	Approximately 10% increase due to damping
Cladding	$100/sq. m.	13–25 mm thick	0.08 kg/sq. m.
Isolation mounts	$300–$1000 per mount	100–150 mm height	14.5–91 kg.
High-imped-ance foundation	$2000 – For materials	50–100 mm tall	Approx. 900 kg.
Passive vibra-tion absorbers	$1000	1 m long cantilever (100 × 100 mm cross-section)	No more than 45 kg.
Active control -intake/exhaust	$15,000 for trans-ducers and control system	Minimal impact	Minimal impact
Active control – vibration	$50,000 to $100 k per system	15 cm tall 15 cm wide 0.3 m across	90 kg. Per actuator
Hydraulic mounts	$500 to $2000	Slightly smaller than regular mounts	14–90 kg.
Distributed vibration absorbers	$10,000 to $20,000	10 cm tall	450 kg.
Enclosure	$15,000	Volume of equip +0.6 m	Approximately 0.63 kg/sq. m.

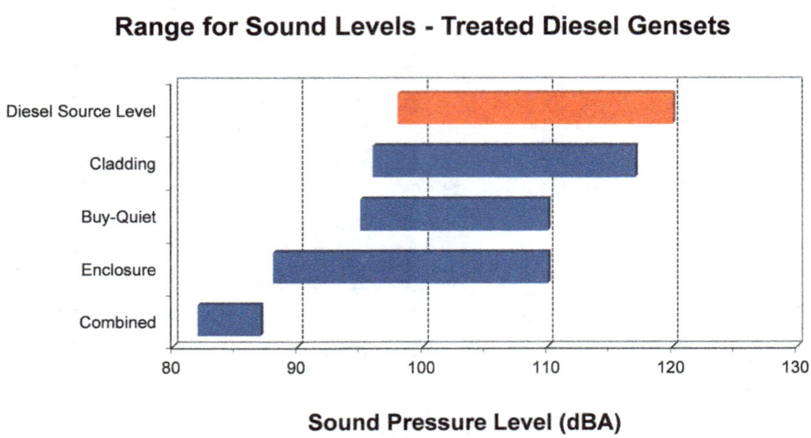

Fig. 12.1 Range of sound level for exposure to diesel noise

Table 12.4 Rough estimate of cost to retrofit 20,000 dwt containership (Renilson 2009)

Item	Approximate cost (US$k)	
	Low estimate[a]	Upper estimate[b]
Detailed design studies, including CFD and model testing	100–200	100–200
Design and construction of new propeller	20–50	300–400
PBCF/PCT	100	100
Installation of propeller, vortex generators and PBCT[c]/PCT[d]	30–50	100–200
Total	250–400	400–700

[a]assumes that the propeller only needs modifying, not replacing, and that all work can be done in a scheduled dry dock period
[b]assumes that a new specialist propeller is needed (such as high skew, or CLT) and that this needs to be fitted at a dedicated dry dock period
[c]*PBCF* Propeller Boss Cap Fins
[d]*PCT* Propeller Cap Turbine

Table 12.5 Rough estimate of cost to retrofit 250,000 dwt large tanker (Renilson 2009)

Item	Approximate cost (US$k)	
	Low estimate[a]	Upper estimate[b]
Detailed design studies, including CFD and model testing	100–200	100–200
Design and construction of new propeller	20–50	1400–1700
PBCF/PCT	200	200
Installation of propeller, vortex generators and PBCT[c]/PCT[d]	80–100	300–400
Total	600–850	2200–2800

[a], [b], [c], and [d] see Tables 12.4

Table 12.6 Rough estimate of cost to retrofit 180 m passenger ferry (Renilson 2009)

Item	Approximate cost (US$k)	
	Low estimate[a]	Upper estimate[b]
Detailed design studies, including CFD and model testing	100–200	100–200
Design and construction of new propeller	40–100	300–1,600[e]
PBCF/PCT	200	200
Installation of propeller, vortex generators and PBCT[c]/PCT[d]	60–100	200–300
Total	400–600	800–2300

[a], [b], [c], and [d] see Tables 12.4
[e] The lower estimate is based on new propeller blades only, utilizing the existing hub, whereas the upper estimate assumes that a completely new hub is needed

1000 Hz. The cost for a PBCF system ranges from $40 k for a cargo ship up to $185 k for a container ship (Renilson 2009).

Yatsenko 2012, provides approximate costs and associated noise reduction for various propeller types. These are summarized below:

- Water Jet or Pump Jet
  - Low-frequency sound reduction (LFS) of 15 dB, approximate cost of 6–10x that of standard propeller if replacing ($600 k–$1000 k)

- Skewed propeller
  - LFS of 10 dB, approximate cost of 2x to replace ($200 k)
- Forward-skew blades in Duct
  - Approximately 20 dB new v old, no cost data
- Kort Nozzle, Ducted Propellers
  - LFS of 5 dB, approximate cost of 3–5x to replace ($300 k–$500 k)
- Kappel Propellers
  - Approximate 4% increase in efficiency

12.4 Conclusions

The acoustic environment in and around ships can be abated with a reasonable ROI. The cost of reducing hearing loss has a relatively short pay-back period. There are other advantages such as health and crew retention that are not included in this metric. The pay-back or ROI with respect to underwater noise is much harder to quantify. However, the costs for various approaches have been provided. New information should be considered as it becomes available.

References

Alamgir, H., Tucker, D., et al.: Economic burden of hearing loss for the U.S. military: a proposed framework for estimation. The Society of Federal Health Professional (AMSUS). **181**(4) (April 2016)

Berglund, B., Lindvall, T. (eds.): *Community Noise. Archives of the Center for Sensory Research* Chapter 7 Effects of Noise on Humans. Jannes Snabbtryck, Stockholm (1995)

Berry, B., Flindell, I.: *Estimating Dose–Response Relationships between Noise Exposure and Human Health Impacts in the UK,* BEL Technical Report, BEL 2009-2. (July 2009)

Collins, B.: Comfort & Silent Class, Novel Class Notations from DNV, Presentation at Silent Ocean, (March 2010)

Fischer, R., Page, C.: Acoustical engineering controls and estimated return on investment for DOD selected high noise sources: a roadmap for future noise control acquisition. DTIC AD 1036036, April 2013. https://apps.dtic.mil/sti/citations/AD1036036

Griffin, R.: Sound Exposure of Southern Resident Killer Whales. (April 2006). https://vdocuments.net/sound-exposure-of-southern-resident-killer-whales-rachael-m-griffin-bsc-april.html

IMO Guidelines for the Reduction of Underwater Noise from Commercial Shipping to Address Adverse Impacts on Marine Life, MEPC.1/Circ. 833, (April 2014) (due to be updated in 2023)

M&I Power Technology Solutions: (Private Communications), *Feasibility Report on LM2500 Intake and Exhaust Retrofit,* (Dec. 2018)

Merchant, N.: Underwater noise abatement: economic factors and policy options. Environ Sci Policy. **92** (2019)

Mewis, F., Hollenbach, U.: Special measures for improving propulsive efficiency. HSVA NewsWave, the Hamburg Ship Model Basin Newsletter, (2006)

Renilson Marine Consulting Pty Ltd.: *Reducing Underwater Noise Pollution From Large Commercial Vessels*, Commissioned by the International Fund for Animal Welfare, March (2009)

Spence, Jesse, Fischer, R., Bahtiarian, M., et al.,: *Review of Existing and Future Potential Treatments for Reducing Underwater Sound from Oil and Gas Industry Activities*, Prepared for Joint Industry Program on E&P Sound and Marine Life, NCE Report 07-001, (Dec. 2007)

Technology for a Quieter America: National Academy of Engineering of the National Academies. National Academies Press, Washington DC (2010)

World Health Organization (WHO), Europe, Quantifying Burden of Disease from Environmental Noise:. *Second technical meeting report*, Copenhagen, Denmark, 2005

Yatsenko, S.: Appendix D: Reducing Underwater Sound to Protect Marine Life. CSL Australia (2012)

Chapter 13
Inspection and Testing

Inspection and testing are the 'insurance' needed to ensure that the previous engineering and construction efforts pay off. The inspections would generally be assigned to the Quality Assurance (QA) group. An important last step is to demonstrate compliance with the acoustic requirements imposed by the owner and/or regulatory agencies. Appropriate test methods, instrumentation, calibration, and documentation need to be addressed. The type of data needed to assist in determining the cause of acoustic excesses will be covered in this chapter.

13.1 Inspection

A key variable in the success of the Noise and Vibration Control Program Plan is the difference between the assumed model parameters and the actual installation. For instance, field fitting of an HVAC duct may include several extra 90-degree elbow turns (see Fig. 13.1), changes in duct size (thus velocity) or other changes that usually have an adverse impact on the noise. Fig. 13.1 shows that ducts at the bulkhead are twisted relative to the design, adding unnecessary turns. While QA checks per the Noise Control Plan may find and account for some of these changes, they may not all be detected. Thus, on-site training of the various shipyard trades proves to provide a significant payback.

To ensure that the trials reflect the finished vessel, the following should be inspected and verified prior to trials. The best way to minimize final minute or even major adjustments/fixes is to schedule inspections at intervals throughout construction. In between inspections, pictures or videos should be taken of suspected poor/incorrect installations to determine if corrective actions are needed. The output of each inspection should be a deficiencies form with recommended actions and dates for completion, if needed.

R. Fischer, L. Boroditsky, *Noise and Vibration Control on Ships*,
https://doi.org/10.1007/978-3-031-55170-3_13

Fig. 13.1 Duct, as installed – not designed

Systems and equipment to be inspected with respect to acoustic 'completeness' are as follows:

- Acoustic insulation and ceilings– type, thickness, attachment method – between frames and beam wrapping, type of vapor barrier and sheathing, taping of edges, paint on surface?
- High transmission treatments – area of coverage, minimized penetrations and clearance around penetration, paint on surface; overlapped joints to avoid gaps?
- Joiner material – unnecessary holes and penetrations; isolation material around perimeter if required?
- Damping treatments – area of coverage, thickness, sufficient attachment to structure, coverage per acoustic drawings?
- Floating deck or room – potential short circuits, clearance around penetrations?
- Isolation mounts correctly installed?
- Pipe hangers on resiliently mounted systems?
  - Attached to stiffeners and not plate?
  - Hanger does not torque isolator or not used for alignment?
  - Correct isolator per acoustic calculations/drawings?
- Flexible hose (see Fig. 13.2)
  - in appropriate locations?
  - not misaligned?
  - not vibration shorted?
  - correct lengths of flexible hose on dog-legs, if used?
  - appropriately supported by isolation mounts, as designed?

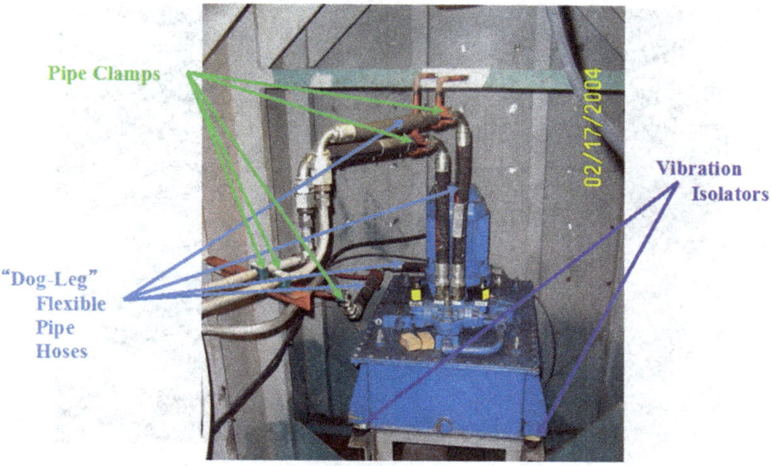

Fig. 13.2 Dog-leg flex hose with pipe clamps

13.1.1 General Inspection Steps for Isolation Mount Systems

The following provides some general inspection guidelines and specific "installation procedure" recommendations with attendant forms.

- Isolation mounts – check all equipment scheduled to be isolation mounted
  - Correct type and number of mounts, including mount hardware
  - Over/under load based on predicted static displacement, measure mount height at four corners if system is deemed critical to program
  - Mounts are oriented properly, loaded in the proper direction, and exhibit acceptable angular alignment; see Figs. 13.3 and 13.4.
  - Mounts are fabricated using hardware and proper materials as indicated on the design drawings and are not damaged by oil, paint, or hot work.
  - Mounts are not older than that allowed by mount vendor
  - Mounts are protected against paint, oil, and inadvertent damage
  - Snubber clearance, if utilized to limit motions
  - Meets clearance envelope for sea-state and/or shock, see Fig. 13.5 for example
  - Is held captive?
  - Checked with respect to vendor recommended 'replacement height'
  - Flexible connections aligned correctly and correct material used – exhaust, piping, and cables; correct installation shown below
  - No vibration short circuits; touching of mounted equipment to ship's structure
  - Grounding straps in place
  - Mounts over stiffened foundations and not on cantilevered plate/foundation
  - Isolation mount (Fig. 13.3) or raft not misaligned
  - Shipping brackets removed (Fig. 13.6)

Fig. 13.3 Spring mount over compressed and misaligned

Fig. 13.4 Mount cocked relative to foundation and machinery foot

13.1.2 Vibration Isolator Inspection Report

This inspection report will document the salient data that should become part of the Test Report. Critical information includes installation data and final loaded heights.

1. Fill out a Vibration Isolator Inspection Report (Table 13.1) for each separate piece of vibration-isolated equipment/machinery.
2. Install vibration isolators onto equipment foundation according to appropriate drawing instructions and/or standard shipyard practices. Grossly uneven mating surfaces should be corrected by grinding, machining, or shimming.

Fig. 13.5 Isolation mounted motor shorted against vertical post

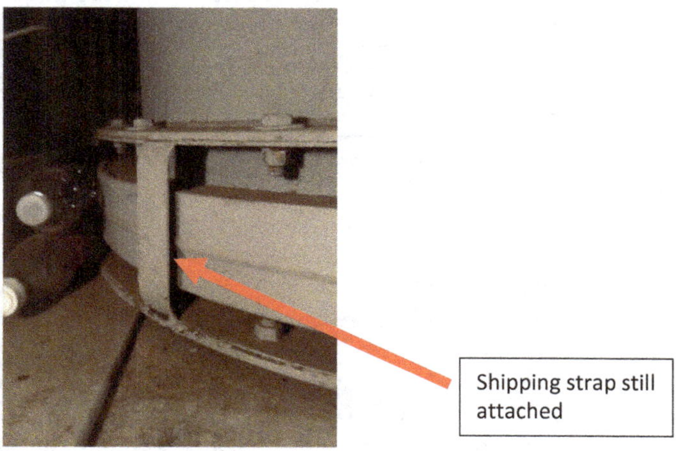

Fig. 13.6 Shipping brackets still installed on isolation mounted fan

Table 13.1 Vibration isolator inspection report – Page 1 of 2

Vibration Isolator Inspection Report Page 1 of 2

MACHINE: ____________________

UNIT ID#: ____________________
(i.e., port, starboard, forward, aft)

COMPARTMENT: ____________________

SYSTEM/SERVICE: ____________________

MOUNT INFO
MFG & MODEL: ____________________
(Take off mount ID Plate)

DESIGN HEIGHT: ____________________
(Take off mount data sheet)

MOLD DATE: ____________________
(Take off mount ID Plate)

INSTALLATION DATE: ____________________

Machinery/Isolation System:						Check when Inspected (See below for Details)				
MOUNT	Undeflected Height (mm)	Initial Loaded Height (mm)	Height After 48 Hours (mm)	Height After Piping & Cables Installed (mm)	Shim Height If required (mm)	Alignment	Clearances	Orientation	Painting	Snubber
1										
2										
3										
4										
5										
6										
7										
8										
9										
10										

3. Before the equipment is placed on the vibration isolators, the height (undeflected) shall be measured and recorded on the attached Vibration Isolator Inspection Report – Table 13.2.
4. The mounts shall be shimmed (or aligned with Chock-fast Orange) so that all the undeflected mount heights are within 0.7 mm (0.028 in).
5. Install equipment/machinery on vibration isolators according to appropriate drawing instructions and/or standard shipyard practices. Remove all shipping blocks and temporary chocks.
6. Measure mount heights from the same reference locations as used in step 3 above.
7. Visually review mounting for general, alignment, clearance, orientation, painting and snubber conditions listed on the Check if okay or make notes of conditions on second page of the Vibration Report.

Table 13.2 Vibration isolator inspection report – Page 2 of 2

Vibration Isolator Inspection Report Page 2 of 2

MACHINE: ______________________________

UNIT ID#: ______________________________

EQUIPMENT DIAGRAM SHOWING MOUNT LOCATIONS

PROBLEMS:

COMMENTS:

PREPARED BY: ______________________ DATE: __________

Approved by: ______________________ DATE: __________

8. After equipment has been on the mounts for 48 hours, measure the mount heights using the same reference positions and report on the Vibration Isolator Inspection Report.
9. After all piping and cabling is connected to the equipment/machinery measure mount heights using same reference positions and report on Vibration Isolator Inspection Report. Any significant height differences from the previous measurement should be investigated. Visually inspect.
10. Submit complete reports as identified in the Noise Control Plan.

- Alignment: After loads are applied, the top and bottom shall remain parallel. Mounts shall not be twisted. The tops of the mounts shall be within 0.7 mm (0.028″) of each other.
- Clearances: All mounts shall have a minimum of 2.54 cm (1″) clearance to any other un-isolated area of the ship.
- Orientation: Check that all mounts are oriented as shown on the installation drawings. Some mounts may be rotated 90° relative to others.
- Painting: Electrometric elements must never be painted. Rubber elements are to be masked if paint is to be applied. If painting is completed, make sure masking is removed and rubber elements have not been painted.
- Snubber: When appropriate, check that the snubbers are in place and properly loaded. See the isolator data sheets for details.
- Visual: Visually check condition of mounts, interferences with other component systems, locations per drawings, flexibility, tightness of all attachment bolts, no welding damage, etc.

13.1.3 Damping Treatments

Damping treatments require close inspection of both tiles and spray-on treatments. Tiles, especially when glued to curved sections, sometimes have voids. Tapping can generally find these voids. The depth of spray-on treatments should be checked with gauges to ensure that the desired thickness is achieved, without apparent sagging.

13.1.4 Safety

Normal shipyard safety procedures should be followed at all times. Potential tests that may require special safety steps should be identified in the Test Plan. This includes the operation of equipment by shipyard or owner personnel. Noise and vibration test specialists should not turn machinery on and off or operate unless advance notice and guidance has been provided by the shipyard or owner personnel. All Close Out notices should be observed.

Cabling from sensors to measurement locations should be tie wrapped in such a way as to avoid tripping hazards. Cabling should not pass through a water tight door – transits through the water-tight bulkhead should be utilized.

Prior to testing on masts and antennas, it is important to secure radar equipment and ensure that it is locked out until testing is completed.

Life jackets should be worn while on-deck, during at-sea transfers, and while using RIHB or lifeboats to conduct in-water testing.

13.2 Testing

The ship's specification would address any acoustic qualifications imposed by the owner, operator, or regulatory agency such as IMO or an NGO. This section should also identify the vessel and equipment operating parameters during which these limits are to be met. Another important parameter is environmental conditions, such as wind speed, water depth, and ambient noise, as appropriate. Finally, the state of completion of the vessel's construction should be identified, e.g., completely balanced HVAC systems, berth outfitted, machinery operable, etc.

Most of the noise and vibration specifications identify the types and methods that need to be utilized to achieve compliance and obtain certification or recertification, as appropriate. In all cases, citing a relevant and current test standard is preferable to leaving the test procedure to chance or misinterpretation.

A test plan, whether for noise or vibration, should address the following components:

1. Requirements
2. Instrumentation and Personnel
3. Schedule
4. Pre-trial inspection
5. Safety requirement
6. Measurement Parameters, Test Procedures, Data Collection and Reduction
7. Conclusions/Report
8. Recommendations for correcting measured excesses or future changes for follow ships

As a result of a properly executed test plan, one should be able to determine compliance with any specified acoustic criteria. Comprehensive approaches to testing are provided in the Mil-Std-1474E, ISO,[1] DI-HFAC,[2] IEC,[3] and ANSI[4] test standards. Many marine regulatory agencies also have their own test requirements for noise and vibration. Specialized test labs and test procedures are also covered in this chapter.

[1] ISO- 10055, Mechanical vibration – Vibration testing requirement for shipboard equipment and machinery components: ISO-13332, Reciprocating internal combustion engines – test codes for measurement of structureborne noise emitted from high-speed and medium-speed reciprocating internal combustion engines measured at the engine feet, ISO 20283, Mechanical vibration – Measurement of vibration on ships; ISO 21984, Ships and marine technology – Guidelines for measurement, evaluation and reporting of vibration with regard to habitability on specific ships; ISO 17208 – Underwater acoustics – Quantities and procedures for description and measurement of underwater sound from ships;.

[2] DI-HFAC 80274 Equipment structureborne vibration acceleration measurements test report.

[3] OVE EN IEC 60034-14, Rotating electrical machines, mechanical vibration

[4] ANSI S2.26 – Vibration testing requirements and acceptance criteria for shipboard equipment, ANSI S12.64, Measurement of underwater noise from vessels.

If no test standard is cited, the testing should be conducted by a knowledgeable marine acoustic consultant. This usually provides a defensible test methodology and competent report.

13.2.1 Instrumentation and Personnel

Specialized equipment is generally needed to undertake noise and vibration measurements. This list includes sensors such as microphones for airborne noise, accelerometers for vibration, and hydrophones for underwater noise measurement. The sensors are attached to spectrum analyzers via amplifiers or conditioning equipment if necessary. The output can be in either digital or analog formats. Modern equipment can either log or output digital data into spreadsheets. Force gauges are needed if mechanical impedance measurements are needed, and torque sensors are necessary for shafting measurements.

Depending on the complexity of the measurement, the use of a single-channel analyzer such as a Sound Level Meter (SLM) may be appropriate. In other cases, to efficiently collect the needed data, a multi-channel system may be more suitable. In all these cases, how the sensor is to be utilized, i.e., mic with wind screen or accelerometer glued to structure, needs to be addressed.

It is a good practice to epoxy accelerometers, or their attachment pads. The surface should be sanded smooth to ensure good contact. Magnetic mounts will significantly limit the valid frequency range. A guide for the useful frequency range for accelerometer attachment methods is provided in Table 13.3.

Calibration requirements for the instrumentation, such as laboratory and field calibration, should follow any test requirement imposed in the ship's specification or test requirement. If none are cited, the testing equipment should have been calibrated by an approved acoustic equipment test laboratory to National Institute of Standards and Technology (NIST) standards within the past year or two at most. The equipment should also be field calibrated before and after testing is conducted. Copies should be provided of the calibration certificates. The field calibrations should be witnessed and signed by the owner or ship's representative and provided with the final test report. The transducer sensitivity, any amplification settings, applicable frequency range, and serial number should all be reported.

Table 13.3 Frequency for various accelerometer attachment methods

Probe tip	Flat till 130 Hz, +10 dB at 1.4 kHz
Flat magnet	Flat till 600 Hz, +20 dB at 9 kHz
Adhesive mounting pad	Flat till 2200 Hz, +28 dB at 15 kHz
Adhesive	Flat till 3200 Hz, +28 dB peak at 24 kHz
Stud	Flat till 3400 Hz, +30 dB peak at 40 k Hz

The need for non-acoustic instruments should be considered. For example, this includes flow meters to measure the flow rate through ventilation terminations. The test team should have appropriate communications equipment to coordinate test number and type, equipment line-ups, etc.

As a minimum, the person in charge of conducting the testing should be either certified[5] or have direct training or relevant experience. They should be familiar with standards set by ANSI, SAE, NEMA[6] and/or ISO for conducting the test.

13.2.2 Schedule

Planning for acoustic tests should be performed well in advance of the anticipated test date. This is needed to assist in minimizing transit time to any test range and meet any special preparations, such as underwater signature testing. Scheduling is also needed in preparation for obtaining and loading onboard any specialized acoustic test equipment, such as shakers or shaft torsional sensors and equipment. The schedule should reflect the construction schedule, pre-trial inspection periods, time needed to instrument the vessel, personnel availability, test pre- and final-approval lead time, sequencing of tests (noise v. vibration), time table for any data analyses that need to be done on-site, and required reports. Reviewing or analyzing the data on-site will assist in ensuring that all the data needed have been collected. If needed, an additional diagnostic investigation can be conducted while the vessel is available.

If divers or support vessels are needed to support the tests, the schedule should identify when and what specific capabilities are needed. It is especially important to be aware of and schedule any special test facility, such as one that would measure underwater radiated noise. For some tests, it is important to have an equipment vendor onboard to assist in operating and testing their equipment or systems. This is often true for sonar self-noise measurements.

13.2.3 Measurement Parameters and Test Procedures

Test instrumentation should be set based on expected or test standards invoked. For instance, this usually is the 'slow response' on a sound level meter, a rms or instantaneous response, or a set measurement time for vibration depending on the needed bandwidth.

[5] ABS has a certification process for an Ambient Environmental Testing Specialist for whole-body vibration and noise measurement; see Appendix 2 of ABS Guide for Passenger Comfort on Ships, Updated 2015.

[6] NEMA – National Electrical Manufactures Assoc.

The test procedure should identify the following parameters:

- Coordination between test personnel and ship operations
- Equipment line-up and operating conditions, i.e., half or full power, rpm, etc.
- Data acquisition system, bandwidths and lower/upper frequency range of interest
- Transducer mounting methods, i.e., tripod and accelerometer mounting methods
- Number of locations/measurement positions at each location
- Placement of sensors (i.e., not within 1.5 m of a surface for airborne noise, accelerometer in middle of plating, etc.)
- Microphones should always be covered by windscreens, especially on-deck
- Signal-to-noise requirement(s), typically 10 dB or greater; if lower than 10 dB, corrections to be applied, if allowed
- Number, type and how often background measurements are to be conducted
- Number of data averages, bandwidth, and identification of frequency filtering, if needed, e.g., human vibration limits
- Coordination with trial personnel for the operation of machinery line-ups needed for testing, securing other 'nonessential' equipment/machinery
- Machinery line-ups, power settings, rpm, etc.
- Environmental conditions, i.e., background noise and/or vibration, temperature, humidity, barometric pressure, sea state, water depth, with/against wind and current, etc.
- Test duration for each type of test and for testing overall
- Test run identification approach and test logs to be retained
- Data reduction and analysis techniques
- Reports/graphs, with recommendations/conclusions wrt criteria
- Type of documentation and photographs needed, if any
- Extent of diagnostic testing, when and if needed

Test procedures should define the methodology, procedures, instrumentation, personnel, and criteria to which the measured data should be compared. For noise critical vessels, the testing should also assist in defining and identifying the sources of noise and/or vibration that control the acoustic environment, both within the hull and in the ocean. This can lead to the establishment of an optimum machinery operating list to achieve as low an acoustic environment as possible. Finally, the test may result in technical information to improve the follow vessels of the class or be used to improve acoustic modeling.

Separate test plans are normally needed for noise, vibration, sonar self-noise, and underwater radiated noise due to the various operating scenarios needed. Depending on the number of test personnel, some of these tests may be taken concurrently. For airborne tests in particular, a sign should be posted on the entrances noting Test in Progress – Do Not Enter.

The procedure should identify data to be reduced on-site as opposed to back at the office. The data reduction should keep records of sensor serial number, sensitivity, run number, data formats – graphs or tables, and sensor gain settings. Graphs should appropriately label axes and provide legends for each separate plot. Where possible, criterion levels should be included in graphs.

13.2.4 Airborne Noise Compliance Testing

ISO 2923 – Measurement of noise on board vessels – addresses compliance testing of airborne noise on ships. This procedure should be considered the gold standard for measuring and accessing noise onboard ships. It covers instrumentation, the test environment, quantities to be measured, operating conditions for the vessel, test procedures, and test reports. Section 8.3 of IMO 2923 states not to measure closer than 1 m from bulkheads and 1 m off the deck (see Sect. 7.5 for discussion of near field).

Regulatory agencies, such as ABS, DNV, BV, and ClassNK, often have their own airborne noise measurement procedures that go along with their classification requirements.

13.2.5 Structureborne & Habitability Compliance Testing

ISO 4867 – Code for the measurement of and reporting of shipboard vibration data and 4868 – Code for the measurement of local vibration data of ships structures and equipment – address compliance testing of structureborne (habitability) noise on ships. These procedures should be considered the gold standard for measuring and accessing vibrations onboard ships. They cover environmental conditions, instrumentation, the test environment, quantities to be measured, operating conditions for the vessel, test procedures, and test reports.

Again, regulatory agencies, such as ABS, DNV, BV, and ClassNK, often have their own airborne noise measurement procedures that go along with their classification requirements.

13.2.6 Test and Conclusions Report

Most Standards list the information to be included in a Trials Report. As a minimum this should include the following:

- Vessel name
- Locations and dates of the acoustic trials
- Test personnel and witnesses, if required
- Lab calibration sheets for each sensor and instrumentation
- Vessel operating condition(s)
- Identification of measurement locations, preferably on a ship's layout drawing or with each individual test log
- Acoustic criteria
- Measurement methodology
- Results and conclusions, including any identified excesses to the criteria and future design changes based on intelligence collected during the testing, if appropriate
- References

13.2.7 Factory Acceptance Testing

In some instances, factory testing of the noise and/or vibration of machinery or equipment is needed, either by the ship's specification or the need for additional information. These tests may be needed to show compliance with a specific acoustic level or provide source-level data for the prediction process. If a Factory Acceptance Test (FAT) is necessary, the instrumentation, calibration, sensors, test procedures, etc., are similar to those discussed above. The major differences are (1) the machinery/equipment will be tested on a different foundation than that provided by the ship, (2) the factory provides a different acoustic environment than the ship's compartment, especially with respect to ambient noise and sufficient signal-to-noise, and (3) the limited ability, in some cases, to properly load and operate the machinery/equipment.

Propulsion diesels are tested on a dynamo, which may increase ambient noise and vibration unless this machinery is specifically isolated from the test compartment. Diesel generators require a load cell rated to the output of the system. Electric motors are sometimes tested back-to-back to provide proper loading. The receive motor needs to be physically spaced away from the tested motor (not on the same foundation), and both must be isolation mounted. Pumps need to be operated at the correct pressure and flow rates. Inlet and outlet piping require flexible couples. Testing the acoustic source levels of fans can be complicated. Testing of often needed to be in accordance with procedures defined in AMCA 300-96, "Reverberant Room Method for Sound Testing of Fans." 1996.

In a limited number of cases, it may be needed to adjust the measured source vibration level data by the difference in the impedance of the factory foundation versus the expected or measured impedance of the ship's foundation. In this case, either an impedance hammer or an electromagnetic shaker or both are often used to measure the foundation's impedance.

13.3 Special Trials – Underwater Radiated Noise

In some cases, particularly with respect to underwater noise, it may be necessary to enlist the support of a specialized test lab or facility.[7] The Test Procedure should identify this effort early in the program. Often, a long lead time is usually needed to schedule such tests. It should also identify the type of data needed, data processing requirements, and report formatting. The Procedure should also identify when the report is due after the trials. If these facilities are operated by the military, special classification issues may need to be addressed. Portable systems are available from

[7] The U.S., Canadian, and Norwegian Royal Navies maintain special underwater radiated noise test facilities.

vendors that allow on-site testing in a quiet area. This means that the known or expected ambient noise is at least 10 dB below the radiated noise requirement. In some cases there is a minimum depth requirement. The travel time to and from this quiet test area may become a factor.

Compliance with any radiated noise criteria needs to be demonstrated by underwater noise testing. In the U.S. some of the earliest radiated noise measurements were undertaken on National Oceanic and Atmospheric Administration (NOAA) fisheries survey vessels (Bahtiarian and Fischer 2006). The American Acoustical Society (ASA) developed some early measurement standards (Bahtiarian 2009). These eventually morphed into an international standard (Bahtiarian 2012).

ISO[8] recently[9] developed test procedures to measure a ship's underwater radiated noise level and then used that data to determine its acoustic source level. The documents are:

- ISO 17208-1:2016 Underwater Acoustics provides the requirements for the 'precision' measurements in deep water (more than 150 m or 1.5 times the overall length of the ship)
- ISO 17208-3 cover measurement in shallow water (less than 150 m and 1.5 times the ship length)
- ISO 17208-2:2019 Underwater Acoustic identifies methodology needed to determine the 'source level' of the ship based on deep water measurements.

These documents spell out measurement requirements, frequency range, test measurement site, instrumentation, ambient (self-noise) limits, calibration, hydrophones and their deployment, testing configuration and course, background adjustments, data processing, and other parameters needed to conduct a good test.

As noted for the other acoustic tests, regulatory agencies often have their own test requirements. This includes ABS – Guide for the Classification Notation – Underwater Noise and External Airborne Noise;

Other references and measurement procedures for underwater noise are provided in NPL Good Practice Guide No. 133, 2014.

An underwater measurement system requires two major components – hydrophones that can measure the in-water sound pressure level and a system to measure the distance from the ship to the hydrophones. The hydrophones need to have a flat response over the entire frequency range of interest and have sufficient 'sensitivity' – produce a high enough voltage in response to the incident pressure that they can effectively pick up the signal. Usually, multiple hydrophones are arrayed at different depths to improve accuracy and provide an effective 'average' response. There are

[8]ANSI S16.64-2009 has been replaced by ISO 17208. The ANSI standard provide 3 different grades of precision, A through C, with B being deemed the most appropriated for commercial vessel.

[9]The latest standard should always be used unless the ship's specification calls for a test standard with a specific date.

many ways to deploy the hydrophone array, ranging from a surface supported system from a buoy or other support boat to a bottom mounted system. Each type of deployment has advantages and disadvantages and to a large extent depends on the water depth and ability to launch and retrieve sensitive equipment and cables.

Other important factors are as follows:

- a test site with low ambient noise to test to a good signal-to-noise ratio, requiring the test site to be far from shipping lanes, other high underwater noise sources, and interested onlookers in noisy power boats
- tests need to be conducted in a low sea state so that the natural ambient noise is sufficiently low to avoid contaminating the data
- if a support vessel is used, it must be dead-in-water, so its signal does not contaminate the measurement; battery power would be needed for instrumentation.
- measurement need to be collected at the prescribed operating speed(s) and with the proper (normally operating) machinery and ship aspect (abeam, astern or bow-on)
- it would be beneficial to drydock the vessel or clean the hull and propeller blades of marine growth a short time before testing.

As with all acoustic testing, the hydrophones should be calibrated in the past year and field calibrated on-site to ensure an accurate measurement. The data are usually recorded both as time waveforms and as spectra over the frequency range of interest and in the bandwidth(s) of interest. The range data are used to determine the spreading losses from the ship to each hydrophone, and the measured data are then 'corrected' to a one-meter source level.

A summary of assumptions, measurement uncertainty, test site requirements, weather conditions, etc., for Bureau Veritas (BV) Class Notation are provided in Palomo et al. 2015. One should always check for the latest BV or IMO measurement requirements.

The components of the measurement system[10] used on the NOAA FRVs are shown in Fig. 13.7.

The measurement system deployed in the water to measure the radiated noise on NOAA's quiet Fisheries Research Vessels is shown in Fig. 13.8.

The vessel under test typically makes numerous passes by the array to average the port and starboard aspect radiated noise abeam of the vessel. The typical test course configuration path is shown in Fig. 13.9.

Various test procedures and test measurement approaches are documented by individual regulatory groups promulgating radiated noise requirements. For instance, DNV Silent-R has different methods identified for testing vessels with thrusters and different allowances for water depth and distance from the vessel to the hydrophone.

[10]Pre implementation and acceptance of ISO 17208

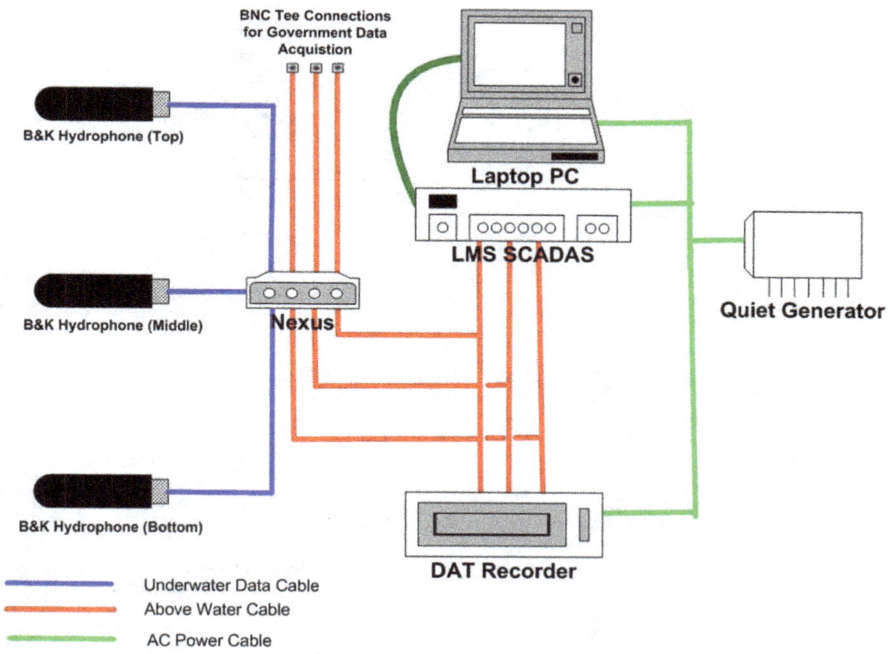

Fig. 13.7 Underwater noise measurement system components

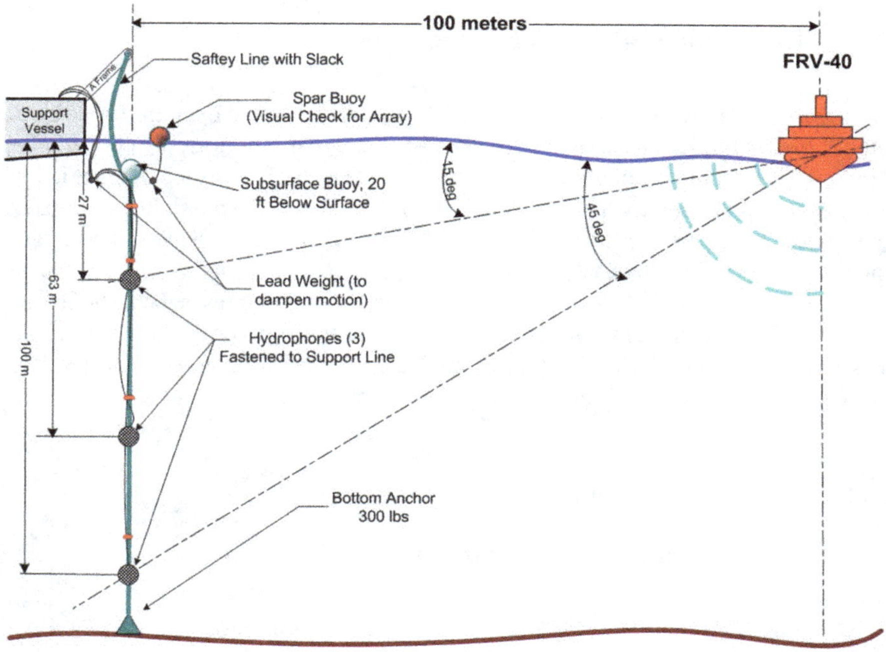

Fig. 13.8 Diagram of hydrophone array deployment

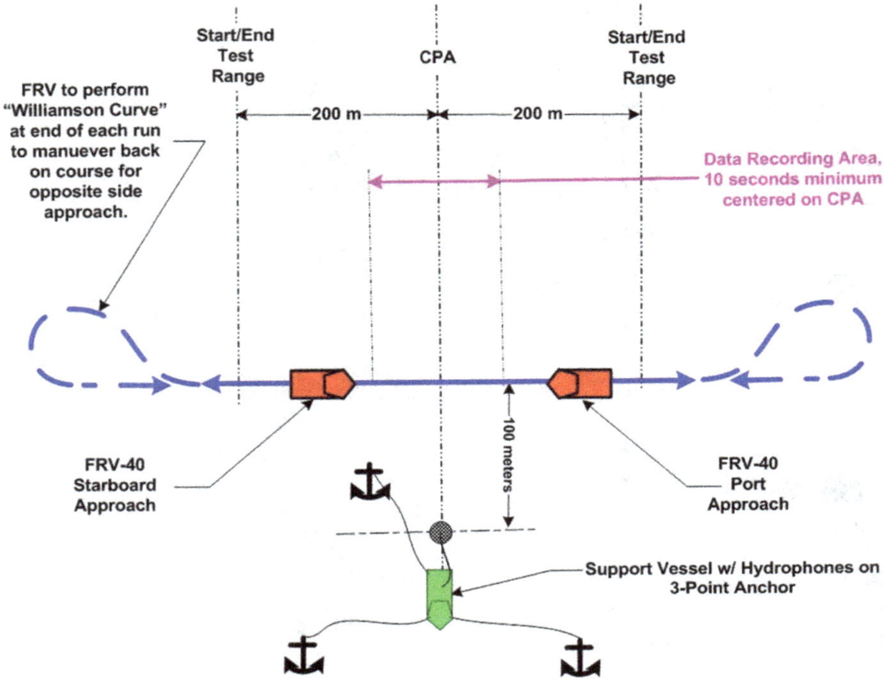

Fig. 13.9 Underwater test course showing Williamson Curves at end of each pass-by

13.4 Recommendations for Changes

First, the extent of the problem, in terms of level, frequency and operating condition, should be identified succinctly. If possible, the contribution of airborne or structureborne sources should be separated (see Sect. 13.5). Any recommendation for change should identify the acoustic impact if not treated, the possible cost impact, and alternative approaches to correct the deficiency or excess. These alternatives should be documented by analytical or software predictions. In some cases, a change in the acoustic criteria may be justified; if so, supporting evidence should be provided. For instance, meeting a low habitability noise criterion in a washroom may not be as important as meeting it in a berth. It should be noted if no feasible (or reasonable) noise or vibration control treatments exist.

13.5 Specialized Tests

13.5.1 Vibration/Diagnostic Surveys

1. Machinery-induced vibration – Vibration measurements should be taken on the feet or surfaces of machinery or equipment to determine whether the vibrations

are excessive or within acceptable limits, e.g., smooth, rough, or defective. Vibration assessment can assist in determining whether bearings are a problem (see #8). These tests can cover a broad frequency range and generally require that the unit be operated separately if possible to reduce interference from other nearby machinery sources. Accelerometers are typically used as sensors.

2. Shafting vibration survey – to measure torsional, lateral or longitudinal response of shaft to determine if there is excessive vibration or stress in the shafting system. Requires underway measurements usually with strain sensors attached to the shaft and proximity probes near flanges for the lateral/longitudinal measurements. Lasers (laser doppler vibrometers) have been used for torsional measurements (Xiang et al. 2012).
3. Structural vibration – to determine if the vibration of a particular ship structure – foundation, deck, bulkhead, etc. – is excessive with respect to structural integrity (cyclical fatigue is usually a concern). Using strain gauges and/or accelerometers, measurements should be taken on the structure to determine the overall response. If necessary, the 'source' and path of energy driving the structure could be identified.
4. Resonant response – survey to determine if a structural resonance exists where the structure vibrates at a particular frequency that matches an excitation frequency. For instance, a mast vibration where the vibration is caused by propeller blade rate excitation or a piping system where the excitation may be associated with pump vane passage rate. Survey conducted with accelerometers and an impact hammer. Mechanical shaker, impact hammer, or 'bump tests' can be used to determine the natural frequency of a structure or system of concern. In some cases, a modal analysis is used to determine modes of vibration – where multiple measurement points are used and the system is excited by an impact hammer.
5. Habitability vibration survey – to determine the low-frequency environment on the vessel with respect to allowed crew exposure to vibration. Measurement of vibration on decks should be in three directions – vertical, athwartships, and longitudinal.
6. It is beneficial during diagnostic testing to collect both narrow band and broadband data. The narrow band data can be used to identify tones, while the broadband data are used to determine excesses relative to any broadband criterion.
7. Modal testing, with a mechanical or pneumatic shaker or impedance hammer, can assist in determining impedances.
8. Bearing frequencies can be identified based on frequency analyses. These can show imbalance, misalignment, looseness, roundness, and local bearing defects. These defects typically appear as multiples of the rotation rate. For instance, imbalance has a dominant frequency at the rotation rate. Misalignment appears at 1× and 2× rotation rates. Race defects show multiple peaks between 4× and 15×. ISO 2372 provides examples of these conditions. Bearing vendors often supply vibration diagnostic guides.

All of these tests involve instrumenting the equipment/system of concern and operating the system – either dockside or underway – as appropriate. If acoustic problems are anticipated or have been identified as a possible risk, pre-trials may be warranted once equipment is operational. In some cases, an impact hammer or shaker can be utilized to provide additional diagnostic support. Usually, a multi-channel acoustic analyzer is used to take data at multiple locations simultaneously if phase information is needed to determine mode shapes.

With respect to the underwater requirements, it would be helpful to conduct noise and vibration tests concurrently on the vessel as the radiated noise tests are being conducted. Data from accelerometers on the hull would assist in the identification of tones related to operating machinery.

If airborne noise excesses exist, it would be valuable to conduct testing both dockside and underway. For instance, HVAC testing can be carried out dockside in those compartments shown to have a noise excess. Similarly, auxiliary machinery can be cycled on and off while dockside to identify if it is a significant contributor and whether the primary path is airborne or structureborne. If time and costs permit, it is always helpful to compare measured to predicted acoustic levels to see where improvements are needed.

In some cases, artificial noise sources or vibration shakers can be used to document airborne or structureborne path transfer functions. Artificial underwater sources can be used to identify the response of the hull to incident underwater noise and calibrate transmission loss at the underwater test site.

Diagnostic of the underwater radiated noise can be carried out pier side, assuming that the ambient noise is very low.[11] Hydrophone(s) suspended over the side can measure the noise while equipment of interest is being cycled on and off. The near field at 16 Hz is nominally approximately 59 m, and at 125 Hz, it is 19 m (Fischer and Boroditsky 2004)

These data can be used directly to determine the 'relative' noise and frequency content from different systems.

13.6 Vessel Operation

In addition to testing to see that noise and vibration criteria are met, acoustic data from the trials should be used to:

- Identify hazardous noise compartments or duty stations and under what operating conditions the noise is considered hazardous.
- Develop a 'quiet equipment' line up for when and if a radiated noise signature is to be met

[11] Nighttime may be quieter than during the daytime.

- Determine the point (pitch/rpm) on a CCP where onboard or underwater noise is minimized
- Ascertain those equipment items
  - With poor vibration isolation
  - Resonant operating conditions
- Identify HVAC systems
  - Dominated by flow noise
  - Unbalanced systems
  - Flow measurements at terminals

13.6.1 Test Standards

ABS:

Guide for the Classification Notation – Underwater Noise and External Airborne Noise

Guidance Notes on noise and vibration control for inhabited spaces

ABS Guide for Crew Habitability on:

Ships
Workboats
Offshore Installations
Mobile Offshore Drilling Units (MODUs)

Guidance Notes on Ship Vibration

ANSI:

ANSI/ASA S2.16-1997 (R2016), American National Standard Vibratory Noise Measurements and Acceptance Criteria of Shipboard Equipment.

ANSI/ASA S2.25-2004 (R2014), American National Standard Guide for the Measurement, Reporting, and Evaluation of Hull and Superstructure Vibration in Ships.

ANSI/ASA S2.26-2001 (R2016), American National Standard Vibration Testing Requirements and Acceptance Criteria for Shipboard Equipment.

ANSI/ASA S2.27-2002 (R2014), American National Standard Guidelines for the Measurement and Evaluation of Vibration of Ship Propulsion Machinery.

ANSI/ASA S2.28-2009 (R2014), American National Standard Guide for the Measurement and Evaluation of Broadband Vibration of Surface Ship Auxiliary Rotating Machinery.

ANSI/ASA S2.29-2003 (R2013), American National Standard Guide for the Measurement and Evaluation of Vibration of Machine Shafts on Shipboard Machinery.

ANSI/ASA S12.64-2009/Part 1 (R2014), American National Standard Quantities and Procedures for Description and Measurement of Underwater Sound from Ships—Part 1: General Requirements.

ANSI/ASA S12.67-2008 (R2013), American National Standard Pre-Installation Airborne Sound Measurements and Acceptance Criteria of Shipboard Equipment.

BV

Part E. Additional Class Notations: Chapter 6, Comfort On Board (COMF)

Rule Note: Underwater Radiated Noise; NR 614 DT R01 E

Rule Note: Comfort and Health on-board Offshore Units, NR 636 DT R00 E

ClassNK

Guideline for the Mandatory Code on Noise Levels on Board Ships

Noise and Vibration Guidelines

DNV

Rules for Classification of Ships; Part 6, Chapter 15 – Vibration Class

Rules for Classification of Ships; Part 6, Chapter 24 – Silent Class Notation

Rules for Classification of Ships; Part 6, Chapter 33 – Comfort Class

Rules for the Classification of Steel Ships, NR 467:

COMF-NOISE, COMF-VIB, CIMF+, COMF-NOISE-Pax
COMF-NOISE-Crew, COMF-VIB-PAX
COMF-VIB-Crew, Accommodations

ISO:

1680 – Test code for the measurement of airborne noise emitted by rotating electrical machines.

2922 – Acoustics – Measurement of airborne sound emitted by vessels on inland waterways and harbours.

2923 – Acoustics – Measurement of noise on board vessels.

3740 – Acoustics – Determination of sound power levels of noise sources – Guidelines for the use of basic standards.

3945 – Mechanical vibration of large rotating machines with speed range from 10 to 200 r/s – Measurement and evaluation of vibration severity in situ.

4867 – Code for the measurement of and reporting of shipboard vibration data.

4868 – Code for the measurement of local vibration data of ships structures and equipment.

6954 – Mechanical vibration – Guidelines for the measurement, reporting and evaluation of vibration with regard to habitability on passenger and merchant ships

8528 – Reciprocating internal combustion engine driven alternating current generating sets – Part 9: Measurement and evaluation of mechanical vibrations

10,816 – Mechanical vibration – Evaluation of machine vibration by measurements on non-rotating parts
13,332 – Reciprocating internal combustion engines – Test code for the measurement of structureborne noise emitted from high-speed and medium-speed reciprocating internal combustion engines measured at the engine feet
20,816 – Mechanical vibration – Measurement and evaluation of machine vibration
20,283 – Mechanical vibration – Measurement of vibration on ships

References

Bahtiarian, M.: ASA Standard Goes Underwater. Acoust. Today. **5**, 30 (2009)

Bahtiarian, M.: Taking American National Standard to the international level. Acoust. Today. **8**, 25 (2012)

Bahtiarian, M., Fischer, R.: Underwater radiated noise of the NOAA ship *OSCAR DYSON*. Noise Control Eng. J. **54**(4), 224 (2006)

Fischer, R., Boroditsky, L.: Development of Ship Radiated Noise Transfer Functions – Near/Far Field Measured Signature Comparison. In:148th Meeting of the American Society of America, 2004

National Physical Laboratory (NPL): Good Practice Guide No. 133, Underwater Noise Measurement, Marine Scotland, 2014

Palomo, P., Mullor, R., et al.: A Comprehensive Framework to Address Ship Underwater Radiated Noise: from Bureau Veritas Class Notation to Validations of Numerical Prediction Tools, ICSV 22, Florence, Italy, 2015

Xiang, L., Yang, S., Gan, C.: Torsional vibration measurements on rotating shaft system using laser doppler vibrometer. Opt. Lasers Eng. **50**(11), 1596–1601 (2012)

Chapter 14
Operation and Care of Treatments

Some acoustic treatments are 'install and forget.' These include many acoustic insulations and damping treatments. Many, however, require some monitoring over their life and replacement at certain intervals. These include isolation mounted equipment and floating decks. If these systems are not maintained properly, the acoustic environment will degrade over time. These features are covered in this chapter.

Maintenance and care of treatments is the 'insurance' needed to warrant that the previous engineering and construction efforts continue to pay off. To carry this out, maintenance and care of treatments, along with scheduling of treatment replacement, should be detailed in ship retained documents. Here again, 'buy quiet' may prove beneficial in reducing the needed care and maintenance. Since maintenance in noisy machinery space can be hazardous to the crew, it is suggested to schedule maintenance shore-side or when it is possible to secure operation of other noisy machinery in the compartment. Properly maintained equipment will eliminate rattling of pipes and ducts and reduce air and fluid leaks and blockages in ventilation systems. Replacement parts and machinery should meet or exceed the noise and vibration control approach designs/used on the original equipment/systems.

One of the first concerns is that equipment and treatments be properly shipped. All vendor treatments, including machinery as appropriate, should be shipped with special handling and storage requirements. For instance, elastomeric mounts shipped and subjected to low temperatures may lose their needed stiffness characteristic and/or may be delivered with cracks, making them unusable. Before installing vendor-supplied equipment or systems, the shipyard should check and remove clamps holding the vendor-supplied isolation mounted equipment in place for shipping purposes only. This is often the case with fans and small motors bought with vendor-supplied mounts. Packages or enclosures need to be sized to allow for maximum deflection of mount units without contact with the shipping package.

R. Fischer, L. Boroditsky, *Noise and Vibration Control on Ships*,
https://doi.org/10.1007/978-3-031-55170-3_14

14.1 Maintenance of Treatments and Equipment

Part of any approach to maintenance is to consider access requirements. Permanent fittings, pipes, ducts, wireways, and structures should be kept clear of equipment or treatment removal routes. This is especially true with respect to isolation mounts. Access is needed to measure the 'loaded height' and for replacement of the mounts. The 'loaded height' should be checked periodically to determine if the mounts need replacement per the vendor specifications. Another factor in maintenance is re-alignment after changes to the mounting system or accounting for mount creep. This includes shafting, flex hose, and wiring connections. Elastomeric mounts should be kept clean and protected from solvents or paint. This can be accomplished using soap and water or isopropyl alcohol. It is recommended to follow vendor recommend 'life-to-replacement' guidelines. For instance, elastomeric mounts generally have a life span of 7–10 years. Some vendors have a longer time span, and these should be followed to ensure no degradation of the acoustic environment.

Spare parts, particularly flex connections, should be maintained for any ship critical system. The ship should have additional acoustic insulation, tape, etc., to repair sections of damaged insulation. Additional acoustic ceiling tiles are always useful since access to the overhead usually damages them. Extra damping tiles should be kept onboard to replace damaged tiles.

These steps will help ensure that the mission is safely and quietly conducted over the life of the vessel. These steps should also be followed during overhaul periods.

To safeguard the acoustic environment of the vessel, it is recommended that the following steps be taken:

- Replace thermal/acoustic insulation after damage or removal for access
- Minimize imbalance by checking alignment and balance a minimum of every 6 months
- Lubricate machinery periodically
- Replace worn bearing
- Tighten hold-down bolts to prevent 'looseness', especially for isolation mounted systems.
- Check for ventilation dampers that are out of position and whistling
- Inspect isolation mounts and other noise control measures, particularly after any repairs or replacement, e.g., vibration shorts, paint, melted sections due to welding, etc.
- Replace control treatments with like materials
- Check for corrosion behind/under treatments such as floating rooms/floors, A/C duct insulation, etc.
- Look for soot and blockages in propulsion machinery intake/exhaust systems
- Examine for blockage of grills on tunnel thrusters or water chests
- View for smooth propeller blade surfaces, locate and repair dents in the blade surface and leading and trailing edges; check for cavitation errosion
- Inspect and remove marine fouling on propellers

- Inspect for smooth hull surfaces to prevent flow separation
- Complete sealing around access doors to machinery spaces or enclosures
- Replace cladding treatment on valves and other equipment after inspection

If a machinery vibration monitoring system is utilized, some of the above issues will be detected via appropriate alarms.

To ensure proper upkeep, a maintenance plan should be developed. The maintenance philosophy for installed acoustic systems and machinery, such as Buy-Quiet, should be defined with respect to planned maintenance and for predictive maintenance for those equipment/systems that have condition monitoring. It should lay out maintenance routines performed on acoustic treatment on a predetermined schedule, using the manufacturer's recommendations as available. The plan should identify maintenance and replacement of treatment routines to be performed over time. The plan should also identify applicable technical manuals from acoustic treatment vendors. The acoustic treatments should be identified on ship drawings and kept as part of the Noise and Vibration Control Plan.

The plan should address regulatory standards or vendor-supplied information on alignment, balance—both static and dynamic. This should include instrumentation requirements and measurement methods. Shaft runout should be limited to a total indicator reading (TIR) of 0.0254 mm. Similarly, the allowable lateral and longitudinal alignment of flexible couplings should be identified and checked, especially after mount changes. Spare parts, if needed, should also be identified and kept on-board, especially for flex hose connections. The plan should also identify the lifting gear needed for isolation mounted equipment, especially large units such as propulsion equipment.

Figure 14.1 shows how the signature of a vessel can change over time if there is only limited maintenance. The broadband noise has increased significantly, especially above 1000 Hz. As a speculation, this is due to propeller cavitation.

14.2 Care of Treatments

As stated in Chap. 13, some of the acoustic treatments are 'install and forget'. Many, however, require some monitoring over their life and replacement at certain intervals. There is a MIL-STD[1] that covers the maintenance and repair of damping materials. As noted above, these include isolation mounted equipment and floating decks. If these systems are not maintained properly, the acoustic environment will degrade over time. Vendors should be able to provide information on the inspection process and life cycle of their material.

[1] MIL-STD-2148(SH) Rev. A. "DoD Standard Practice. Vibration Damping Materials, Procedures for Installation, Maintenance and Repair".

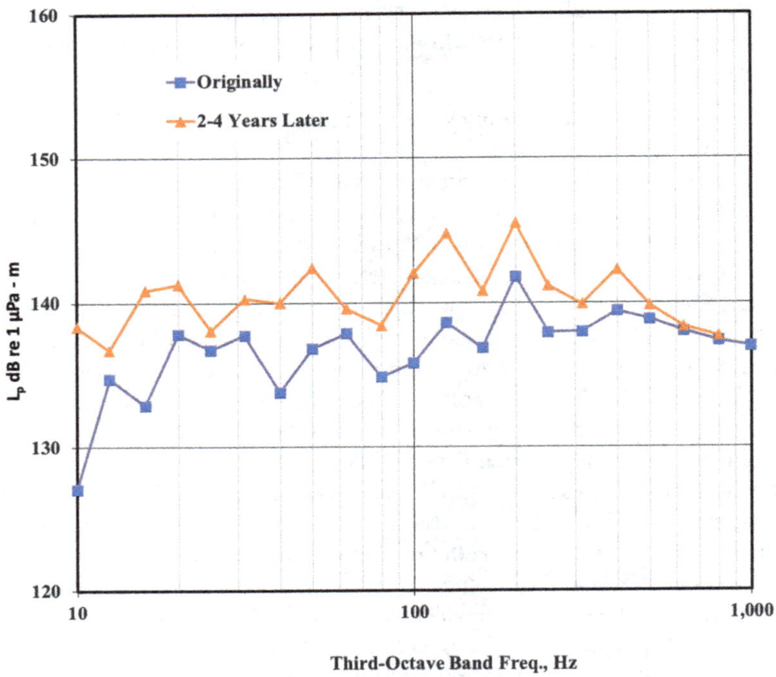

Fig. 14.1 Change in radiated spectrum after several years

14.3 Re-testing

For noise-sensitive vessels, re-testing of the radiated signature should be conducted at set intervals, usually every 2 years. Regulatory agencies often have recertification requirements.

A thorough machinery vibration survey during commissioning of a vessel can also form a baseline for monitoring their health. In this case, sensor pads for the accelerometers can be identified in drawing (or pictures) and left in place for future surveys. The equipment, instrumentation, and measurement standard used during the original testing should be documented.

For vessels whose operation depends on a reduced underwater noise signature, sensors and their monitoring could lead to early identification of machinery or systems that are increasing in their contribution to the signature. Monitoring of the hull vibration or hydrophones onboard the vessel can be used to determine whether the propeller is cavitating or not (Walrod 1999). A general discussion of the benefits of ship maintenance is provided in Dlugokecki et al. 2014.

References

Dlugokecki, V., Fanguy, D., Hepinstall, L., Tedesco, M.: Reducing the cost of ship maintenance and repair. Conference Paper, (Oct. 2014). https://www.researchgate.net/publication/354492188_Reducing_the_Cost_of_Ship_Maintenance_and_Repair

Walrod, J.: Sensor and Actuator Networks for Acoustic Signature and Control. Undersea Defense Technology, Nice (June 1999)

Index

R. Fischer, L. Boroditsky, *Noise and Vibration Control on Ships*,
https://doi.org/10.1007/978-3-031-55170-3

www.ingramcontent.com/pod-product-compliance
Lightning Source LLC
Chambersburg PA
CBHW061556310826
50743CB00003B/38
* 9 7 8 3 0 3 1 5 5 1 6 9 7 *